Global Volcanic Hazards and Risk

Approximately 800 million people live within 100 km of active volcanoes worldwide, and with ever-growing populations, the likelihood of volcanic emergencies is increasing. Volcanic eruptions can cause extreme societal and economic disruption through loss of life and livelihoods, and damage to critical infrastructure.

Originally prepared for the United Nations Office for Disaster Risk Reduction, this is the first comprehensive assessment of global volcanic hazard and risk, drawing on a wide range of international expertise. It presents the state of the art in our understanding of global volcanic activity, as well as a thorough introduction to volcanology, accessible to a broad audience. It also looks at our assessment and management capabilities, and considers the preparedness of the global scientific community and government agencies to manage volcanic hazards and risk.

Volcanic hazard profiles and local case studies are provided online for all countries with active volcanoes, with invaluable information on volcanic hazard and risk at the local, national and global scale. Particular attention is paid to volcanic ash, the most frequent and wide-ranging volcanic hazard. The first global ash fall hazard map is presented along with a discussion of the characteristics and impacts associated with volcanic ash fall.

Of interest to all those concerned with reducing the impact of natural hazards and disaster risk reduction, including government officials, the private sector, students, researchers and professional scientists, this book is a key resource for the disaster risk reduction community and for those interested in volcanology and natural hazards. A non-technical summary report is also included for policy makers and general interest readers. This title is also available as Open Access via www.cambridge.org/volcano.

Dr Susan Loughlin is the Head of Volcanology at the British Geological Survey (BGS) and joint leader of the Global Volcano Model (GVM). Her research interests include volcanic processes, hazards and risk, communication, social and environmental impacts of eruptions and the interaction of scientists and decision makers. Dr. Loughlin spent several years at Montserrat Volcano Observatory and was Director for two years. She has provided advice to governments and communities during volcanic unrest and eruptions (e.g. Montserrat and Iceland/UK) and provided scientific evidence for longer-term planning.

Professor Steve Sparks is a volcanologist at the University of Bristol and joint leader of the Global Volcano Model (GVM). With expertise in many aspects of volcanology, he is the most highly cited scientist in this field. His interests include volcanic hazards and risk, the physics of volcanic eruptions and fluid dynamics of hazardous flows. Professor Sparks has provided advice to governments during ongoing and developing volcanic emergencies in Montserrat and Iceland.

Dr Sarah Brown is a researcher in volcanology at the University of Bristol. Her interests lie in physical volcanology with an emphasis on the assessment of hazard and risk. Dr. Brown works on combining and developing volcanological datasets including the Large Magnitude Explosive Volcanic Eruptions database (LaMEVE) to investigate the global eruption record with an aim towards developing a better understanding of volcanic risk.

Dr Susanna Jenkins is a volcanologist at the University of Bristol. Her research focuses on the assessment of hazards and risks associated with explosive volcanism. Dr Jenkins has worked with research, government and civil protection agencies, particularly in south-east Asia and the Lesser Antilles, in quantifying the risk from future eruptions and assessing the impact of recent damaging eruptions.

Dr Charlotte Vye-Brown is a volcanologist at the British Geological Survey (BGS). She applies a multi-disciplinary approach of field studies, geochemistry and remote sensing to her research. Her interests include volcanic geology, formation of continental flood basalts, lava flow emplacement, rift volcanism and communication of science to support planning and response to volcanic activity.

Global Volcanic Hazards and Risk

Edited by

SUSAN C. LOUGHLIN
British Geological Survey, Edinburgh, UK

STEVE SPARKS
University of Bristol, UK

SARAH K. BROWN
University of Bristol, UK

SUSANNA F. JENKINS
University of Bristol, UK

CHARLOTTE VYE-BROWN
British Geological Survey, Edinburgh, UK

CAMBRIDGE
UNIVERSITY PRESS

University Printing House, Cambridge CB2 8BS, United Kingdom

Cambridge University Press is part of the University of Cambridge.

It furthers the University's mission by disseminating knowledge in the pursuit of
education, learning and research at the highest international levels of excellence.

www.cambridge.org
Information on this title: www.cambridge.org/9781107111752

© Susan C. Loughlin, Steve Sparks, Sarah K. Brown, Susanna F. Jenkins and Charlotte Vye-Brown 2015

This work is in copyright. It is subject to statutory exceptions and to the provisions of
relevant licensing agreements; with the exception of the Creative Commons version
the link for which is provided below, no reproduction of any part of this work may take
place without the written permission of Cambridge University Press.

An online version of this work is published at http://dx.doi.org/10.1017/CBO9781316276273 under a
Creative Commons Open Access license CC-BY-NC-ND 3.0 which permits re-use, distribution and
reproduction in any medium for non-commercial purposes providing appropriate credit to the original
work is given. You may not distribute derivative works without permission. To view a copy of
this license, visit https://creativecommons.org/licenses/by-nc-nd/3.0.

All versions of this work may contain content reproduced under license from third parties.
Permission to reproduce this third-party content must be obtained from these third-parties directly.

When citing this work, please include a reference to the DOI 10.1017/CBO9781316276273.

First published 2015

A catalogue record for this publication is available from the British Library

Library of Congress Cataloguing in Publication data
Global volcanic hazards and risk / edited by Susan C. Loughlin, British Geological Survey, Edinburgh, UK,
Steve Sparks, University of Bristol, UK, Sarah K. Brown,
University of Bristol, UK, Susanna F. Jenkins, University of Bristol, UK.
pages cm
Includes bibliographical references and index.
ISBN 978-1-107-11175-2 (Hardback : alk. paper)
1. Volcanic hazard analysis. 2. Volcanoes. I. Loughlin, Susan C., editor. II. Sparks, R. S. J. (Robert
Stephen John), 1949– editor. III. Brown, Sarah K., editor. IV. Jenkins, Susanna F., editor.
QE527.6.G56 2015
363.34′95–dc23   2015011193

ISBN 978-1-107-11175-2 Hardback

Additional resources for this publication at www.cambridge.org/volcano

Cambridge University Press has no responsibility for the persistence or accuracy
of URLs for external or third-party internet websites referred to in this publication,
and does not guarantee that any content on such websites is, or will remain,
accurate or appropriate.

Contents

	List of contributors	viii
	Foreword	x
	Preface	xii
	Acknowledgements	xiv
1	An introduction to global volcanic hazard and risk S.C. Loughlin, C. Vye-Brown, R.S.J. Sparks, S.K. Brown, J. Barclay, E. Calder, E. Cottrell, G. Jolly, J-C. Komorowski, C. Mandeville, C. Newhall, J. Palma, S. Potter and G. Valentine	1
	Appendix: Summaries of Chapters 4-26 and Supplementary Case Studies 1-3	41
2	Global volcanic hazard and risk S.K. Brown, S.C. Loughlin, R.S.J. Sparks, C. Vye-Brown, J. Barclay, E. Calder, E. Cottrell, G. Jolly, J-C. Komorowski, C. Mandeville, C. Newhall, J. Palma, S. Potter and G. Valentine	81
3	Volcanic ash fall hazard and risk S.F. Jenkins, T.M. Wilson, C. Magill, V. Miller, C. Stewart, R. Blong, W. Marzocchi, M. Boulton, C. Bonadonna and A. Costa	173
	Appendix A: Global average recurrence intervals	Online
4	Populations around Holocene volcanoes and development of a Population Exposure Index S.K. Brown, M.R. Auker and R.S.J. Sparks	223
5	An integrated approach to Determining Volcanic Risk in Auckland, New Zealand: the multi-disciplinary DEVORA project N.I. Deligne, J.M. Lindsay and E. Smid	233
6	Tephra fall hazard for the Neapolitan area W. Marzocchi, J. Selva, A. Costa, L. Sandri, R. Tonini and G. Macedonio	239
7	Eruptions and lahars of Mount Pinatubo, 1991-2000 C.G. Newhall and R. Solidum	249
8	Improving crisis decision-making at times of uncertain volcanic unrest (Guadeloupe, 1976) J-C. Komorowski, T. Hincks, R.S.J. Sparks, W. Aspinall And CASAVA ANR Project Consortium	255

9	Forecasting the November 2010 eruption of Merapi, Indonesia J. Pallister and Surono	263
10	The importance of communication in hazard zone areas: case study during and after 2010 Merapi eruption, Indonesia S. Andreastuti, J. Subandriyo, S. Sumarti and D. Sayudi	267
11	Nyiragongo (Democratic Republic of Congo), January 2002: a major eruption in the midst of a complex humanitarian emergency J-C. Komorowski and K. Karume	273
12	Volcanic ash fall impacts T.M. Wilson, S.F. Jenkins and C. Stewart	281
13	Health impacts of volcanic eruptions C.J. Horwell, P.J. Baxter and R. Kamanyire	289
14	Volcanoes and the aviation industry P.W. Webley	295
15	The role of volcano observatories in risk reduction G. Jolly	299
16	Developing effective communication tools for volcanic hazards in New Zealand, using social science G. Leonard and S. Potter	305
17	Volcano monitoring from space M. Poland	311
18	Volcanic unrest and short-term forecasting capacity J. Gottsmann	317
19	Global monitoring capacity: development of the Global Volcano Research and Monitoring Institutions Database and analysis of monitoring in Latin America N. Ortiz Guerrero, S.K. Brown, H. Delgado Granados and C. Lombana Criollo	323
20	Volcanic hazard maps E. Calder, K. Wagner And S.E. Ogburn	335
21	Risk assessment case history: the Soufrière Hills Volcano, Montserrat W. Aspinall And G. Wadge	343
22	Development of a new global Volcanic Hazard Index (VHI) M.R. Auker, R.S.J. Sparks, S.F. Jenkins, W. Aspinall, S.K. Brown, N.I. Deligne, G. Jolly, S.C. Loughlin, W. Marzocchi, C.G. Newhall and J.L. Palma	349

23	Global distribution of volcanic threat S.K. Brown, R.S.J. Sparks and S.F. Jenkins	359
24	Scientific communication of uncertainty during volcanic emergencies J. Marti	371
25	Volcano Disaster Assistance Program: Preventing volcanic crises from becoming disasters and advancing science diplomacy J. Pallister	379
26	Communities coping with uncertainty and reducing their risk: the collaborative monitoring and management of volcanic activity with the vigías of Tungurahua J. Stone, J.Barclay, P. Ramon, P. Mothes and STREVA	385
	Index	389
	Appendix B: Country and regional profiles of volcanic hazard and risk S.K. Brown, R.S.J. Sparks, K.Mee, C. Vye-Brown, E.Ilyinskaya, S.F. Jenkins, S.C. Loughlin, et al.[*]	Online[†]

---

[*] The contributors to this report are listed separately within Appendix B.
[†] See www.cambridge.org/volcano for Appendix B which comprises a short discussion of the global distribution of volcanic hazard and risk and individual profiles of volcanism for all countries and regions with volcanic activity within the last 10,000 years.

Contributors

Andreastuti, S.	Center for Volcanology and Geological Hazard Mitigation, Indonesia
Aspinall, W.	University of Bristol, UK
Auker, M.R.	University of Bristol, UK
Baptie, B.	British Geological Survey, UK
Barclay, J.	University of East Anglia, UK
Baxter, P.J.	University of Cambridge, UK
Biggs, J.	University of Bristol, UK
Blong, R.	Aon Benfield, Australia
Bonadonna, C.	University of Geneva, Switzerland
Boulton, M.	University of Bristol, UK
Brown, S.K.	University of Bristol, UK
Calder, E.	University of Edinburgh, UK
Costa, A.	Istituto Nazionale di Geofisica e Vulcanologia, Italy
Cottrell, E.	Smithsonian Institution, USA
Crosweller, H.S.	University of Bristol, UK
Daud, S.	Civil Contingencies Secretariat, Cabinet Office, UK
Delgado-Granados, H.	Universidad Nacional Autónoma de México, México
Deligne, N.I.	GNS Science, New Zealand
Felton, C.	Civil Contingencies Secretariat, Cabinet Office, UK
Gilbert, J.S.	Lancaster University, UK
Gottsmann, J.	University of Bristol, UK
Hincks, T.	University of Bristol, UK
Hobbs, L.K.	Lancaster University, UK
Horwell, C.J.	Durham University, UK
Ilyinskaya, E.	British Geological Survey, UK
Jenkins, S.F.	University of Bristol, UK
Jolly, G.	GNS Science, New Zealand
Kamanyire, R.	Public Health England, UK
Karume, K.	Goma Volcano Observatory, Democratic Republic of Congo
Kilburn, C.	University College London, UK
Komorowski, J-C.	Institut de Physique du Globe de Paris, France
Lane, S.J.	Lancaster University, UK
Leonard, G.	GNS Science, New Zealand
Lindsay, J.M.	University of Auckland, New Zealand
Lombana-Criollo, C.	Universidad Mariana, Colombia
Loughlin, S.C.	British Geological Survey, UK
Macedonio, G.	Istituto Nazionale di Geofisica e Vulcanologia, Italy
Magill, C.R.	Macquarie University, Australia
Mandeville, C.	US Geological Survey, USA
Marti, J.	Consejo Superior de Investigaciones Científicas, Spain
Marzocchi, W.	Istituto Nazionale di Geofisica e Vulcanologia, Italy
Mee, K.	British Geological Survey, UK
Miller, V.	Geoscience Australia, Australia
Mothes, P.	Instituto Geofísico Escuela Politécnica Nacional, Ecuador
Newhall, C.	Earth Observatory of Singapore, Singapore

Oddsson, B.	Department of Civil Protection and Emergency Management, Iceland
Ogburn, S.E.	University at Buffalo, USA
Ortiz Guerrero, N.	Universidad Mariana, Colombia; Universidad Nacional Autónoma de México, México
Pallister, J.	Volcano Disaster Assistance Program, US Geological Survey, USA
Palma, J.	University of Concepcion, Chile
Poland, M.	Hawaiian Volcano Observatory, US Geological Survey, USA
Potter, S.	GNS Science, New Zealand
Pritchard, M.	Cornell University, USA
Ramon, P.	Instituo Geofísico EPN, Ecuador
Sandri, L.	Istituto Nazionale di Geofisica e Vulcanologia, Italy
Sayudi, D.	Geological Agency of Indonesia, Indonesia
Selva, J.	Istituto Nazionale di Geofisica e Vulcanologia, Italy
Smid, E.	University of Auckland, New Zealand
Solidum, R.U.	Philippine Institute of Volcanology and Seismology, Philippines
Sparks, R.S.J.	University of Bristol, UK
Stewart, C.	Massey University, New Zealand
Stone, J.	University of East Anglia, UK
Subandriyo, J.	Geological Agency of Indonesia, Indonesia
Sumarti, S.	Geological Agency of Indonesia, Indonesia
Surono	Geological Agency of Indonesia, Indonesia
Tonini, R.	Istituto Nazionale di Geofisica e Vulcanologia, Italy
Valentine, G.	University at Buffalo, USA
Vye-Brown, C.	British Geological Survey, UK
Wadge, G.	University of Reading, UK
Wagner, K.	University at Buffalo, USA
Webley, P.	University of Alaska Fairbanks, USA
Wilson, T.M.	University of Canterbury, New Zealand

The editors would also like to thank Nick Barnard and Sue Mahony for their help in preparing the index for this volume.

Foreword

Ray Cas
President, International Association for Volcanology and Chemistry of the Earth's Interior (IAVCEI)
October 2014.

This contribution from the Global Volcano Model Network (GVM) and the International Association for Volcanology and Chemistry of the Earth's Interior (IAVCEI), on the status of global volcanic hazards and risk assessment capability for the United Nations Office for Disaster Reduction (UNISDR) Global Assessment Report for Risk Reduction 2015 (GAR15 Report) is an extremely timely and important reminder that there is still a huge amount of work to be done. GVM is a collaborative international initiative, involving multiple research and government institutions, in collaboration with IAVCEI, and has as its mandate *"to create a sustainable, accessible information platform on volcanic hazard and risk"*. This task would be difficult for any learned association or institution by itself, and has required funding and logistic support from multiple international sources.

Over 130 scientists from 86 institutions in nearly 50 countries worldwide have contributed to this work, representing a remarkable collaborative effort of the volcanological community. The World Organisation of Volcano Observatories (WOVO) is a key Commission of IAVCEI and has contributed to profiles of volcanism for the 95 countries or territories with active volcanoes.

This book provides a state-of-the-art assessment of the preparedness of the global scientific community and government agencies to manage volcano hazards and risks globally. It demonstrates alarmingly that adequate information to make informed hazard and threat assessment exists for only 328 (about 20%) of the Earth's 1,551 "active" volcanoes that are known to have erupted during the Holocene (<10,000 years). The situation is even more concerning when considering that there are many dormant volcanoes that have not erupted in the Holocene, but could still erupt.

This situation clearly indicates that much more needs to be done by governments worldwide to improve both the monitoring capabilities for all the known active volcanoes, and as importantly, undertake detailed investigations of the geological histories of all known active and dormant volcanoes.

Monitoring provides only a modern snapshot of the level of activity or unrest of volcanoes, which is crucial to assessing if volcanic eruption is imminent. Seismic and geodetic networks are core to such monitoring, as is gas sampling and analysis. Development of modern airborne and ground-based remote sensing technologies and data sets are now also enhancing our abilities to assess unrest at volcanoes.

However, even if an eruption is imminent, without a database on the eruption history, the frequency and magnitude of eruptions, and the previous eruption styles of a volcano, trying to predict the most likely hazards and their magnitude, becomes poorly constrained guesswork. *Understanding the geological history of volcanoes is one of the most important tools in modern volcano hazard and risk assessment.* Understanding the previous behaviour of a volcano requires a programme of careful geological mapping, providing data on the dispersal patterns and stratigraphic occurrence of the spectrum of deposit types and their magnitude. Together with knowledge of the geochemistry and geochemical evolution, and a well-constrained geochronological framework of events, factually-based hazard and risk assessment is only then possible.

Sadly it seems that such basic and essential geological knowledge is lacking for almost 80% of the world's active volcanoes! Is this a function of inadequate funding, or an assumption that geological and stratigraphic fieldwork is old fashioned and no longer relevant, or both? This requires urgent attention.

Undertaking geological mapping of volcanoes need no longer be tedious and require covering every square metre of a volcano. Modern remote sensing databases such as Aster, radiometrics, aeromagnetics, LiDAR, etc, offer fast, smart ways of producing first-order maps of volcanoes, that can then be ground-truthed in strategic areas to confirm apparent stratigraphic superposition relationships, evaluate deposit types, collect samples for geochemistry and geochronology, and efficiently produce an assessment of the geological history, eruption styles, deposit types, eruption magnitudes, hazards and risks.

Having compiled a geological database through collaboration with the Smithsonian Institute's Global Volcano Program (GVP), GVM has introduced a Volcano Hazard Index (VHI) for each volcano for which there is an adequate geological record. This important new innovation begins to provide an overview of the range of possible hazards for a particular volcano, the likelihood of specific hazards occurring, and their magnitude, based on the previous history of the volcano. I am pleased to note that just this year to emphasise the importance of understanding the geology of volcanoes, Secretary-General of IAVCEI, Joan Marti, organised an international workshop on the theme of "The geology of volcanoes" on the volcanic island of Madeira. A proposal to form a new IAVCEI Research Commission on this theme is now being prepared.

In addition, measures of the populace exposed to volcanic hazards are introduced to better understand the volcanic threat. A significant statistic of the report is that 800 million people live within 100 km of active volcanoes, 226 live million within 30 km and 29 million live within 10 km. This again highlights the importance of developing a better understanding of volcanic hazards and their impact.

The report also briefly addresses the potential economic impacts of volcanic events, which as global populations increase are just likely to rise. The 2010 Eyjafjallajökull eruption in Iceland was a startling wake-up call on this.

In summary, the GAR15 Report on Global Volcanic Hazards and Risks is a stark reminder that there is still a huge amount of work to be done in understanding the hazards and risks of the world's volcanoes. Major investments are required not only in acquiring and deploying more monitoring equipment on more volcanoes, but also for undertaking ongoing geological mapping and fieldwork to improve understanding of hazards and risks on all active volcanoes.

On behalf of IAVCEI, I congratulate GVM and everyone who has contributed to the GAR15 Report, most of whom are members of IAVCEI. The GAR15 Report will provide UNISDR, governments, IAVCEI and its members with much to consider.

Preface

Volcanic hazards and risk have not been considered in previous global assessments by UNISDR as part of the biennial reports on disaster risk reduction. This book developed as a consequence of Global Volcano Model (GVM) being invited to make such an assessment by UNISDR for its 2015 report. GVM worked in close collaboration with the International Association of Volcanology and Chemistry of the Earth's Interior (IAVCEI) to contribute four background papers for the 2015 Global Assessment Report (GAR15) of UNISDR. These background papers contain a lot more information than could be included in GAR15 and can be construed as the evidence on which UN ISDR have been able to include volcanic risk into their report. Although the background papers were placed on the UNISDR website they would have become part of the ephemeral grey literature that increasingly pervades scientific publication. Thus the decision was made to publish the reports together as an open access e-book with the support of UNISDR.

The book represents the efforts of the global volcanological community to provide a synthesis of what we understand about volcanoes, volcanic hazards and the attendant risks. The book owes its existence to the efforts of many scientists from many countries. There are over 130 authors from 47 countries. Members of the World Organisation of Volcano Observatories (WOVO) have been immensely helpful and collaborative in providing information for the country profiles and making sure that the facts are correct. Outside of those who have directly contributed are many thousands of scientists throughout the world who have provided the data and scientific analysis within the peer-reviewed literature to contribute to the collective knowledge, which we have tried to synthesise. There will be shortcomings and omissions in any endeavour of this kind. GVM and IAVCEI have the ambition to carry out future global analyses to reflect advances in knowledge and to address shortcomings and omissions in this inaugural attempt at a global synthesis.

The book is organised and presented in a rather unconventional way, reflecting that it represents four different background papers for the GAR15. Each background paper has a different and complementary purpose and may also attract different readers. We decided not to change the reports in any significant way apart from some minor re-formatting and cross-referencing. The reader will likely notice some repetition between the main chapters, which reflects the logic of the reporting to UNISDR. Chapter 1 is a summary of our findings and key issues designed for a non-technical readership. We hope that a wide range of people within the disaster risk reduction community will find this chapter accessible. Our findings are evidence-based and draw from the scientific literature as well as some new analysis. We also utilise case studies to illustrate the issues or provide a more detailed analysis of certain key topics. Thus Chapter 2 is essentially a much longer version of Chapter 1 containing much more technical detail and the evidence base on which Chapter 1 draws, including references to the peer-reviewed scientific literature and authoritative sources. This chapter is written more for a technical audience or for those who want to understand the science and evidence in more detail. We do not though assume any expertise in geoscience disciplines so the chapter is reaching out to a wide technical audience within the disaster risk reduction (DRR) and natural hazards communities. Chapter 3 is a more specialist study of volcanic ash fall hazard based on the work of the GVM ash hazard task force. Ash hazard has risen to prominence in recent years due to the impacts on aviation and is the volcanic hazard where probabilistic methods have advanced the most. There are 23 case studies, each of which constitutes a short chapter. Brief synopses of these short case studies are included in Chapter 1 for the non-technical readership, with three supplementary short case studies. These case studies were chosen to illustrate the wide range of scientific and risk management issues related to volcanoes. Finally there is supplementary material, which consists of profiles of each of the 95 countries and territories with active volcanoes. Most of these profiles were written in collaboration with members of the World Organisation of Volcano Observatories (WOVO). The intention is to update these profiles as new information becomes available and it is anticipated that these updates will be a collaboration between GVM and WOVO members.

Acknowledgements

We are indebted to colleagues around the world in the volcanological community who have generated the contemporary understanding of volcanoes on which this book draws. We are very grateful to colleagues and reviewers, including Dave Ramsey, Victoria Avery and Christina Neal of the US Geological Survey, for their valued input and suggestions, which greatly improved this book. Support was provided by the European Research Council and the Natural Environment Research Council of the UK (NERC) through their International Opportunities Fund. We are also thankful for the support provided by the UNISDR for the publication of this work through Cambridge University Press.

Chapter 1

An introduction to global volcanic hazard and risk

S.C. Loughlin, C. Vye-Brown, R.S.J. Sparks, S.K. Brown, J. Barclay, E. Calder, E. Cottrell, G. Jolly, J-C. Komorowski, C. Mandeville, C. Newhall, J. Palma, S. Potter and G. Valentine

with contributions from B. Baptie, J. Biggs, H.S. Crosweller, E. Ilyinskaya, C. Kilburn, K. Mee, M. Pritchard and authors of Chapters 3-26

Contents

1.1 Introduction
1.2 Background
1.3 Volcanoes in space and time
1.4 Volcanic hazards and impacts
1.5 Monitoring and forecasting
1.6 Assessing volcanic hazards and risk
1.7 Volcanic emergencies and DRR
1.8 The way forward

1.1 Introduction

The aim of this book is provide a broad synopsis of global volcanic hazards and risk with a focus on the impact of eruptions on society and to provide the first comprehensive global assessment of volcanic hazard and risk. The work was originally undertaken by the Global Volcano Model (GVM, http://globalvolcanomodel.org/) in collaboration with the International Association of Volcanology and Chemistry of the Earth's Interior (IAVCEI, http://www.iavcei.org/) as a contribution to the Global Assessment Report on Disaster Risk Reduction, 2015 (GAR15), produced by the United Nations Office for Disaster Risk Reduction (UN ISDR). The Volcanoes of the World database collated by the Smithsonian Institution (Siebert et al., 2010, Smithsonian, 2014) is regarded as the authoritative source of information on Earth's volcanism and is the main resource for this study (data cited in this report are from version VOTW4.22).

Chapter 1 provides a short summary of global volcanic hazards and risks intended for a non-technical readership. Chapter 2 provides a more detailed analysis of global volcanic hazards and risks. Chapter 3 focuses on volcanic ash fall hazard and risk. Chapters 4 to 26 provide additional detail and case studies about subjects covered in Chapters 1 and 2. These case studies, along with published literature, provide the evidence base for this work. Summaries of Chapters 4 to 26, and additional case studies 1-3 are provided as an appendix to this chapter.

A complementary report comprising country profiles of volcanism, is provided online in support of this book (Appendix B). The country-by-country analysis of volcanoes, hazards, vulnerabilities and technical coping capacity is provided to give a snapshot of the current state of volcanic risk across the world.

Loughlin, S.C., Vye-Brown, C., Sparks, R.S.J., Brown, S.K., Barclay, J., Calder, E., Cottrell, E., Jolly, G., Komorowski, J-C., Mandeville, C., Newhall, C., Palma, J., Potter, S., Valentine, G. (2015) An introduction to global volcanic hazard and risk. In: S.C. Loughlin, R.S.J. Sparks, S.K. Brown, S.F. Jenkins & C. Vye-Brown (eds) *Global Volcanic Hazards and Risk,* Cambridge: Cambridge University Press.

1.2 Background

Volcanic eruptions can cause loss of life and livelihoods in exposed communities, damage critical infrastructure, displace populations, disrupt business and add stress to already fragile environments (Blong, 1984). Currently, an estimated 800 million people live within 100 km of a volcano that has the potential to erupt [Chapter 4]. These volcanoes are located in 86 countries and additional overseas territories worldwide [see Appendix B]*.

The total documented loss of life from volcanic eruptions has been modest compared to other natural hazards (~280,000 since 1600 AD, Auker et al., 2013). However, a small number of eruptions are responsible for a large proportion of these fatalities, demonstrating the potential for devastating mass casualties in a single event (Figure 1.1). Importantly, these eruptions are not all large and the impacts are not all proximal to the volcano. For example, the moderate-sized eruption of Nevado del Ruiz, (Colombia) in 1985 triggered lahars (volcanic mudflows), which resulted in the deaths of more than 23,000 people tens of kilometres from the volcano (Voight, 1990).

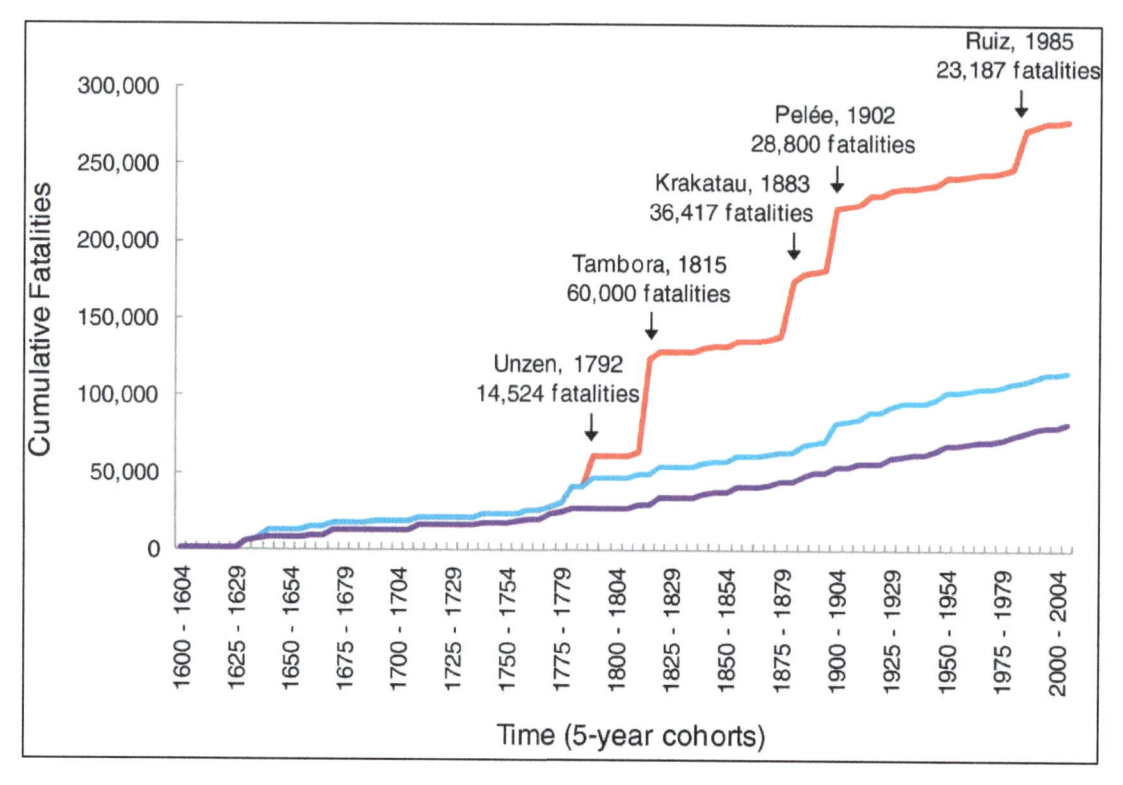

Figure 1.1 Cumulative number of fatalities directly resulting from volcanic eruptions (Auker et al., 2013). Shown using all 533 fatal volcanic incidents (red line), with the five largest disasters removed (blue line), and with the largest ten disasters removed (purple line). The largest five disasters are: Tambora, Indonesia in 1815 (60,000 fatalities); Krakatau, Indonesia in 1883 (36,417 fatalities); Pelée, Martinique in 1902 (28,800 fatalities); Nevado del Ruiz, Colombia in 1985 (23,187 fatalities); Unzen, Japan in 1792 (14,524 fatalities). The sixth to tenth largest disasters are: Grímsvötn, Iceland, in 1783 (9,350 fatalities); Santa María, Guatemala, in 1902 (8,700 fatalities); Kilauea, Hawaii, in 1790 (5,405 fatalities); Kelut, Indonesia, in 1919 (5,099 fatalities); Tungurahua, Ecuador, in 1640 (5,000 fatalities). Counts are calculated in five-year cohorts. This figure is reproduced as Figure 2.13 in Chapter 2.

---

* Appendix B (www.cambridge.org/volcano) comprises country profiles of volcanism.

Despite exponential population growth, the number of fatalities per eruption has declined markedly in the last few decades, suggesting that risk reduction measures are working to some extent (Auker et al., 2013). There has been an increase in volcano monitoring and resultant improvements in hazard assessments, early warnings, short-term forecasts, hazard awareness, communication and preparedness around specific volcanoes (Leonard et al., 2008, Solana et al., 2008, Lindsay, 2010, Larson et al., 2010, Roberts et al., 2011, Marzocchi & Bebbington, 2012, Wadge et al., 2014). Many volcano observatories are active in vulnerable communities, helping to build awareness of volcanic hazards and risk. They now have a key role in building resilience and reducing risk. It is conservatively estimated that at least 50,000 lives have been saved over the last century (Auker et al., 2013) probably as a consequence of these developments. Unfortunately, many volcanoes worldwide are either unmonitored or not sufficiently monitored to result in effective risk mitigation and therefore when they re-awaken the losses may be considerable. The inequalities in monitoring capacity worldwide and the lack of basic geological information at some volcanoes is demonstrated in the country and regional profiles of volcanism in Appendix B.

Volcanic eruptions are almost always preceded by 'unrest' (Potter et al., 2012, Barberi et al., 1984) including volcanic earthquakes and ground movements which can in themselves be hazardous. Volcanic unrest can allow scientists at volcano observatories to provide early warnings if there is a good monitoring network (Phillipson et al., 2013) [Chapters 15 and 18]. Increasingly, effective monitoring from both the ground and space is enabling volcano observatories to provide good short-term forecasts of the onset of eruptions or changing hazards situations (Sparks, 2003, Segall, 2013; Chapter 17). Such forecasts and early warnings can support timely decision-making and risk mitigation measures by civil authorities (Newhall & Punongbayan, 1996, Lockwood & Hazlett, 2013). For example, nearly 400,000 people were evacuated during the November 2010 eruption of Merapi, Indonesia and it is estimated that 10,000 to 20,000 thousand lives were saved as a result (Surono et al., 2012). Nevertheless, there were 386 fatalities reflecting in part the complex contexts in which individuals receive information and make decisions.

Long-lived or frequent eruptions pose particular challenges for communities and there are good examples of social adaptation in response to these difficult situations (e.g. Sword-Daniels, 2011). For example, the long-lived but intermittent eruption of Soufrière Hills Volcano in Montserrat (Lesser Antilles), comprised five phases of lava extrusion between 1995 and 2010 (Wadge et al., 2014). The eruption caused severe social and economic disruption, with 19 fatalities on 25 June 1997(Loughlin et al., 2002), and the subsequent loss of the capital, port and airport. The progressive off-island evacuation of more than 7,500 people (two thirds of the pre-eruption population), left a population of less than 3,000 in 1998 (Clay et al., 1999). A strong cultural identity has helped islanders to cope and a state-of-the-art volcano observatory has become established that continues to support development of new methodologies in hazard and risk assessment [Chapter 21]. Tungurahua in Ecuador has erupted since 1999 and innovative incentives to encourage rapid evacuation have been developed. A system of community 'vigías' (watchers) support scientists, civil defence and their communities by observing the volcano and organising evacuations of their communities if necessary (Stone et al., 2014). Some of the farmers at highest risk have been allocated additional fields away from the volcano, providing options for retreat in times of threat and uncertainty [Chapter 26]. The preservation or rebuilding of livelihoods, critical infrastructure systems and social capital is essential to successful adaptation under these conditions.

The economic impact of volcanic eruptions has recently become more apparent at local, regional and global scales. The 2010 eruption of the Eyjafjallajökull volcano in Iceland caused serious disruption to air traffic in the north Atlantic and Europe as fine volcanic ash in the atmosphere drifted thousands of kilometres from the volcano (Þorkelsson, 2012). The resulting global economic losses from this modest-sized eruption accumulated to about US$ 5 billion (Ragona et al., 2011) as global businesses and supply chains were affected. In the eruption of Merapi, Indonesia in 2010, losses were estimated at US$ 300 million (BNPB., 2011)[Chapters 9 and 10]. Economic losses due to damage of exposed critical infrastructure are unavoidable, but the goal is to minimise them as far as possible through effective long-term planning.

There is often a lack of awareness of volcanic risk both in the proximity of a volcano and further afield, and indeed the risk may not have been assessed at all (Lockwood & Hazlett, 2013). In part this is due to the long duration between eruptions at some volcanoes. Understanding the risks posed by a volcano first requires a thorough understanding of the eruptive history of that volcano, ideally through both geological and historical research (Sparks & Aspinall, 2004). There is still significant uncertainty about the eruption history at many of the world's volcanoes so understanding of potential future hazards, and their likely frequency and magnitude is limited. For example, before the 2008 eruption of Chaitén volcano, Chile, the few studies available suggested that the last major eruption occurred thousands of years ago and little was known of any historical eruptions. The threat appeared low and so the closest monitoring station operated by the national monitoring institution was more than 200 km away. It was only after the 2008 eruption, which resulted in the rapid evacuation of Chaitén town, that new dating was undertaken showing that in fact Chaitén volcano has been more active than previously thought. Had the research been done first, an eruption may have been anticipated (e.g. Lara et al. 2013).

Although volcanoes do pose risks during unrest and eruption, they also provide benefits to society during their much longer periods of repose (Lane et al., 2003, Kelman & Mather, 2008, Bird et al., 2010, Witter, 2012). Volcanic environments are typically appealing: soils are fertile; elevated topography provides good living and agricultural conditions, especially in the equatorial regions (Small & Naumann, 2001); water resources are commonly plentiful; volcano tourism can provide livelihoods; some volcanoes have geothermal systems that can be exploited (Witter, 2012) and some have religious or spiritual significance. These benefits mean that providing equivalent alternatives if evacuation/resettlement is advised can be challenging.

1.3 Volcanoes in space and time

Most active volcanoes (Figure 1.2) occur at the boundaries between tectonic plates (Schmincke, 2004, Cottrell, 2014) where the Earth's crust is either created in rift zones (where tectonic plates move slowly apart) or destroyed in subduction zones (where plates collide and one is pushed below the other). Most volcanoes along rift zones are deep in the oceans along mid-ocean ridges. Some rift zones extend from the oceans and seas onto land, for example in Iceland and the East African Rift valley. The Pacific 'ring of fire' comprises chains of island volcanoes (e.g. Aleutians, Indonesia, Philippines) and continental volcanoes (e.g. in the Andes) that have formed above subduction zones. These volcanoes have the potential to be highly explosive. Other notable subduction zone volcanic chains include the Lesser Antilles in the Caribbean and the South Sandwich Islands in the Southern Atlantic. Some active volcanoes occur in the interiors of tectonic plates above mantle 'hot spots', the Hawaiian volcanic chain and Yellowstone in the USA being the best-known examples.

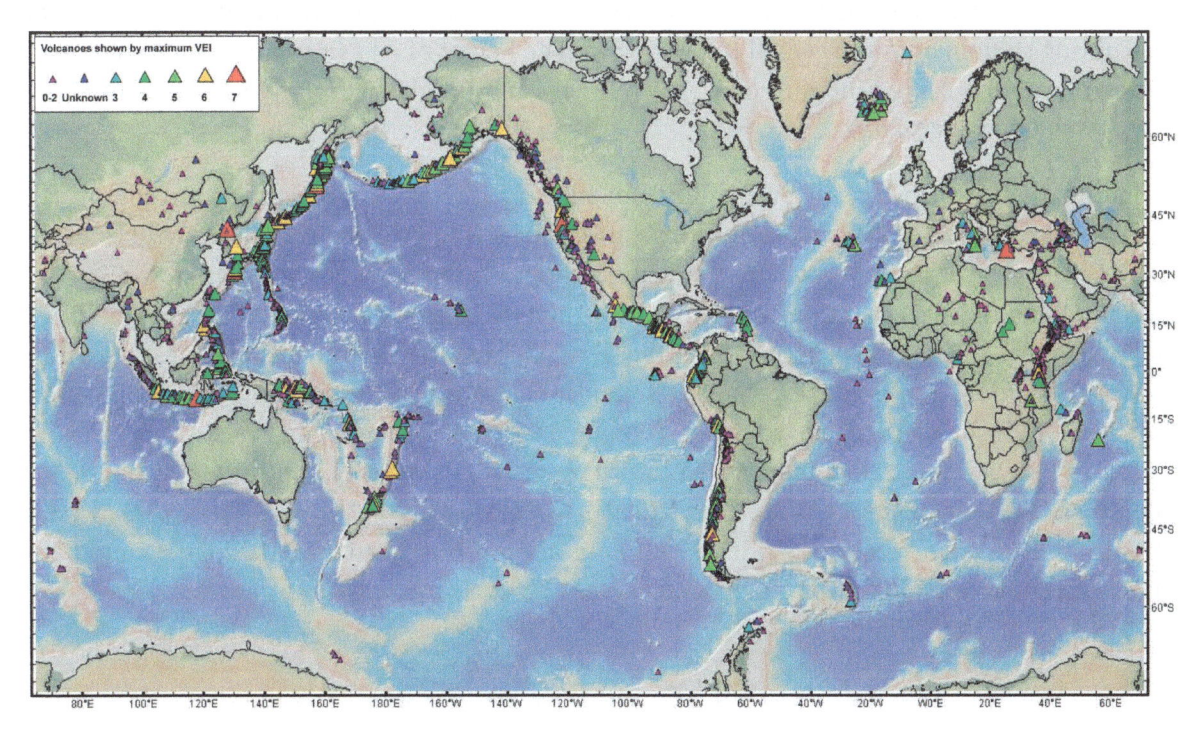

Figure 1.2 Potentially hazardous volcanoes are shown with their maximum recorded Volcanic Explosivity Index (VEI) – a measure of explosive eruption size. Small eruptions (VEI 0-2) and eruptions of unknown size are shown in purple and dark blue. The warming of the colours and the increase in size of the triangles represents increasing VEI. Volcanoes mostly occur along plate boundaries with a few exceptions. There may be thousands of additional active submarine volcanoes along mid-ocean ridges but they don't threaten populated areas. Records are for the Holocene (the last ~10,000 years).

There are many different types of volcanoes in each of these settings, some are typical steep-sided cones, some are broad shields, some of the larger caldera volcanoes are almost indistinguishable on the ground and can only be seen clearly from space (Siebert et al., 2010, Cottrell, 2014). Each volcano may demonstrate diverse eruption styles from large explosions that send buoyant plumes of ash high into the atmosphere to flowing lavas. Each eruption evolves over time, resulting in a variety of different hazards and a wide range of consequent impacts. This variety in behaviours arises because of the complex and non-linear processes involved in the generation and supply of magma to the Earth's surface (Cashman et al., 2013). The subsequent interaction of erupting magma with surface environments such as water or ice may further alter the characteristics of eruptions and thus their impacts. This great diversity of behaviours and consequent hazards means that each volcano needs to be assessed and monitored individually by a volcano observatory.

Volcanic eruptions are usually measured by magnitude and/or intensity (Pyle, 2015) but neither is easy to measure, particularly for explosive eruptions. The magnitude of an eruption is defined as total erupted mass (kg), while intensity is defined as the rate of eruption, or mass flux (kg per second). In order to compare the size of different types of eruptions, a magnitude scale is commonly used. A widely used alternative to characterise and compare the size of purely explosive eruptions is the *Volcanic Explosivity Index* (VEI) which comprises a scale from 0 to 8 (Figure 1.3). The VEI is usually based on the volume of material erupted during an explosive eruption (which can be estimated based on fieldwork after an eruption) and also the height of the erupting column of ash (Newhall & Self, 1982). The height of an ash column generated in an explosive eruption can be measured relatively easily and is related to intensity (Mastin et al., 2009, Bonadonna et al., 2012).

In general, there is an increasing probability of fatalities with increasing eruption magnitude, for example, all recorded VEI 6 and 7 eruptions since 1600 AD have caused fatalities (Auker et al., 2013). Five major disasters dominate the historical dataset on fatalities accounting for 58% of all recorded fatalities since 4350 BC (Figure 1.1). The two largest disasters in terms of fatalities were caused by the largest eruptions (Tambora 1850; Krakatau 1883). Nevertheless, small to moderate eruptions can be devastating, the modest eruptions of Nevado del Ruiz (VEI 3) and Mont Pelée (VEI 4) being good examples (Voight, 1990). A statistical analysis of all volcanic incidents (any volcanic event that has caused human fatalities), excluding the five dominant major disasters, highlights the fact that VEI 2-3 eruptions are most likely to cause a fatal volcanic incident of any scale and VEI 3-4 eruptions are most likely to have the highest numbers of fatalities (Auker et al., 2013).

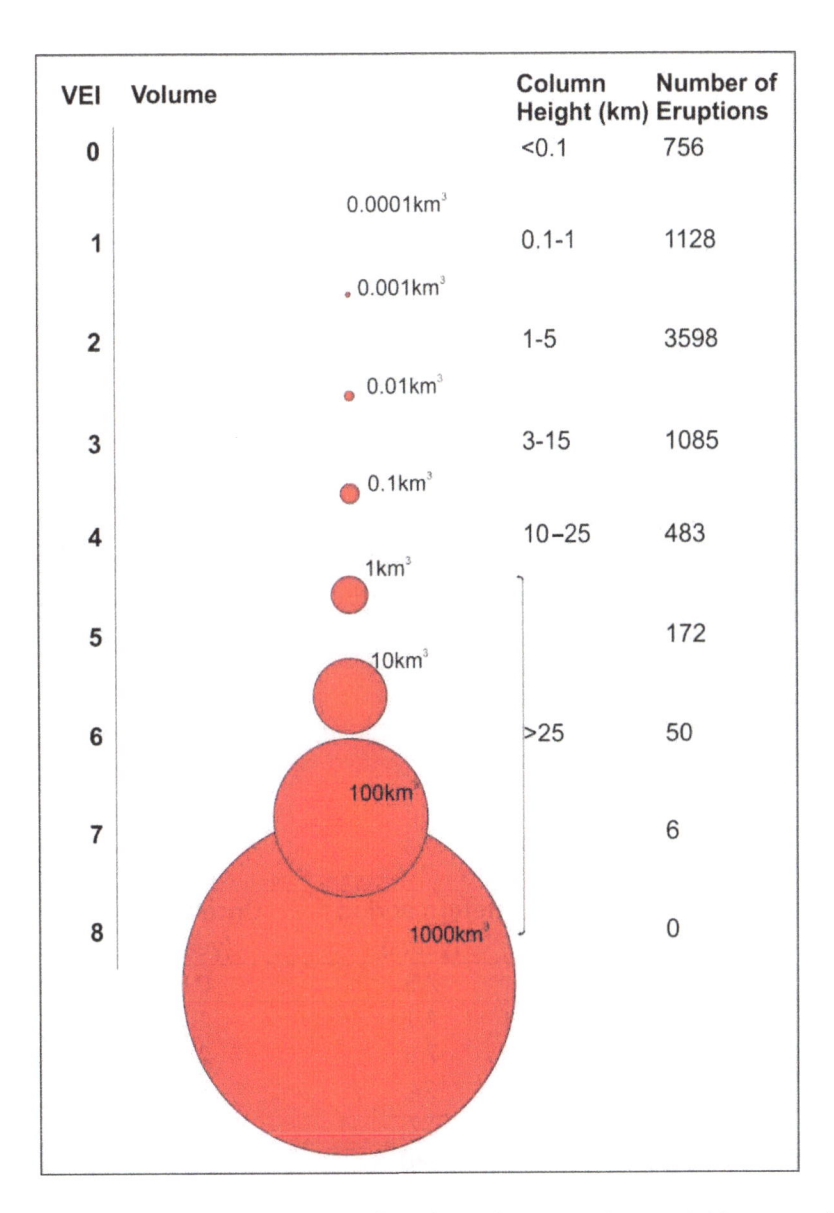

VEI	Volume	Column Height (km)	Number of Eruptions
0		<0.1	756
1	0.0001 km³	0.1–1	1128
2	0.001 km³	1–5	3598
3	0.01 km³	3–15	1085
4	0.1 km³	10–25	483
5	1 km³		172
6	10 km³	>25	50
7	100 km³		6
8	1000 km³		0

Figure 1.3 VEI is best estimated from volume of explosively erupted material but can also be estimated from column height. The typical eruption column heights and number of confirmed Holocene eruptions with an attributed VEI in VOTW4.22 are shown (Siebert et al., 2010).

In total there are 1,551 volcanoes in the Smithsonian Institution database VOTW4.22, of which 866 are known to have erupted in the last 10,000 years (the Holocene). Since 1500 AD, there are 596 volcanoes that are known to have erupted. Only about 30% of the world's Holocene volcanoes have any published information about eruptions before 1500 AD, while 38% have no records earlier than 1900 AD. Geological, historical and dating records become less complete further back in time. Statistical studies of the available records (Deligne et al., 2010, Furlan, 2010, Brown et al., 2014) suggest that only about 40% of explosive eruptions are known between 1500 and 1900 AD, while only 15% of large Holocene explosive eruptions are known prior to 1 AD.

The record since 1950 is believed to be almost complete with 2,208 eruptions recorded from 347 volcanoes. The average number of eruptions ongoing per year since 1950 is 63, with a minimum of 46 and maximum of 85 eruptions recorded per year. On average 34 of these are new eruptions beginning each year.

Going further back in time, the Large Magnitude Explosive Volcanic Eruptions (LaMEVE) database (Crosweller et al., 2012) lists 3,130 volcanoes that have been active in the last 2.58 million years (Quaternary period), and some of these may well be dormant rather than extinct. Many of these volcanoes remain unstudied and much more information is needed to understand fully the threat posed by all of the world's volcanoes. There are also thousands of submarine volcanoes, but the great majority of these (with one or two exceptions) do not constitute a major threat.

Estimating the global frequency and magnitude of volcanic eruptions requires this under-recording to be taken into account (Deligne et al., 2010, Furlan, 2010, Brown et al., 2014). Statistical analysis of global data for explosive eruptions (with under-recording accounted for) shows that as eruption magnitude increases, the frequency of eruptions decreases (Table 1.1).

Table 1.1 Global return periods for explosive eruptions of magnitude M (where M = Log_{10}m -7 and m is the mass erupted in kilograms (Pyle, 2015)). The estimates are based on a statistical analysis of data from VOTW4.22 and the Large Magnitude Explosive Volcanic Eruptions database (LaMEVE) version 2 (http://www.bgs.ac.uk/vogripa/)(Crosweller et al., 2012). The analysis method takes account of the decrease of event reporting back in time (Deligne et al., 2010). Note that the data are for M ≥ 4. This table is reproduced as Table 2.1 in Chapter 2.

Magnitude	Return period (years)	Uncertainty (years)
≥4.0	2.5	0.9
≥4.5	4.1	1.3
≥5.0	7.8	2.5
≥5.5	24	5.0
≥6.0	72	10
≥6.5	380	18
≥7.0	2,925	190
≥7.5	39,500	2,500
≥8.0	133,350	16,000

Volcanoes that erupt infrequently may surprise nearby populations if monitoring is not in place, and eruptions may be large. For example, Pinatubo, Philippines, (Newhall & Punongbayan, 1996) was dormant for a few hundred years before the large eruption in 1991 [Chapter 7], so populations, civil protection services and government authorities had no previous experience or even expectation of activity at the volcano. Conversely, some volcanoes are frequently active and local communities have learned to adapt to these modest eruptions (e.g. Sakurajima, Japan; Etna, Italy; Tungurahua, Ecuador [Chapter 26]; Soufrière Hills volcano, Montserrat (Sword-Daniels, 2011)). Very infrequent, extremely large volcanic eruptions (i.e. VEI 7-8+) have the potential for regional and global consequences and yet we have no experience of such events in recent historical time (Self & Blake, 2008). The super-eruptions that took place at Yellowstone (Magnitude M=8 or more) have a very low probability of occurrence in the context of human society (Table 1.1).

1.4 Volcanic hazards and their impacts

Volcanoes produce multiple primary and secondary hazards (Blong, 1984, Papale, 2014) that must each be recognised and assessed in order to mitigate their impacts. Depending upon volcano type, magma composition, eruption style, scale and intensity at any given time, these hazards will have different characteristics and may occur in different combinations at different times. The major volcanic hazards that create risks for communities include those outlined below:

Ballistics. Ballistics (also referred to as volcanic bombs) are rocks ejected on ballistic trajectories by volcanic explosions. In most cases the range of ballistics is a few hundred metres to about two kilometres from the vent, but they can be blasted to distances of more than 5 km in the most powerful explosions. Fatalities, injuries and structural damage result from direct impacts of ballistics, and those which are very hot on impact can start fires.

Volcanic ash and tephra. Explosive eruptions and pyroclastic density currents (see below) produce large quantities of intensely fragmented rock, referred to as tephra. The very finest fragments from 2 mm down to nanoparticles are known as 'volcanic ash' and can be produced in huge volumes. The physical and chemical properties of volcanic ash are highly variable and this has implications for impacts on health, environment and critical infrastructure [Chapters 12 and 13], and also for the detection of ash in the atmosphere using remote sensing. Falling volcanic ash may cause darkness and very hazardous driving conditions, while concurrent rainfall leads to raining mud. Even relatively thin ash fall deposits (≥ 1 mm) may threaten public health (Horwell & Baxter, 2006, Carlsen et al., 2012) damage crops and vegetation, disrupt critical infrastructure systems (Spence et al., 2005, Sword-Daniels, 2011, Wilson et al., 2012, Wilson et al., 2014), transport, primary production and other socio-economic activities over potentially very large areas. Ash fall creates major clean-up demands (Blong, 1984) [Chapter 12], which need to be planned for (e.g. the availability of large volumes of water for hosing, trucks and sites to dump ash). The accumulation of ash on roofs can be hazardous especially if it is wet; for example, the collapse of roofs during the 1991 Mount Pinatubo eruption killed about 300 people [Chapter 7]. Unfortunately, volcanic ash fall can also be persistent during long-lived eruptions, giving crops, the environment and impacted communities limited chance to recover (Cronin & Sharp, 2002). Remobilisation of volcanic ash by wind can continue for many months or even years after an eruption, prolonging exposure (Carlsen et al., 2012, Wilson et al., 2012).

Volcanic explosions inject volcanic ash into the atmosphere and it may be transported by prevailing winds hundreds or even thousands of kilometres away from a volcano. Airborne ash is a major hazard for aviation (Guffanti et al., 2010) [Chapter 14]. For example, eruptions at Galunggung volcano, Indonesia, in 1982 and Redoubt volcano, Alaska, in 1989 caused engine failure of two airliners that encountered the drifting volcanic ash clouds. Forecasting the dispersal of volcanic ash in the atmosphere for civil aviation (Bonadonna et al., 2012) is a major challenge during eruptions and is the role of Volcanic Ash Advisory Centres supported by volcano observatories [Chapter 12].

The potentially wide geographic reach of volcanic ash, the relatively high frequency of explosive volcanic eruptions and the variety of potential impacts make volcanic ash the hazard most likely to affect the greatest number of people [Chapter 3].

Pyroclastic flows, surges and blasts. These are hot, fast-moving flows (Figure 1.4) that may originate from explosive lateral blasts, the collapse of explosive eruption columns or the collapse of lava domes (Calder et al., 2002). *Pyroclastic flows* are concentrated avalanches of volcanic rocks, ash and gases that are typically confined to valleys, and *pyroclastic surges* are more dilute turbulent clouds of ash and gases that can rapidly spread across the landscape and even travel uphill or across water (Carey et al., 1996). A *volcanic blast* is a term commonly used to describe a very energetic kind of pyroclastic density current which is not controlled by topography and is characterised by very high velocities (more than 100 m/s in some cases) and dynamic pressures (Jenkins et al., 2013). Volcanic blasts can destroy or cause severe damage to infrastructure, vegetation and agricultural land (Blong, 1984, Jenkins et al., 2013, Charbonnier et al., 2013), and can even remove soil from the bedrock (Wadge et al., 2014). The spectrum of flow types are sometimes collectively referred to as *pyroclastic density currents.* They are the most lethal volcanic hazard accounting for one third of all known volcanic fatalities. They travel at velocities of tens to hundreds of kilometres per hour and have temperatures of hundreds of degrees centigrade.

Figure 1.4 Pyroclastic flows from the 1984 explosive eruption of Mayon, Philippines (C. Newhall).

Eyewitnesses have reported that pyroclastic flows and surges make little sound so may offer no warning of their advance if they are not seen (Loughlin et al., 2002). Surviving a pyroclastic density current is very unlikely. Those who have survived in buildings at the margins of dilute currents have been very badly burned, thus the only appropriate response to the threat of an imminent pyroclastic density current is evacuation. Pyroclastic density currents account for one third of all historical volcanic fatalities (Auker et al. 2013).

Lahars and floods. Lahars (volcanic mudflows) are fast-moving mixtures of volcanic debris and water that can destroy bridges and roads, bury buildings and cut off escape routes (Figure 1.5). Lahars can directly affect areas tens of kilometres from a volcano and may cause flooding hazards at even greater distances. They may occur when intense rain falls on unconsolidated volcanic ash and debris, but they may also result from volcanic activity melting summit ice caps/glaciers or from eruptions in crater lakes.

Figure 1.5 a) Only the roofs of 2-storey buildings are visible after repeated inundation by lahars following the 1991 eruption of Pinatubo, Philippines (C. Newhall). b) Lahars during the 1991 eruption of Pinatubo in the Philippines caused the destruction of concrete bridges (USGS archive).

Geothermal activity beneath ice or the breaching of crater lakes and reservoirs can also trigger lahars between eruptions. Following explosive eruptions the potential for lahars during heavy rainfall can persist for years or even decades if there are significant thicknesses of loose deposits, as was the case following the 1991 eruption of Pinatubo in the Philippines [Chapter 7]. Such long-term disruption can seriously impact recovery. Lahars account for 15% of all historical volcanic fatalities (Auker et al., 2013).

Debris avalanches, landslides and tsunamis. *Debris avalanches* can be large and remarkably mobile flows formed during the major collapse of volcanic edifices. They are commonly associated with volcanic eruptions or magmatic intrusions and may be a particular issue in edifices which have been weakened by active hydrothermal systems (Siebert, 1984, Voight, 2000). Debris avalanches can lead to *lateral volcanic blasts* as the highly pressurised interior of a volcano is exposed (e.g. Mount St. Helens, USA, 1980). The rapid entry of voluminous debris avalanches into the sea displaces large volumes of water and may cause tsunamis. In 1792 a debris avalanche from Mount Unzen, Japan, caused a tsunami resulting in over 32,000 fatalities. Most of the 36,417 fatalities reported during the 1883 eruption of Krakatau, Indonesia, were the result of tsunamis generated by pyroclastic flows entering the sea (Mandeville et al., 1996). Most volcanoes are steep-sided mountains partly built of poorly consolidated volcanic deposits and many are in multiple hazard environments. Volcanic landslides and debris avalanches can be caused by intense rainfall or regional tectonic earthquakes. Hurricane Mitch in 1998 triggered a major landslide on Casita volcano in Nicaragua, causing at least 3,800 fatalities. Landslides are common on many volcanoes, whether active or not.

Volcanic gases and aerosols. Volcanic gases can directly cause fatalities, health impacts and damage to vegetation and property [Chapters 10, 11 and 13]. Although the main component of gases released during most eruptions is water vapour, there are many other gas species and aerosols released, including carbon dioxide, sulfur dioxide, hydrogen sulphide and halogens (hydrogen fluoride and chloride). The impact of volcanic gases on people depends on the concentrations present in the atmosphere and the duration of exposure. Volcanic gases tend to be more dense than air and may accumulate in depressions or confined spaces (such as basements and work trenches), or flow along valleys. In 1986, a sudden overturn of Lake Nyos in Cameroon (Oku Volcanic Field) released a silent and invisible cloud of carbon dioxide that flowed into surrounding villages, causing 1,800 fatalities as a result of asphyxiation (Kling et al., 1987). Such lake overturns may occur without eruptive activity, for example following earthquakes or landslides into lakes (e.g. Lake Kivu (Baxter et al., 2003) [Chapter 11]).

Fluorine- and chlorine-bearing gases can also be hazardous and may adhere to the surfaces of erupting volcanic ash which subsequently falls to the ground. If people and/or animals consume affected water, soil, vegetation or crops they can be affected by fluoride poisoning. Volcanic gases emitted by a volcano may combine with rainfall to produce acid rain, which damages sensitive vegetation and ecosystems. Sulfur dioxide gas converts in the atmosphere to sulfate aerosols, a major cause of air pollution (Schmidt et al., 2011).

Lava. Anything in the path of a lava flow will be damaged or destroyed, including buildings, vegetation and infrastructure. They usually advance sufficiently slowly to allow people and animals time to evacuate. Nevertheless, unusual chemical compositions found at a small number of volcanoes can produce rapidly flowing lavas. For example, Nyiragongo in the Democratic Republic of Congo has a summit crater containing a lake of very fluid lava. In 1977, the crater wall fractured releasing the lava which flowed downhill at speeds of more than 60 km/h. An estimated 70 people were killed (Komorowski et al., 2002-2003). Another exceptionally mobile lava flow in 2002 [Chapter 11] destroyed about 13% of Goma city, 80% of its economic assets, part of the international airport runway and the homes of 120,000 people (Komorowski, 2002-2003). These losses combined with felt earthquakes and fear of death caused severe psychological distress (Baxter et al., 2003).

In contrast, very viscous lava will pile up to form a *lava dome* above a vent. Domes can be extremely hazardous with high pressure, gas-rich interiors and a tendency for partial or total collapse leading to pyroclastic flows and surges (pyroclastic density currents) [Chapter 9].

Volcanic earthquakes. Volcanic earthquakes are typically small in magnitude (≤M5) and relatively shallow, but they may be felt and may cause structural damage. They may be particularly strong before a volcanic eruption as magma is forcing a path through the Earth's crust. Most volcanoes are in tectonically active environments prone to larger and more destructive earthquakes.

Lightning. Lightning occurs during explosive eruptions in volcanic ash clouds and has caused a number of fatalities (Auker et al., 2013).

Each volcanic hazard is a controlled by different physical and chemical processes that may occur at varying intensities and for different durations over time. Different hazards may occur concurrently (e.g. pyroclastic density currents and volcanic gas) or sequentially (ash fall followed by generation of lahars during intense rainfall). Some hazards are short-lived (e.g. ballistics associated with an explosion) or long-lived (e.g. repeated volcanic ash fall over weeks and months).

Secondary hazards such as disease or famine arising from evacuation, contaminated water, crop failure, loss of livestock, pollution and environmental degradation for example, can be widespread and account for over 65,000 fatalities since 1600 AD (Auker et al., 2013). If a volcanic eruption is superimposed on an existing humanitarian crisis, as occurred in Goma, Democratic Republic of Congo, in 2002, the likelihood of cascading impacts is much higher (Baxter et al., 2003).

Consideration for the short- and long-term health consequences of various volcanic hazards has been a focus of attention for many years, resulting in a compilation of resources (including recommended sampling and analysis protocols) and a network of experts known as the International Volcanic Health Hazard Network [Chapter 13]. Concentration thresholds and durations of exposure to volcanic gases, for example, are available to enable quantitative risk assessments to be developed for particular hazards scenarios [Chapter 21].

1.5 Monitoring and forecasting

1.5.1 Monitoring

A volcano observatory is an institution (e.g. geological survey, university, national research institute, meteorological office, or dedicated observatory) whose role it is to monitor active volcanoes and provide early warnings of anticipated volcanic activity to the authorities and usually also the public [Chapter 15]. There are more than 100 volcano observatories worldwide and many have responsibility for multiple volcanoes. Indeed, many have responsibility for multiple hazards including earthquakes and tsunami. For each country, the exact constitution and responsibilities of a volcano observatory may differ, but it is typically the source of authoritative short-term forecasts of volcanic activity as well as scientific advice about hazards and in some cases risk. They therefore have a key role in building resilience and reducing risk. They also have a critical role in ensuring aviation safety around the world working collaboratively with the world's Volcanic Ash Advisory Centres (VAACs; Chapter 14).

Volcanic eruptions are usually preceded by days to months or even years of precursory activity or 'unrest' (Siebert et al., 2010, Phillipson et al., 2013), unlike other natural hazards such as earthquakes. Detecting and recognising these signs provides the best means to anticipate eruptions, and to mitigate against potential risks [Chapter 18]. Unfortunately, only about 35% of Earth's historically active (those with eruptions since 1500 AD) volcanoes are continuously monitored, which is essential if scientists are to identify and act upon such warning signs. Based on reports from volcano observatories between 2000-2011 as summarised by the Global Volcanism Program of the Smithsonian Institution, 228 monitored volcanoes experienced unrest (Phillipson et al., 2013) and approximately half of them went on to experience eruptions within an 11 year time period.

Ground-based monitoring programmes for active volcanoes typically include (Sparks et al., 2012): a network of seismometers to detect volcanic earthquakes caused by magma movement (Chouet, 1996, McNutt, 2005); a ground deformation network (e.g. Global Positioning System) to measure the rise and fall of the ground surface as magma migrates in the subsurface (Dzurisin, 2003, Larson et al., 2010); remote sensing assessment of gas emissions into the atmosphere (Nadeau et al., 2011, Edmonds, 2008); sampling and analysis of gases and water emitted from the summit and flanks of a volcano (Aiuppa et al., 2010); observations of volcanic activity using webcams and thermal imagery; measurements of other geophysical properties (e.g. strainmeters (Roberts et al., 2011), infrasound (Johnson & Ripepe, 2011)) and environmental indicators (e.g. groundwater levels). Volcano observatories may have telemetry that enables real-time analysis of monitoring data, particularly seismicity, or staff may undertake campaigns to collect data from sensors on a regular basis (e.g. daily, weekly).

Near real-time automatically processed monitoring data are increasingly being made available online by volcano observatories. Real-time monitoring allows the public and civil authorities to improve their understanding of monitoring methods and gain awareness of background activity during quiescence. Monitoring then facilitates real-time decision-making. For example, in Iceland before the Eyjafjallajökull eruption in 2010, some individuals self-evacuated before the official evacuation was announced when they saw the rapidly increasing numbers of earthquakes (Icelandic Meteorological Office: http://en.vedur.is/earthquakes-and-volcanism/earthquakes/).

Ground-based monitoring instrumentation can be vulnerable to destruction by volcanic activity or other threats, such as weather, theft or fire, so resources to maintain and restore monitoring are required. There are excellent examples of monitoring capability being developed very quickly and effectively and even improved after losses. For example the Vanuatu Geohazards Observatory was completely destroyed by fire in 2007, leaving Vanuatu with no monitoring capacity. Following this, Vanuatu Geohazards and GNS Science, New Zealand, formed a partnership installing new monitoring equipment and improving the monitoring capabilities (Todman et al., 2010).

Information derived from satellite earth observation can be a valuable addition to monitoring. High temporal and spatial resolution satellite remote sensing of volumetric changes in topography (of a growing lava dome) complemented ground monitoring and contributed to the rapid and timely evacuation at Merapi volcano, Indonesia in 2010 (Surono et al., 2012) [Chapter 10]. Radar (InSAR) is able to detect unrest at volcanoes previously thought to be dormant or extinct (Biggs et al., 2009), but whether this unrest is caused by magmatic movement or other processes requires validation using ground-based methods (Larson et al., 2010). Thermal anomalies can be correlated with eruption rate of magma, and ash and sulfur dioxide can also be detected in the atmosphere (Bonadonna et al., 2012). Only a few volcano observatories have the capacity to process satellite data in-house. However, moves by the space agencies to contribute to post-Hyogo Framework for Action initiatives signal that satellite remote sensing has significant potential in disaster risk reduction [Chapter 17]. A wider participation in the International Charter for Space and Major Disasters and greater access to data and free and open-source software will undoubtedly contribute to further effective risk mitigation actions [Chapter 9].

The Global Volcano Research and Monitoring Institutions Database (GLOVOREMID, [Chapter 19]) is in development. This will allow an understanding of global monitoring capabilities, equipment and expertise distribution to be developed and will highlight gaps. GLOVOREMID began as a study of monitoring in Latin America, comprising 314 Holocene volcanoes across Mexico, Central and South America [Chapter 19]. Efforts to expand GLOVOREMID to a global dataset are ongoing, but it is not yet complete.

A useful objective globally is to establish a minimum of baseline monitoring (e.g. seismometers) at all active volcanoes. Such monitoring levels will at least detect some signs of unrest so that enhanced monitoring networks can be rapidly deployed if necessary. There are nevertheless many locations where *rapid* deployment is not possible, a situation that should be considered in contingency planning.

1.5.2 Forecasting and early warning

An ability to forecast the onset of an eruption and significant changes during an eruption, are key components of an effective early warning system (Sparks & Aspinall, 2004, Marzocchi & Bebbington, 2012). Intensive monitoring of recent eruptions has generated integrated time-series of data, which have resulted in several successful examples of warnings being issued on impending eruptions [Chapters 7 and 9].

Real-time analysis of multi-parameter time-series datasets is necessary to make reliable and robust forecasts at volcanoes (Nadeau et al., 2011, Sparks et al., 2012). It has become evident that some signals or combinations of signals have more diagnostic value than others. Volcanic earthquakes, in particular long period earthquakes have been used to make short-term forecasts of eruptions (Chouet, 1996), for example at Popocatepetl, Mexico, in 2000 when thousands were evacuated 48 hours before a large eruption. Such earthquakes were also a strong indicator of imminent eruption at Soufrière Hills volcano, Montserrat, and elsewhere.

The ability of a volcano observatory to effectively make short-term forecasts about the onset of a volcanic eruption or an increase in hazardous behaviour during an eruption is dependent on many things. They include having functioning monitoring equipment and telemetry, real-time data acquisition and processing, as well as some knowledge of the past behaviour of the volcano and a conceptual model for how the volcano works. There needs to be a team that includes skilled research scientists and technicians, with sufficient resources to respond when necessary, maintain equipment, acquire, process and interpret data, as well as disseminate knowledge and information on hazard (and possibly risk) to multiple stakeholders in a timely and effective way. Increasingly the ability to acquire and process Earth Observation data is necessary.

The great complexity of natural systems means that we cannot in most cases give exact time and place predictions of volcanic eruptions and their consequences. There have been a few exceptions, for example, before the 1991 and 2000 eruptions of Hekla, Iceland, public warnings were issued tens of minutes before each eruption began with the likely time of eruption indicated (Sparks, 2003, Roberts et al., 2011). The predictions were correct to within a few minutes. In general though, forecasting the outcomes of volcanic unrest and ongoing eruptions is inherently uncertain. Forecasts are becoming increasingly quantitative, evolving from empirical pattern recognition to forecasting based on models of the underlying eruption dynamics. This quantitative approach has led to the development and use of models for forecasting volcanic ash fall and pyroclastic flows, for example. Forecasting requires the use of quantitative probabilistic models to address aleatory uncertainty (irreducible uncertainties relating to the inherent complexity of volcanoes), as well as epistemic uncertainty (data- or knowledge-limited uncertainties). Forecasts of eruptions and hazards can be developed in a manner similar to weather forecasting [Chapter 24] (Sparks & Aspinall, 2004).

Tools can be developed to support scientists in hazards analysis (e.g. modelling tools) and also to support consistent decision-making, such as raising and lowering alert levels. Event trees have been successfully used at many eruptions worldwide since the 1980s (Newhall & Hoblitt, 2002, Lockwood & Hazlett, 2013)[Chapter 7]. Bayesian Belief Network analysis is another method (Sparks et al., 2013, Hincks et al., 2014, Marzocchi & Bebbington, 2012), which provides logical frameworks for discussing probabilities of possible outcomes at volcanoes showing unrest or already in eruption (Sparks & Aspinall, 2004, Newhall & Hoblitt, 2002) [Chapter 8]. Other Bayesian tools are particularly useful for short-term forecasting. They take account of available monitoring information [Chapters 6 and 8] and patterns of previous volcanic behaviour and can help to ensure consistency (Lockwood & Hazlett, 2013) of scientific advice, thereby assisting public officials in making urgent evacuation decisions and policy choices [Chapter 10].

Such tools can be valuable for discussion between scientific teams, but also can facilitate communication with authorities and the public. The probability estimates might be based on past and current activity (empirical), expert elicitation (Aspinall, 2010), numerical simulations, or a combination of methods. The probabilities can be revised regularly as knowledge or methodologies improve or when volcanic activity changes.

Short-term forecasting and recognition of the very dynamic nature of risk is essential for rapid response actions such as evacuation. Longer term forecasts over years or decades will be based mainly upon geological and geochronological data. Probabilistic forecast models for major hazards should ideally be used for managing risk at identified high-risk volcanoes, where both long-term mitigation actions such as moving critical infrastructure or short-term mitigation actions, such as evacuation, incur considerable costs. Long-term forecasts of the likelihood of volcanic activity over a given period of time (e.g. 100 years) can be extremely useful for mitigation actions such as land use planning.

1.6 Assessing volcanic hazards and risk

In order to make a thorough risk assessment, hazard, exposure and vulnerability must all be accounted for. Indeed, there are many factors that contribute to risk. In practice, most volcano observatories have focused on hazard assessments and where risk assessments are made there has been a tendency to focus only on hazard and exposure, and to consider only loss of life. Methods to quantify different aspects of vulnerability to volcanic hazards are improving and there are examples of detailed and comprehensive qualitative and semi-quantitative assessments of vulnerability to volcanic hazards (Spence et al., 2005), leading to risk mitigation recommendations. There is considerable potential to develop quantitative risk assessment methodologies to include loss of livelihoods, loss of critical infrastructure and economic losses for example. There is also future potential in risk monitoring.

1.6.1 Hazards assessments and maps

Given the large number of individual volcanic hazards, each of which has different characteristics, hazard assessment is inevitably complex and multi-faceted and reliable hazard assessment requires volcano-by-volcano investigation. In most countries, the volcano observatory (or official institution) provides scientific advice about hazards to the local and national authorities who hold the responsibility to take mitigation measures (e.g. evacuation). The actual mechanism for provision of this advice differs from country to country, depending on the relevant legislation.

There is scientific consensus that any hazard analysis should be based on understanding of a volcano's past eruptive activity through time combining field geology, geochemical characterisation and dating. The next step requires modelling and statistical approaches but based on a thorough understanding of the data.

An important concept in natural hazards is the *hazard footprint*, which can be defined as the area likely to be adversely affected by a hazard over a given time period. Hazards assessments thus usually take the form of maps. They are typically based upon one or more volcanic hazards and knowledge of past eruptions from geological studies and historical records over a given period of time. Hazard maps take many forms, from circles of a given radius around a volcano, or different zones likely to be impacted by different hazards, to probabilistic maps based on hazard modelling. 'Risk management' maps integrate hazards and identify zones of overall increasing or decreasing hazard. Thus they show communities at highest risk. There are also a variety of probabilistic maps that depend on the nature of the hazard. For volcanic flows (pyroclastic density currents, lahars and lavas) the map typically displays the spatial variation of inundation probability over some suitable time period or given that the flow event takes place [Chapter 20]. For volcanic ash fall hazard the probability of exceeding some thickness or loading threshold is typically presented (Jenkins et al., 2012). Hazards maps and derivative risk management maps can be used for multiple purposes, such as raising awareness of hazards and identifying likely impacts to enable effective land use planning and to help emergency managers mitigate risks (Lockwood & Hazlett, 2013).

Once a volcanic eruption has begun, hazards maps may become rapidly obsolete as topography is changed. For example, valleys extending from a volcano's summit may fill with hot pyroclastic deposits enabling subsequent pyroclastic density currents to travel further (Loughlin et al., 2002). Frequent updates of some hazards maps may therefore be necessary.

Most hazard assessments focus at the volcano scale, but probabilistic methods can be now applied to ash fall hazards at regional (Jenkins et al., 2012) and global scales (Chapter 3). Given that ash fall is the hazard that affects most people through a variety of different impacts, this approach provides a valuable way to manage and mitigate a number of risks.

1.6.2 Exposure and vulnerability

There can be many different kinds of loss as a consequence of volcanic eruptions including: loss of life and livelihoods (Kelman & Mather, 2008, Usamah & Haynes, 2012); detrimental effects on health [Chapter 13]; destruction or damage to assets (e.g. buildings, bridges, electrical lines and power stations, potable water systems, sewer systems, agricultural land) (Blong, 1984, Wilson et al., 2012, Wilson et al., 2014); economic losses (Ragona et al., 2011); threats to natural resources including geothermal energy (Witter, 2012); systemic vulnerability; and loss of social capital. Each of these will have its own specific characteristics in terms of exposure and vulnerability, which, like hazards, will vary in space and time (Adger, 2006). Therefore, moving from hazard to risk ideally requires an assessment of exposed populations and assets, as well as their vulnerability.

In the vicinity of volcanoes, the potential for loss of life has been the priority, and hazard 'footprints' are traditionally superimposed on census data to identify 'exposed' populations for preliminary societal risk calculations. Similarly hazard footprints can be used to identify exposed assets, such as buildings, critical infrastructure, environment, ecosystems and so on.

Vulnerability has many forms which may include physical, social, organisational, economic and environmental. In terms of social vulnerability, geographically, socially or politically marginalised communities are typically the most vulnerable. Within these communities the young, elderly and sick are some of the more vulnerable individuals. The resilience of livelihoods is increasingly recognised as a key factor that plays a role in the vulnerability and exposure of communities and individuals. For example, if subsistence farmers are evacuated, the longer the period of evacuation, the more likely it is that attempts will be made to return to evacuated at-risk areas to harvest crops and care for livestock and this has been documented many times around volcanoes (e.g. Philippines (Seitz, 2004); Ecuador (Lane et al., 2003); Indonesia (Laksono, 1988), Tonga (Lewis, 1999)). Providing options (e.g. alternative farmland) has proven an effective risk mitigation technique in several places (e.g. Ecuador (Lane et al., 2003)). The same issues apply to all scales of private enterprise and there are examples of individuals and businesses trying to retrieve capital assets from high-risk evacuated areas. Physical vulnerabilities are typically closely associated with social vulnerabilities and may include, for example, the type and quality of roofing, and the quality of evacuation routes and transport. Assessing the vulnerability of critical systems which support communities specifically addresses the complex nature of vulnerability with its many variables and enables the analysis of resilience (Sword-Daniels, 2011). Vulnerabilities are ideally assessed at a community level and with a strong understanding of the local social, cultural, economic and political landscape.

Nevertheless, this should always be considered in a wider context. For example, tourists have been recognised as a vulnerable group unlikely to be aware of evacuation procedures or how to receive emergency communications when volcanic activity escalates (Bird et al., 2010). Volcanic eruptions can lead to populations being evacuated and displaced for considerable periods of time and may ultimately lead in some cases to permanent resettlement (Usamah & Haynes, 2012). If the conditions under which evacuees must live are poor, individuals are more likely to return to their homes in at-risk areas. For example, in Montserrat, Lesser Antilles, evacuated families were living in temporary shelters for months and ultimately years (Clay et al., 1999), and some individuals sought peace and quiet at their homes in the evacuated zone or continued to farm, resulting in 19 unnecessary deaths in 1997 (Loughlin et al., 2002). Concerns about looting also cause people to delay evacuation or return to at-risk areas.

A health and vulnerability study for the Goma volcanic crisis in 2002 considered human, infrastructural, geo-environmental and political vulnerability following the spontaneous and temporary evacuation of 400,000 people at the onset of the eruption (Baxter et al., 2003). The area was already in the grip of a humanitarian crisis and a chronic complex emergency involving armies and armed groups of at least six countries. The potential for cascading health impacts (e.g. cholera epidemic) as a result of such a large displaced and vulnerable population was extremely high, however in the case of Goma, the response was remarkable and catastrophic losses were averted [Chapter 11].

The forensic analysis of past volcanic disasters offers an opportunity to identify and investigate risk factors in different situations and also to identify evidence of good practice (Integrated Research on Disaster Risk Forensic Investigations of Disasters: http://www.irdrinternational.org/projects/forin/). Long-lived eruptions such as Soufrière Hills volcano, Montserrat, and Tungurahua, Ecuador, offer opportunities to assess adaptation to extensive risks, for example coping with the cascading impacts of repeated ash fall (Sword-Daniels, 2011).

Like natural hazards, understanding all the factors that contribute to vulnerability and exposure at any particular place at a particular moment in time is challenging. Nevertheless, growing knowledge, improved methodologies and an increasing willingness to integrate information across disciplines should contribute to increased understanding of risk drivers.

1.6.3 Volcanic risk

The priority in the vicinity of volcanoes has been risk to life and only in recent years have volcanologists started to try to quantify such risks. The great value of quantification is that it allows risks to be measured, ranked and compared. Quantifying vulnerability in particular is challenging and is only beginning to be applied for volcanic risk analysis (Kelman & Mather, 2008, Marzocchi & Woo, 2009). To facilitate semi-quantitative approaches to risk, vulnerability is commonly converted to indices. For example the vulnerability of roofs to collapse following ash fall (physical vulnerability) can be assessed using an index of different roof types and thresholds for collapse under different conditions (Spence et al., 2005).

A common means of representing volcanic risk, following methods used for industrial accidents, is to consider the societal risk in terms of the probability of exceeding a given number of fatalities N and the cumulative frequency F of events having N or more fatalities. The resulting

F-N curves have been used successfully in Montserrat [Chapter 21]. Also in Montserrat, a study on the exposure of the population to very fine respirable ash (Hincks et al., 2006) combined volcanology, sedimentology, meteorology and epidemiology to assess the probability of exposure to ash of different population groups over a 20-year period. The study illustrates the multi-disciplinary character of risk assessments, where diverse experts are needed. Quantitative risk assessments are also being developed for cities exposed to particularly high-risk volcanoes [Chapters 5 and 6] where rigorous, repeatable and defendable analysis is essential.

Other potential losses, such as livelihoods, infrastructure, buildings, agriculture and environmental assets, would all benefit from rigorous hazard and risk assessment approaches. In most cases though, despite the considerable potential of quantitative risk assessment approaches, volcanic risks have so far been managed without being quantified. Where vulnerabilities have been identified and assessed in a qualitative manner, they can be addressed. For example, communities identified as vulnerable can be engaged in participatory risk reduction activities. A good example is the system of community '*vigías*' (volcano watchers) in place in Ecuador to support the volcano observatory and to ensure rapid communication between at-risk communities and civil authorities in the event of a sudden escalation in volcanic activity [Chapter 26]. The communities themselves take account of the most vulnerable individuals in their evacuation planning.

More participation of communities in risk assessment, risk management and risk reduction can have considerable benefits to the community and can influence the psychological and sociological aspects of risk. For example, there is evidence that uncertainties may be better understood and there is more acceptance of risk reduction actions taken in the face of uncertainty. Participatory approaches can also benefit scientists and civil authorities through an increase in trust and greater awareness of local knowledge (Haynes et al., 2008a).

The temporal and spatial scales of risk assessments brings in different uncertainties and assumptions due to data availability. Care is needed that assessments do not appear contradictory at different scales. There is a need for harmonisation of methods and data sources. Exposure is largely dealt with through population data and vulnerabilities to various volcanic hazards are usually expressed using proxies, such as the Human Development Index (HDI). Building inventories including roof types could allow the application of established indices for structural vulnerability to ash fall.

For example, in SE Asia, volcanic ash fall is the volcanic hazard most likely to have widespread impacts since a single location may receive ash fall at different times from different volcanoes. Tephra fall thickness exceedance probability curves can be calculated using volcanic histories and simulations of eruption characteristics, eruption column height, tephra volume and wind directions at multiple levels in the atmosphere (Jenkins et al., 2012). Exposure can be calculated using urban population density based on LandScan data and the HDI to contribute towards an estimate of risk across a region. Analysis shows the influence of each of the risk components to total risk for each city from a 1 mm or greater fall of tephra, highlighting the different contributions made by hazard, exposure and vulnerability [Chapter 12].

Increasing the opportunities to integrate knowledge and experience from scientists (of all disciplines), authorities and communities at risk should enable improvements in understanding risk, enhancing resilience, supporting adaptation and reducing risk.

1.6.4 A new global assessment of volcanic threat

The UN Global Assessment Report (2015) required a new assessment of volcanic hazard and risk at global, regional and country scales in order to identify countries and regions at significant risk, to identify gaps in knowledge and to enable prioritisation of resources. A standardised and simple approach was needed and so indices were developed for hazard and exposure. The supplementary online report (Appendix B) provides a compendium of regional and country profiles, which use these indices, where sufficient data allows, to identify high-threat volcanoes.

The Volcano Hazard Index (VHI) characterises hazard at volcanoes based on their recorded eruption frequency, modal and maximum recorded VEI levels and occurrence of pyroclastic density currents, lahars and lava flows. The full methodology is given in Chapter 22. The index builds on previous similar approaches (Ewert et al., 2005, Aspinall et al., 2011).

The VHI is too coarse for local use, but is a useful indicator of regional and global threat. The VHI can change for volcanoes as more information becomes available and if there are new occurrences of either volcanic unrest or eruptions or both. Unfortunately, lack of data for many of the world's volcanoes precludes the possibility of assessing all volcanoes in this way. 328 volcanoes have eruptive histories judged sufficiently comprehensive to calculate VHI and most of these volcanoes (305) have had documented historical eruptions since 1500 AD. There are 596 volcanoes with post-1500 AD eruptions, so the VHI can currently be applied to just over half the world's recently active volcanoes. A meaningful VHI cannot currently be calculated for the remaining 1,223 volcanoes due to lack of information. The absence of thorough eruptive histories (based on geological, geochronological and historical research) for most of the world's volcanoes makes hazard assessments at these sites particularly difficult. This knowledge gap must be addressed with urgency.

The Population Exposure Index (PEI) is based on populations within 10, 30 and 100 km of a volcano, which are then weighted according to evidence on historical distributions of fatalities with distance from volcanoes. The methodology extends previous concepts (Ewert & Harpel, 2004) and is given in Chapter 4.

Volcano population data derived from VOTW4.0 are used to calculate PEI, which is divided into seven levels from sparsely to very densely populated areas. The PEI is an indicator of relative threat to life and can be used as a proxy for economic impact based on the distance from the volcano. This method does not account for secondary losses, such as disease or famine, or far-field losses due to business disruption as a result of volcanic ash and gas dispersion.

The VHI is here combined with the PEI to provide an indicator of risk, which is divided into Risk Levels I to III with increasing risk. The aim is to identify volcanoes which are high risk due to a combination of high hazard and population density. 156, 110 and 62 volcanoes classify as Risk Levels I, II and III respectively. In the country profiles of Appendix B, plots of VHI versus PEI provide a way of understanding volcanic risk. Indonesia and the Philippines are plotted as an example (Figure 1.6). Volcanoes with insufficient information to calculate VHI should be given serious attention and their relative threat should be assessed through PEI.

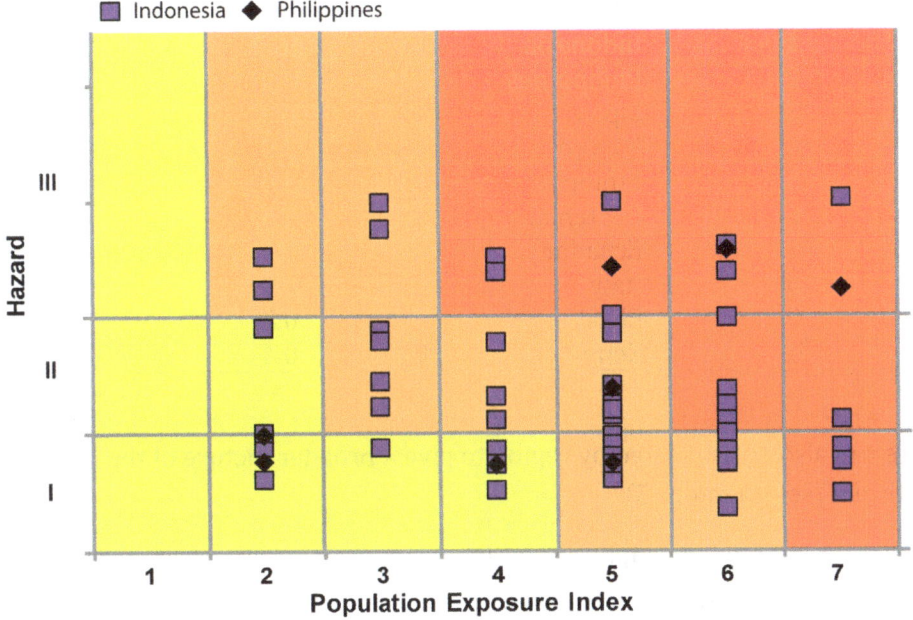

Figure 1.6 Plot of Volcanic Hazard Index (VHI) and Population Exposure Index (PEI) for Indonesia and the Philippines, including only those volcanoes with sufficient eruptive history data to calculate VHI. The warming of the background colours is representative of increasing risk through Risk Levels I-III. This figure is reproduced as Figure 2.28 in Chapter 2.

1.6.5 Distribution of volcanic threat between countries

In this section we investigate the distribution of volcanic threat (potential loss of life) in order to identify countries where threat is relatively high. The full methodology and results are presented in Chapter 23.

The term 'threat' is used simply as a combination of hazard and exposure because we do not consider vulnerability or value. We have developed two measures that combine the number of volcanoes in a country, the size of the population living within 30 km of active volcanoes (Pop30) and the mean hazard index score (VHI). Population exposure is determined using LandScan data (Bright et al., 2012) to calculate the total population living within 30 km of one or more volcanoes with known, or suspected, Holocene activity. We then rank countries using the two measures. Each measure deliberately focuses on a different perspective of threat.

Measure 1 gives the overall volcanic threat country by country based on the number of active volcanoes, an estimate of exposed population and average hazard index of the volcanoes. Table 1.2 shows the distribution of this measure between the 10 highest scoring countries. Indonesia clearly stands out as the country with two thirds of the share of global volcanic threat due to the large number of active volcanoes and high population density.

Measure 1 $= mean\ VHI\ x\ number\ of\ volcanoes\ x\ Pop30$

Table 1.2 The top 10 countries with highest overall volcanic threat. The normalised percentage represents the country's threat as a percentage of the total global threat.

Rank	Country	Normalised %
1	Indonesia	66.0
2	Philippines	10.6
3	Japan	6.9
4	Mexico	3.9
5	Ethiopia	3.9
6	Guatemala	1.5
7	Ecuador	1.1
8	Italy	0.9
9	El Salvador	0.8
10	Kenya	0.4

The measure can also be calculated by region to give a broader picture of the global distribution of volcanic threat (see Chapter 23).

Measure 1 may be misleading because individual countries may vary considerably in the proportion of their population that is exposed to volcanic threat. Nation states vary greatly in their size and populations, from, for example, China with 1.3 billion people (<1% exposed) to St. Kitts and Nevis in the Caribbean with only 54,000 people (100% exposed).

To address this point, Measure 2 ranks the importance of threat in each country. This measure is independent of the country's size, so numbers of volcanoes and exposed population numbers are not included in the calculation. The focus is on the proportion of the population exposed. Measure 2 is defined as follows:

$$\textbf{Measure 2} = \frac{Pop30}{TPop} \times Mean\,VHI$$

The countries that rank highest using this measure are completely different to the rankings using Measure 1. They are a collection of small island states and small countries (Table 1.3).

Table 1.3 The top 10 countries or territories ranked by proportional threat: the product of the proportion of the population exposed per country and the mean VHI.

Rank	Country
1	UK-Montserrat
2	St. Vincent & the Grenadines
3	France – West Indies
4	St. Kitts & Nevis
5	Dominica
6	Portugal – Azores
7	St. Lucia
8	UK – Atlantic
9	El Salvador
10	Costa Rica

These measures and rankings simply provide contexts and answers to different perspectives and questions. There is no suggestion which of these different country and regional rankings should be preferred. If the objective is to identify where most volcanic threat is concentrated, then SE Asia and East Asian countries, such as Indonesia, the Philippines and Japan, have a large share of the total global volcanic threat. If the question is in which countries and regions, irrespective of size, could potential losses be disproportionately high in the context of the country's size, then the West Indies and small nation states are indicated.

There is great potential to enhance and refine the indices and measures of threat. Different measures can be developed in future to answer different questions.

1.7 Volcanic emergencies and disaster risk reduction

The role of scientists at volcano monitoring institutions is to provide volcanic hazards assessments, timely and impartial information, short- and long-term forecasts and early warning to civil authorities so they can make effective risk-based decisions, for example about evacuation or land use planning. In practice, many monitoring institutions must also respond to other natural hazards including earthquakes or tsunami.

Volcanic eruptions are somewhat unique, in that they are usually preceded by 'unrest' which can be detected if monitoring networks are in place (Chapter 18). Some signs of unrest such as felt earthquakes, increased degassing and changes in the hydrothermal and groundwater systems may be evident to local communities and observers. However, not all episodes of unrest lead to an eruption and so scientists must address this uncertainty when advising civil authorities. This can be particularly challenging if there is limited monitoring (Chapter 8). Volcanic Alert Levels are a common way for volcano observatories to characterise the level of unrest or volcanic activity at a volcano and are designed primarily for people on the ground, to support communication and decision-making [Chapter 16]. Such systems can be useful, especially if supported by an agreed common understanding and recognised procedures by authorities and the public [Chapter 10]. However, they also need to be flexible to account for local context and uncertainty. The international aviation colour code system introduced by the International Civil Aviation Organisation provides a framework for notifications to the aviation sector [Chapter 14] and aids communication between volcano observatories and VAACs.

Short-term forecasts of the start of eruptions or increases in hazardous activity can be made by scientists if real-time monitoring is in place. High resolution earth observation products (such as radar) can be highly complementary to ground monitoring networks facilitating timely forecasts and mitigation actions (Chapter 17). The 2010 eruption of Merapi, Indonesia, showed rapid escalation of monitoring signals leading to an increased alert level and a series of evacuations saving the lives of 10,000-20,000 people [Chapter 9] (Surono et al., 2012). Scientists at volcano observatories commonly work collaboratively with networks of international researchers, thus enhancing their access to new methods, research and ideas. However, the observatory itself should be the source of definitive scientific advice. Scientists are often involved in educational activities, so that authorities and the communities can better understand the potential hazards and risks from their volcano(es). This involvement may also involve regular exercises with civil protection agencies (and VAACs) to test planning for eruption response. All of these activities require effective communication and long-term relationships between scientists and authorities, the public, non-governmental organisations (NGOs) and the private sector [Chapters 10 and 24]. The understanding, communication networks and trust, which are built up over time, underpin effective eruption response and risk reduction (Barclay et al., 2008, Haynes et al., 2008a, Haynes et al., 2008b, Solana et al., 2008).

Volcanic risk management and risk reduction at a societal level is the official responsibility of civil authorities, but to be effective also relies on the engagement of communities, individuals, non-governmental organisations and the private sector. In practice, the scientists are likely to have useful knowledge and experience about the potential impacts of volcanic eruptions, and are thus also well-placed to offer advice on risk-based lessons learned at previous eruptions [Chapters 8, 24 and 26].

Several eruptions in recent years have resulted in significant scientific and risk management advances as a result of focused post-event analysis and consideration of lessons learnt. A key example was the installation of extensive monitoring at Nevado del Huila volcano in Colombia after the Nevado del Ruiz disaster, even though Huila had been dormant for more than 500 years. Early warning systems and emergency response activities were practiced between scientists, authorities, NGOs and communities, reportedly leading to timely evacuations and preventing many fatalities during eruptions in 2007-8. A more recent example is the Eyjafjallajökull eruption in Iceland, where significant progress in volcanic ash dispersal modelling and forecasting (Mastin et al., 2009, Woodhouse et al., 2013), data assimilation and observational methods has been achieved since the eruption as a result of cross-disciplinary efforts focused on clear scientific challenges and stakeholder needs (Bonadonna et al., 2012). In order to act on lessons learnt, take full advantage of opportunities and respond effectively to future eruptions, scientists are beginning to engage in formal collaborative and coordinated activities, and research across regions and internationally. Such collaborative and cross-disciplinary research is facilitating progress and has helped to ensure volcano observatories are able to draw useful research into operational activities. Following the controversial management of the 1976 eruption of La Soufrière in Guadeloupe (a large-scale evacuation of the capital city with no subsequent major eruption), a major effort in disaster risk reduction began in the area around the volcano. A dedicated volcano observatory was established and new methods in hazard and risk assessment are being developed alongside cost-benefit analysis in support of pragmatic long-term development and risk mitigation.

During a volcanic crisis, civil authorities and scientists are under immense pressure and must make decisions in short time-frames and often with limited information. Commonly an 'emergency committee' will meet and consider scientific advice before taking official action. Effective official response during an emergency is underpinned by long-term relationships, trust and mutual understanding of different institutional needs, priorities and contexts (Barclay et al., 2008, Haynes et al., 2008a, Haynes et al., 2008b, Solana et al., 2008).

There are a variety of different disaster risk management options open to authorities. Attempts to reduce the hazard are rare, reflecting that this is in many cases not possible, but there have been some examples of lava flow diversion and lahar barriers which have had some effect. Short-term exposure can be reduced directly through evacuation of people and long-term exposure can be reduced by transferring existing assets to geographical areas of lower risk. Improved connectivity between risk management and development is very much needed so that new assets are built in areas of relatively low risk.

Where a known high-risk volcano may erupt in the near future threatening large urban populations, for example Auckland, New Zealand [Chapter 5], and Naples, Italy [Chapter 6], the attention is on planning for the evacuation of large numbers of people in short periods of time. Planning typically assumes an effective short-term alert or forecast is received. During some long-lived eruptions evacuations may become regular occurrences as populations continue to live and work alongside a sporadically active volcano (e.g. Tungurahua, Ecuador) or there may be permanent large scale movements of populations (e.g. Montserrat in 1997). Once a permanent evacuation has occurred, risk assessments are needed to manage access into evacuated areas, to manage access and land use in marginal zones (e.g. Montserrat), and to consider the potential for hazards of even greater impact than previously experienced. At White Island, New Zealand, risk assessments have been used to enable land managers to make decisions on the timing of access to a popular hiking trail that was impacted in the 2012 eruptions. Risk assessments have also been used by the Volcano Observatory to guide decisions on when scientists can access areas for monitoring tasks. In Indonesia, provision is now made for farmers to move animals during some evacuations.

Tools are needed to support scientific and risk management decision-making and there are good examples already available. One effective way to build a bridge between civil authorities and scientists is to combine hazards and risk assessments with cost-benefit analysis, for example an analysis of the costs and benefits of an evacuation [Chapter 5]. Recently, the argument for studying the trade-offs involved in taking mitigating action in the interests of public safety within the economic decision framework of cost-benefit analysis (Leonard et al., 2008, Marzocchi & Woo, 2009) has gained traction [Chapter 6]. These trade-offs may be important to ensure populations are not at more risk when evacuated (e.g. from disease, conflict, security). Cost-benefit analysis does in some cases raise some difficult issues, such as the value of human life, but can be used to support any aspect of decision-making not just evacuation, such as land use planning and the establishment of monitoring capability. Importantly cost-benefit analysis can be done before any crisis develops. Response decisions, about evacuation for example, may be based on pre-defined thresholds and probabilities. Such methods can also be applied retrospectively to examine decision-making in the past, for example the controversial evacuations in Guadeloupe (Hincks et al., 2014) in 1976, which may in fact have been justified.

The desire to attract visitors to support livelihoods in the tourism sector (e.g. in spa towns associated with geothermal areas) can lead to a lack of transparency in terms of making information about hazards and risk available. Tourists often come to volcanic areas because of the volcanoes (Bird et al., 2010) and require appropriate information on the potential hazards, impacts and appropriate response to warnings. Ensuring tourists and tourism employees are aware of early warning and information systems and how to respond if a warning is issued is essential to reduce vulnerability. For example, at White Island, New Zealand, the Volcano Observatory is working in close partnership with regional and national civil protection to develop an understanding of the volcanic risks for both tourists and tourism employees alike.

The UN 'Hyogo Framework for Action 2005-2015' has been a good blueprint for risk reduction activities and the five priorities for action remain highly relevant to volcanic risk:

1. Ensure that disaster risk reduction is a national and local priority with a strong institutional basis for implementation.

2. Identify, assess and monitor disaster risks and enhance early warning.

3. Use knowledge, innovation and education to build a culture of safety and resilience at all levels.

4. Reduce the underlying risk factors.

5. Strengthen disaster preparedness for effective response at all levels.

The reduction in fatalities caused by volcanic eruptions through recent decades demonstrates how the application of science and technology largely coordinated through volcano observatories can lead to anticipation of hazards, increased societal resilience and can effectively reduce risk.

1.8 The way forward

Many aspects of volcanic hazards are localised around a particular volcano and each volcano is to some extent unique, as indeed are the communities that live around them. Thus monitoring institutions (e.g. volcano observatories) and their staff, where they exist, are a very important component of disaster risk reduction. These institutions can help emergency managers, civil authorities and communities understand potential future eruption scenarios and volcanic hazards, and can provide monitoring, forecasts and early warning when a volcano threatens to erupt or change its behaviour. Ideally, a monitoring institution can be at the heart of a 'people-centred early warning system' (Leonard et al., 2008) to support informed decision-making by individuals and authorities. Scientific advisory groups, including scientists from monitoring institutions as well as other national or regional institutions and universities, are an excellent resource for emergency managers and civil authorities before, during and after volcanic crises.

Scientific research across disciplines has a very significant role to play in enhancing resilience, improving the knowledge and evidence base, harnessing resources such as big data and new technologies, developing hazard and risk assessment approaches and carrying out analyses of past eruptions to establish lessons learnt. Some research funding opportunities have been very effective for facilitating international scientific cooperation and collaboration by funding partners in multiple countries. Where research funding is available to work overseas, it's essential that in-country scientists are fully engaged in the research design and process. Volcanic risk and resilience research projects should ideally also be developed in partnership with civil protection/emergency managers to ensure full integration into the disaster risk reduction (DRR) process.

Building resilience and reducing risk alongside an active volcano requires good communication between scientists, civil authorities, emergency managers and the public. In addition, understanding of the hazards and risks, effective planning, exercises of emergency responses, development of trust, understanding of cultural factors that affect community responses are some of the factors that need to be taken into account.

This book highlights some of the wide range of hazards posed by volcanoes, describes their diverse impacts on communities and provides a new global analysis of volcanic hazards and risks. Based on this analysis we identify three key pillars for the reduction of risks associated with volcanic hazards worldwide and list recommended actions (see Chapter 2).

Pillar 1: Identify areas and assets at risk, and quantify the hazard and the risk

Systematic geological, geochronological and historical studies are required to compile quality-assessed data on which rigorous hazard and risk assessments can be based. There is a fundamental need to characterise hazards and risk at many volcanoes worldwide where existing information is incomplete or lacking altogether.

Action 1.1 Those volcanoes shown to be poorly known with major knowledge gaps regarding their past activity and with a high population exposure index (in this study) should be prioritised for geological studies that document recent volcanic history with a hazard assessment context. Recommended studies include stratigraphy, geochronology, petrology, geochemistry and physical volcanology. Such studies greatly enhance the ability of volcanologists to interpret volcanic unrest and respond effectively when activity begins. In some cases, findings are likely to increase the currently known risk.

Action 1.2 Probabilistic assessment of hazard and risk that fully characterises uncertainty is becoming mandatory to inform robust decision-making. Assessments and forecasts are typically combinations of interpreting geological and monitoring data, and various kinds of modelling. Probabilistic event trees and hazard maps for individual volcanoes are best made by local or national scientists, with priority given to high-risk volcanoes. Some data from beyond the specific volcano in question are also needed for these trees and maps, especially if the volcano in question is poorly known.

Action 1.3 Global databases can serve as references for local scientists, providing analogue data and distributions of likely eruption parameters. Creation and maintenance of global databases on volcanoes, volcanic unrest and volcanic hazards, and quality assurance on data, hazard assessment methods, forecast models, and monitoring capacity are best done through international co-operation. Funding the compilation of such databases does not fit easily into national and regional research funding and needs stronger international support.

Action 1.4 Forensic assessments of volcanic hazards, their impact and risk drivers are needed during and after eruptions. Such studies are essential to improve knowledge of hazards and vulnerability in particular and to improve and test methodologies, such as forecast modelling based on real observational data. National Governments should be encouraged to support their institutions to include timeline-based analysis of their actions and subsequent impacts, and to report successes and shortcomings of crisis responses. Evaluations of "lessons learnt" from past emergencies are important to improve future responses and avoid repetition of mistakes.

Action 1.5 Risks from volcanic ash fall associated with a particular volcano or region can be characterised by detailed probabilistic modelling, taking into account the range of physical processes (atmospheric and volcanic) and associated uncertainties. There is also a need to better understand the impacts of volcanic ash, and define thresholds of atmospheric concentration and deposit thickness for various levels of damage to different sectors. We recommend that further analysis be performed for all high-risk volcanoes, to enable more conclusive statements to be made about expected losses and disruption and to support resilience and future adaptation measures.

Pillar 2: Strengthen local to national coping capacity and implement risk mitigation measures

Mitigation means implementing activities that prevent or reduce the adverse effects of extreme natural events. Broadly, mitigation includes: volcano monitoring, reliable and effective early warning systems, active engineering measures, effective political, legal and administrative frameworks. Mitigation also includes land-use planning, careful siting of key infrastructure in low risk areas, and efforts to influence the behaviour of at-risk populations in order to increase resilience. Good communication, education and community participation are critical ingredients to successful strategies. All these measures can help minimise losses, increase societal resilience and assure long-term success.

Action 2.1 Many active volcanoes are either not monitored at all, or have only rudimentary monitoring. Some of these volcanoes are classified in this study as high risk. A major advance for hazard mitigation would be if all active volcanoes had at least one volcano-dedicated seismic station with continuous telemetry to a nominated responsible institution (volcano observatory) combined with a plan for use of satellite services. For volcanoes in repose there are two suggested responses, namely implementation of low-cost systems for monitoring and raising awareness of volcanic hazards and risk among vulnerable populations. Provision of funding to purchase equipment must be complemented by support for scientific monitoring, training and development of staff and long-term equipment maintenance. We recommend this action as a high priority to address volcanic risk.

Action 2.2 Volcanoes identified as high-risk should ideally be monitored by a combination of complementary multi-parameter techniques, including volcano-seismic networks, ground deformation, gas measurements and near real-time satellite remote sensing services and products. This should be maintained, interpreted and responded to by a nominated institution (volcano observatory). Donations of equipment and knowledge transfer schemes need to be sustainable long-term with respect to equipment maintenance and consumables. Support for monitoring institutions and investment in local expertise is essential.

Action 2.3 Technological innovation should strive towards reducing costs of instrumentation and making application of state-of-the-art science as easy as possible so more volcanoes can be monitored effectively. For example, satellite observation offers a new and promising approach, but lower costs, easier access, technological training, and better and more timely sharing of data are needed to realise the potential. Many of the new models derived from research of volcanic processes and hazardous phenomena for forecasting can be made into accessible and easy-to-apply operational tools to support observatory work and decision-making. More resources need to be put into converting potentially useful research into effective and accessible tools.

Action 2.4 Volcanic hazards, monitoring capacity, early warning capability and the quality of communication by volcanologists are key risk factors. The behaviour, attitudes and perceptions of scientists, decision-makers and communities also influence risk. Reducing risk is thus possible with better assessment and awareness of the hazards, effective communication by scientific institutions and authorities, well-practiced response protocols, participatory activities with communities and a greater awareness by all of key risk factors and how they can be managed/reduced. We recommend open, transparent interaction and communication with effective exchange of knowledge. In addition well-thought-out contingency plans for emergencies are essential in all sectors of society.

Pillar 3: Strengthen national and international coping capacity

Efforts should be made to increase coping capacity to address a wide range of hazards, especially relatively infrequent events like major volcanic eruptions. Many countries are enhancing their own disaster preparedness as suggested in the Hyogo Framework for Action.

Some volcanic emergencies cross borders and have regional or global impacts. Coordinated planning, mitigation, regulation and response from different countries are needed in these situations. A key challenge with all projects from donor countries is to be assured that they are needs-based, sustainable and well anchored in the host countries' own development plans. Another challenge is coordination between different projects and sectors.

Action 3.1 Exchange visits, workshops, summer schools and international research collaboration are good ways to share experience and expertise in volcano monitoring, appraisal of unrest, assessment of hazard and risk, and communication. The value of interdisciplinary science is becoming more evident and an understanding of methodologies available in other disciplines can greatly strengthen effective collaboration. Collaborative regional networks of countries are an efficient way to build capacity, carry out research, undertake coordinated monitoring and planning and make effective use of leveraged resources.

Action 3.2 There needs to be much more effort to integrate volcanic hazard and risk assessments with sustainable development and land use planning activities, preferably before eruptions occur, so issues around livelihood, evacuation and potential resettlement are considered as part of resilience building and risk reduction activities.

Action 3.3 Free and easy access to the most advanced science and data will greatly enhance the ability to manage and reduce volcanic risk. Access to knowledge is globally very uneven between the developed and developing nations. For volcanic hazards, easy and reliable access to the internet, high-resolution digital elevation data and satellite remote sensing data, together with appropriate training would significantly improve the scientific capacity of many countries. We encourage ISDR to promote open access of scientific knowledge to all and support the deployment of advanced technologies and information wherever it is needed. Equally important, ground-based data need to be shared among volcano observatories and with the Earth Observation (EO) community (for validation purposes).

Action 3.4 Index-based methods to characterise hazard, exposure, threat and monitoring capacity used in this study are straightforward, and are intended to provide a basic broad overview of volcanic hazard and risk across the world as well as highlight knowledge gaps. The Volcanic Hazards Index and Population Exposure Index should not be used to assess or portray hazard and risk in detail at individual volcanoes, which is the responsibility of national institutions and volcano observatories.

References

Adger, W. N. 2006. Vulnerability. *Global Environmental Change,* 16, 268-281.

Aiuppa, A., Burton, M., Caltabiano, T., Giudice, G., Guerrieri, S., Liuzzo, M., Mur, F. & Salerno, G. 2010. Unusually large magmatic CO_2 gas emissions prior to a basaltic paroxysm. *Geophysical Research Letters,* 37, L17303.

Aspinall, W. 2010. A route to more tractable expert advice. *Nature,* 463, 294-295.

Auker, M. R., Sparks, R. S. J., Siebert, L., Crosweller, H. S. & Ewert, J. 2013. A statistical analysis of the global historical volcanic fatalities record. *Journal of Applied Volcanology,* 2, 1-24.

Barberi, F., Corrado, G., Innocenti, F. & Luongo, G. 1984. Phlegraean Fields 1982–1984: Brief chronicle of a volcano emergency in a densely populated area. *Bulletin Volcanologique,* 47, 175-185.

Barclay, J., Haynes, K., Mitchell, T., Solana, C., Teeuw, R., Darnell, A., Crosweller, H. S., Cole, P., Pyle, D. & Lowe, C. 2008. Framing volcanic risk communication within disaster risk reduction: finding ways for the social and physical sciences to work together. *Geological Society, London, Special Publications,* 305, 163-177.

Baxter, P., Allard, P., Halbwachs, M., Komorowski, J., Andrew, W. & Ancia, A. 2003. Human health and vulnerability in the Nyiragongo volcano eruption and humanitarian crisis at Goma, Democratic Republic of Congo. *Acta Vulcanologica,* 14, 109.

Biggs, J., Anthony, E. Y. & Ebinger, C. J. 2009. Multiple inflation and deflation events at Kenyan volcanoes, East African Rift. *Geology,* 37, 979-982.

Bird, D. K., Gisladottir, G. & Dominey-Howes, D. 2010. Volcanic risk and tourism in southern Iceland: Implications for hazard, risk and emergency response education and training. *Journal of Volcanology and Geothermal Research,* 189, 33-48.

Blong, R. J. 1984. *Volcanic Hazards. A Sourcebook on the Effects of Eruptions,* Australia, Academic Press.

BNPB - Badan Nasional Penaggulangan Bencana (Indonesian National Board for Disaster Management). 2011. Ketangguhan Bangsa Dalam Menghadapi Bencana: dari Wasior, Mentawai, hingga Merapi. *GEMA BNPB,* 2, 48-48.

Bonadonna, C., Folch, A., Loughlin, S. & Puempel, H. 2012. Future developments in modelling and monitoring of volcanic ash clouds: Outcomes from the first IAVCEI-WMO workshop on Ash Dispersal Forecast and Civil Aviation. *Bulletin of Volcanology,* 74, 1-10.

Bright, E. A., Coleman, P. R., Rose, A. N. & Urban, M. L. 2012. *RE: LandScan 2011.* Oak Ridge National Laboratory, Oak Ridge, TN, USA, digital raster data. Available: http://web.ornl.gov/sci/landscan/

Brown, S. K., Crosweller, H. S., Sparks, R. S. J., Cottrell, E., Deligne, N. I., Ortiz Guerrero, N., Hobbs, L., Kiyosugi, K., Loughlin, S. C., Siebert, L. & Takarada, S. 2014. Characterisation of the Quaternary eruption record: analysis of the Large Magnitude Explosive Volcanic Eruptions (LaMEVE) database. *Journal of Applied Volcanology,* 3:5, pp.22.

Calder, E., Sparks, R. S. J., Luckett, R. & Voight, B. 2002. Mechanisms of lava dome instability and generation of rockfalls and pyroclastic flows at Soufriere Hills Volcano, Montserrat. *Geological Society of London Memoirs,* 21, 173-190.

Carey, S., Sigurdsson, H., Mandeville, C. & Bronto, S. 1996. Pyroclastic flows and surges over water: an example from the 1883 Krakatau eruption. *Bulletin of Volcanology,* 57, 493-511.

Carlsen, H. K., Hauksdottir, A., Valdimarsdottir, U. A., Gíslason, T., Einarsdottir, G., Runolfsson, H., Briem, H., Finnbjornsdottir, R. G., Gudmundsson, S. & Kolbeinsson, T. B. 2012. Health effects following the Eyjafjallajökull volcanic eruption: a cohort study. *BMJ Open,* 2, e001851.

Cashman, K. V., Stephen, R. & Sparks, J. 2013. How volcanoes work: A 25 year perspective. *Bulletin of the Geological Society of America,* 125, 664-690.

Charbonnier, S. J., Germa, A., Connor, C. B., Gertisser, R., Preece, K., Komorowski, J. C., Lavigne, F., Dixon, T. & Connor, L. 2013. Evaluation of the impact of the 2010 pyroclastic density currents at Merapi volcano from high-resolution satellite imagery, field investigations and numerical simulations. *Journal of Volcanology and Geothermal Research,* 261, 295-315.

Chouet, B. A. 1996. Long-period volcano seismicity: its source and use in eruption forecasting. *Nature,* 380, 309-316.

Clay, E., Barrow, C., Benson, C., Dempster, J., Kokelaar, P., Pillai, N. & Seaman, J. 1999. An evaluation of HMG's response to the Montserrat Volcanic Emergency. Part 1. DFID (UK Government) Evaluation Report EV5635.

Cottrell, E. 2014. Global Distribution of Active Volcanoes. *In:* Papale, P. (ed.)*Volcanic Hazards, Risks and Disasters.* Academic Press.

Cronin, S. J. & Sharp, D. S. 2002. Environmental impacts on health from continuous volcanic activity at Yasur (Tanna) and Ambrym, Vanuatu. *International Journal of Environmental Health Research,* 12, 109-123.

Crosweller, H. S., Arora, B., Brown, S. K., Cottrell, E., Deligne, N. I., Guerrero, N. O., Hobbs, L., Kiyosugi, K., Loughlin, S. C. & Lowndes, J. 2012. Global database on large magnitude explosive volcanic eruptions (LaMEVE). *Journal of Applied Volcanology,* 1:4, pp.13.

Decker, R. & Decker, B. 2006. *Volcanoes,* W. H. Freeman.

Deligne, N. I., Coles, S. G. & Sparks, R. S. J. 2010. Recurrence rates of large explosive volcanic eruptions. *Journal of Geophysical Research: Solid Earth,* 115, B06203.

Dzurisin, D. 2003. A comprehensive approach to monitoring volcano deformation as a window on the eruption cycle. *Reviews of Geophysics,* 41.

Edmonds, M. 2008. New geochemical insights into volcanic degassing. *Philosophical transactions. Series A, Mathematical, Physical, and Engineering Sciences,* 366, 4559-4579.

Furlan, C. 2010. Extreme value methods for modelling historical series of large volcanic magnitudes. *Statistical Modelling,* 10, 113-132.

Guffanti, M., Casadevall, T. J. & Budding, K. 2010. Encounters of aircraft with volcanic ash clouds: a compilation of known incidents, 1953-2009. *US Geological Survey Data Series,* 545, 12-12.

Haynes, K., Barclay, J. & Pidgeon, N. 2008a. The issue of trust and its influence on risk communication during a volcanic crisis. *Bulletin of Volcanology,* 70, 605-621.

Haynes, K., Barclay, J. & Pidgeon, N. 2008b. Whose reality counts? Factors affecting the perception of volcanic risk. *Journal of Volcanology and Geothermal Research,* 172, 259-272.

Hincks, T., Aspinall, W., Baxter, P., Searl, A., Sparks, R. & Woo, G. 2006. Long term exposure to respirable volcanic ash on Montserrat: a time series simulation. *Bulletin of Volcanology,* 68, 266-284.

Hincks, T. K., Komorowski, J.-C., Sparks, S. R. & Aspinall, W. P. 2014. Retrospective analysis of uncertain eruption precursors at La Soufrière volcano, Guadeloupe, 1975–77: volcanic hazard assessment using a Bayesian Belief Network approach. *Journal of Applied Volcanology,* 3:3, pp.26.

Horwell, C. J. & Baxter, P. J. 2006. The respiratory health hazards of volcanic ash: a review for volcanic risk mitigation. *Bulletin of Volcanology,* 69, 1-24.

Jenkins, S., Komorowski, J. C., Baxter, P. J., Spence, R., Picquout, A., Lavigne, F. & Surono 2013. The Merapi 2010 eruption: An interdisciplinary impact assessment methodology for studying pyroclastic density current dynamics. *Journal of Volcanology and Geothermal Research,* 261, 316-329.

Jenkins, S., McAneney, J., Magill, C. & Blong, R. 2012. Regional ash fall hazard II: Asia-Pacific modelling results and implications. *Bulletin of Volcanology,* 74, 1713-1727.

Johnson, J. B. & Ripepe, M. 2011. Volcano infrasound: A review. *Journal of Volcanology and Geothermal Research,* 206, 61-69.

Kelman, I. & Mather, T. A. 2008. Living with volcanoes: The sustainable livelihoods approach for volcano-related opportunities. *Journal of Volcanology and Geothermal Research,* 172, 189-198.

Kling, G. W., Clark, M. A., Wagner, G. N., Compton, H. R., Humphrey, A. M., Devine, J. D., Evans, W. C., Lockwood, J. P., Tuttle, M. L. & Koenigsberg, E. J. 1987. The 1986 Lake Nyos gas disaster in Cameroon, West Africa. *Science (New York, N.Y.),* 236, 169-175.

Komorowski, J. 2002-2003. The January 2002 flank eruption of Nyiragongo volcano (Democratic Republic of Congo): Chronology, evidence for a tectonic rift trigger, and impact of lava flows on the city of Goma. *Acta vulcanologica,* 14-15 27-62.

Laksono, P. M. 1988. Perception of volcanic hazards: villagers versus government officials in Central Java. *The Real and Imagined Role of Culture in Development: Case Studies from Indonesia.* Honolulu: University of Hawaii Press.

Lane, L. R., Tobin, G. a. & Whiteford, L. M. 2003. Volcanic hazard or economic destitution: hard choices in Baños, Ecuador. *Environmental Hazards,* 5, 23-34.

Lara, L. E., Moreno, R., Amigo, Á., Hoblitt, R. P. & Pierson, T. C. 2013. Late Holocene history of Chaitén Volcano: New evidence for a 17th century eruption. *Andean Geology,* 40, 249-261.

Larson, K. M., Poland, M. & Miklius, a. 2010. Volcano monitoring using {GPS:} Developing data analysis strategies based on the June 2007 Kilauea Volcano intrusion and eruption. *Journal of Geophysical Research-Solid Earth,* 115, B07406-B07406.

Leonard, G. S., Johnston, D. M., Paton, D., Christianson, A., Becker, J. & Keys, H. 2008. Developing effective warning systems: ongoing research at Ruapehu volcano, New Zealand. *Journal of Volcanology and Geothermal Research,* 172, 199-215.

Lewis, J. 1999. *Development in Disaster-prone Places: Studies of Vulnerability,* London, Intermediate Technology Publications, pp. 174.

Lindsay, J. M. Volcanoes in the big smoke: a review of hazard and risk in the Auckland Volcanic Field. Geologically Active. Delegate Papers of the 11th Congress of the International Association for Engineering Geology and the Environment (IAEG), 2010.

Lockwood, J. P. & Hazlett, R. W. 2013. *Volcanoes: Global Perspectives,* John Wiley & Sons: Chichester. pp.552

Loughlin, S., Baxter, P., Aspinall, W., Darroux, B., Harford, C. & Miller, A. 2002. Eyewitness accounts of the 25 June 1997 pyroclastic flows and surges at Soufrière Hills Volcano, Montserrat, and implications for disaster mitigation. *Geological Society, London, Memoirs,* 21, 211-230.

Mandeville, C. W., Carey, S. & Sigurdsson, H. 1996. Sedimentology of the Krakatau 1883 submarine pyroclastic deposits. *Bulletin of Volcanology,* 57, 512-529.

Marzocchi, W. & Bebbington, M. S. 2012. Probabilistic eruption forecasting at short and long time scales. *Bulletin of Volcanology,* 74, 1777-1805.

Marzocchi, W. & Woo, G. 2009. Principles of volcanic risk metrics: Theory and the case study of Mount Vesuvius and Campi Flegrei, Italy. *Journal of Geophysical Research,* 114, B03213.

Mastin, L. G., Guffanti, M., Servranckx, R., Webley, P., Barsotti, S., Dean, K., Durant, A., Ewert, J. W., Neri, A., Rose, W. I., Schneider, D., Siebert, L., Stunder, B., Swanson, G., Tupper, A., Volentik, A. & Waythomas, C. F. 2009. A multidisciplinary effort to assign realistic source parameters to models of volcanic ash-cloud transport and dispersion during eruptions. *Journal of Volcanology and Geothermal Research,* 186, 10-21.

McNutt, S. R. 2005. Volcanic Seismology. *Annual Review of Earth and Planetary Sciences,* 33, 461-491.

Nadeau, P. A., Palma, J. L. & Waite, G. P. 2011. Linking volcanic tremor, degassing, and eruption dynamics via SO 2 imaging. *Geophysical Research Letters,* 38.

Newhall, C. & Hoblitt, R. 2002. Constructing event trees for volcanic crises. *Bulletin of Volcanology,* 64, 3-20.

Newhall, C. G. & Punongbayan, R. 1996. *Fire and Mud: Eruptions and Lahars of Mount Pinatubo, Philippines,* Philippine Institute of Volcanology and Seismology Quezon City.

Newhall, C. G. & Self, S. 1982. The volcanic explosivity index (VEI) an estimate of explosive magnitude for historical volcanism. *Journal of Geophysical Research,* 87, 1231-1231.

Papale, P. & Shroder, J.F. 2014. *Volcanic Hazards, Risks and Disasters*, Elsevier: Amsterdam. pp.532

Phillipson, G., Sobradelo, R. & Gottsmann, J. 2013. Global volcanic unrest in the 21st century: an analysis of the first decade. *Journal of Volcanology and Geothermal Research,* 264, 183-196.

Potter, S. H., Scott, B. J. & Jolly, G. E. 2012. Caldera Unrest Management Sourcebook. *GNS Science Report.* GNS Science.

Pyle, D. 2015. Sizes of volcanic eruptions. *In:* Sigurdsson, H., Houghton, B., McNutt, S., Rymer, H. & Stix, J. *The Encyclopedia of Volcanoes.* Second ed.: Academic Press.

Ragona, M., Hannstein, F. & Mazzocchi, M. 2011. The impact of volcanic ash crisis on the European Airline industry. *In:* Alemanno, A. (ed.) *Governing Disasters: The Challenges of Emergency Risk regulations.* Edward Elgar Publishing.

Roberts, M. R., Linde, A. T., Vogfjord, K. S. & Sacks, S. Forecasting Eruptions of Hekla Volcano, Iceland, using Borehole Strain Observations, EGU2011-14208. 2011.

Schmidt, A., Ostro, B., Carslaw, K. S., Wilson, M., Thordarson, T., Mann, G. W. & Simmons, A. J. 2011. Excess mortality in Europe following a future Laki-style Icelandic eruption. *Proceedings of the National Academy of Sciences,* 108, 15710-15715.

Schmincke, H. U. 2004. *Volcanism,* Berlin Heidelberg, Springer-Verlag, pp. 324.

Segall, P. 2013. Volcano deformation and eruption forecasting. *Geological Society, London, Special Publications,* 380, 85-106.

Seitz, S. 2004. *The Aeta at the Mount Pinatubo, Philippines: A Minority Group Coping with Disaster,* Quezon City, New Day Publishers.

Self, S. & Blake, S. 2008. Supervolcanoes: Consequences of explosive supereruptions. *Elements,* 4, 41-46.

Siebert, L. 1984. Large volcanic debris avalanches: characteristics of source areas, deposits, and associated eruptions. *Journal of Volcanology and Geothermal Research,* 22, 163-197.

Siebert, L., Simkin, T. & Kimberly, P. 2010. *Volcanoes of the World: Third Edition*, University of California Press.

Sigurdsson, H., Houghton, B., Rymer, H., Stix, J. & McNutt, S. 2015. *Encyclopedia of Volcanoes*, Academic Press.

Small, C. & Naumann, T. 2001. The global distribution of human population and recent volcanism. *Global Environmental Change Part B: Environmental Hazards,* 3, 93-109.

Smithsonian. 2014. *Global Volcanism Program: Volcanoes of the World 4.0.* [Online]. Washington D.C. Available: http://www.volcano.si.edu/.

Solana, M. C., Kilburn, C. R. J. & Rolandi, G. 2008. Communicating eruption and hazard forecasts on Vesuvius, Southern Italy. *Journal of Volcanology and Geothermal Research,* 172, 308-314.

Sparks, R., Biggs, J. & Neuberg, J. 2012. Monitoring volcanoes. *Science,* 335, 1310-1311.

Sparks, R. S. J. 2003. Forecasting volcanic eruptions. *Earth and Planetary Science Letters,* 210, 1-15.

Sparks, R. S. J. & Aspinall, W. P. 2004. Volcanic activity: frontiers and challenges in forecasting, prediction and risk assessment. *The State of the Planet: Frontiers and Challenges in Geophysics.* American Geophysical Union.

Sparks, R. S. J., Aspinall, W. P., Crosweller, H. S. & Hincks, T. K. 2013. Risk and uncertainty assessment of volcanic hazards. *In:* Rougier, J., Sparks, R. S. J. & Hill, L. *Risk and Uncertainty Assessment for Natural Hazards.* Cambridge: Cambridge University Press, 364-397.

Spence, R., Kelman, I., Baxter, P., Zuccaro, G. & Petrazzuoli, S. 2005. Residential building and occupant vulnerability to tephra fall. *Natural Hazards and Earth System Science,* 5, 477-494.

Stone, J., Barclay, J., Simmons, P., Cole, P. D., Loughlin, S. C., Ramón, P. & Mothes, P. 2014. Risk reduction through community-based monitoring: the vigías of Tungurahua, Ecuador. *Journal of Applied Volcanology,* 3:11, 1-14.

Surono, Jousset, P., Pallister, J., Boichu, M., Buongiorno, M. F., Budisantoso, A., Costa, F., Andreastuti, S., Prata, F., Schneider, D., Clarisse, L., Humaida, H., Sumarti, S., Bignami, C., Griswold, J., Carn, S., Oppenheimer, C. & Lavigne, F. 2012. The 2010 explosive eruption of Java's Merapi volcano-A '100-year' event. *Journal of Volcanology and Geothermal Research,* 241-242, 121-135.

Sword-Daniels, V. 2011. Living with volcanic risk: The consequences of, and response to, ongoing volcanic ashfall from a social infrastructure systems perspective on Montserrat. *New Zealand Journal of Psychology,* 40, 131-138.

Þorkelsson, B. 2012. The 2010 Eyjafjallajokull Eruption, Iceland: Report to ICAO. Icelandic Meteorological Office.

Todman, S., Garaebiti, E., Jolly, G. E., Sherburn, S., Scott, B., Jolly, A. D., Fournier, N. & MIller, C. A. 2010. Developing monitoring capability of a volcano observatory: the example of the Vanuatu Geohazards Observatory. *AGU.* San Francisco.

Usamah, M. & Haynes, K. 2012. An examination of the resettlement program at Mayon Volcano: What can we learn for sustainable volcanic risk reduction? *Bulletin of Volcanology,* 74, 839-859.

Voight, B. 1990. The 1985 Nevado del Ruiz volcano catastrophe: anatomy and retrospection. *Journal of Volcanology and Geothermal Research,* 42, 151-188.

Voight, B. 2000. Structural stability of andesite volcanoes and lava domes. *Philosophical Transactions of the Royal Society A: Mathematical, Physical and Engineering Sciences,* 358, 1663-1703.

Wadge, G., Voight, B., Sparks, R. S. J., Cole, P. D., Loughlin, S. C. & Robertson, R. E. A. 2014. An overview of the eruption of Soufriere Hills Volcano, Montserrat from 2000 to 2010. *Geological Society, London, Memoirs,* 39, 1-40.

Wilson, G., Wilson, T. M., Deligne, N. I. & Cole, J. W. 2014. Volcanic hazard impacts to critical infrastructure: A review. *Journal of Volcanology and Geothermal Research,* 286, 148-182.

Wilson, T. M., Stewart, C., Sword-Daniels, V., Leonard, G. S., Johnston, D. M., Cole, J. W., Wardman, J., Wilson, G. & Barnard, S. T. 2012. Volcanic ash impacts on critical infrastructure. *Physics and Chemistry of the Earth, Parts A/B/C,* 45, 5-23.

Witter, J. B. 2012. Volcanic hazards and geothermal development. *Geothermal Resources Council Transactions,* 36, 965-971. l eruption, Iceland. *Journal of Geophysical Research: Solid Earth,* 118, 92-109.

Woodhouse, M. J., Hogg, A. J., Phillips, J. C. & Sparks, R. S. J. 2013. Interaction between volcanic plumes and wind during the 2010 Eyjafjallajökull eruption, Iceland. *Journal of Geophysical Research: Solid Earth,* 118, 92-109.

Summaries

Chapters 4-26 provide an evidence base for both Chapter 1 (this chapter) and Chapter 2. The relevant chapters are indicated to the reader. In this appendix short summaries of Chapters 4-26 are provided along with Supplementary Case Studies 1 to 3.

4	Populations around Holocene volcanoes and development of a Population Exposure Index S.K. Brown, M.R. Auker and R.S.J. Sparks	43
5	An integrated approach to Determining Volcanic Risk in Auckland, New Zealand: the multi-disciplinary DEVORA project N.I. Deligne, J.M. Lindsay and E. Smid	44
6	Tephra fall hazard for the Neapolitan area W. Marzocchi, J. Selva, A. Costa, L. Sandri, R. Tonini and G. Macedonio	45
7	Eruptions and lahars of Mount Pinatubo, 1991-2000 C.G. Newhall and R. Solidum	47
8	Improving crisis decision-making at times of uncertain volcanic unrest (Guadeloupe, 1976) J.-C. Komorowski, T. Hincks, R.S.J. Sparks, W. Aspinall and CASAVA ANR project consortium	48
9	Forecasting the November 2010 eruption of Merapi, Indonesia J. Pallister and Surono	49
10	The importance of communication in hazard zone areas: case study during and after 2010 Merapi eruption, Indonesia S. Andreastuti, J. Subandriyo, S. Sumarti and D. Sayudi	50
11	Nyiragongo (Democratic Republic of Congo), January 2002: a major eruption in the midst of a complex humanitarian emergency J.-C. Komorowski and K. Karume	52
12	Volcanic ash fall impacts T.M. Wilson, S.F. Jenkins and C. Stewart	53
13	Health impacts of volcanic eruptions C.J. Horwell, P.J. Baxter and R. Kamanyire	55
14	Volcanoes and the aviation industry P.W. Webley	57
15	The role of volcano observatories in risk reduction G. Jolly	59
16	Developing effective communication tools for volcanic hazards in New Zealand, using social science G. Leonard and S. Potter	61

17	Volcano monitoring from space M. Poland	62
18	Volcanic unrest and short-term forecasting capacity J. Gottsmann	63
19	Global monitoring capacity: development of the Global Volcano Research and Monitoring Institutions Database and analysis of monitoring in Latin America N. Ortiz Guerrero, S.K. Brown, H. Delgado Granados and C. Lombana Criollo	64
20	Volcanic hazard maps E. Calder, K. Wagner and S.E. Ogburn	65
21	Risk assessment case history: the Soufrière Hills Volcano, Montserrat W. Aspinall and G. Wadge	67
22	Development of a new global Volcanic Hazard Index (VHI) M.R. Auker, R.S.J. Sparks, S.F. Jenkins, W. Aspinall, S.K. Brown, N.I. Deligne, G. Jolly, S.C. Loughlin, W. Marzocchi, C.G. Newhall and J.L. Palma	69
23	Global distribution of volcanic threat S.K. Brown, R.S.J. Sparks and S.F. Jenkins	71
24	Scientific communication of uncertainty during volcanic emergencies J. Marti	72
25	Volcano Disaster Assistance Program: Preventing volcanic crises from becoming disasters and advancing science diplomacy J. Pallister	74
26	Communities coping with uncertainty and reducing their risk: the collaborative monitoring and management of volcanic activity with the vigías of Tungurahua J. Stone, J. Barclay, P. Ramon, P. Mothes and STREVA	75
S.CS1	Multi-agency response to eruptions with cross-border impacts B. Oddsson	76
S.CS2	Planning and preparedness for an effusive volcanic eruption: the Laki scenario C. Vye-Brown, S.C. Loughlin, S. Daud and C. Felton	77
S.CS3	Interactions of volcanic airfall and glaciers L.K. Hobbs, J.S. Gilbert, S.J. Lane and S.C. Loughlin	78

Chapter 4 Summary: Populations around Holocene volcanoes and development of a Population Exposure Index

S.K. Brown, M.R. Auker and R.S.J. Sparks

Population exposure provides an indication of direct risk to life from volcanic hazards such as pyroclastic density currents and lahars and can be used as a proxy for threat to livelihoods, infrastructure and economic assets. This index doesn't account for indirect fatalities from famine and disease or far-field losses in the aviation and agriculture industries caused by the distribution of volcanic ash, gas and aerosols. The direct threat to the population is affected by the distance from the volcano. More than 800 million people live within 100 km of active volcanoes in 86 countries. Indonesia, the Philippines and Japan top the list for the greatest number of people living close to volcanoes; however, some countries have a higher proportion of their total population within 100 km of a volcano (e.g. Guatemala and Iceland with >90%). Eruptions can produce hazardous flows that extend for tens of kilometres. The Population Exposure Index (PEI 1-7) is therefore determined from the population within 100 km, weighted for circle area and fatality incidence within radii of 10, 30 and 100 km.

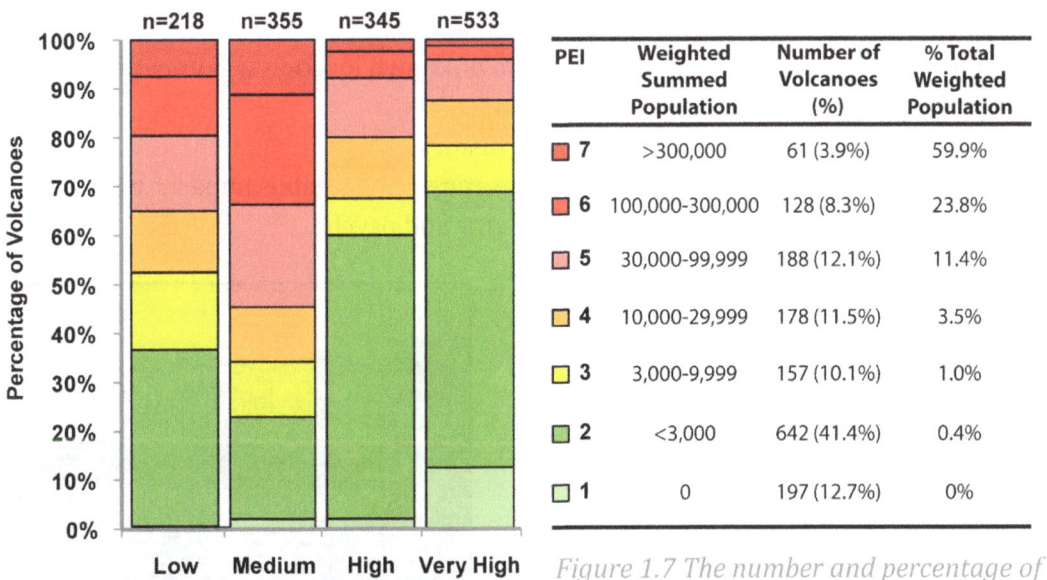

PEI	Weighted Summed Population	Number of Volcanoes (%)	% Total Weighted Population
7	>300,000	61 (3.9%)	59.9%
6	100,000-300,000	128 (8.3%)	23.8%
5	30,000-99,999	188 (12.1%)	11.4%
4	10,000-29,999	178 (11.5%)	3.5%
3	3,000-9,999	157 (10.1%)	1.0%
2	<3,000	642 (41.4%)	0.4%
1	0	197 (12.7%)	0%

Figure 1.7 The number and percentage of volcanoes at each PEI level shown with the HDI.

Most volcanoes classify as PEI 2, accounting for <1% of the total population under threat. Just 4% of volcanoes are ranked at PEI 7, but these account for 60% of that total population. The greatest numbers of high PEI (5-7) volcanoes are in the Indonesia, Mexico & Central America and Africa & Red Sea regions, however as a proportion of its volcanoes, the Philippines and SE Asia ranks highest, with ~70% of volcanoes classified as PEI 5-7. More volcanoes are located in countries of Very High HDI than Low; however only <15% of volcanoes in High and Very High HDI countries classify with PEI≥5, rising to 45% in Low and Medium HDI countries, indicating a broad relationship between a lower level of development and a higher percentage of volcanoes with high proximal populations. These countries may have fewer resources to dedicate to disaster mitigation and may experience greater relative losses in the event of volcanic activity. PEI provides a first-order method of identifying volcanoes close to large populations, which might therefore have priority in resource allocation. Full assessment based on local factors such as volcano morphology may lead to different conclusions about priorities.

Chapter 5 Summary: An integrated approach to Determining Volcanic Risk in Auckland, New Zealand: the multi-disciplinary DEVORA project

N.I. Deligne, J.M. Lindsay and E. Smid

Auckland, New Zealand, home to 1.4 million people and over a third of New Zealand's population, is built on top of the Auckland Volcanic Field (AVF). The AVF covers 360 km², has over 50 eruptive centres (vents), and has erupted over 55 times in the past 250,000 years. The most recent eruption, Rangitoto, was only 550 years ago. Most vents are monogenetic, i.e. they only erupt once. This poses a considerable problem for emergency and risk managers, as it is unknown where or when the next eruption will occur. The DEterming VOlcanic Risk for Auckland (DEVORA) program is a 7-year multi-agency research programme primarily funded by the government, and has a mandate to investigate the geologic underpinnings, volcanic hazards and risk posed by the AVF. DEVORA researchers work in collaboration with Auckland Council (local government) and Civil Defence (crisis responders) to implement findings into policy. The main challenges facing Auckland and other populated areas coinciding with volcanic fields include:

- uncertainty of where and when the next eruption will be;
- communicating to the public how an eruption of unknown location will impact them and how they can best prepare;
- planning for an event which hasn't occurred in historic time;
- foreseeing and appropriately planning for the range of possible impacts to the built environment, local, regional, and national economy and psyche.

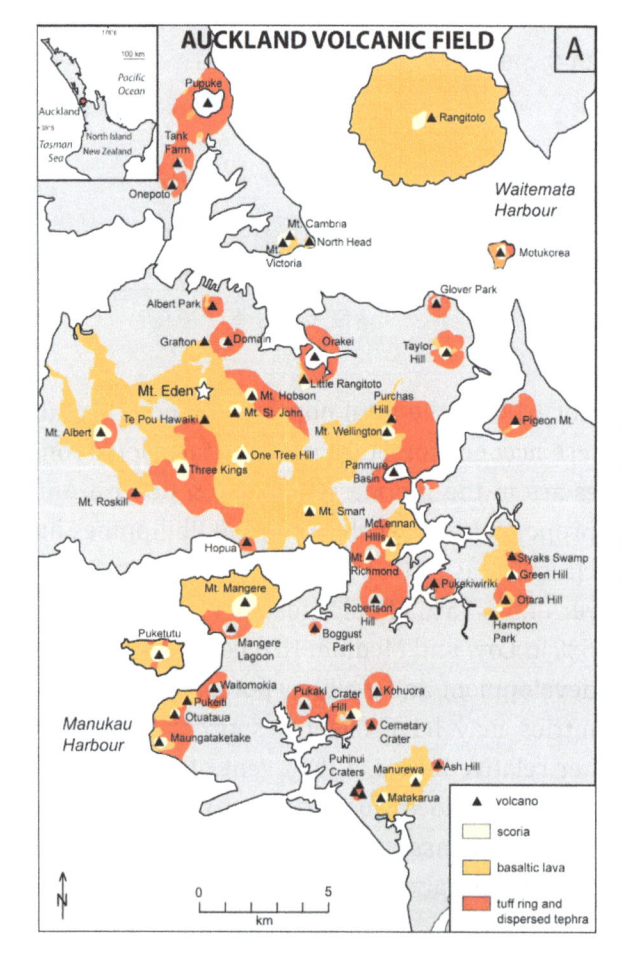

Figure 1.8 a) Map of Auckland Volcanic Field; star indicates location of Mt Eden. b) View of Mt Eden looking to the north highlighting the complete overlap of AVF and city (© Auckland Council).

Chapter 6 Summary: Tephra fall hazard for the Neapolitan area

W. Marzocchi, J. Selva, A. Costa, L. Sandri, R. Tonini and G. Macedonio

The Neapolitan area represents one of the highest volcanic risk areas in the world, both for the presence of three potentially explosive and active volcanoes (Vesuvius, Campi Flegrei and Ischia), and for the extremely high exposure (over a million people located in a very large and important metropolitan area). Risk management has to be based on the evaluation of the long-term impact of the volcanoes (long-term volcanic hazard), and on tracking the space and time evolution of potential pre-eruptive signals. The Osservatorio Vesuviano (INGV-OV) of the Istituto Nazionale di Geofisica e Vulcanologia is continuously monitoring these volcanoes using advanced techniques to record the evolution of seismic activity, ground deformation, geochemical signals and of many other potential pre-eruptive indicators. Moreover, INGV-OV provides updated hazard information to the Italian Civil Protection Department that is responsible for planning risk mitigation actions.

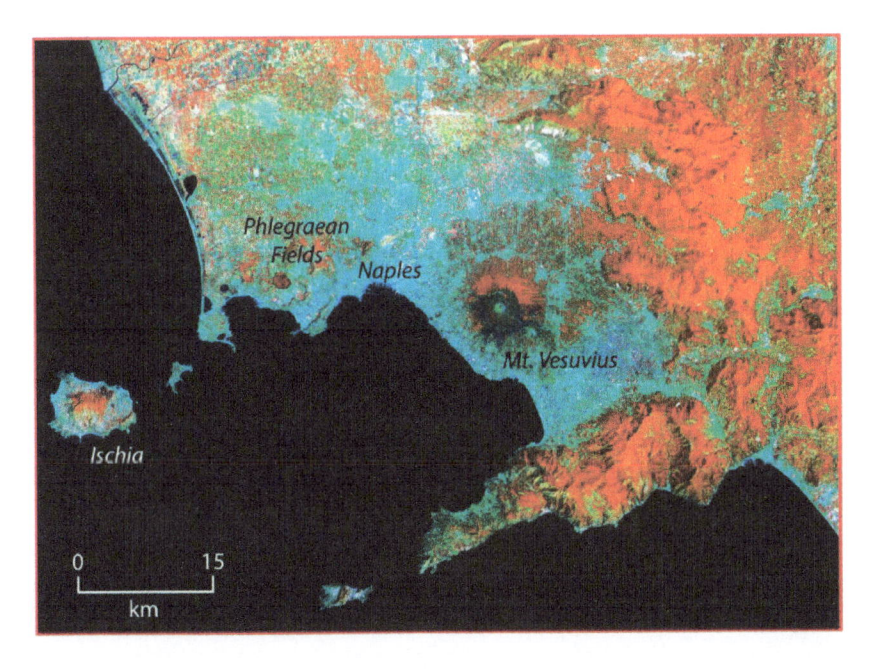

Figure 1.9 Satellite map of the Neopolitan area. Modified from Laboratorio di Geomatica, INGV-OV.

Because of the large and ubiquitous uncertainties in the knowledge of pre-eruptive processes, hazard information essentially consists of the probabilistic assessment of different types of threatening events. The presence of such uncertainties poses several major challenges to scientists and decision makers.

- Volcanologists have to articulate scientific information including all known uncertainties, and merge different types of knowledge including: data, expert opinion and models.
- Naples illustrates the importance of multi-hazard analysis, because it is threatened by three volcanoes that may produce diverse hazards such as ash fall, pyroclastic flows and lavas flows, as well as related threats like earthquakes, ground deformation and

tsunamis; this requires study of different physical processes and understanding of cascading events that can amplify the overall risk.
- Decision makers have to plan risk mitigation strategies with uncertain scientific information. Since the societal and economic costs of most feasible mitigation actions may be extremely high, a sound risk mitigation strategy requires a careful evaluation of what is feasible, and what is affordable accounting for costs and benefits.
- Any kind of risk mitigation plan in high-risk areas requires an efficient risk communication strategy during volcanic unrest, and a strong educational program during quiescence to improve the preparedness of the population and their resilience.
- There are no past monitored eruptions in the Neapolitan area. This encourages volcanologists and decision makers to share their knowledge and to learn from experience gained from other analogue cases from around the world.

Chapter 7 Summary: Eruptions and lahars of Mount Pinatubo, 1991-2000

C. Newhall and R. Solidum

After sleeping for ~ 500 years, Mount Pinatubo (Philippines) began to stir in mid March 1991, and produced a giant eruption on 15 June 1991, the second largest of the twentieth century. About 20,000 indigenous Aetas lived on the volcano, and ~1,000,000 lowland Filipinos lived around it. Two large American military bases, Clark Air Base and Subic Bay Naval Station, were also at risk.

- Despite considerable uncertainties, the eruption was correctly forecast and more than 85,000 were evacuated by 14 June. Many aircraft were also protected from the eruption.
- About 300 lowlanders died from roof collapse during the eruption, but nearly all of the Aetas survived. At least 10,000 and perhaps as many as 20,000 were saved by timely warnings and evacuations.
- Regrettably, ~500 Aeta children died of measles in evacuation camps, because their parents distrusted Western-trained doctors and refused help.
- The hazard lasted far beyond the eruption – and, indeed, continues today though at a much-reduced level. Voluminous rain-induced lahars continued for more than 10 years, and sediment-clogged channels still overflow today during heavy rains.
- Although about 200,000 were "permanently displaced" by lahars, only about 400 died from lahars. Timely warnings from scientists and police helped to keep most people safe.
- Warnings and evacuations before the eruptions were clearly cost effective; lahar warnings and evacuations were also cost effective. Construction of sediment control structures might or might not have been cost effective, depending on how one counts costs and benefits.

Figure 1.10 Lahars repeatedly buried the town of Bacolor from 1991-1995. Only roofs of 2-storey buildings are visible. Photo by Chris Newhall, USGS.

Chapter 8 Summary: Improving crisis decision-making at times of uncertain volcanic unrest (Guadeloupe, 1976)

J-C. Komorowski, T. Hincks, R.S.J. Sparks and W. Aspinall

Scientists monitoring active volcanoes are increasingly required to provide decision support to civil authorities during periods of unrest. As the extent and resolution of monitoring improves, the process of jointly interpreting multiple strands of indirect evidence becomes increasingly complex. During a volcanic crisis, decisions typically have to be made with limited information and high uncertainty, on short time scales. The primary goal is to minimise loss and damage from any event, but social and economic loss resulting from false alarms and evacuations must also be considered. Although it is not the responsibility of the scientist to call an evacuation or manage a crisis, there is an increasing requirement to assess risks and present scientific information and associated uncertainties in ways that enable public officials to make urgent evacuation decisions or other mitigation policy choices.

Increasingly intense seismicity was recorded and felt at La Soufrière 1 year prior to the eruption which began with an unexpected explosion on 8 July 1976. Ash-venting associated with sulfur (H_2S, SO_2) and halogen-rich (HCl, HF, Br) gases released during the eruption led to moderate environmental impact with short-term public health implications. Given evidence of continued escalating pressurisation and the uncertain transition to a devastating eruption, authorities declared a 4-6-month evacuation of ca. 70,000 people on 15 August. The evacuation resulted in severe socio-economic consequences until long after the crisis had subsided. The costs have been estimated as 60% of the total annual per capita Gross Domestic Product of Guadeloupe in 1976, excluding losses of uninsured personal assets and open-grazing livestock. There were no fatalities, but this eruption stills ranks amongst the most costly of the twentieth century. Hence analysis, forecast and crisis response were highly challenging for scientists and authorities in the context of markedly escalating and fluctuating activity as well as the societal pressures cast in an insular setting.

As the extent and resolution of monitoring improves, the process of jointly interpreting multiple strands of indirect evidence becomes increasingly complex. The use of new probabilistic formalism for decision-making (e.g. Bayesian Belief Network analysis, Bayesian event decision trees) can significantly reduce scientific uncertainty and better assist public officials in making urgent evacuation decisions and policy choices when facing volcanic unrest.

A recent retrospective Bayesian Belief Network analysis of this crisis demonstrates that a formal evidential case would have supported the authorities' concerns about public safety and their decision to evacuate in 1976.

At present, following the controversial management of the 1976 eruption, a major effort in infrastructural development has begun in the area potentially at risk from volcanic activity. Hence, risk assessment, monitoring and cost-benefit analysis must continue to be enhanced in support of pragmatic long-term development and risk mitigation policies.

Chapter 9 Summary: Forecasting the November 2010 eruption of Merapi, Indonesia

J. Pallister and Surono

Merapi volcano (Indonesia) is one of the most active and hazardous volcanoes in the world. It is known for frequent small to moderate eruptions, pyroclastic flows produced by lava dome collapse and the large population settled on and around the flanks of the volcano that is at risk. Its usual behaviour for the last decades abruptly changed in late October and early November 2010, when the volcano produced its largest and most explosive eruptions in more than a century, displacing about 400,000 people, and claiming nearly 400 lives. Despite the challenges involved in forecasting this 'hundred year eruption', the magnitude of precursory signals (seismicity, ground deformation, gas emissions) was proportional to the large size and intensity of the eruption. In addition and for the first time, near-real-time satellite radar imagery played a major role along with seismic, geodetic and gas observations in monitoring and forecasting eruptive activity during a major volcanic crisis. The Indonesian Center of Volcanology and Geological Hazard Mitigation (CVGHM) was able to issue timely forecasts of the magnitude of the eruption phases, saving an estimated 10,000–20,000 lives.

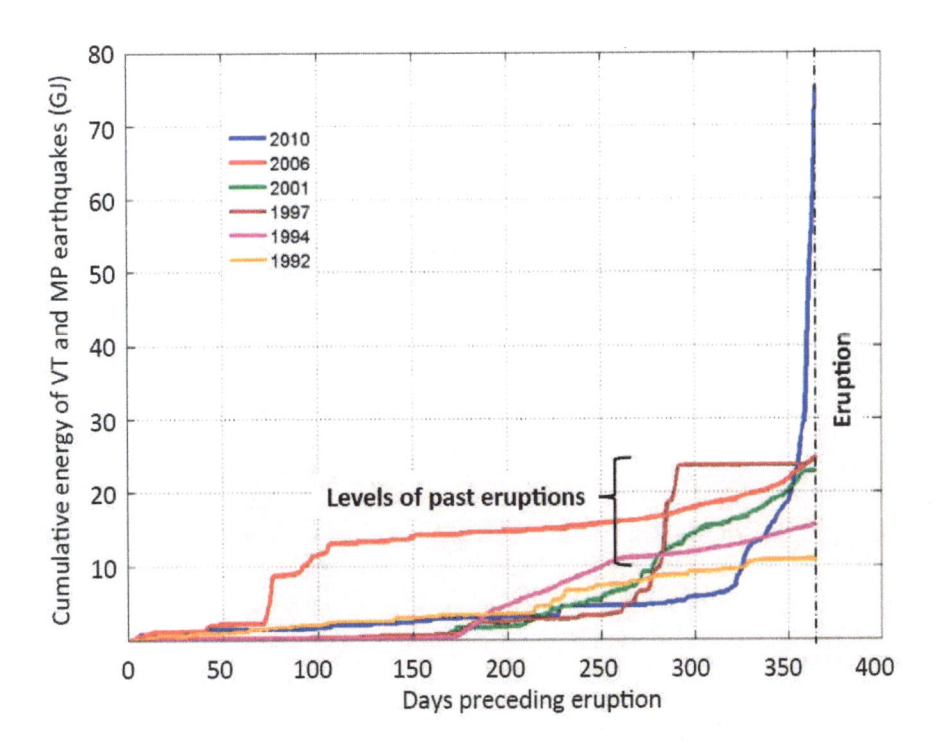

Figure 1.11 Cumulative seismic energy release of volcano-tectonic (VT) and multiphase (MP) earthquakes for eruptions of Merapi in 1997, 2001, 2006 and 26 October 2010. Modified from Budi-Santoso et al. (2013).

Chapter 10 Summary: The importance of communication in hazard zone areas: case study during and after 2010 Merapi eruption, Indonesia

S. Andreastuti, J. Subandriyo, S. Sumarti and D. Sayudi

Merapi is one of the most active volcanoes in Indonesia. Eruptions during the twentieth and twenty-first centuries resulted in: 1369 casualties (Thouret et al., 2000) (1930-1931), 66 casualties (1994) and 386 casualties (2010). The 2010 eruption had impacts that were similar to unusually large 1872 eruption, which had widespread impacts and resulted in approximately 200 casualties (Hartmann 1934): a large number given the relatively sparse population in the late nineteenth century compared to today.

The 2010 Merapi eruption affected two provinces and four regencies, namely Magelang (west-southwest flank), Sleman (south flank), Klaten (southeast-east flank and Boyolali (northern flank). The eruption led to evacuation of 399,000 people and resulted in a total loss of US$ 3.12 billion (National Planning Agency).

Indonesia applies four levels of warnings for volcano activity. From the lowest to highest: at Level I (Normal), the volcano shows a normal (background) state of activity; at Level II (Advisory) visual and seismic data show significant activity that is above normal levels; at Level III (Watch) the volcano shows a trend of increasing activity that is likely to lead to eruption; and at Level IV there are obvious changes that indicate an imminent and hazardous eruption, or a small eruption has already started and may lead to a larger and more hazardous eruption. At Level III people must be prepared for evacuation and at Level IV evacuations are required.

ALERT LEVEL	DATES	RADIUS	ERUPTION
NORMAL (DECREASING)	15-9-2011		
ADVISORY	30-12-2010		
WATCH	3-12-2010		
	4-11-2010	20 KM (11:00 UTC)	4 Nov. 17:05 UTC (16,5 km)
	3-11-2010	15 KM (08:05 UTC)	3 Nov. 08:30 UTC (9 km)
WARNING (INCREASING)	25-10-2010	10 KM (11:00 UTC)	26-10-2010 (10:02 UTC)
WATCH	21-10-2010		
ADVISORY	20-9-2010		
NORMAL	17-9-2010		

Figure 1.12 Chronology of warnings and radius of evacuations during the Merapi eruption in 2010 (time increases from the bottom of the diagram upwards).

During the time of the 2010 crisis, there was rapid escalation of seismicity, deformation and rates of initial lava extrusion. All of these monitoring parameters exceeded levels observed during previous eruptions of the late twentieth century. This raised concerns of an impending much larger eruption. Consequently, a Level IV warning was issued and evacuations were carried out and then extended progressively to greater distances as the activity escalated. The exclusion zone was extended from 10 to 15 and then to 20 km from Merapi's summit.

The 2010 Merapi eruption offers an excellent lesson in dealing with eruption uncertainties, crises management and public communication. Good decision making depends not only on good leadership, but also on the capabilities of scientists, good communication and coordination amongst stakeholders, public communication and on the capacity of the community to respond. All of these factors were in place before the 2010 eruption and contributed to the saving of many thousands of lives.

Impacts of Merapi eruptions on the human and cultural environment, livelihood and properties provide a lesson that in dense-populated areas around a volcano there is a need for regular review of hazard mitigation strategy, including spatial planning, mandatory disaster training, contingency planning and for regular evacuation drills. Merapi is well known for a capacity building programme named 'wajib latih' (mandatory training) required for people living near the volcano. The aim of this activity is to improve hazard knowledge, awareness and skill to protect self, family and community. In addition to the wajib latih, people also learn from direct experience with volcano hazards, which at Merapi occur frequently. However, the 2010 Merapi eruption showed that well-trained and experienced people must also be supported by good management, and that training and mitigation programmes must consider not only "normal" but also unusually large eruptions (Mei et al., 2013).

Chapter 11 Summary: Nyiragongo (Democratic Republic of Congo), January 2002: a major eruption in the midst of a complex humanitarian emergency

J.-C. Komorowski and K. Karume

Nyiragongo is a 3,470 m high volcano located in the western branch of the East African Rift in the Democratic Republic of Congo (DRC), close to the border with Rwanda. It has a 1.3 km wide summit crater that has been filled with an active lava lake since 1894. The area is affected by permanent passive degassing of carbon dioxide (CO_2). Fatal concentrations of CO_2 can accumulate in low-lying areas, threatening the permanent population and internally displaced persons (IDPs) in refugee evacuation centres. Nyiragongo volcano is responsible for 92% of global lava-flow related fatalities (ca. 824) since 1900.

On 17 January 2002, fractures opened on Nyiragongo's upper southern flanks triggering a catastrophic drainage of the lava lake. Two main flows entered the city producing major devastation, and forcing the rapid exodus of most of Goma's 300,000–400,000 inhabitants across the border into neighbouring Rwanda. There were international concerns about the evacuation causing an additional humanitarian catastrophe exacerbating the ongoing regional ethnic and military conflict. Lava flows destroyed about 13% of Goma, 21% of the electricity network, 80 % of its economic assets, 1/3 of the international airport runway and the housing of 120,000 people. The eruption caused about 470 injuries and about 140 to 160 deaths mostly from CO_2 asphyxiation and from the explosion of a petrol station near the active hot lava flow.

This was the first time in history that a city of such a size had been so severely impacted by lava flows. The eruption caused a major humanitarian emergency that further weakened the already fragile lifelines of the population in an area subjected to many years of regional instability and military conflicts. The medical and humanitarian community feared a renewal of cholera epidemics that caused a high mortality in refugee evacuations centres after the 1994 genocide. However, rapid and efficient response by relief workers from UN agencies, numerous non-governmental organisations (NGOs), and local utility agencies prevented major epidemics.

The limited number of fatalities in 2002 is attributed to:

- timely recognition by the Goma Volcano Observatory (GVO) of the reactivation of the volcano about 1 year prior to the eruption and their efficient communication with authorities once the eruption began;
- memory of the devastating 1977 eruption which triggered life-saving actions by villagers;
- panic-less self-evacuation of the population;
- presence of a large humanitarian community in Goma;
- occurrence of the eruption in the morning, and the relatively slow progression of eruptive vents towards Goma with the dike and fractures stopping before the water-saturated zone and the lake.

Had any one of these parameters been negatively exaggerated, the death told would have been much greater and potentially catastrophic.

Chapter 12 Summary: Volcanic ash fall impacts

T.M. Wilson, S.F. Jenkins and C. Stewart

All explosive eruptions produce volcanic ash (fragments of volcanic rock < 2 mm), which is then dispersed by prevailing winds and deposited as ash falls hundreds or even thousands of kilometres away. The wide geographic reach of ash falls, and their high frequency, makes them the volcanic hazard most likely to affect the greatest numbers of people. However, forecasting how much ash will fall, where and with what characteristics is a major challenge. In addition, ash fall impacts are wide-ranging, influenced by environmental agents such as wind and rain, and often not well understood. As a very general rule, three zones of impact may be broadly expected; these are summarised in Figure 1.13 where physical ash impacts to selected societal assets are depicted against deposit thickness, which generally decreases with distance from the source volcano. Thick ash falls (>100 mm) may damage infrastructure, crops and vegetation, damage buildings and create major clean-up demands, but are typically confined to within tens of kilometres of the vent. Relatively thin falls (<10 mm) may cause adverse health effects for vulnerable individuals and can disrupt critical infrastructure services, aviation and other socio-economic activities over potentially very large areas.

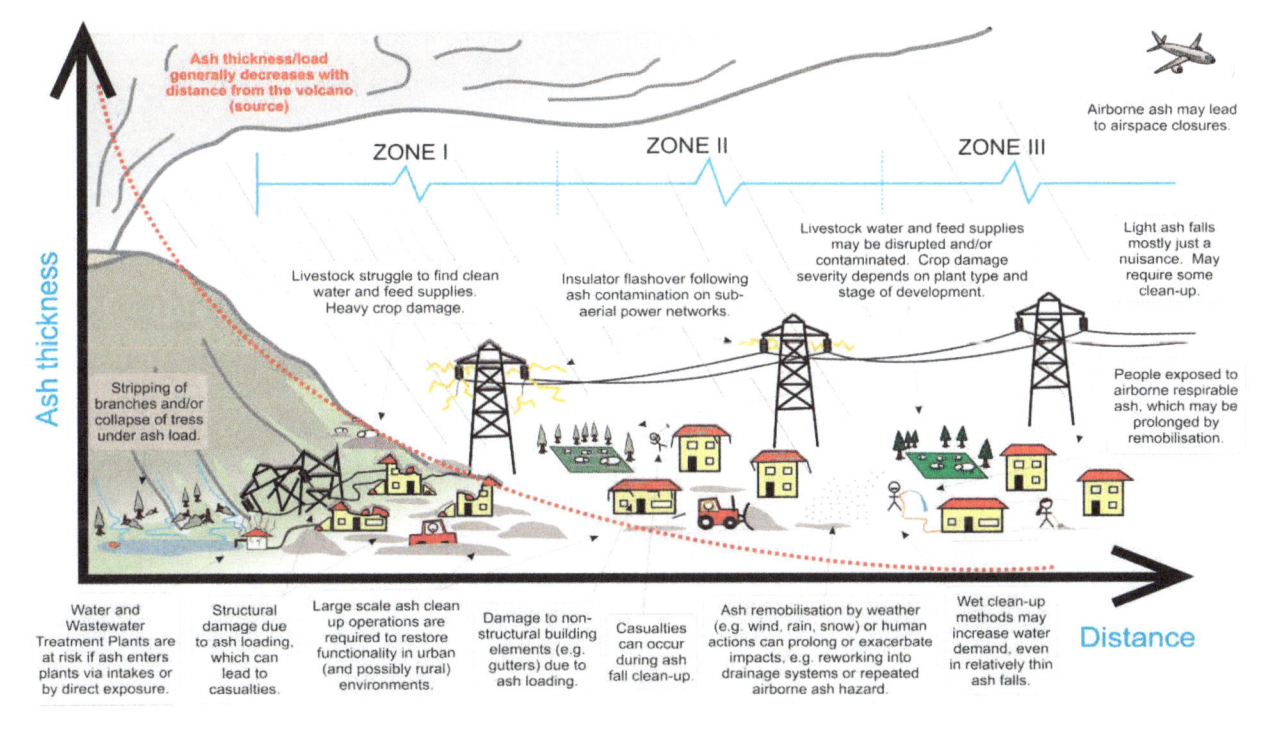

Figure 1.13 Schematic of some ash fall impacts with distance from a volcano. This assumes a large explosive eruption with significant ash fall thicknesses in the proximal zone and is intended to be illustrative rather than prescriptive. Three main zones of ash fall impact are defined: 1) Destructive and immediately life-threatening (Zone I); 2) Damaging and/or disruptive (Zone II); 3) Disruptive and/or a nuisance (Zone III).

Impacts depend not only upon the amount of volcanic ash deposited and its characteristics (hazard), but also the numbers and distribution of people and assets (exposure), and the ability of people and assets to cope with ash fall impacts (vulnerability). While volcanic eruptions cannot be prevented, the exposure and vulnerability of the population to their impacts may, in theory, be reduced, through the considerable tasks of hazard and risk assessment, improved land use planning, risk education and communication and increasing economic development.

Chapter 13 Summary: Health impacts of volcanic eruptions

C.J. Horwell, P.J. Baxter and R. Kamanyire

Volcanoes emit a variety of products which may be harmful to human and animal health. Some cause traumatic injury or death; others may trigger disease or stress, particularly in the respiratory and cardiovascular systems.

Injury agents. Injury and death are caused by a range of volcanic hazards, which can be summarised by their impact on the body: 1) mechanical injury (lahars, rock avalanches, ballistics and tephra falls) where the body is crushed; 2) thermal injury (pyroclastic flows and surges, lava flows) where the body is burned; 3) toxicological effects (gases, ash and aerosols) where emissions react with the body; 4) electrical impact (lightning).

Volcanic gases. Volcanoes emit hazardous gases (e.g. CO_2, SO_2, H_2S and radon). Gas exposures occur during and following eruptions, and during periods of quiescence, and may be proximal or distal to the vent, depending on the size of eruption. Most gas-related deaths occur by asphyxiation near the volcano, but large eruptions may generate mega-tonnes of SO_2 which can be transported globally, potentially triggering acute respiratory diseases, such as asthma, where populations are exposed.

Volcanic ash. Whilst ash may cause skin and eye irritation, the primary concern for humans is ash inhalation; the style of eruption and composition of the magma govern the size and composition of the particles which, in turn, control their pathogenic potential when inhaled. The most hazardous eruptions generate fine-grained, crystalline silica-rich ash which has the potential to cause silicosis. Inhalation of fine particles (sub-2.5 μm diameter) affects both cardiovascular and respiratory mortality and morbidity.

Secondary effects. Large populations brought together in evacuation camps may contract diseases through poor sanitation. Some evacuees may suffer mental stress and other psychological disorders related to displacement. Widespread ashfall or gas impact (acid rain) may lead to crop failure, loss of livestock and contamination of water supplies which, in turn, may trigger famine and related diseases. Heavy ashfall can cause roof collapse and is slippery, making clean-up and driving hazardous. Infrastructure may be impacted, affecting healthcare responses.

Figure 1.14 Ash mobilisation in Yogyakarta following the 2014 Kelud eruption. Photo: Tri Wahyudi.

Hazard/Impact planning and response. A key aspect of public health planning and response is the assessment of population exposure to ash and gas through air quality monitoring networks, which should provide real-time data and be set up in advance. Syndromic surveillance of respiratory symptoms can also inform public health advice. The International Volcanic Health Hazard Network (www.ivhhn.org), the umbrella organisation for volcanic health-related research and dissemination, has produced pamphlets and guidelines on volcanic health issues for the public, scientists, governmental bodies and agencies. IVHHN has also developed protocols for rapid characterisation of ash (such as particle size, crystalline silica content and basic toxicology) giving timely information to hazard managers during, or soon after, an eruption, to facilitate informed decision-making on health interventions.

Chapter 14 Summary: Volcanoes and the aviation industry

P.W. Webley

Since the start of commercial airline travel in the 1950s, 247 volcanoes have been active, some with multiple eruptions. Volcanic ash encounters from 1953-2009 have been documented by Guffanti et al. (2010). Two of the most significant encounters occurred in the 1980s which resulted in total engine shut-down (Casadevall, 1994) and, along with those from the 1991 eruption of Mount Pinatubo (Casadevall et al., 1996), led the International Civil Aviation Organization (ICAO) to set up nine regional volcanic ash advisory centres or VAACs (ICAO, 2007). They provide volcanic ash advisories to the aviation community for their own area of responsibility.

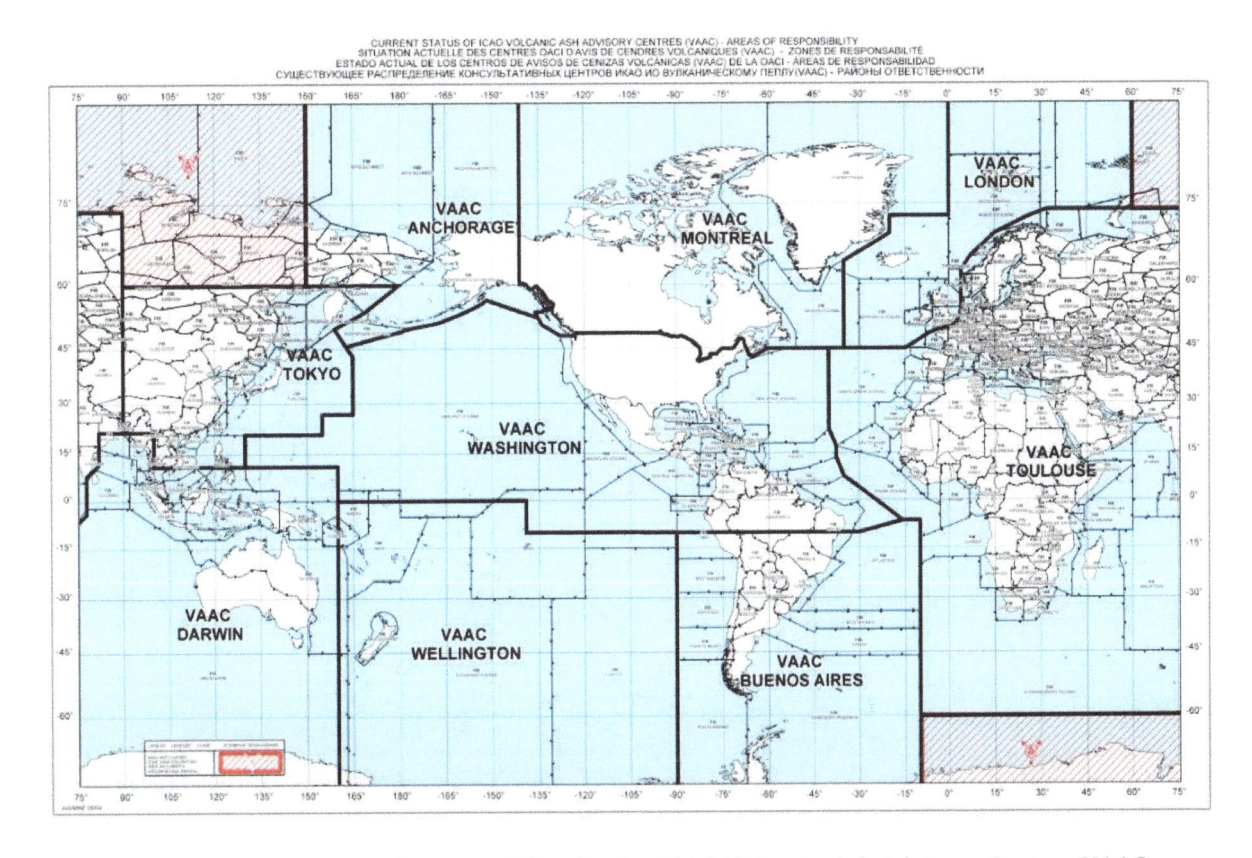

Figure 1.15 Map of the areas of responsibility for the ICAO Volcanic Ash Advisory Centres VAACs.

There are several different alerting systems used worldwide, each with the aim to update both local population centres close to the volcano and the aviation community. One common system used across the North Pacific is the United States Geological Survey (USGS) colour code system, see Gardner and Guffanti (2006). This uses a green-yellow-orange-red system for aviation alerts, which with its corresponding text (USGS, 2014), allows the aviation community to stay informed on the activity levels of the volcano. Risk mitigation to minimise aviation impact is dependent on real-time monitoring of volcano activity, detection and tracking of ash clouds using satellite data, dispersion modelling to forecast ash movement and global communication of timely information. International working groups, task forces and meetings have been assembled to tackle the questions related to volcanic ash in the atmosphere. The World

Meteorological Organization (WMO) and International Union of Geology and Geophysics (IUGG) held workshops on ash dispersal forecast and civil aviation in 2010 and 2013 (WMO, 2013). Additionally, ICAO assembled the International Volcanic Ash Task Force (IVATF) as a focal point and coordinating body of work related to volcanic ash at global and regional levels.

Globally, there can be many volcanoes active and potentially hazardous to the aviation industry. Therefore, the VAACs and local volcano observatories work closely together to provide the most effective advisory system and ensure the safety of all those on the ground and in the air.

Chapter 15 Summary: The role of volcano observatories in risk reduction

G. Jolly

Volcanic risk reduction is a partnership between science, responding agencies and the affected communities. A critical organisation in the volcanic risk reduction cycle is a volcano observatory (VO), which is an institute or group of institutes whose role it is to monitor active volcanoes and provide early warnings of future activity to the authorities. For each country, the exact constitution and responsibilities of a VO may differ, but that establishment is the source of authoritative short-term forecasts of volcanic activity. There are over 100 VOs around the world to monitor ca. 1500 volcanoes considered to be active or potentially active. Some of these VOs have responsibility for multiple volcanoes. In some countries an academic institute may have to fulfil both the monitoring and research function for a volcano.

To be able to effectively monitor their volcanoes, VOs potentially have a very wide suite of tools available to them; however, the range of the capability and capacity of VOs globally is enormous. Many active volcanoes have no monitoring whatsoever, whereas some VOs in developed countries may have hundreds of sensors on a single volcano. This leads to major gaps in provision of warnings of volcanic activity, particularly in developing countries.

Monitoring programmes typically include: tracking the location and type of earthquake activity under a volcano; measuring the deformation of the ground surface as magma intrudes a volcano; sampling and analysing gases and water being emitted from the summit and flanks of a volcano; observing volcanic activity using webcams and thermal imagery; measurements of other geophysical properties such as electrical conductivity, magnetism or gravity. VOs may have ground-based sensors measuring these data in real-time or they may have staff undertaking campaigns to collect data on a regular basis (e.g. weekly, monthly, annually). Some VOs may also the capability to collect and analyse satellite data.

VOs play a critical role in all parts of the risk management cycle. VOs are often involved in outreach activities in times of volcanic quiet so that the authorities and the communities can better understand the potential risk from their volcano(es); this may also involve regular exercising with civil protection agencies to test planning for eruption responses. During the lead up to an eruption, VOs may provide regular updates on activity which inform decisions on evacuations or mitigation actions to reduce risk to people or to critical infrastructure. For example, power transmission companies may choose to shut off high-voltage lines if there is a high probability of ashfall. During an eruption, VOs will then provide up-to-date information about the progression of activity. For an explosive eruption, information might include the duration, the height that ash reaches in the atmosphere and areas being impacted on the ground. This can inform decisions such as search and rescue attempts or provide input to ash dispersion forecasts for aviation. After an eruption has ceased, VOs can aid recovery through advice about ongoing hazards such as remobilisation of ash deposits during heavy rainfall.

The World Organisation of Volcano Observatories (WOVO) is an IAVCEI commission that aims to co-ordinate communication between VOs and to advocate enhancing volcano monitoring around the globe. WOVO is an organisation of and for VOs of the world (www.wovo.org). One of the main recent roles of WOVO has been to link VOs with Volcanic Ash Advisory Centres for enhancing communication between VOs and the aviation sector. Early notification of eruptions is critical for air traffic controllers and airlines so that they can undertake appropriate mitigation of risk to aircraft.

The role of VOs is critical in reducing risk from volcanoes, both on the ground and in the air. Volcanic risk reduction can only improve if VOs are adequately resourced by national governments.

Chapter 16 Summary: Developing effective communication tools for volcanic hazards in New Zealand using social science

S. Potter and G. Leonard

New Zealand has a number of active volcanoes in a wide range of risk and geological settings. The effective communication of information about volcanic hazards to society is important to reduce the risk from these volcanoes, and is achieved by integrating the disciplines of social science and volcanology. This includes:

- The development of a new Volcanic Alert Level system for New Zealand. Qualitative research methods allowed the needs of stakeholders to be incorporated into the new system, resulting in a more effective communication tool to inform their decision-making (Potter et al., 2014).
- The improvement of lahar warnings and hazard information for visitors to the ski areas on Mt Ruapehu (Figure 1.16). The observation of responses to multiple simulated events indicated changes to education and procedures to improve future responses (Leonard et al., 2008). This is supported by longitudinal surveys of hazard perception and safety action recall.
- The creation of a crisis volcanic hazard map for eruptions at Mt. Tongariro in 2012 (Figure 1.16; Leonard et al., 2014). The area impacted by the eruptions included a section of the popular Tongariro Alpine Crossing walking track. Requirements of stakeholders were considered alongside scientific modelling and geological information to develop an effective communication product.

By incorporating social science, information derived from volcano monitoring and data interpretation can be used more effectively to reduce the risk of volcanic hazards to society.

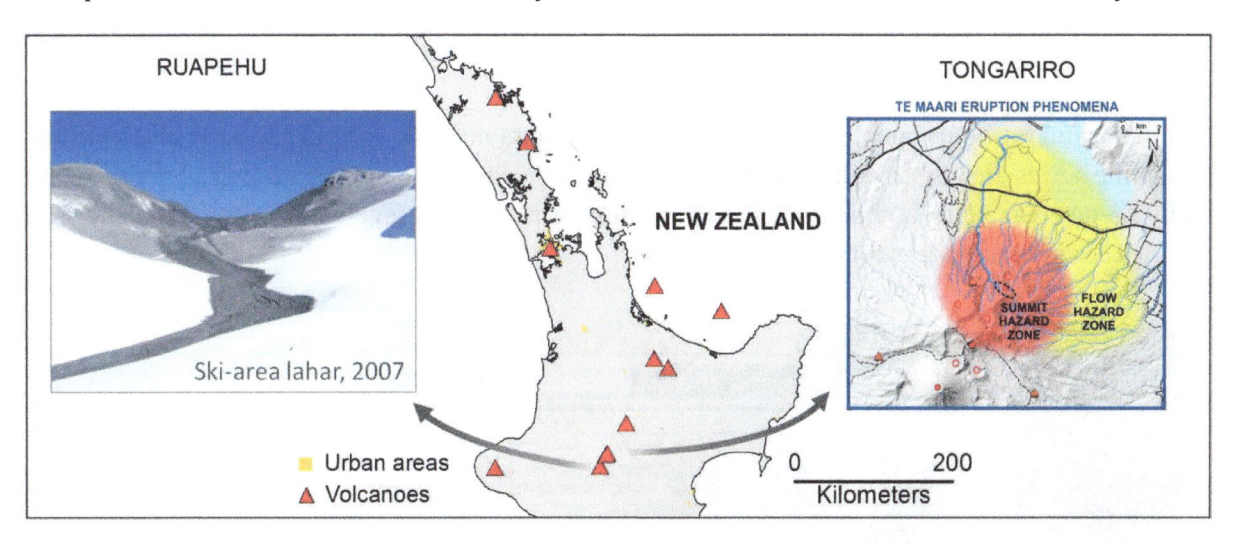

Figure 1.16 Volcanoes in New Zealand. The comprehensive Tongariro hazard map can be found at www.gns.cri.nz/volcano.

Chapter 17 Summary: Volcano monitoring from Space

M. Poland

Unfortunately, only some of Earth's active volcanoes are continuously monitored; the others are too remote or lack of infrastructure (often due to limited financial resources in the host country) for systematic observation. This lack of monitoring is a critical gap in hazards assessment and risk management. Volcanic eruptions are usually preceded by days to months of precursory activity, unlike other natural processes like earthquakes and tornados. Detecting such warning signs at an early stage thus provides the best means to plan and mitigate against potential hazards.

Satellite-based Earth Observation (EO) provides the best means of bridging the currently existing volcano-monitoring gap. EO data are global in coverage and provide information on some of the most common eruption precursors, including ground deformation, thermal anomalies, and gas emissions. Once an eruption is in progress, continued tracking of these parameters, as well as ash emission and dispersal, is critical for modelling the temporal and spatial evolution of the hazards and the likely future course of the eruption. The need for volcano-monitoring EO data is demonstrated by a number of international projects, including:

- the 2012 the International Forum on Satellite EO and Geohazards, which articulated the vision for EO volcano monitoring (http://www.int-eo-geo-hazard-forum-esa.org/);
- the Geohazard Supersites and Natural Laboratories initiative, which aims to reduce loss of life from geological disasters through research using improved access to multi-disciplinary Earth science data (http://supersites.earthobservations.org/);
- the European Volcano Observatory Space Services (EVOSS), which has the goal of providing near-real-time access to gas, thermal, and deformation data from satellites at a number of volcanoes around the world (http://www.evoss-project.eu/);
- the Disaster Risk Management volcano pilot project of the Committee on Earth Observation Satellites (CEOS), which is designed to demonstrate how free access to a diversity of remote sensing data over volcanoes can benefit hazards mitigation efforts .

To be useful for operational volcano monitoring, EO data must be temporally extensive to allow for time series analysis, available with low latency to facilitate rapid utilization by scientists and emergency managers, and be available at minimal or no cost, as few countries and agencies can afford commercial prices for satellite imagery.

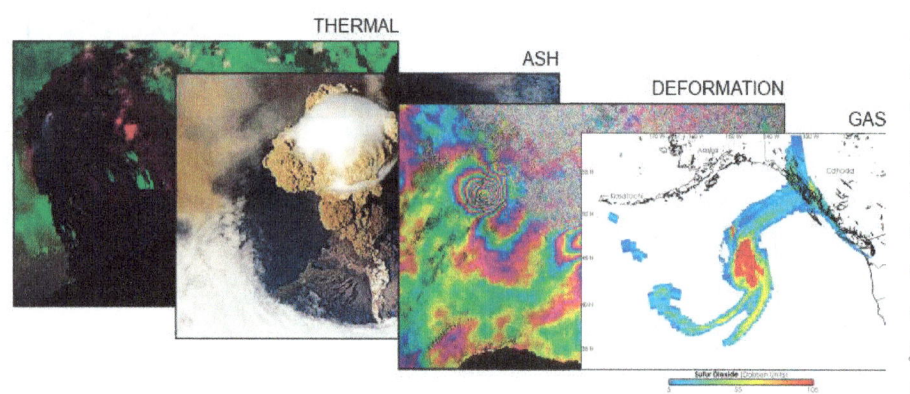

Figure 1.17 Examples of space-based volcano-monitoring products, to detect thermal anomalies, ash emissions, deformation of Earth's surface, and gas emissions. Images: NASA.

Chapter 18 Summary: Volcanic unrest and short-term forecasting capacity

J. Gottsmann

It is important that early on in a developing unrest crisis scientists are able to decipher the nature, timescale and likely outcome of volcano reawakening following long periods of quiescence. There are major challenges when assessing whether unrest will actually lead to an eruption or wane with time. An analysis of reported volcanic unrest between 2000 and 2011 (Figure 1.18) showed that that the median pre-eruptive unrest duration was different across different volcano types (Phillipson et al., 2013) lasting between a few weeks to few months. The same study also showed that volcanoes with long periods of quiescence between eruptions will not necessarily undergo prolonged periods of unrest before their next eruption.

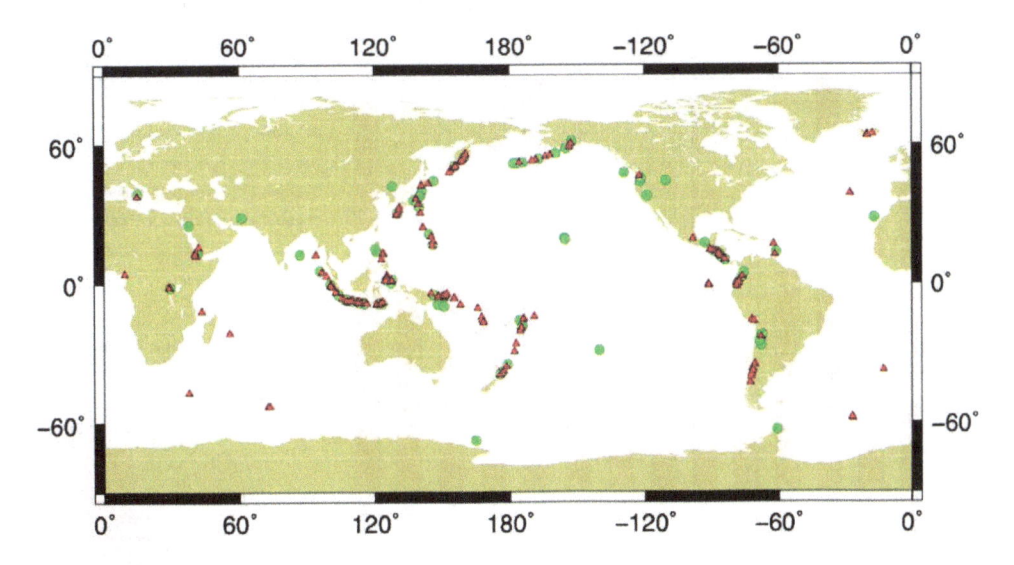

Figure 1.18 Location maps of 228 volcanoes with reported unrest between January 2000 and July 2011. Green circles show volcanoes with unrest not followed by eruption within reporting period, while red triangles show those with eruption.

Forecasting the outcomes of volcanic unrest requires the use of quantitative probabilistic models (Marzocchi and Bebbington, 2012) to adequately address intrinsic (epistemic) uncertainty as to how an unrest process may evolve as well as aleatory uncertainty regarding the limited knowledge about the process. To improve the knowledge-base on volcanic unrest, a globally validated protocol for the reporting of volcanic unrest and archiving of unrest data is needed. Such data are important for the short-term forecasting of volcanic activity amid technological and scientific uncertainty and the inherent complexity of volcanic systems. Selection of appropriate mitigation actions based on informed societal decision-making using probabilistic forecast models and properly addressing uncertainties is particularly critical for managing the evolution of a volcanic unrest episode in high-risk volcanoes, where mitigation actions require advance warning and incur considerable costs.

Chapter 19 Summary: Global monitoring capacity: development of the Global Volcano Research and Monitoring Institutions Database and analysis of monitoring in Latin America

N. Ortiz Guerrero, S.K. Brown, H. Delgado Granados and C. Lombana Criollo

Volcano observatories and monitoring institutions play a critical role in real-time information, providing hazard assessments and enabling timely evacuations. Their monitoring capacity is fundamental in disaster risk reduction. The Global Volcano Research and Monitoring Institutions Database (GLOVOREMID) has been developed to collate data on institutional capacity including techniques used, and instrumental and laboratory capabilities. This is being expanded to a global dataset, but began as a study of monitoring capacity across 314 volcanoes through Mexico, Central and South America. Monitoring Levels of 0 to 5 are assigned to volcanoes based on the use of seismic, deformation and gas monitoring.

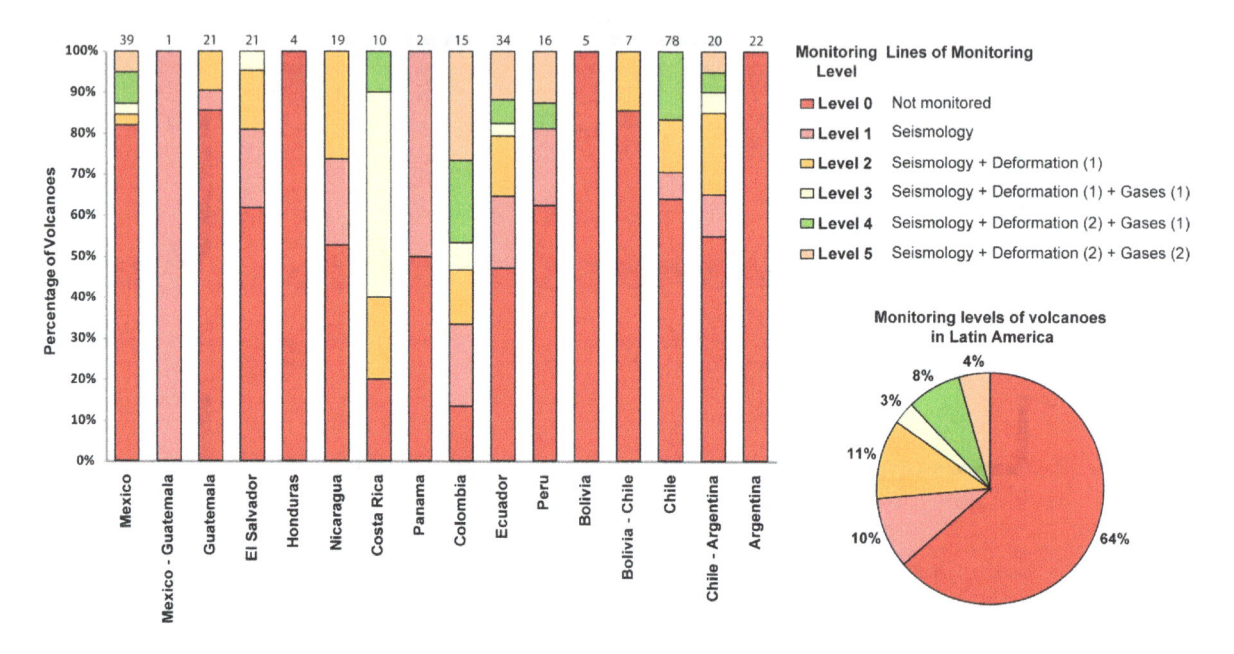

Figure 1.19 The percentage of volcanoes in each country of Latin America with different monitoring levels. The levels and their defining characteristics are shown (top).

A total of 200 Latin American volcanoes classify as Level 0 as they are not continuously monitored using these techniques. Several countries have no monitoring systems in place; however, of these few have confirmed Holocene eruptions. There are, however, 30 unmonitored volcanoes with recorded historical eruptions. Their presence suggests that resources may be required to better equip the region for anticipation and monitoring of volcanic activity. Of the monitored volcanoes, most are Level 2, with dedicated seismic and deformation stations. 15% of Latin American volcanoes are monitored using these and gas analysis. With just 13% and 20%, respectively, of Colombian and Costa Rican volcanoes being unmonitored and 100% of their historically active volcanoes being monitored, these countries are proportionally best for having at least minimal monitoring. Coupled with monitoring Levels 3-5 at over 50% of their volcanoes, these countries show the most comprehensive monitoring regimes. As expected, there is an overall positive correlation between the monitoring of volcanoes and their hazard and risk levels.

Chapter 20 Summary: Volcanic hazard maps

E. Calder, K. Wagner and S.E. Ogburn

Generating hazard maps for active or potentially active volcanoes is recognised as a fundamental step towards the mitigation of risk to vulnerable communities. The responsibility for generating such maps most commonly lies with government institutions but in many cases input from the academic community is solicited. A wide variety of methods are currently employed to generate such maps, and the respective philosophies on which they are based varies; there is also acknowledgement of the notion that one model cannot fit all situations. Some hazard maps are based solely on the distribution of prior erupted products, others take into account estimated recurrence intervals of past events, or use computer models of volcanic processes to gauge potential future extents of impact. Those that are based on modelling generally use empirical, or relatively simple models that capture the essence of a complex process. Simulations are then used to indicate the outcome of an eruptive scenario, or set of scenarios, or, less frequently, are applied probabilistically.

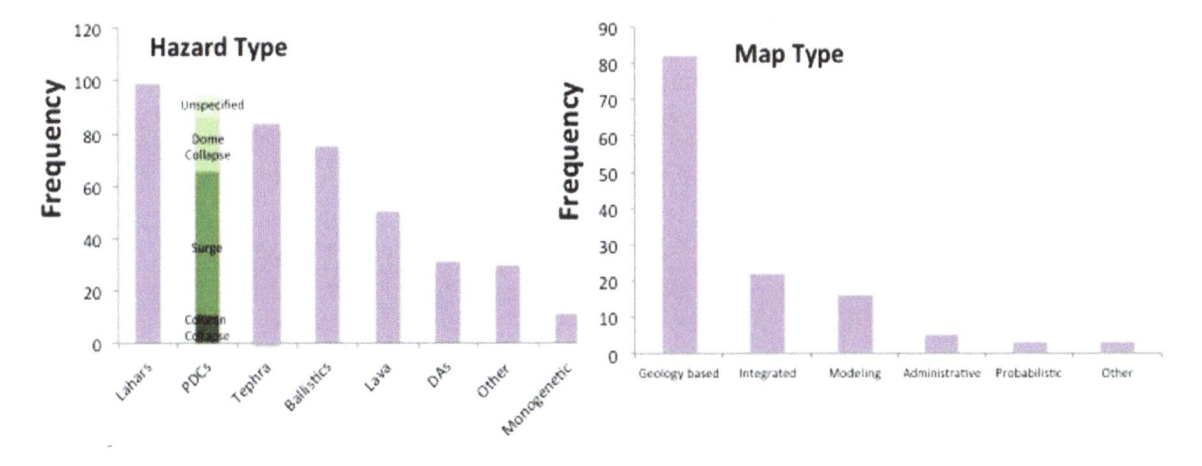

Figure 1.20 a) Types of hazards in the 120 maps reviewed, including: lahars, PDCs, tephra fall, lava flows, debris avalanches and monogenetic volcanism. PDCs were further distinguished based on specific type (column collapse, surge, dome collapse, or unspecified). Some 75% of maps include lahars and/or PDCs and 63% include tephra. Less than half include lava and/or debris avalanches, while less than 10% include hazards associated with unknown source locations, such as monogenetic eruptions. b) Hazard maps can be subdivided into categories based on how and what information is conveyed. Those based solely on the geologic history of the area are significantly more common (63%) than all other map types. Integrated qualitative maps make up a further 17% of maps. Map complexity increases to the right as the number of maps in that category decreases.

A recent review undertaken of 120 volcanic hazard maps provides the following information: The hazards of most widespread concern, as indicated by frequency of occurrence on hazards maps are: lahars (volcanic mudflows), pyroclastic density currents (PDCs), tephra fall, ballistics, lava flows, debris avalanches (volcanic landslides) and monogenetic eruptions (Figure 1.20a). Hazard maps can be categorised into five main types, which, in order of decreasing frequency, are: Geology-based maps: Indicate hazard footprints for the relevant suite of hazards based on the distribution of past eruptive products; Integrated qualitative maps: Display integrated information on the hazards, usually as zones of high, medium, low hazard levels; Modelling based maps: Involve scenario-based application of simulation tools often for a single hazard type; Administrative maps: Combine hazard zones with administrative needs to generate a zonation map used for crisis management; Probabilistic hazard maps: Involve probabilistic application of simulation tools usually for a single hazard type (Figure 1.20b).

The volcanology community currently lacks a coherent approach for hazard mapping but there is consensus that improved quantification is necessary. The variation in currently utilised approaches results in part from differences in the extent of understanding and capability of modelling the respective physical processes (for example tephra fall hazards are currently better quantified than other hazards). Probabilistic hazard maps, in particular, are highly variable in terms of what they represent. Yet there is the need for probabilistic approaches to be fully transparent; they are used to communicate and inform stakeholders, for whom an understanding of the significance of the uncertainties involved is crucial. A recent initiative through the newly formed IAVCEI Commission on Volcanic Hazards and Risk, will focus on hazard mapping with the objective of constructing a framework for a classification scheme for hazard maps, promoting the harmonisation of terminology and providing guidelines for best practices. Driven by the needs of today's stakeholders there is also a need for future research efforts to advance the science that would aid in the production of a new generation of robust, fully quantitative, accountable and defendable hazard maps.

Chapter 21 Summary: Risk assessment case history: The Soufrière Hills Volcano, Montserrat

W. Aspinall and G. Wadge

The Soufrière Hills Volcano (SHV), Montserrat, has been erupting episodically since 1995, with life-threatening pyroclastic flows generated by dome collapse and explosive events. Volcanic activity is monitored by the Montserrat Volcano Observatory (MVO), with an international panel - the Scientific Advisory Committee on Montserrat Volcanic Activity (SAC) - providing regular hazard and risk assessments. Advanced quantitative risk analysis techniques have been developed, forming an important basis for mitigation decisions.

Over 18 years, the SAC has used the following sources of information and methods: MVO data on current activity at the SHV; knowledge of other dome volcanoes; computer models of hazardous volcanic processes; formalised elicitations of probabilities of future hazards scenarios; probabilistic event trees; Bayesian belief networks; census data on population numbers and distribution, and Monte Carlo modelling of risk levels faced by individuals, communities and the island population.

Important findings of the SAC's work are outlined below:

- For hazards, the performance of probabilistic event forecasts against actual outcomes has been measured using the Brier Skill Score: more than 80% of life-critical forecasts had positive scores indicating dependable hazard anticipation. These hazard assessments are crucial for risk estimation and mitigation decisions.

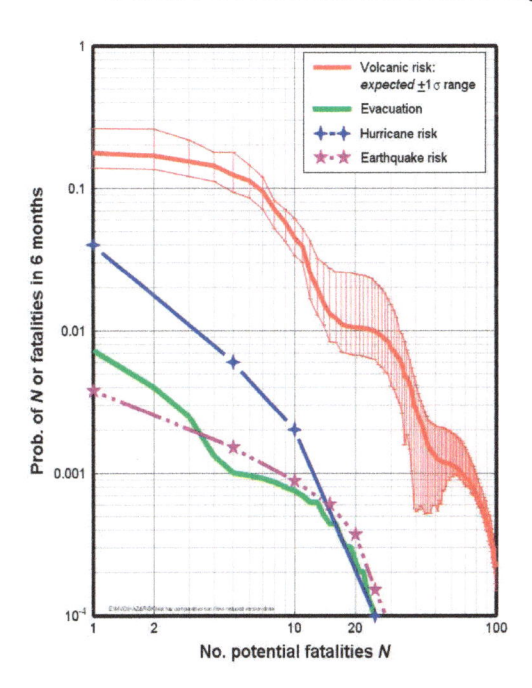

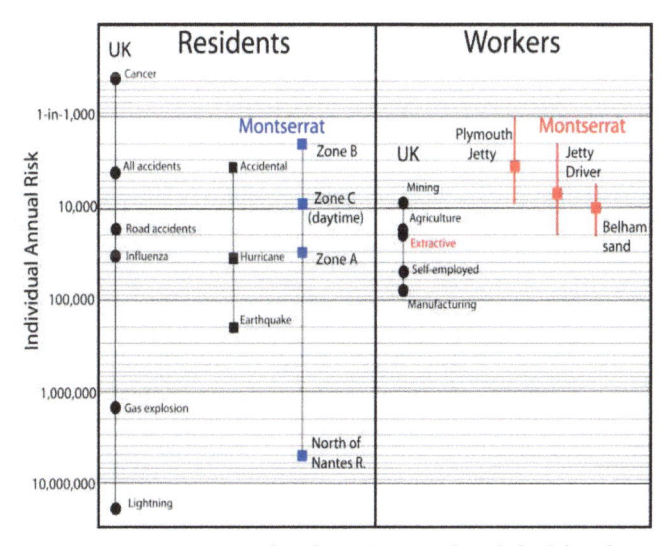

Figure 1.21 F-N plot for 2003 and risk ladder for 2011. See text and Figure 21.1

- It is vital that risk assessments are presented to the authorities and public via open reports in a manner that is understandable. Societal casualty risks and individual risk of death are both calculated. The *F-N* plot from 2003 (left) shows the probability of *N* or more fatalities due to the volcano (red, with uncertainty), the reduced risk if the main at-risk area is evacuated (green) and comparative hurricane and earthquake risks. An individual risk ladder from 2011 is shown (right) with both residential zone risk levels and work-related risk levels plotted, with uncertainties. Comparative values from familiar circumstances are shown for reference.

- Appraising how the authorities respond to specific risk assessments and evaluating outcomes in societal terms has proved difficult, partly because there is no formal feedback mechanism.
- Whilst observatory operations, political aspects and social contexts have changed greatly over this drawn-out episode, the SAC has adopted a uniform approach to risk assessment. This continuity has ensured a consistent approach to scientific advice and helped build public trust. Since risk assessments began in late 1997 there have been no further casualties from volcanic activity, even though it escalated significantly in subsequent years.

SAC risk assessment reports are available from www.mvo.ms.

Chapter 22 Summary: Development of a new global Volcanic Hazard Index

M.R. Auker, R.S.J. Sparks, S.F. Jenkins, W. Aspinall, S.K. Brown, N.I. Deligne, G. Jolly, S.C. Loughlin, W. Marzocchi, C. Newhall and J.L. Palma

A Volcano Hazard Index (VHI) has been developed to characterise the hazard level of volcanoes based on their recorded eruption frequency, modal and maximum recorded VEI levels and occurrence of pyroclastic density currents, lahars and lava flows. VHI is based on a scoring of these hazards indicators with subsequent use of these scores to classify volcanoes into three levels (I, II and III). There are 596 historically active volcanoes, 305 of which have sufficiently detailed eruptive histories to calculate VHI; VHI can be applied to about half the world's recently active volcanoes. A further 23 Holocene volcanoes have a valid VHI score. A meaningful VHI cannot be calculated for the remaining volcanoes due to sparse records.

The volcanoes with an assigned VHI divide between the three levels: I (41%), II (32%) and III (27%). The levels indicate the relative hazard of individual volcanoes. However, **all volcanoes pose significant hazards,** so Level I volcanoes should not be regarded as benign. Scores should not be used as precise numerical values: e.g. a Level III volcano with a score of 24 should not be considered as twice as hazardous as a Level II volcano with a score of 12. VHI is an ordinal characterisation and should not be used for spurious quantification. Volcanoes with the same score may pose quite different hazards. These indices cannot be used for specific hazard assessment. The VHI can change as more data become available and if there are new occurrences of either unrest or eruptions.

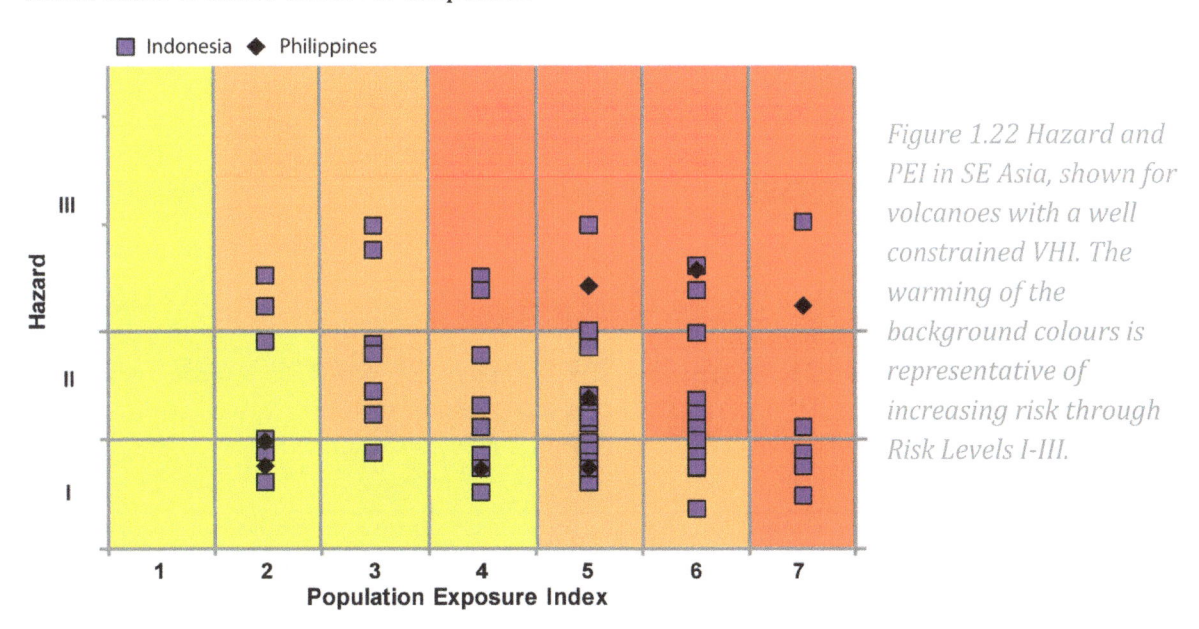

Figure 1.22 Hazard and PEI in SE Asia, shown for volcanoes with a well constrained VHI. The warming of the background colours is representative of increasing risk through Risk Levels I-III.

The Population Exposure Index (PEI) is derived from a population at 10, 30 and 100 km from the volcano, weighted according to the historic occurrence of fatalities and area (Chapter 4). PEI is divided into seven levels from sparsely to very densely populated areas. VHI is combined with the PEI to provide an indicator of risk, which is described as Risk Levels I to III with increasing risk at individual volcanoes. The essential aim of the scheme is to identify volcanoes which are high risk due to a combination of high hazard and population density. A total of 156, 110 and 62 volcanoes classify as Risk Levels I, II and III, respectively. In the country profiles plots of VHI versus PEI provide a way of understanding volcanic risk. Indonesia and the Philippines are plotted as an example. Relative threat can be assessed through PEI where VHI cannot be calculated. The absence of thorough eruptive histories for most of the world's volcanoes and hence absence of VHI is a knowledge gap that must be addressed.

Chapter 23 Summary: Global distribution of volcanic threat

S.K. Brown, R.S.J. Sparks and S.F. Jenkins

An understanding of the total volcanic threat born by each country is gained through the calculation of two measures, combining the number of volcanoes per country, the total population living within 30 km of active volcanoes within the country (Pop30), the total population (Tpop) and the mean hazard score (VHI). The mean VHI per country is determined from the hazard scores of the classified volcanoes and proxy hazard scores derived by volcano type for unclassified volcanoes, permitting a global analysis of the volcanic threat.

The first measure developed here considers the overall threat to life, identifying those countries with the highest threat due to a combination of large numbers of people living within 30 km of active volcanoes, large numbers of volcanoes and high hazard scores.

$$Overall\ threat = mean\ VHI\ x\ number\ of\ volcanoes\ x\ Pop30$$

Indonesia, the Philippines and Japan rank most highly using this measure, all with large populations living within 30 km distance and numerous volcanoes. The sum of the resultant risk scores from the global dataset provides the total global threat and as a proportion of this Indonesia has an astounding dominance, with about two-thirds of the global threat within its borders. As expected, some correlation is observed between threat and the occurrence of fatalities.

The second measure considers the proportion of the population within a country exposed to the volcanic threat, disregarding the numbers of volcanoes.

$$Proportional\ threat = \frac{Pop30}{TPop}\ x\ Mean\ VHI$$

The countries in which volcanic threat is highly significant in terms of the proportion of population exposed are dominantly the small-area nations and island states, with much of the West Indies and Central America ranking most highly.

Both measures provide quite crude assessments of threat and do not take any important local controls on risk into account, such as monitoring capabilities or hazard mitigation measures. However, the differences between the two measures illustrate how, in the event of volcanic activity without advance mitigation measures, losses could be greatest in absolute terms in some countries ranked highly through Measure 1, while the relative social and economic losses could be much greater in smaller countries where a larger proportion of the population would be affected (Measure 2).

Chapter 24 Summary: Scientific communication of uncertainty during volcanic crises

J. Marti

One of the most challenging aspects when managing a volcanic crisis is scientific communication. Volcanology is by its nature an inexact science, such that an appropriate scientific communication should convey information not only on the volcanic activity itself, but also on the uncertainties that always accompany any estimate or prediction. Deciphering the nature of unrest signals (volcanic reactivation) and determining whether or not an unrest episode may be precursory to a new eruption requires knowledge on the volcano's past, current and future behaviour. In order to achieve such a complex objective it is necessary to have different specialists involved in information exchange including those from disciplines such as field studies, volcano monitoring, experimentation, modelling and probabilistic forecasting. It is hence important that these stakeholders communicate on a level that caters for needs and expectations of all disciplines; i.e. to share a common technical language. This is particularly relevant when volcano monitoring is carried out on a systematic survey basis without continuous scientific scrutiny of monitoring protocols or interpretation of data. In an emerging unrest situation, difficulties may arise with communication between different stakeholders with different levels of involvement from different disciplines.

Of particular importance is the communication link between scientists with Civil Protection agents and decision makers during evolving volcanic crises. In this case, it is necessary to translate the scientific understanding of volcanic activity into a series of clearly explained scenarios that are accessible to the decision-making authorities. Also, direct interaction between volcanologists and the general public is rather common both during times of quiescence and activity. Information coming directly from the scientific community has a special influence on risk perception and on the confidence that people put in scientific information. Therefore, effective volcanic crisis management requires identification of feasible actions to improve communication strategies at different levels including: scientists to scientists, scientists to technicians, scientists to Civil Protection, scientists to decision makers, and scientists to general public.

The main goal of eruption forecasting is to identify how, where and when an eruption will occur. To answer these questions we need to use probabilities, which is a way to quantify the intrinsic uncertainty of each parameter. However, communicating probabilities and, in particular, the degree of uncertainty they may have, is not an easy task, and may require a very different approach depending on who is the receiver of such information. Making predictions on what is going to be the future of a volcano follows basically the same reasoning as in other natural hazards (storms, landslides, earthquakes, tsunamis etc.), but does not necessarily have the same level of understanding by the population and decision-makers. This is in part due to lack of experience in making predictions on the behaviour of volcanoes. Compared to meteorologists who have much more data and observations, volcanologists have to deal with a higher degree of uncertainty, mainly derived from this lack of observational data. It is also important to consider that all volcanoes behave in a different way, so a universal model to understand the behaviour of volcanoes does not exist. Each volcano has its own particularities depending on magma composition and physics, rock rheology, stress field, geodynamic environment, local geology, etc., which make them unique, so that what is indicative in one volcano may be not relevant in another. All this makes volcano forecasting very challenging and even more difficult to communicate such high degrees of uncertainty to the population and decision makers. In order to improve scientific communication during volcanic crises comparisons between communication protocols and procedures adopted by different volcano observatories and scientific advisory committees are recommended, in order to identify difficulties and best practice at all levels of communication: scientist to scientist, scientist to technician, scientist to Civil Protection, scientist to general public. Experience from the management and communication of other natural hazards should be brought in and common communication protocols should be defined based on clear and effective ways of showing probabilities and associated uncertainties. Although each cultural and socio-economic situation will have different communication requirements, comparison between different experiences will help to improve each particular communication approach, thus reducing uncertainty in communicating eruption forecasts.

Chapter 25 Summary: Volcano Disaster Assistance Program: Preventing volcanic crises from becoming disasters and advancing science diplomacy

J. Pallister

The Volcano Disaster Assistance Program is a cooperative partnership of the USAID Office of US Foreign Disaster Assistance (OFDA) and the US Geological Survey (USGS). Founded in 1986 in the wake of the Nevado del Ruiz catastrophe wherein more than 23,000 people perished needlessly in a volcanic eruption, VDAP works by invitation to reduce volcanic risk, primarily in developing nations with substantial volcano hazards. The majority of emergency responses and capacity building projects occur in, but are not limited to, Pacific Rim nations. The single most successful VDAP operation was its response with the Philippine Institute of Volcanology and Seismology to the reawakening and subsequent eruption of Mount Pinatubo in 1991. This response alone saved 20,000 lives, including US military personnel at Clark Air Base, and a conservative estimate indicates that at least 250 million dollars in tangible assets were removed from harm's way ahead of the eruption (Newhall et al., 1997). More recently, in late 2010 VDAP assisted Indonesia's Center for Volcanology and Geologic Hazard Mitigation respond to the eruption of Merapi volcano, which saved 10,000 to 20,000 lives.

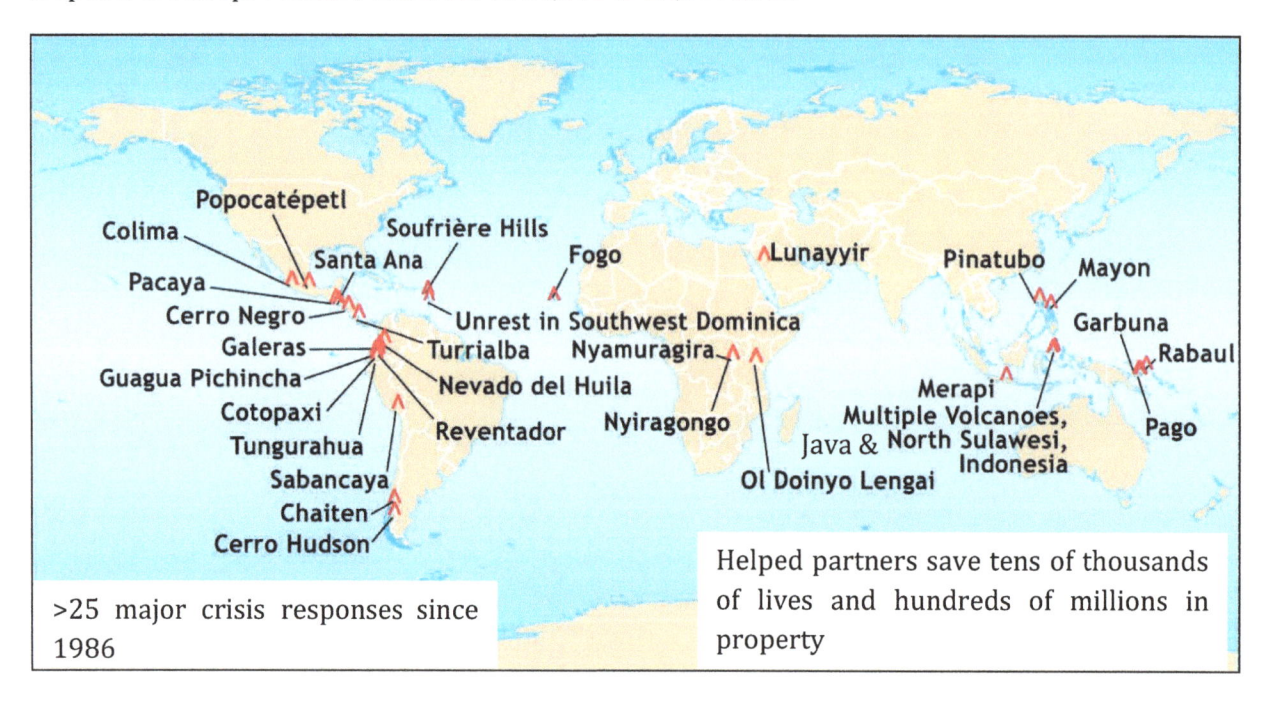

Figure 1.23 Map of VDAP deployments 1986-2012

Over the past 25 years, the VDAP program has served as a development and proving ground for much of the volcano monitoring technology and eruption forecasting science that is applied at US volcanoes. International experience in crisis response and risk mitigation has informed, strengthened and helped guide development of domestic capabilities.

Chapter 26 Summary: Communities coping with uncertainty and reducing their risk: the collaborative monitoring and management of volcanic activity with the *vigías* of Tungurahua

J. Stone, J. Barclay, P. Ramon, P. Mothes and STREVA

Volcán Tungurahua in the Ecuadorian Andes has been in eruption since 1999. Enforced evacuations ended with acrimonious re-occupation within 3 months and the management of risk has been more collaborative ever since.

A network, formed from volunteers already living in the communities at risk, was created with two main goals in mind: (i) to facilitate timely evacuations as part of the Civil Defence communication network, including the management of sirens, and (ii) to communicate observations about the volcano to the scientists. They are called '*vigias*' and around 25 of them are equipped with VHF radios to communicate regularly with observatory scientists and local civil protection.

Since 2000 the *vigías* have provided early warnings to and effective evacuations of their communities (Stone et al., 2014). They also provide detailed updates of increases in activity and hazardous flows to the scientists. In combination this has helped to minimise loss of life and enabled the communities to maintain their lives and livelihoods in the face of dynamic risk. The network has been sustained for >14 years resulting in improved communication pathways and an active involvement in risk reduction at a community level. *Vigías* also maintain scientific instruments and have been able to coordinate the response to fires, road traffic accidents, medical emergencies, thefts, assaults and to plan for future earthquakes and landslides. Motivation to continue the network is provided by its strong value to the community and the mutually beneficial trust-based relationships that it brings, particularly between the scientists and the *vigías*.

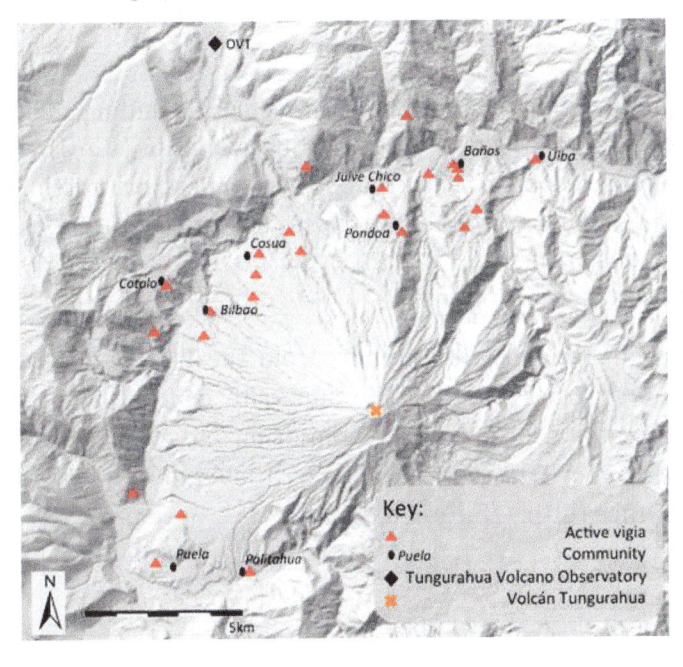

Figure 1.24 Map showing the location of the vigías and significant communities affected by volcanic hazards (adapted from Stone et al., 2014).

Supplementary Case Study 1: Multi-agency response to eruptions with cross-border impacts

B. Oddsson

Iceland lies on the Mid-Atlantic Ridge, the spreading boundary between the Eurasian and North American tectonic plates. In this dynamic environment there are more than 30 volcanic systems, the most frequently active of which lie under Vatnajökull, Europe's largest ice sheet. Since the settlement of Iceland in the late ninth century, over 200 eruptions have been documented, with three in the last 4 years. The eruption of Eyjafjallajökull in 2010 significantly disrupted aviation in Europe and the north Atlantic causing global financial losses. Locally, the sustained ashfall from the Eyjafjallajökull eruption had severe effects on farming in southern Iceland. The fissure eruption at the Barðarbunga volcanic system (ongoing at the time of writing - 2014) has at times resulted in high concentrations of volcanic gases in populated areas of Iceland and sulfur dioxide from the eruption has been detected in the UK.

The Icelandic Meteorological Office (IMO) is responsible for monitoring and warning of natural hazards in Iceland (http://en.vedur.is/), while The National Commissioner of the Icelandic Police, Department of Civil Protection and Emergency Management (DCPEM) is responsible for general emergency coordination, first response in a crisis, communications with the public and mitigation action and recovery (http://www.almannavarnir.is/).

The IMO, DCEPM, University of Iceland and other relevant institutes in Iceland work together during volcanic emergencies at the National Crisis Coordination Center. Two innovative and major initiatives are now underway in Iceland supported by national and international funding to develop risk products and to enhance multi-agency collaboration and data/information sharing.

The first is supported by the national Government and the International Civil Aviation Organisation (ICAO) and aims are to:

- build an online accessible Catalogue for all active volcanoes in Iceland including their main characteristics, eruption histories and possible future eruption scenarios (ICAO);
- develop an interagency plan and general response for the public in case of an eruption;
- develop risk assessments and plans with communities close to active volcanoes, including mitigation actions and response plans;
- develop risk assessments for large, explosive eruptions.

The second is development of a 'Supersite' in Iceland with support from the EUFP7 project 'FUTUREVOLC', a consortium of 26 partners across Europe. The supersite concept implies integration of space and ground-based observations for improved monitoring and evaluation of volcanic hazards, and there is an open data policy. The project is led by University of Iceland together with the Icelandic Meteorological Office (http://futurevolc.hi.is/).

Supplementary Case Study 2: Planning and preparedness for an effusive volcanic eruption: the Laki scenario

C. Vye-Brown, S.C. Loughlin, S. Daud and C. Felton

Following the eruption of Eyjafjallajökull (Iceland) in 2010 the government department handling civil protection in the UK, the Civil Contingencies Secretariat (CCS) of the Cabinet Office, introduced volcanic risks into the National Risk Register (NRR) for the first time. In order to enhance UK preparedness for, and increase resilience to, most types of eruption in Iceland and their distal impacts two scenarios were included in the NRR based on past events: a small to moderate explosive eruption of several weeks duration (the Eyjafjallajökull eruption) and a large fissure eruption of several months duration (the 'Laki' eruption of Grimsvötn volcano).

The Laki eruption occurred over a period of ~8 months in 1783-84 from a fissure in south-eastern Iceland and is the second largest such eruption in Iceland in historical time with huge outpourings of mainly lava, gases and aerosols (atmospheric particles). Hazards might include sulfur dioxide or other gases at flight and ground levels, particulate matter including sulfates ($PM_{2.5}$ and PM_{10}) and plume contents reaching the ground. There are good historical accounts of the eruption and its impact both in Iceland and across Europe and such eruptions are known to cause regional to hemispheric-scale impacts on multiple sectors from health and transport to environment and economy. However, assessing the potential impacts on the UK of such an eruption now is challenging but most risks could be mitigated with effective planning. Therefore, planning and preparedness for volcanic eruptions involves co-ordination and working across multiple departments both within and across national boundaries.

Since the incorporation of the Laki scenario in the NRR, cross-cutting work coordinated by the CCS has brought together government, research institutions and academia to investigate volcanic risks to the UK, better understand uncertainties, build UK resilience to volcanic risks and prepare our response to them. Collaboration across disciplines is essential to understand the likely impact with variability in the eruption dynamics and meteorology as well as interaction with modern systems including the potential timescales, intensities and concentrations of these hazards. Whilst the Laki scenario is a relatively low-frequency event, it is high magnitude and models of the distal impact of volcanic hazards are needed to ensure proportionate planning and to enable government departments to consider likely societal impacts and response strategies.

Supplementary Case Study 3: Interactions of volcanic airfall and glaciers

L.K. Hobbs, J.S. Gilbert, S.J. Lane and S.C. Loughlin

Volcanic airfall (defined here as any material, such as ash, that falls from an eruption plume and cools before deposition) may land on glaciers and snowfields and the interactions between these deposits and underlying glaciers have a range of possible outcomes, depending on the nature of the airfall. If the deposit thickness exceeds a local 'critical thickness', ablation rate (removal of ice by melting or sublimation processes) will be reduced relative to the bare surface; conversely, thinner deposits enhance ablation of the underlying glacier surface. Both scenarios have potentially hazardous consequences. Reduced ablation can lead to shortages in water supplies as well as release of physical contaminants and accumulation of leachates from the deposits. Relatively thin airfall deposits and enhanced ablation can lead to contamination of water supplies by leaching and facilitate production of lahars and avalanches by increasing available meltwater and providing failure surfaces. In both cases, glacial mass balance is affected and increased gravitational loading by deposits can make failure hazards more likely to occur, while the presence of debris on the glacier surface provides additional material for incorporation into lahars.

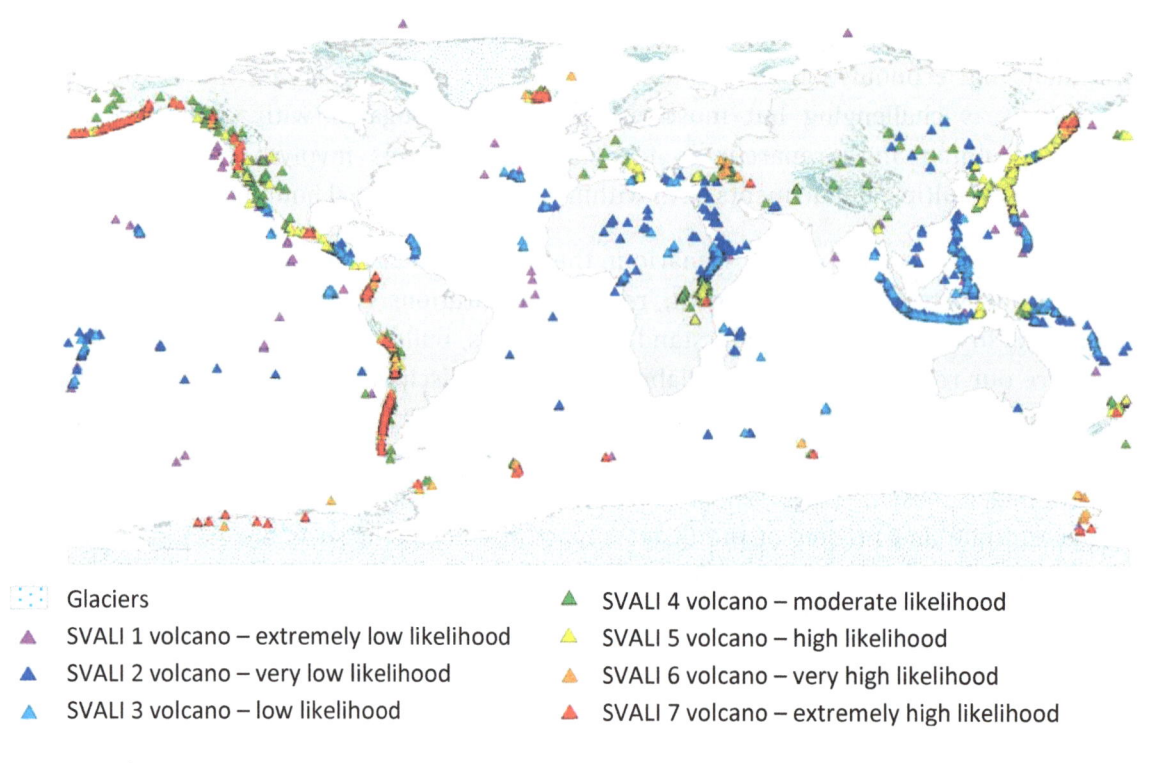

- Glaciers
- SVALI 1 volcano – extremely low likelihood
- SVALI 2 volcano – very low likelihood
- SVALI 3 volcano – low likelihood
- SVALI 4 volcano – moderate likelihood
- SVALI 5 volcano – high likelihood
- SVALI 6 volcano – very high likelihood
- SVALI 7 volcano – extremely high likelihood

Figure 1.25 Likelihood of deposition of supraglacial airfall by active volcanoes. Symbols are not representative of scale of volcanoes or glaciers. Volcano location data are from the Global Volcanism Program (Simkin and Siebert, 2002-2013); glacier data are from the World Glacier Inventory (WGMS and NSIDC (1989, updated 2012), the Global Land Ice Measurements from Space Glacier Database (Armstrong et al., 2012), Morales-Arnao (1998), UNEP and WGMS (2008), Moussavi et al. (2010) and Fukui and Iida (2012).

It is important to assess the potential for deposition of volcanic airfall on glaciers where such supraglacial deposition may pose a hazard to life or economy. The Supraglacial Volcanic Airfall Likelihood Index (SVALI) provides a framework for making such assessments based on eruption characteristics and the geographical location of the source volcano relative to the locations of glaciers (Figure 1.25).

References

Armstrong, R., Raup., B., Khalsa, S.J.S, Barry, R., Kargel, J. and Helm, C., 2012. GLIMS glacier database. Boulder, Colorado, USA: National Snow and Ice Data Center. Digital Media. Accessed regularly.

Fukui, K. and Iida, H., 2012. Identifying active glaciers in Mt. Tateyama and Mt. Tsurugi. *Journal of the Japanese Society of Ice and Snow (Seppyo)*, 74(3), 213-222.

Morales-Arnao, B., 1998. Glaciers of South America – Glaciers of Perú. In: Williams, R.S. Jr. and Ferrigno, J.G., (Eds). *Satellite image atlas of glaciers of the world, South America*, USGS Professional Paper 1386I, I, pp. 151-157.

Moussavi, M.S., Valadan Zoej, M.J., Vaziri, F., Sahebi, M.R. and Rezaei, Y., 2010. A new glacier inventory of Iran. *Annals of Glaciology*, 50, 93-103.

UNEP and WGMS, 2008. *Global Glacier Changes: facts and figures*. Joint UNEP and WGMS Publication, 88 pp.

Simkin, T. and Siebert, L., 2002-2013. *Volcanoes of the World: an Illustrated Catalog of Holocene Volcanoes and their Eruptions. Smithsonian Institution, Global Volcanism Program Digital Information Series, GVP-3*. www.volcano.si.edu Accessed regularly.

WGMS and NSIDC, 1989, updated 2012. *World Glacier Inventory*. Compiled and made available by the World Glacier Monitoring Service, Zurich, Switzerland, and the National Snow and Ice Data Center, Boulder CO, USA. Digital media. Accessed regularly.

Chapter 2

Global volcanic hazard and risk

S.K. Brown, S.C. Loughlin, R.S.J. Sparks, C. Vye-Brown, J. Barclay, E. Calder, E. Cottrell, G. Jolly, J-C. Komorowski, C. Mandeville, C. Newhall, J. Palma, S. Potter and G. Valentine

with contributions from B. Baptie, J. Biggs, H.S. Crosweller, E. Ilyinskaya, C. Kilburn, K. Mee, M. Pritchard and authors of Chapters 3-26

Contents

2.1 Introduction
2.2 Background on volcanoes
2.3 Volcanoes in space and time
2.4 Volcanic hazards and impacts

2.5 Monitoring and forecasting
2.6 Assessing hazards and risk
2.7 Volcanic emergencies and DRR
2.8 The way forward

2.1 Introduction

An estimated 800 million people live within 100 km of an active volcano in 86 countries and additional overseas territories worldwide [see Chapter 4 and Appendix B][1]. Volcanoes are compelling evidence that the Earth is a dynamic planet characterised by endless change and renewal. Humans have always found volcanic activity fascinating and have often chosen to live close to volcanoes, which commonly provide favourable environments for life. Volcanoes bring many benefits to society: eruptions fertilise soils; elevated topography provides good sites for infrastructure (e.g. telecommunications on elevated ground); water resources are commonly plentiful; volcano tourism can be lucrative; and volcanoes can acquire spiritual, aesthetic or religious significance. Some volcanoes are also associated with geothermal resources, making them a target for exploration and a potential energy resource.

Much of the time volcanoes are not a threat because they erupt very infrequently or because communities have become resilient to frequently erupting volcanoes. However, there is an ever-present danger of a long-dormant volcano re-awakening or of volcanoes producing anomalously large or unexpected eruptions. Volcanic eruptions can cause loss of life and livelihoods in exposed communities, damage or disrupt critical infrastructure and add stress to already fragile

---

[1] Chapters 4 to 26 provide additional detail and case studies about subjects covered in this chapter. An electronic supplementary report (Appendix B, www.cambridge.org/volcano) is provided comprising country and regional profiles of volcanism.

Brown, S.K., Loughlin, S.C., Sparks, R.S.J., Vye-Brown, C., Barclay, J., Calder, E., Cottrell, E., Jolly, G., Komorowski, J-C., Mandeville, C., Newhall, C., Palma, J., Potter, S., Valentine, G. (2015) Global volcanic hazard and risk. In: S.C. Loughlin, R.S.J. Sparks, S.K. Brown, S.F. Jenkins & C. Vye-Brown (eds) *Global Volcanic Hazards and Risk,* Cambridge: Cambridge University Press.

environments. Their impacts can be both short-term, e.g. physical damage, and long-term, e.g. sustained or permanent displacement of populations. The risk from volcanic eruptions and their attendant hazards is often underestimated beyond areas within the immediate proximity of a volcano. For example, volcanic ash hazards can have effects hundreds of kilometres away from the vent and have an adverse impact on human and animal health, infrastructure, transport, agriculture and horticulture, the environment and economies. The products of volcanism and their impacts can extend beyond country borders, to be regional and even global in scale.

Although known historical loss of life from volcanic eruptions (since 1600 AD about 280,000 fatalities are recorded, Auker et al. (2013)) is modest compared to other major natural hazards, volcanic eruptions can be catastrophic for exposed communities. In 1985 the town of Armero in Colombia was buried by lahars (volcanic mudflows) with more than 21,000 fatalities due to relatively small explosive eruptions at the summit of Nevado del Ruiz volcano that partially melted a glacier (Voight, 1990). Since 1985 an estimated 2 million people have been evacuated due to eruptions or threats of eruption. Some of these people have been permanently relocated. The 2010 eruption of Merapi volcano in Indonesia caused the evacuation of approximately 400,000 people, 386 fatalities (Surono et al., 2012) and an estimated loss of US$ 300 million (IDR 3.56 trillion) (BNPB., 2011). Timely evacuations saved an estimated 10,000 to 20,000 lives. More recently the economic impact of volcanic eruptions has become more apparent on local, regional and global scales. A modest-sized eruption of the Eyjafjallajökull volcano, Iceland, in 2010 caused havoc when air traffic was restricted due to an extensive ash cloud, demonstrating the regulatory challenges for the aviation sector. The global financial losses approximated US$5bn as almost all parts of the world were affected by disrupted global business and supply chains (Ragona et al., 2011). Managing volcanic risk is thus a worldwide problem. Very large magnitude eruptions are the only natural phenomenon, apart from meteor impacts, with the potential for global disaster (Self & Blake, 2008).

Volcanic eruptions are difficult to predict accurately. However, progress has been made in forecasting the onset of an eruption by using scientific interpretation of volcanic unrest (Sparks, 2003, Segall, 2013). Volcanic unrest usually precedes eruptions, and may consist of earthquakes, ground deformation, gas release and other manifestations caused by rock fracturing or magma movement below the Earth's surface. The ability to issue early warnings is improving with advances in methods of detection and scientific knowledge. Volcanic unrest may only be detected if there is a good monitoring network in place, but many volcanoes worldwide are not monitored sufficiently or at all. As some volcanic hazards can develop rapidly once an eruption begins, precautionary responses such as evacuations are commonly undertaken prior to the eruption starting or in periods of heightened activity. However, volcanic unrest does not necessarily lead to an eruption, and unrest can be hazardous even without a resulting eruption (Barberi et al., 1984, Potter et al., 2012). About half of historically active volcanoes have reawakened after a repose interval of a century or more. Some of these volcanoes have subsequently erupted more frequently, while some return to dormancy. Inexperienced communities living on long-dormant volcanoes tend to be sceptical of the level of hazard posed by their volcano when it threatens to erupt.

Volcanoes present many different hazards and eruptions are often complex sequences of hazardous phenomena. Each hazard has different characteristics and can cause a wide range of impacts distributed across small to large areas. External factors can influence the occurrence and distribution of these hazards, with wind, for example, determining the direction and extent of hazardous volcanic ash fall, and rain potentially causing volcanic mudflows and landslides. Volcanic eruptions can last minutes to decades (Siebert et al., 2010). These attributes provide challenges for successful emergency management and disaster risk reduction. Volcano observatories dedicated to monitoring high-risk volcanoes are crucial for effective mitigation and emergency management; they support resilient communities and systems. There are many factors that contribute towards exposure (i.e. the number and distribution of threatened people and assets) and vulnerability (i.e. their response) to volcanic hazards, and these require integration with hazards assessments to produce local, regional and global assessment of volcanic risk.

The aim of this book is provide a broad synopsis of global volcanic hazards and risk with a focus on the impact of eruptions on society and to provide the first comprehensive global assessment of volcanic hazard and risk. This work was originally undertaken by the Global Volcano Model (GVM, http://globalvolcanomodel.org/) in collaboration with the International Association of Volcanology and Chemistry of the Earth's Interior (IAVCEI, http://www.iavcei.org/) as a contribution to the Global Assessment Report on Disaster Risk Reduction, 2015 (GAR15), produced by the United Nations Office for Disaster Risk Reduction (UN ISDR).

This chapter complements Chapter 1, which provided a non-technical summary of the topics covered through the rest of the book. This chapter presents a more technical synopsis of volcanism from a hazard and risk perspective, with selected references and links to online resources that will enable a reader to learn about particular topics in more detail. Although more technical from this point, this work does not assume a readership specialising in geosciences.

Chapter 2 comprises eight sections as follows:

- Section 1: Introductory section
- Section 2: Background on volcanoes, the cause of eruptions and the processes driving them
- Section 3: Volcanic eruptions in space and time
- Section 4: Volcanic hazards and their impacts
- Section 5: Monitoring and forecasting of volcanic eruptions
- Section 6: Methods of assessing volcanic hazards and risk
- Section 7: Management of volcanic emergencies and disaster risk reduction
- Section 8: Prognosis on the ways to improve knowledge, emergency management and risk reduction

Here, the readers are frequently pointed towards Chapters 3 through 26. These chapters are designed to illustrate key concepts, methodologies and approaches to the assessment and management of volcanic hazards and risk. These chapters, along with published literature, provide the evidence base for this work.

A complementary report, Appendix B, is provided online in support of this book, in which a country-by-country analysis of volcanoes, hazards, vulnerabilities and technical coping capacity is provided to give a snapshot of the current state of volcanic risk across the world.

2.2 Background on volcanoes and volcanic eruptions

This section provides a basic background on volcanoes and hazards for those unfamiliar with the basic ideas and contemporary understanding. There are numerous books, publications and websites devoted to volcanoes, contemporary theories of volcanism and volcanic hazards. Selected books (Blong, 1984, Schmincke, 2004, Decker & Decker, 2006, Lockwood & Hazlett, 2013, Papale, 2014), review papers (Sparks, 2003, Newhall, 2007, Cashman et al., 2013, Sparks & Aspinall, 2013) and the Encyclopedia of Volcanoes (Sigurdsson et al., 2015) are recommended starting points. The US Geological Survey website http://volcanoes.usgs.gov/hazards/index provides comprehensive information on volcanic processes and hazards. The Smithsonian Institution provides comprehensive and authoritative information on the world's volcanoes as well as weekly reports on volcanic activity around the world (http://www.volcano.si.edu/). The Smithsonian Volcanoes of the World database is the source for much of the basic information used in this book (Siebert et al., 2010, Cottrell, 2014). Version 4 of the database (VOTW4) is online (Smithsonian, 2013). The figures cited throughout this book are from VOTW4.22.

2.2.1 Causes of volcanism

Volcanoes are a manifestation of the Earth's internal dynamics related to heat loss. Most volcanoes are located close to the boundaries of tectonic plates and are the consequence of melting the Earth's interior at depths ranging mostly between 10 and 200 km (Figure 2.1). At depths of a few tens of kilometres the solid Earth is very hot (1200°C or more) and close to its melting temperature. Tectonic plate boundaries are regions where the cool rigid carapace of the Earth is disrupted, like a cracked egg shell. Plates are formed where they are rifted apart (mostly in the oceans) and destroyed where plates collide and one of the plates is pushed back into the Earth's interior (a process called *subduction*).

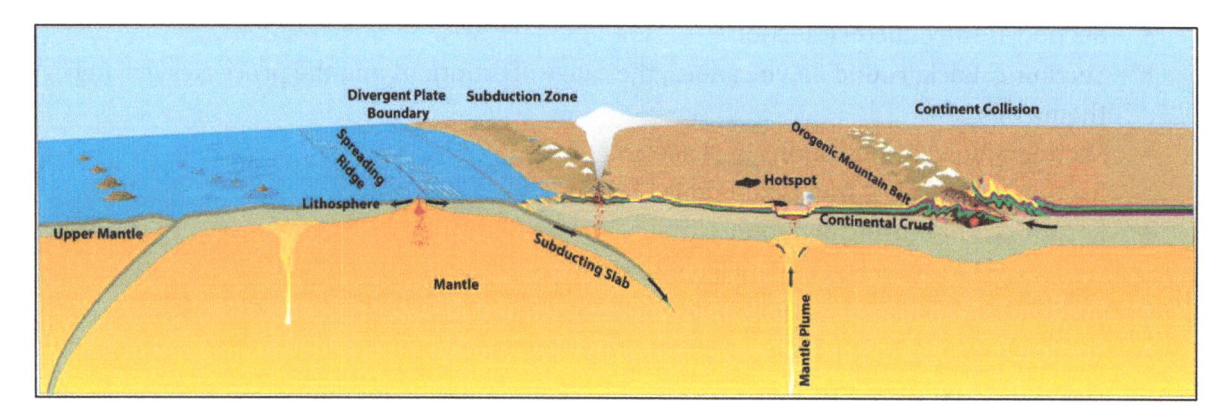

Figure 2.1 A cross-section through the Earth's upper mantle and crust illustrating the plate tectonics and magma generation which gives rise to Earth's volcanoes at subduction zones, spreading ridges and hotspots. (Image courtesy of the Global Volcanism Program, Smithsonian Institution.)

Numerous volcanoes form on the world's rifted plate boundaries, but mostly these are located deep below the ocean surface along submarine ocean ridges. Most active volcanoes that pose hazards are located at subduction zones, forming arc-shaped chains of volcanic islands like the Lesser Antilles in the Caribbean or lines of volcanoes parallel to the coasts of major continents as in the Andes. There are some dangerous volcanoes located where rifting plates form on land

and in the shallow ocean, such as Iceland and the great East African rift valley. There are also active volcanoes within tectonic plates, the Hawaiian volcanoes being the best-known examples.

Where there are large convection currents in the Earth's mantle beneath the plates, hot mantle rock melts as it moves towards the Earth's surface due to a reduction in pressure. This pressure reduction melting process occurs below rifting plates and volcanoes like those in Hawaii in the interior of plates. In a subduction zone, one of the colliding plates is forced back into the Earth's interior. Hydrated crust of the sea floor is subducted to depths of about 100 km where the water is released into the surrounding hot rocks. Water dramatically lowers the melting temperature of these rocks and copious melting results. Regardless of tectonic setting, intergranular melt coalesces and moves along cracks and conduits towards the Earth's surface. Volcanoes form as a consequence.

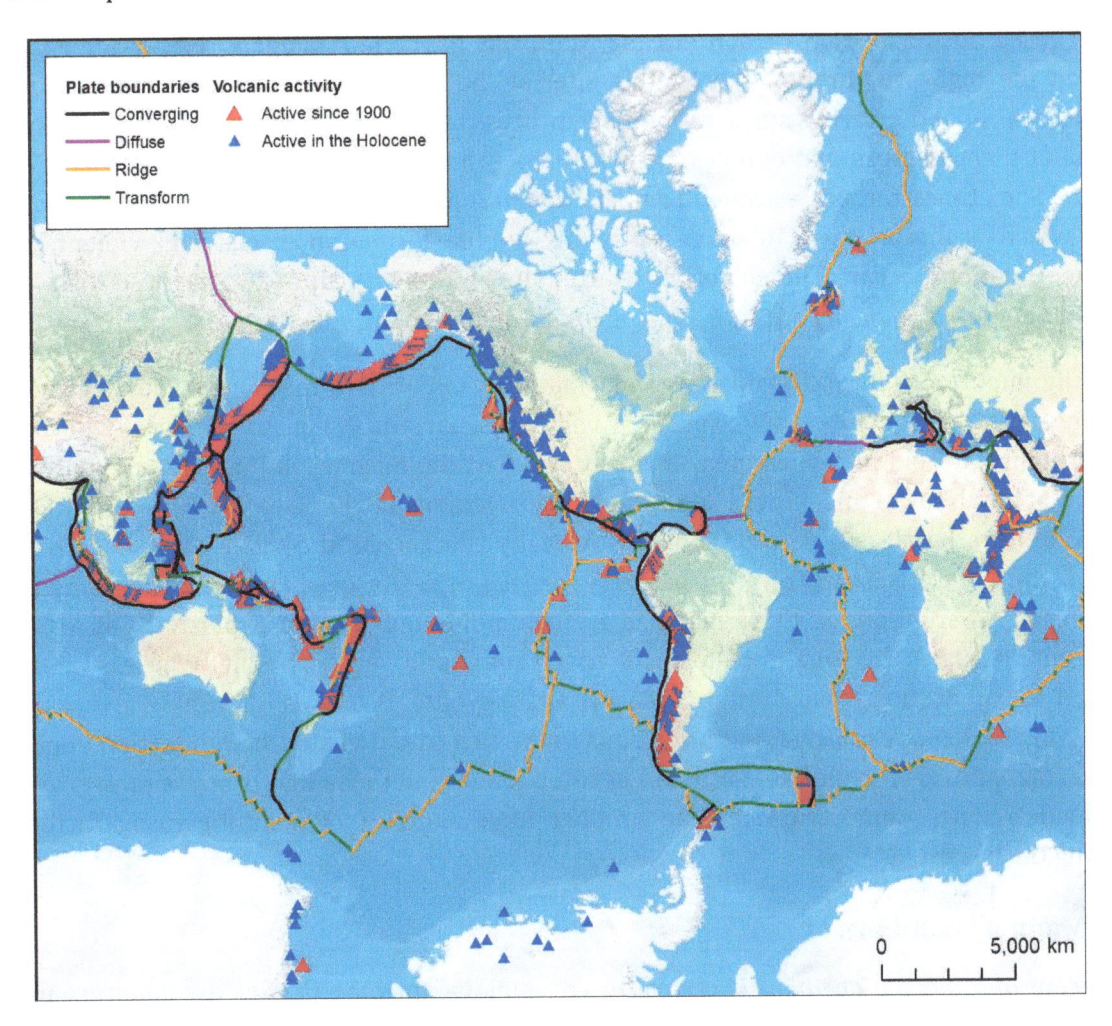

Figure 2.2 Global map of the distribution and status of Holocene volcanoes as listed in VOTW4.22. The distribution of volcanoes also outlines the boundaries of major tectonic plates.

The spatial distribution of volcanoes (Figure 2.2) is now very well understood (Cottrell, 2014) and enables volcanologists to be very confident about where to expect active volcanoes and new volcanoes in the future. There are many parts of the world where active volcanism can be excluded, although they can still be affected by the economic, environmental and climatic impacts of volcanism.

2.2.2 Magma

Magma is subsurface molten rock, commonly mixed with suspended crystals and gas bubbles. Magmas vary in composition from those typically rich in elements such as magnesium, calcium and iron and containing about 50% silica (silicon dioxide) to those rich in alkali elements, such as sodium and potassium, with only minor amounts of magnesium, calcium and iron, and containing as much as 75% silica. The former magmas are known as *basalts* and have temperatures typically in the range 1100 to 1300°C. The latter are called *rhyolites* and have temperatures typically in the range 700 to 900°C. There are many magmas intermediate between basalt and rhyolite, the most common being *andesite*. There are a plethora of other magma types and related nomenclature that relate to variations in chemical compositions and mineralogy (Le Maitre et al., 2002).

Volcanic gases, such as water, carbon dioxide, sulfurous gases and halogens, are dissolved in magma at the high pressures of the Earth's interior, but bubble out of the magma at low pressures near or at the Earth's surface. The same process is familiar in fizzy drinks where gas is dissolved at high pressure and bubbles out when the can or bottle is opened and pressure is released. Sometimes the gas escapes from the magma quietly and slowly to form gas-poor magma which erupts as lava. In other cases gas bubble formation is fast and violent, so explosive eruptions occur. The materials ejected in explosive eruptions are described as *pyroclastic,* and *pyroclastic deposits* are a major constituent of many volcanoes.

A critically important physical property of magma is its *viscosity* (a measure of how easily a liquid flows) as this controls many aspects of volcano behaviour. Hot basalts typically have a viscosity similar to cold honey, whereas andesite and rhyolite magmas are much more viscous by factors of hundreds to millions (i.e. much more viscous than tar). For example, gas can escape quite easily from basalt and its eruptions are typically characterised by lava flows and weak explosions. In contrast escape of gas from very viscous andesite or rhyolite magma is much more difficult so eruptions of these magmas are commonly much more explosive. Volcanoes located within plates or where plates rift apart commonly erupt basalt, whereas many volcanoes in subduction zones erupt andesite and rhyolite. There is therefore a marked tendency for the most explosive and hazardous volcanoes to be located in subduction zones. However, there are very explosive volcanoes at rifted plates and those that produce mostly lava in subduction zones. Some eruptions can produce huge amounts of polluting gases, further increasing the hazard.

2.2.3 Magma chambers

The concept of a *magma chamber* is crucial to volcanology, forming part of the underground plumbing of a volcano. Magma chambers can be defined as a subsurface region or regions where magma accumulates, supplying the volcano during an eruption (Figure 2.3). They commonly form because magma ascending stalls below ground rather than erupts. Typically magma chambers can form at depths of a few kilometres to the base of the Earth's crust, which ranges in thickness from 6 to as much as 70 km. The Earth's crust becomes cold and strong near the surface and magmas can be prevented from reaching the Earth's surface by various mechanisms. Magma can solidify and may have insufficient pressure to break through to the surface. Stagnation of magma can result in cooling, loss of gas and crystallisation, while heating of surrounding rocks can result in them melting and formation of more magma. Magma chambers can also form when melt is squeezed out of partially molten rocks to form lenses and pockets. These melts can merge together to form large eruptible volumes of magma. These complex processes lead to a wide range of magma volumes, compositions, temperatures and volcanic gas contents, which explains why a single volcano can erupt different magmas with a wide range of eruption styles and hazards.

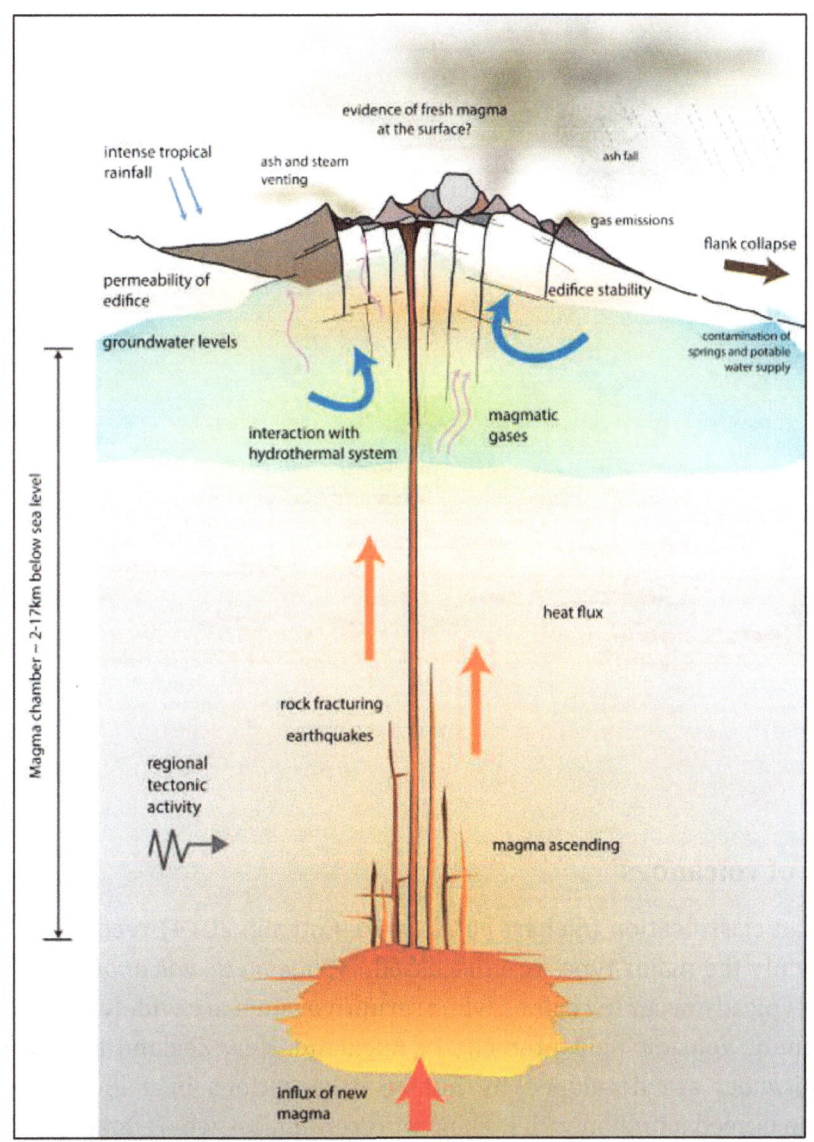

Figure 2.3 Schematic cross-section through a volcano in a tropical setting and its underlying magma chamber illustrating some of the major processes that lead to phenomena that are monitored on active volcanoes. Modified from Hincks et al. (2014). The magma chamber may become pressurised, for example from influx of new magma from depth or build up of internal gas pressure as it cools and crystallises. Typically a narrow volcanic conduit connects the chamber to the surface during an eruption. The rising magma and pressure from the chamber makes the volcano deform and results in many small earthquakes. Volcanic gases are released and ground water is heated, resulting in surface hot springs and fumaroles.

Magma chambers are an important concept in the interpretation of monitoring data at volcano observatories. Earthquakes, ground deformation and anomalous gas emissions, that are commonly precursors to eruptions, are often interpreted in terms of processes within magma chambers and in movement of magma and gases from a magma chambers to the surface (Figure 2.3). Recent research is recognising that many volcanic systems involve multiple regions of magma (Figure 2.4) and that there can be other causes of geophysical phenomena at volcanoes, such as movements of ground water (Figure 2.3), which need to be distinguished from manifestations of magma chambers.

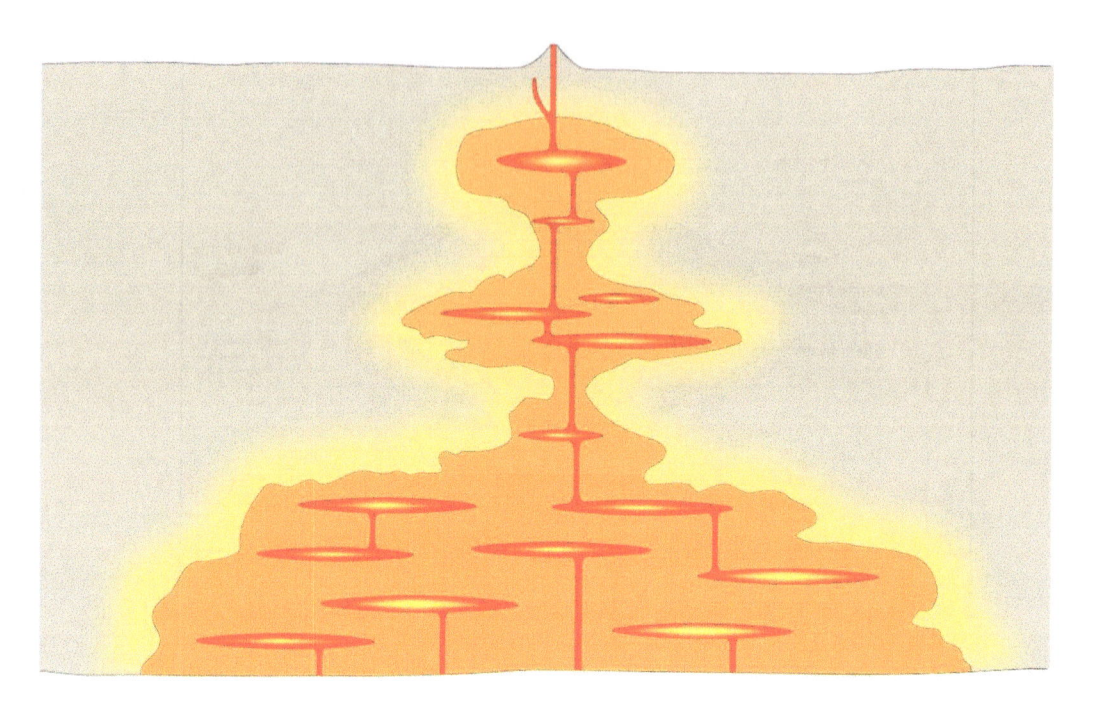

Figure 2.4 Schematic diagram showing a crustal region comprising several magma bodies embedded in hot partially melted rocks at different depths below a volcano. Figure 2.3 shows the uppermost chamber connected to the volcano.

2.2.4 Types of volcanoes

The Smithsonian classification (Siebert et al., 2010, Cottrell, 2014) recognises 26 categories of volcano. Here only the major types are discussed. *Monogenetic volcanoes* are formed by single eruptions and typically occur in regions where eruptive vents are widely distributed in what are called monogenetic volcanic fields; the city of Auckland, New Zealand is built in such an area. *Polygenetic volcanoes* are developed by numerous eruptions in a localised area over time periods that can exceed a million years. They represent places where magma ascent is focussed. Many polygenetic volcanoes are thought to be underlain by large regions of very hot rock, containing small amounts of melt and multiple magma chambers. Polygenetic volcanoes can be broadly classified into different types based on magma chemistry, size and dominant eruptive styles.

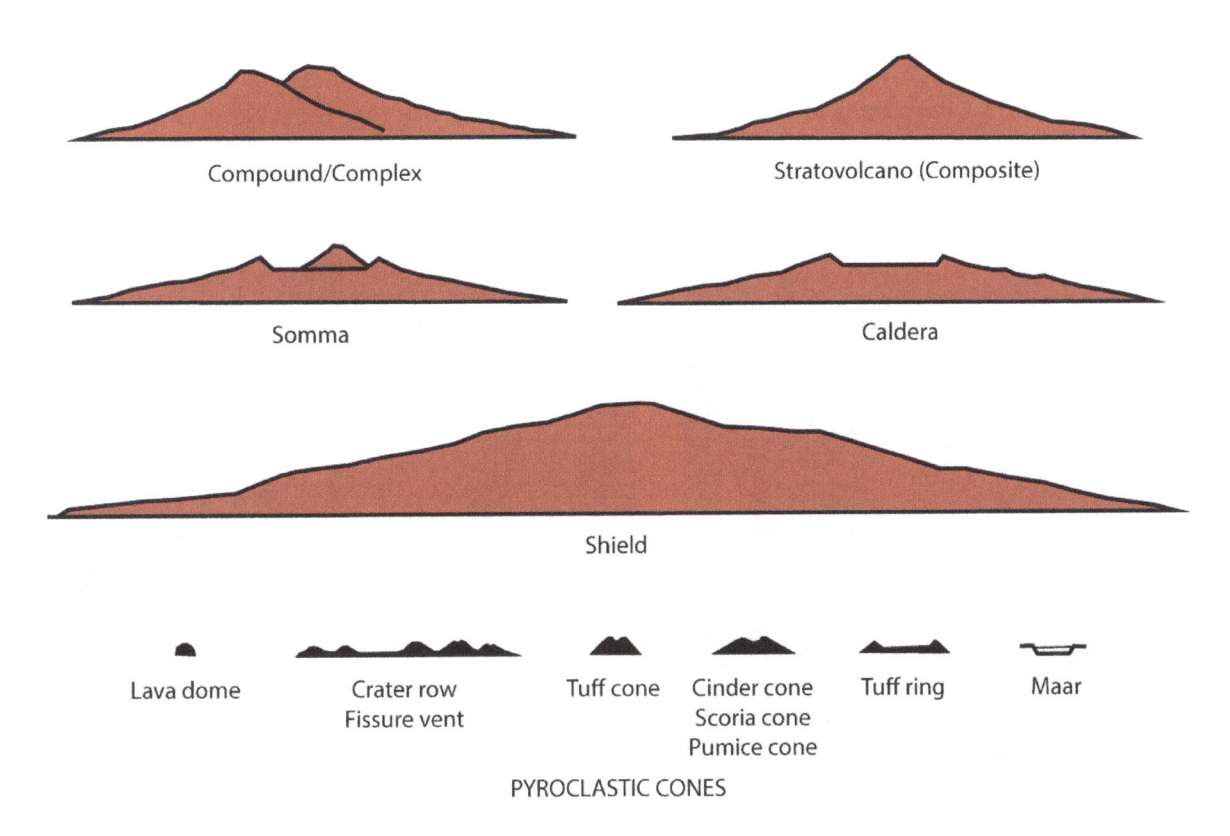

Figure 2.5 Volcanoes take a variety of forms as dictated by their chemistry, eruption style and products. Here the main volcano types are illustrated with a vertical exaggeration of 2:1 for the main edifice constructs and 4:1 for the pyroclastic cones (Siebert et al., 2010). Relative sizes are approximate.

Some common volcano types are illustrated in Figure 2.5. *Fissure volcanoes* are where large fractures (or fissures) form in the Earth's crust and are characterised by eruption of copious lava and gas. The 1783 eruption of Laki in Iceland is a type example (Figure 2.6a), when over 15 km^3 of basalt erupted over 6 months. *Shield volcanoes*, like Kilauea and Mauna Loa (Hawaii), are amalgamations of numerous lava flows and are typically basaltic (Figure 2.6b). *Stratovolcanoes*, like Fuji (Japan) and Colima (Mexico), are typically steep-sided and are mixtures of lava and pyroclastic deposits (Figure 2.6c). *Lava dome volcanoes*, like Soufrière Hills Volcano (Montserrat, Eastern Caribbean) are made of mounds of andesitic or rhyolitic lava known as *domes*, together with pyroclastic deposits (Figure 2.6d). *Calderas* are large volcanic craters (1 to more than 50 km diameter) mostly formed by large magnitude volcanic eruptions (Figure 2.6e). Eruption of large volumes of magma causes the ground above the magma chamber to collapse. The largest of these, like Yellowstone (USA), have been called supervolcanoes (Self & Blake, 2008).

While basic classifications are useful, many volcanoes are very diverse in their styles of eruption, in their magnitudes, intensities and frequency of eruption. For example an active volcano like Santorini (Greece) over a history of 700,000 years has behaved as a shield volcano and stratovolcano at different times, has formed several large calderas from major explosive eruptions, and has erupted basalt, andesite and rhyolite at different times. This variety comes about because the processes of magma generation and the interaction of erupting volcanoes with surface environments are complex. From a hazard perspective every volcano is thus unique in some respects and this means that forecasting of eruptions and assessment of hazards needs to be carried out at a local volcano scale. For this reason a critical aspect of living with an active volcano is to have a dedicated volcano observatory.

Figure 2.6 Examples of volcano forms. a) Fissure from the 1783 Laki, Iceland, eruption (O. Sigurdsson). b) The Mauna Loa shield volcano, Hawaii (US Geological Survey archive). c) Colima stratovolcano, Mexico. (S. Brown). d) The lava dome at Unzen volcano, Japan (S.Jenkins). e) The 10 km diameter Aniakchak caldera, Alaska (US Geological Survey archive).

2.2.5 Styles of eruption

At the most basic level volcanic activity can be divided into effusion of lava and explosive eruptions (Figure 2.7). In some cases eruptions are only explosive, while in others they are dominantly effusive. However, many eruptions are a mixture of explosive and effusive activity, which can sometimes occur simultaneously or in complex alternating sequences. Explosive eruptions can vary from discrete explosions lasting a few minutes to sustained and intense discharges over many hours. Explosions can result from violent release of volcanic gases dissolved under pressure in the magma (Figure 2.8a, b) and by interaction of hot erupting magma with water (Figure 2.8c). Lava represents magma that has lost most of the originally dissolved gases prior to eruption. Basalt magmas with low viscosity can form rapidly moving rivers of thin lava when the effusion rate is high (Figure 2.8d), while more viscous andesite and rhyolite form much thicker lavas and domes (Figure 2.8e).

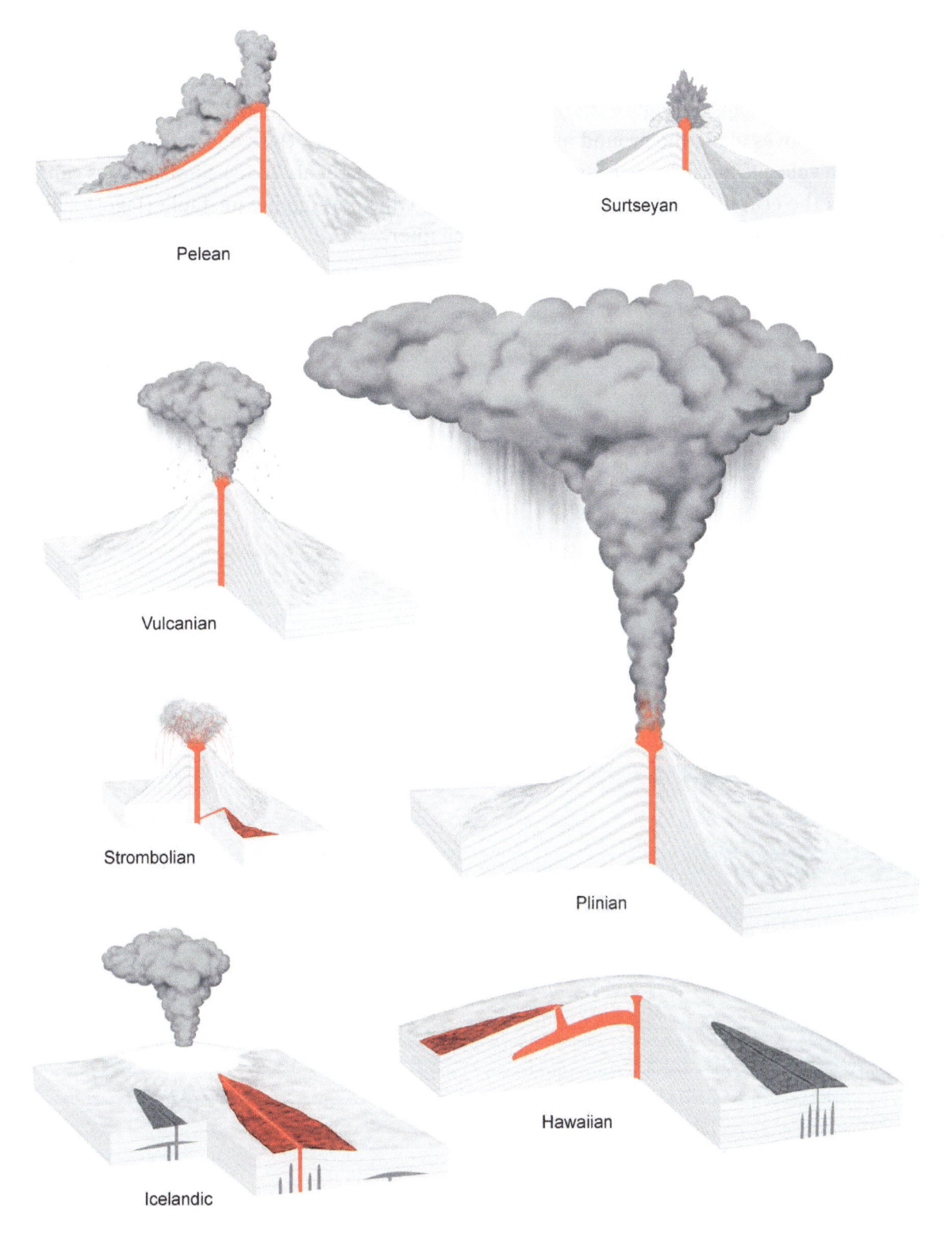

Figure 2.7 Eruption styles vary from effusive lava outflows in Hawaiian style eruptions through to large explosive Plinian eruptions, which can inject ash into the Stratosphere. Here the main eruption styles are illustrated (Artist John Norton in Siebert et al. (2010)). See Figure 2.8 for examples.

Explosive eruptions are responsible for much of the threat to life. Explosive eruptions discharge mixtures of hot volcanic rocks, volcanic gases and sometimes surface-derived water into the atmosphere. These flows are highly turbulent so large amounts of air are engulfed and heated (Figure 2.8b). In some cases this mixture, ejected at speeds of tens to hundreds of kilometres per hour, rises as a plume to great heights in the atmosphere, reaching the stratosphere in the more powerful eruptions (Figure 2.8b). In other cases the mixture is so full of rocks and ash that erupted mixture collapses (Figure 2.8f) and forms a dense flow (called a pyroclastic density current) that moves along the ground under gravity (Figure 2.8g). Over 43% of deaths directly attributed to volcanic hazards are caused by such flows (Auker et al., 2013). Similar flows can be formed with the eruption and collapse of lava domes, which are highly unstable (Figure 2.8h). Volcanic hazards are discussed in Section 4 of this chapter.

(Next page) Figure 2.8 Major styles of volcanic eruptions and their associated hazards. a) Strombolian explosive eruption at the summit of the basaltic volcano Fuego in Guatemala, 2014 (J. Crosby). b) A Plinian explosive eruption of Mount St. Helens, USA, in 1980 displays the characteristic turbulent clouds of ash, gas and hot air rising to heights of over 15 km into the atmosphere. These rising clouds are known as plumes. A pyroclastic flow moves down the flanks of the volcano generating a smaller ash cloud (US Geological Survey archive). c) A Surtseyan explosive eruption of the submarine Nishino-shima volcano, Japan, 2013 (Japan Coast Guard). d) A Hawaiian style eruption-generated basalt lava flow on Kilauea volcano, Hawaii, about 1 km south of the Kupaianaha vent in 1987 (S. Rowland). e) Rhyolite lava dome at Chaitén volcano, Chile in 2009. The 2008-2009 eruption of the volcano occurred after 7,400 years of dormancy and resulted in the evacuation of 900 people from the town of Chaitén (A. Amigo). f) The collapse in a fountain like structure of the eruption column at Mount St. Helens in 1980 due to the density of the column, forming a pyroclastic flow (US Geological Survey archive). g) Pyroclastic flow from the 1984 Pelean explosive eruption of Mayon, Philippines (C. Newhall). h) Pyroclastic flow from a dome collapse on Montserrat (H. Odbert).

Background on volcanoes | 93

2.2.6 Size and intensity of eruptions

There are two main measures of volcanic eruptions, namely magnitude and intensity. The magnitude is defined as an erupted mass while intensity is defined as rate of eruption or mass flux. Magnitude, M, is defined as the base 10 log of the mass erupted in kilograms minus 7. Magnitudes range over 9 orders of magnitude with magnitude 9 eruptions being the largest and the largest Holocene (the last 10,000 years) eruption being magnitude 7.4 (Brown et al., 2014). Intensity is usually expressed as kg/s or m³/s. The range of intensities is likewise very large, from a few kg/s up to a billion kg/s in exceptional rare events.

Neither magnitude nor intensity is easy to measure accurately. Volume is often used rather than mass as it is typically easier to estimate. A widely used index for the size of explosive eruptions is the *Volcanic Explosivity Index* (VEI) (Newhall & Self, 1982), which is used by the Smithsonian Institution to categorise all explosive eruptions based on multiple criteria (Figure 2.9). VEI is on a scale from 0 to 8 and is approximately equivalent to the magnitude (M), which is based on a logarithmic scale of mass. VEI is usually estimated from volumes of volcanic ash, but can also be estimated from eruption column height if volume information is not available. VEI only applies to explosive eruptions and a magnitude scale that includes both effusive (lava) and explosive products is more general.

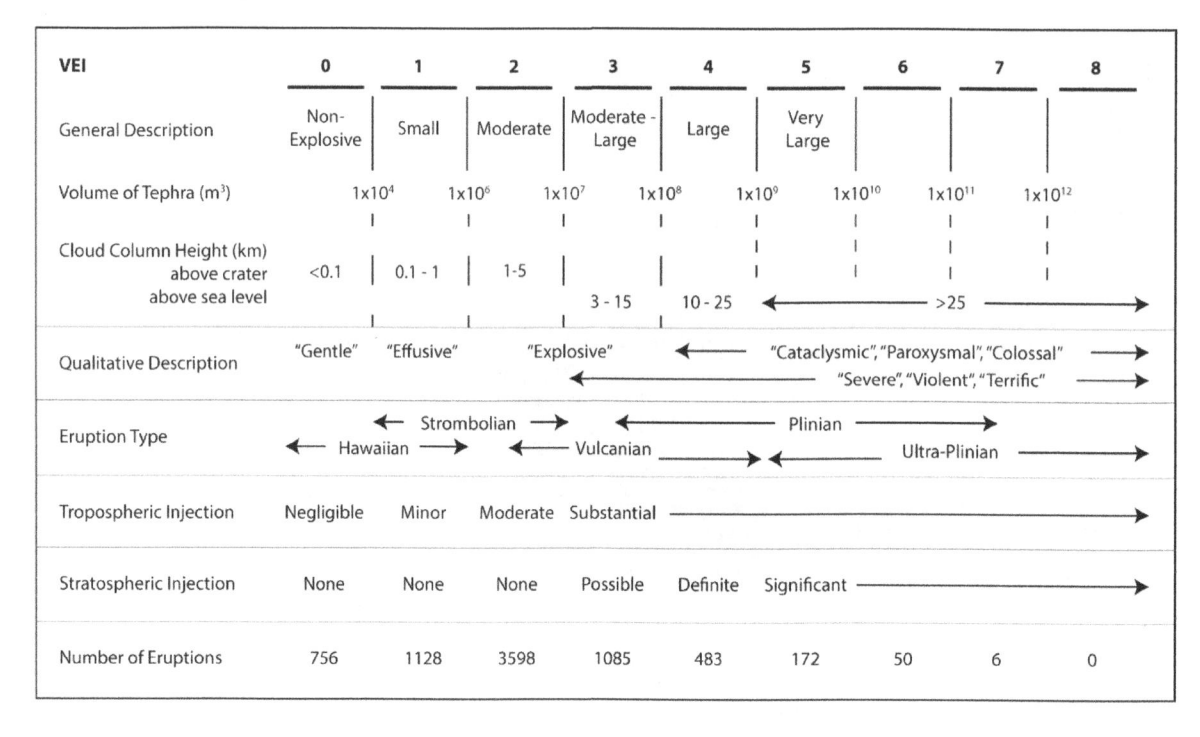

VEI	0	1	2	3	4	5	6	7	8
General Description	Non-Explosive	Small	Moderate	Moderate - Large	Large	Very Large			
Volume of Tephra (m³)		1×10^4	1×10^6	1×10^7	1×10^8	1×10^9	1×10^{10}	1×10^{11}	1×10^{12}
Cloud Column Height (km) above crater / above sea level	<0.1	0.1 - 1	1-5	3 - 15	10 - 25	>25 →			
Qualitative Description	"Gentle"	"Effusive"		"Explosive"		"Cataclysmic","Paroxysmal","Colossal" / "Severe","Violent","Terrific"			
Eruption Type	← Hawaiian →	← Strombolian → / ← Vulcanian →			← Plinian → / Ultra-Plinian →				
Tropospheric Injection	Negligible	Minor	Moderate	Substantial →					
Stratospheric Injection	None	None	None	Possible	Definite	Significant →			
Number of Eruptions	756	1128	3598	1085	483	172	50	6	0

Figure 2.9 Scheme to illustrate the assessment of Volcanic Explosivity Index (VEI) from diverse observations adapted from Newhall & Self (1982) and Siebert et al. (2010). VEI is best estimated from erupted volumes of ash but can also be estimated from column height. The nomenclature of common kinds of explosive eruption and typical duration of the eruptions are indicated. The number of confirmed Holocene eruptions with an attributed VEI in VOTW4.22 are shown.

There is a widely used classification of explosive volcanic eruptions based on the well-known eruptions at type volcanoes, such as Vulcanian, Hawaiian and Strombolian (Figure 2.7). Some of the most common terms are indicated in Figure 2.9 and are qualitatively correlated with VEI and intensity of eruption. The term Plinian comes from the AD79 eruption of Vesuvius and highlights the seminal description of a major powerful explosive eruption by Pliny the Younger. The eruption of Mount Pinatubo (Philippines) in 1991 is a modern example of a Plinian eruption.

The height of a volcanic eruption column generated in an explosive eruption is related to intensity (Mastin et al., 2009, Bonadonna et al., 2012) which cannot be measured directly. Adjustments are required for wind in the case of weak eruptions (Woodhouse et al., 2013).

Many eruptions are sequences of different styles of eruption (e.g. explosions and lava flows) of varying intensity and magnitude. Volcanic eruptions vary greatly in duration from just a few minutes to decades. There are several volcanoes that erupt almost continuously, such as Stromboli in Italy, which has had countless small explosions over at least two millennia.

2.3 Volcanoes in space and time

Since 1960 the Smithsonian Institution has collated data on the world's active Volcanoes. Their Volcanoes of the World database (Siebert et al., 2010, Smithsonian, 2013) (VOTW4.0) is regarded as the authoritative source of information on Earth's volcanism and is the main resource for this study. Eruption data cited in this book are from VOTW4.22.

2.3.1 Volcano inventory

VOTW4.0 contains a catalogue of 1,551 volcanoes. Their distribution is shown in Figure 2.2. There are 596 volcanoes that have an historical eruption record since 1500 AD, and 866 volcanoes with known Holocene eruptions (the last 10,000 years). There are 9,444 eruptions recorded in VOTW4.0. There are many more volcanoes that have been active in the Quaternary period (defined as the last 2.6 million years). The LaMEVE database (Crosweller et al., 2012) lists 3,130 Quaternary volcanoes. Some of those that are not catalogued in VOTW4.0 may well be dormant rather than extinct. Individual volcanoes can change from one of these categories of activity to another as more information becomes available. For example, prior to the 2010 eruption of Sinabung in Indonesia, the volcano had no historical record and was classified as a dormant Holocene volcano. Evidence of Holocene activity in those volcanoes without historical records is based on geological studies. However, many Quaternary volcanoes still remain unstudied, including numerous small monogenetic volcanoes that have not been systematically catalogued. Remote sensing using synthetic aperture radar is recognising unrest in volcanoes previously thought to be long dormant or even extinct (Biggs et al., 2014). There are likely many thousands of active submarine volcanoes along the Earth's ocean ridges, most of which have never been catalogued or explored. From a hazard and risk perspective it is those volcanoes close to communities that are of most concern; however, even remote and uninhabited island volcanoes pose a threat to aviation and distal populations.

2.3.2 Rates of eruption

A key question is how often do volcanoes erupt? This is not a straightforward question to answer as many volcanoes do not have long historical or geological records (Simkin, 1993). Indeed analysis of VOTW4.0 data established that only about 30% of the world's volcanoes have any information before 1500 AD, while 38% have no record earlier than 1900 AD. All of the records that exist are affected by severe under-recording (Deligne et al., 2010, Furlan, 2010, Brown et al., 2014), that is the historical and geological records become less complete back in time. For example statistical studies of the available records (Deligne et al., 2010, Furlan, 2010, Brown et al., 2014) suggest that only about 40% of explosive eruptions are known between 1500 and 1900 AD, while only 15% of large Holocene explosive eruptions are known prior to 1 AD. Most volcanoes alternate between long periods of repose and short bursts of activity. Since the repose periods can be decades to millennia or more there are very few volcanoes with long enough or complete enough records to enable statistical models of eruption frequency to be developed.

Table 2.1 Global return periods for explosive eruptions of magnitude M, where M = $Log_{10}m$ -7 and m is the mass erupted in kilograms. The estimates are based on a statistical analysis of data from VOTW4 and the Large Magnitude Explosive Volcanic Eruptions database (LaMEVE) version 2 (http://www.bgs.ac.uk/vogripa/) (Crosweller et al., 2012). The analysis method takes account of the decrease of event reporting back in time (Deligne et al., 2010). Note that the data are for M ≥ 4.

Magnitude	Return period (years)	Uncertainty (years)
≥4.0	2.5	0.9
≥4.5	4.1	1.3
≥5.0	7.8	2.5
≥5.5	24.0	5.0
≥6.0	72	10
≥6.5	380	18
≥7.0	2925	190
≥7.5	39,500	2500
≥8.0	133,500	16000

Analysis of global data for explosive eruptions shows a decrease in the frequency of eruptions as eruption magnitude increases (Table 2.1), as observed for many other Earth systems (e.g. earthquakes, tropical cyclones and high-latitude winter storms. Up to M 6.5 the data define a comparable decrease of average return period with magnitude to that seen in earthquake data. However, for M ≥6.5 the average return periods become greater than this empirical law and the decrease becomes greater for larger magnitudes. Here we note that super-eruptions (Self & Blake, 2008) like those that took place in Yellowstone are defined as having a magnitude of M = 8 or greater. The estimate of a global average return period of 130,000 years for super-eruptions indicates events of very low probability in the context of human society.

The global eruption record since 1950 is considered largely complete for eruptions on land. Eruptions of submarine volcanoes are largely undocumented, although they likely exceed eruptions on land in number. There are 2,208 confirmed eruptions recorded in the VOTW4.0 database since 1950, from 347 volcanoes. Despite our knowledge of 1,551 volcanoes, the number of individual erupting volcanoes each year varies within a relatively narrow range, from 44 to 77 volcanoes, with on average 57 volcanoes in eruption in any given year. The average number of eruptions ongoing per year since 1950 is 63, with a minimum of 46 and maximum of 85 eruptions recorded per year, including on average 34 new eruptions beginning per year. VOTW4.0 counts all eruptions occurring less than three months after the preceding eruption to be part of the same single eruption; those occurring after three months of repose are counted as new eruptions, unless clearly shown to be otherwise.

2.3.3 Examples of volcanic activity

Examples of volcanoes and their eruptions are presented in this section to illustrate the wide range of behaviours together with implications for risk. A key point is that every volcano and volcanic region is in some respect unique.

In 1943 farmers in the Mexican state of Michoacán witnessed the ground in cornfields break open and a new volcano, named Paricutin, formed (Luhr et al., 1993). There was no official warning, although many small earthquakes had been noticed in the months before. The eruption lasted nine years, generating huge volumes of ash and lava. Paricutin is an example of a monogenetic volcano, i.e. one that erupts only once. Auckland, New Zealand, home to 1.4 million people and over a third of New Zealand's population, is built on top of the Auckland Volcanic Field (AVF) and illustrates the issues in regions of scattered monogenetic volcanoes [Chapter 5] (Lindsay, 2010). The AVF covers 360 km², has over 50 eruptive centres (vents), and over 55 eruptions have occurred here in the past 250,000 years (Figure 2.10). The most recent eruption, Rangitoto, occurred only 550 years ago. Most vents are monogenetic, i.e. they only erupt once. This poses a considerable problem for emergency and risk managers, as it is unknown where or when the next eruption will occur.

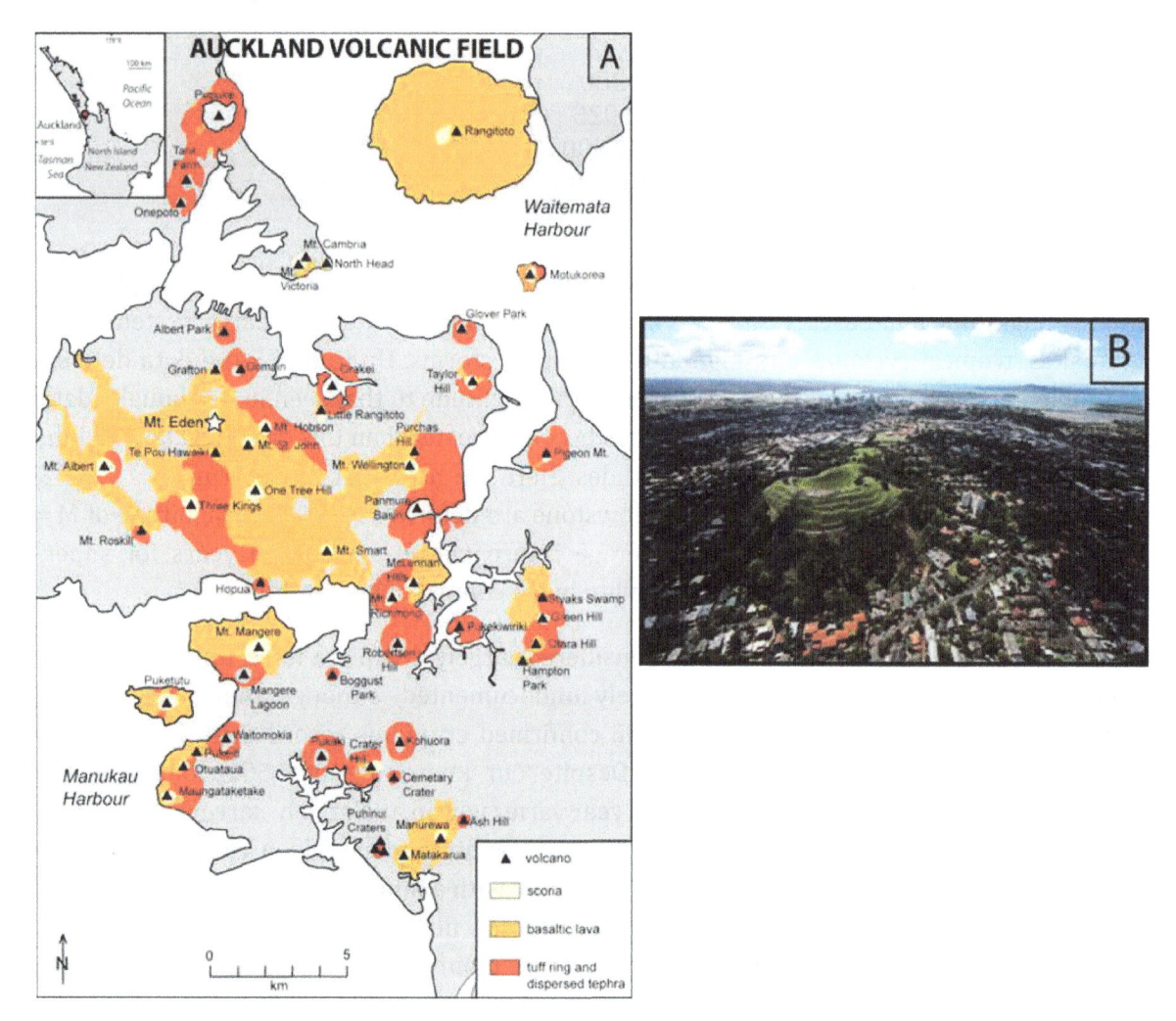

Figure 2.10 a) Map of Auckland Volcanic Field (AVF) showing the distribution of volcanic vents and products in the city of Auckland, New Zealand. Star shows the location of Mount Eden. b) View of Mount Eden looking to the north, highlighting the complete overlap of AVF and city (©Auckland Council). This figure can also be seen in Chapter 5 as Figure 5.1.

With over half of the world's population now in cities, volcanic risk to large urban communities is considerable. Naples is one of the cities in the world with the highest volcanic risk [Chapter 6]. Millions of inhabitants are directly threatened by three active volcanoes, namely Vesuvius, Campi Flegrei caldera and Ischia (Figure 2.11). The Osservatorio Vesuviano (OV) of the Istituto Nazionale di Geofisica e Vulcanologia (INGV) continuously monitors these volcanoes using advanced techniques to record the time and spatial evolution of seismic activity, ground deformation, geochemical signals, and many other potential pre-eruptive indicators. The Osservatorio Vesuviano provides updated hazard information to the Italian Civil Protection Department that is responsible for planning risk mitigation actions. There are great challenges in planning for the evacuation of hundreds of thousands of people from large cities close to active volcanoes.

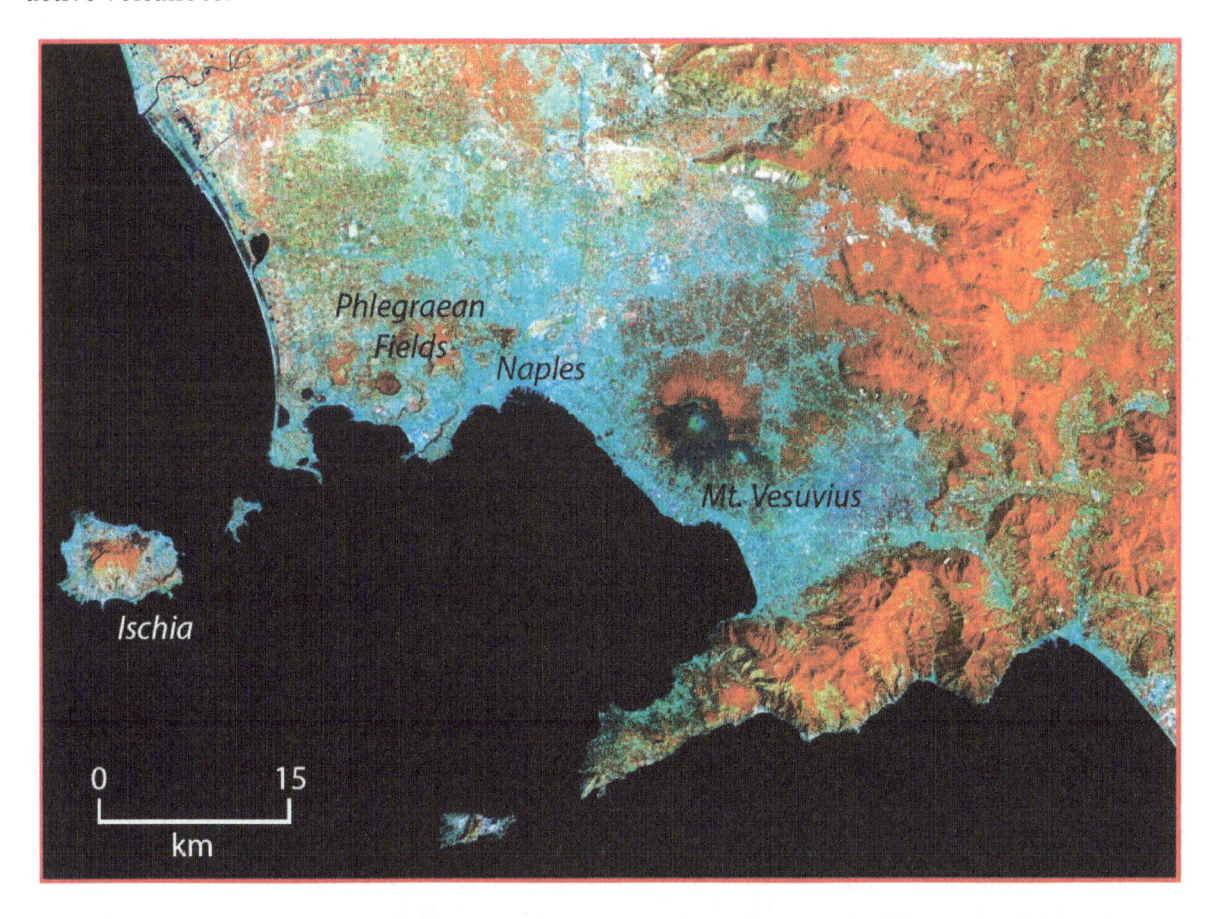

Figure 2.11 Satellite image of the Naples area where over 2 million people live on the flanks of the three active volcanoes of Vesuvius, Campi Flegrei (also known as Phlegraean Fields) and Ischia.

In 1992 small earthquakes were felt on the island of Montserrat in the eastern Caribbean. In 1995 a major andesite eruption of the Soufrière Hills Volcano began, which may still be ongoing (Wadge et al., 2014). There are no records of historical eruptions at Soufrière Hills since the island was colonised in 1642. Periods of small earthquakes in 1896-97, 1933-37 and 1966-67 had been experienced, but they had not led to eruptions. These episodes are an example of volcanic unrest and are sometimes described as failed eruptions. The present eruptive period has been long (1995-2013) and complex with many different hazards. There were 19 fatalities on 25 June 1997 and the population of the island was reduced through evacuations from about 10,000 to about 3,000, with major economic consequences (Clay et al., 1999).

After sleeping for over 500 years, Mount Pinatubo, Philippines, began to stir in mid-March 1991, producing a giant eruption on 15 June 1991, the second largest of the twentieth century [Chapter 7, (Newhall & Punongbayan, 1996a). About ~1,000,000 lowland Filipinos lived around the volcano and 20,000 indigenous Aetas lived on the volcano. Two large American military bases, Clark Air Base and Subic Bay Naval Station, were also at risk. Earthquakes, emissions of sulfur dioxide gas and small- to moderate-sized explosions preceded the paroxysms of 15 June. As many as 20,000 lives were saved as a consequence of the prompt action of authorities to evacuate, based on advice from the Philippine Institute of Volcanology supported by the US Geological Survey Volcano Disaster Assistance Program with funding from US Aid [see Chapter 25]. Despite considerable uncertainties, the eruption was correctly forecast and more than 85,000 were evacuated by 14 June, with many aircraft also protected from the eruption. The duration of the hazards lasted far beyond the eruption and indeed eruption-related hazards continue today, although at a much-reduced level. Voluminous rain-induced volcanic mudflows (lahars) continued for more than 10 years, and sediment-clogged channels still overflow today during heavy rains. Although about 200,000 were "permanently displaced" by lahars, only about 400 fatalities are attributed to lahars. Timely warnings from scientists and police helped to keep most people safe.

Montserrat and Pinatubo are examples of a common situation where a long-dormant volcano with a limited eruption record erupts. Populations, civil protection services and political authorities have had no previous experience of activity at the volcano. In a case like Montserrat the periods of unrest with no eruption may also make a population hesitant to listen to scientists, or to respond appropriately. Chaitén Volcano, Chile, is another example of a volcano that erupted after a period of quiescence of approximately four centuries (Lara et al., 2013).

The emergency in 1976 of La Soufrière Volcano, Guadeloupe (Eastern Caribbean) illustrates the difficulty of assessing the outcome of an eruption [Chapter 8, Hincks et al. (2014)]. A major eruption had occurred in the sixteenth century. However, there had been numerous periods of unrest and very small eruptions related to heating of groundwater by magma in 1690, 1797-1798, 1812, 1836-1837 and 1956 (Komorowski et al., 2005). Increasingly intense earthquakes were recorded at La Soufrière one year prior to the eruption, which began with an explosion on 8 July 1976 followed by 25 explosions over a period of nine months. There was no real way of knowing whether a devastating major explosive eruption might occur, or another period of unrest with minor steam explosions. Intense earthquakes and steam explosions in 1976 led to the precautionary evacuation of 73,000 people for four months in August 1976. They returned when three months later the volcano quietened without a major eruption. High levels of uncertainty and evacuations that are retrospectively identified as unnecessary are a major issue for management of volcanic crises. The importance of volcano observatories is that they can refute false reports of activity, and provide warnings and forecasts of hazardous activity.

Some volcanoes are frequently active. The classic andesite Merapi stratovolcano on the island of Java, Indonesia, has been active for much of the last 200 years [Chapters 9 and 10] (Surono et al., 2012). The frequent eruptions have fluctuated in intensity and magnitude, and there have been periods of quiet, which usually do not last more than a few years. In such cases communities become acquainted with and quite knowledgeable about volcanic activity and learn to live with the volcano. Even in such cases the volcano can be dangerous if unexpectedly large eruptions occur that are beyond the range experienced by the population. On the first day of the 2010 eruption, on 26 October, the second most violent explosive paroxysm of the entire eruption occurred and took the lives of 38 people. The eruption continued to escalate to climax 11 days later and on 5 November 2010 a large explosive eruption discharged hot pyroclastic flows down the flanks to distances almost twice as far as in previous historical eruptions (Surono et al., 2012). The volcanologists recognised unusual and threatening behaviour

[Chapter 9]. Although the eruption was considered responsible for 367 deaths (Jenkins et al., 2013), the authorities evacuated approximately 400,000 people prior to the eruption based on the recommendations of scientists from the Centre of Volcanology and Geological Hazard Mitigation (CVGHM) who operate the Merapi Volcano Observatory [Chapter 10]. An estimated 10,000 to 20,000 lives were saved. The case highlights the need for awareness that volcanoes will not always behave as they have in the past and that for long-lived centres the maximum credible event possible has to be determined from geologic field investigations that extend to time periods well beyond the historical period. The 2010 eruption of Merapi also underscores well that in some cases, paroxysmal activity can occur at the onset of the eruption and additional and even more violent activity can occur later in the eruption as well.

Quite small eruptions can cause major problems. In 1985 a medium-sized explosive eruption (VEI 3) took place at the summit of Nevado del Ruiz in Colombia, which was covered by an ice cap (Voight et al., 2013). The eruption melted ice rapidly and intense floods discharged down several major valleys. As they moved they incorporated loose rock, sand and soil and turned into devastating flows of mud and debris which buried the town of Armero, 45 km away with the loss of 23,000 lives and flooded the village of Chinchina with a loss of 1,927 lives. Here there seems to have been a lack of communication between the authorities and these communities with tragic consequences.

Volcanic eruptions can coincide with very difficult political or social situations, exacerbating the problems. In January 2002 a major eruption of Nyiragongo volcano, Democratic Republic of Congo, occurred in the midst of a complex humanitarian emergency [Chapter 11] (Komorowski, 2003). A basalt lava flow erupted from the crater and inundated the city of Goma producing major devastation and forcing the rapid exodus of most of Goma's 300,000-400,000 inhabitants across the border into neighbouring Rwanda. This situation caused international concerns about an additional humanitarian catastrophe that could have worsened the ongoing regional ethnic and military conflict. Lava flows destroyed about 13% of Goma, 21% of the electricity network, 80% of its economic assets, 1/3 of the international airport runway and the housing of 120,000 people. The eruption caused about 470 injuries and about 140-160 deaths, mostly from CO_2 asphyxiation and from the explosion of a petrol station near the active hot lava flow (Baxter et al., 2003). The eruption caused a major humanitarian emergency that further weakened the already fragile lifelines of the vulnerable population. The limited number of fatalities in 2002 is largely a result of the timely recognition by the Goma Volcano Observatory (GVO) of the reactivation of the volcano about 1 year prior the eruption and their efficient communication with authorities once the eruption began, memory of the devastating 1977 eruption which triggered life-saving actions by the population, the presence of a large humanitarian operation in Goma, and the occurrence of the eruption in the morning.

Extreme volcanic eruptions have potential for regional and global consequences. The 1991 eruption of Mount Pinatubo, Philippines, was one of the biggest explosive eruptions of the twentieth century. Sulfur dioxide and sulfuric acid aerosol pollution spread around the equator within 3 weeks and it took over 2 years for the global atmospheric pollution to dissipate. The pollution was so great that the trend of increasing CO_2 in the atmosphere was momentarily halted, there was global cooling and there was a significant reduction in ozone over northern Europe. The two largest eruptions in recent history are the Laki basalt lava eruption (Iceland) in 1783 and the magnitude 7 explosive eruption of Tambora (Indonesia) in 1815. Up to one third of Icelanders (~8,000) died from the magnitude 6.7 Laki eruption, largely through famine due to the environmental catastrophe (Thordarson & Self, 2003) and there is compelling evidence that there were tens of thousands of deaths in England and France related to the resulting sulfur pollution and crop failures (Schmidt et al., 2011). Likewise the Tambora death toll was an estimated 70,000, mostly related to post-eruption famine (Auker et al., 2013). There was major northern hemisphere cooling of about 1°C in the two years following the eruption of Tambora. In 1816 after Tambora erupted, summer frosts destroyed crops in New England and there was the "year without a summer" in Europe.

The critical culprit in the effects of large explosive eruptions on climate is sulfur dioxide (Robock, 2000). Sulfur dioxide gas reacts with atmospheric water to form tiny droplets of sulfuric acid, which can remain in the stratosphere for several years. Solar radiation is reflected back into space and absorbed by the aerosol. Thus the lower atmosphere becomes abnormally cool and the stratosphere is heated. There is about a 1 in 3 chance of an eruption similar in magnitude to Laki or Tambora in the twenty-first century. In the modern globalised and interconnected world the economic and societal impacts of such an eruption would be considerable.

About 75,000 years ago the largest volcanic eruption in the Earth's recent history took place at Toba in Sumatra, Indonesia (Self & Blake, 2008). Thick layers of volcanic ash were spread over the Indian Ocean, south-east Asia and probably most of China. The eruption ejected about 3,000 cubic kilometres of volcanic ash, 10,000 times more ash than Mount St Helens produced in 1980. The biggest volcanic crater on Earth (Lake Toba) formed with a length of 80 km and width of 30 km. Atmospheric pollution from such an eruption could cause major climatic deterioration for a decade or more. Eruptions on this scale (magnitude 8 or above) have been described as super-eruptions (Self & Blake, 2008). Volcanoes capable of super-eruptions include, but are not limited to, Yellowstone (USA), Taupo Volcanic Centre (New Zealand) and Campi Flegrei (Italy). Such eruptions would have severe global impact, but they occur very infrequently, roughly every 130,000 years or so (Table 2.1), although this is about 5 to 10 times more frequently than meteorite impacts that would have comparable global impact.

Recent studies have shown that large magmatic reservoirs that feed VEI or M 7-8+ eruptions can recharge and become critically primed for large eruptions on much shorter time scales (decades to months) than previously thought (Druitt et al., 2012). Moreover, the return period for ≥ VEI 6.5 eruptions might be shorter than currently estimated as recent large eruptions have been newly identified (e.g. the VEI 6-7 Samalas-Rinjani eruption in Indonesia that occurred in 1257AD (Lavigne et al., 2013)). Hence, very infrequent, extreme volcanic events (i.e. ≥ VEI 6.5) that have potential for regional and global consequences must be integrated into long-term risk assessment and yet we have no experience of such events in recent historical time (Self & Blake, 2008).

2.4 Volcanic hazards and their impacts

Volcanoes produce multiple hazards (Blong, 1984, Papale, 2014), that must each be recognised and accounted for in order to mitigate their impacts. Depending upon volcano type, magma composition and eruption style and intensity at any given time, these hazards will have different characteristics. Thus reliable hazard assessment requires volcano-by-volcano investigation. An important concept in natural hazards is the *hazard footprint*, which can be defined as the area likely to be adversely affected by hazard. The following is a brief account of the major kinds of volcanic hazards that create risks for communities with examples to illustrate their impacts.

Figure 2.12 Volcanic hazards and their impacts. a) The turbulent eruption column of the 2011 Grímsvötn eruption, Iceland (Ó. Sigurjónsson). b) The 2010 plume of Eyjafjallajökull, Iceland, which went on to cause mass disruption across Europe as air travel was grounded (S. Jenkins). c) Burial of houses in ash deposits during the eruption of Soufrière Hills, Montserrat (H. Odbert). d) Extensive damage 6 km from the vent after pyroclastic density currents occurred at Merapi, Indonesia in 2010 (S. Jenkins). e) Only the roofs of 2-storey buildings are visible in Bacolor after repeated inundation by lahars following the 1991 eruption of Pinatubo, Philippines (C. Newhall). f) Cars buried in lavas from the 2002 eruption of Nyiragongo in Democratic Republic of Congo (G. Kourounis).

Not all hazards are generated in every eruption or by every volcano. Individual volcanic eruptions are characterised by their magnitude (mass of erupted material), intensity (the rate of mass eruption), duration and eruptive phenomena (e.g. lava flows or explosions). Each eruption will have its own set of "hazard footprints", which can be defined as the areas affected by each of the hazardous processes. These hazard footprints can evolve during an eruption as it progresses.

2.4.1 Ballistics

Ballistics (also referred to as blocks or bombs) are rocks ejected by volcanic explosions on cannon ball-like trajectories. They are typically decimetres to a couple of metres in size. In most cases the range of ballistics is a few hundred metres to perhaps two kilometres, but they can be thrown to distances of five kilometres or more in the most powerful explosions (Blong, 1984) and so the hazard footprint remains close to the volcano. Fatalities, injuries and structural damage result from direct impacts and very hot ballistics can start fires. Tourists and scientists have proved to be particularly vulnerable: the unexpected explosion of Mount Ontake, Japan, on 27 September 2014 resulted in the deaths of 50 hikers. At Aso in Japan, bomb shelters have been built in case of unexpected explosions. Intense volcanic explosions can cause shock and infrasonic waves in the atmosphere, which can shatter windows and damage delicate equipment (e.g. electronic doors) at distances of several kilometres from the volcano.

2.4.2 Volcanic ash and tephra

All explosive volcanic eruptions generate tephra, fragments of rock that are produced when magma or vent material is explosively disintegrated. Volcanic ash (tephra <2 mm diameter) is then convected upwards within the eruption column and carried downwind, falling out of suspension and potentially affecting communities across hundreds, or even thousands, of square kilometres. Ash is the most frequent, and often widespread, volcanic hazard. Although ash falls rarely endanger human life directly, threats to public health and disruption to critical infrastructure services, aviation and primary production can lead to potentially substantial societal impacts and costs, even at thicknesses of only a few millimetres. A comprehensive volcanic hazard assessment must include ash fall in addition to more localised hazards such as pyroclastic density currents. However, the impacts of ash fall are arguably more complex and multi-faceted than for any of the other volcanic hazards and therefore a separate chapter on volcanic ash fall hazard and risk is provided (see Chapter 3).

Forecasting the dispersal of ash in the atmosphere and how much will fall, where, when and with what characteristics are major challenges during eruptions. Volcanic ash may be transported by prevailing winds hundreds or even thousands of kilometres away from a volcano (Figure 2.12b) and very large explosions can inject volcanic ash into the stratosphere. The dispersal of volcanic ash depends principally on meteorological conditions, including wind (speed and direction) and humidity, the grain size distribution of the ash, and the height of the volcanic plume, which depends on the intensity of the eruption. Hazard footprints of both ashfall on the ground and dispersal of hazardous ash concentrations in the atmosphere can be very large (up to millions of square kilometres in the largest eruptions) and can affect many different countries. The dispersal of volcanic ash is therefore of global concern.

Near the volcano, thick accumulations of tephra and ash can cause roofs to collapse and lead to consequent fatalities and injuries (Figure 2.12c). In the 1991 eruption of Pinatubo [Chapter 7] about 300 people died from roof collapse during the eruption. Moderate ash falls of several centimetres may damage infrastructure (e.g. power grids), cause structural damage to buildings and create major clean-up demands [Chapter 12] (Blong, 1984, Spence et al., 2005, Wilson et al., 2012b). Even relatively thin falls (≥ 1 mm) may threaten public health, damage crops and vegetation, and disrupt critical infrastructure services, aviation, primary production and other socio-economic activities over potentially very large areas (Wilson et al., 2012b).

Very fine ash at ground level is a health hazard to both animals and humans [Chapter 13] (Horwell & Baxter, 2006, Durant et al., 2010), and can also be readily remobilised by wind which can prolong exposure to airborne ash (Carlsen et al., 2012, Wilson et al., 2012a). Inhalation of fine ash may trigger asthma and other acute respiratory diseases (Horwell & Baxter, 2006), although these effects are inconsistent between different eruptions. To date, no longer-term diseases such as silicosis have been attributed to exposure to volcanic ash, although this may be due to inadequate case collection (Kar-Purkayastha et al., 2012). Ash can carry a soluble salt burden that is readily released on contact with water or body fluids. This can lead to both beneficial effects (such as the addition of agronomically-useful quantities of plant growth nutrients to pastoral systems (Cronin & Sharp, 2002); and harmful effects (such as fluorine toxicity to livestock). Famines have occurred following some major eruptions due to destruction of food supplies due to ash fall.

Airborne volcanic ash is a major hazard to aviation [Chapter 14] (Guffanti et al., 2010) and other forms of transport, jeopardising supply chains, provision of emergency services, and many essential services. Eruptions at Galunggung volcano, Indonesia in 1982 and Redoubt volcano, Alaska in 1989 caused engine failure of two airliners that encountered the drifting volcanic ash clouds. Concern over aviation safety resulted in the establishment of nine Volcanic Ash Advisory Centres (VAACs) around the world to issue notices, observations and forecasts of volcanic ash dispersal to civil aviation authorities. VAACs, hosted by meteorological services, work closely with volcano observatories.

2.4.3 Pyroclastic flows, surges and volcanic blasts

Volcanic explosions and rockfalls (a type of landslide) from lava domes may generate high velocity mixtures of hot volcanic rocks, ash and gases that flow across the ground (Figure 2.8g, h) called *pyroclastic density currents*. *Pyroclastic flows* are concentrated avalanches of hot ash, gases and blocks that are typically confined to valleys (Figure 2.8g), while *pyroclastic surges* are more dilute turbulent clouds of hot ash, gases and rocks that can spread widely across the landscape. Flows and surges (pyroclastic density currents) typically occur together with a more dilute surge overlaying a more concentrated flow. A *volcanic blast* is a term commonly used to describe a very energetic kind of pyroclastic density current which is not controlled by topography and is characterised by very high velocities (more than 100 m/s in some cases) and dynamic pressures (Jenkins et al., 2013).

Pyroclastic density currents are the most lethal volcanic hazard, travelling at velocities of tens to hundreds of kilometres per hour and with temperatures of hundreds of degrees centigrade. Escape is difficult and survival unlikely. They can cause severe damage to buildings, infrastructure, vegetation and agricultural land (Blong, 1984, Charbonnier et al., 2013, Jenkins et al., 2013). The Roman town of Pompeii was devastated by pyroclastic density currents and buried by their deposits in the AD79 eruption of Vesuvius. A pyroclastic density current from Mont Pelée volcano on the Caribbean island of Martinique destroyed the town of St Pierre in 1902 with the loss of 29,000 people. The current took only three minutes to reach the town 6 km from the volcano summit.

Pyroclastic density current footprints are influenced principally by topography, by the intensity of the explosion, and, in the case of lava domes, by the volume of collapsed lava. Pyroclastic flows are typically confined to valleys, but the associated surges can spill out of the valley and can reach unexpected places. In 1991 a surge from an eruption of Mount Unzen, Japan, killed 43 people on a ridge a few tens of metres above the valley floor. They had judged that they were safe because pyroclastic flows had not reached this location. The distance pyroclastic density currents can travel ranges from a few kilometres in smaller eruptions to over 100 km in the largest and most intense eruptions.

A volcanic or lateral blast is a term commonly used to describe a very energetic kind of pyroclastic density current. These more energetic currents take little account of topography, and may be hundreds of metres thick and travelling at hundreds of kilometres per hour. In 1980 the volcanic blast of Mount St Helens devastated 600 square kilometres in only four minutes, reaching distances of 25 km from the volcano; 56 people were killed. Even the explosion of relatively small pressurised volcanic domes can produce devastating and mobile high-energy pyroclastic density currents such as at Merapi in 2010 (Charbonnier et al., 2013, Jenkins et al., 2013, Komorowski et al., 2013). In large, explosive eruptions, pyroclastic density currents can travel over the sea and cause fatalities on islands and neighbouring coasts at considerable distances from the volcano. During the 1883 eruption of Krakatoa (Indonesia) pyroclastic density currents flowed over 80 km causing 150 deaths on the island of Sebuku, 30 km from the volcano (Carey et al., 1996).

Pyroclastic density currents account for one third of all volcanic fatalities (Auker et al., 2013). There is no plausible protection; shelters or bunkers can become buried in the hot deposits. Thus the only response to the threat of an imminent pyroclastic density current is evacuation.

2.4.4 Lahars and floods

Lahars and floods are a major cause of loss of life in volcanic eruptions, accounting for 15% of all historical fatalities (Auker et al., 2013). *Lahars* (an Indonesian word) are fast-moving mixtures of volcanic debris and water, sometimes referred to as volcanic mudflows. There are many causes of lahars, but they commonly occur when intense rain moves loose volcanic deposits formed during an eruption. Lahars can persist and continue to threaten an area for years or even decades after an eruption if there are significant thicknesses of unconsolidated ash, as was the case after the 1991 eruption of Pinatubo volcano in the Philippines [Chapter 7].

In addition to being triggered by rain, lahars can be caused by volcanic activity melting ice caps and glaciers. The moderate VEI 3 eruption of Nevado del Ruiz (Colombia) in 1985 produced pyroclastic density currents that melted some of the ice cap and generated lahars, causing ~23,000 fatalities in the town of Armero and village of Chinchina (Voight et al., 2013). Breaching of lakes, notably in craters and reservoirs, is another mechanism that can generate lahars. There are also examples of hot mud being discharged directly from fissures and vents on volcanoes, likely the result of magma heating, disturbing and pressurising ground-water. Lahars tend to bulk up and grow in volume as they flow down steep valleys and incorporate loose material (boulders, trees, soil etc.), a process that increases their energy and destructive potential. On flatter ground they lose energy and drop their sediment load, and can turn into muddy floods. They can be hot, for example when pyroclastic flow deposits are mobilised into lahars, and typical speeds of tens of kilometres per hour mean that they cannot be out-run, so moving quickly to higher ground is the best response.

Large energetic lahars are very destructive and are able to move car- and house-sized boulders great distances. Bridges and other structures can be destroyed due to impact and burial (Figure 2.12e), and escape routes may be cut off. Lahars are confined to valleys, and close to volcanoes it is usually straightforward to identify vulnerable areas. Farther from the volcano in more subdued topography and on floodplains they can inundate large areas. Channels shift frequently as sediment fills one channel and erosion opens another. Lahars can directly affect areas at distances of tens of kilometres from a volcano and may cause flooding hazards at even greater distances as channel capacity is lost to sediment fill. Communities in the hazard footprint of an imminent lahar should be evacuated or have identified escape routes to high ground if a warning is given.

2.4.5 Debris avalanches, landslides and tsunamis

Many volcanoes are steep-sided mountains, often partly constructed from poorly consolidated volcanic deposits and rocks weakened by alteration. Volcanic edifices are thus prone to instability (Siebert, 1984, Voight, 2000). Landslides are therefore common on volcanoes, whether currently active or not. *Debris avalanches* are very large and remarkably mobile flows of rock debris formed during the collapse of volcanic edifices, and they are commonly associated with volcanic eruptions or magmatic intrusions. In 1980 an intrusion of magma into the interior of Mount St Helens, USA, caused the steep northern flanks of the volcano to move outwards at about two metres per day creating a large bulge; after six weeks this collapsed generating a huge debris avalanche. This was accompanied by a lateral volcanic blast and followed by a major vertical explosive eruption. Volcanic landslides and debris avalanches can also be caused by hurricanes or regional tectonic earthquakes, and thus are sometimes unrelated to volcanic activity. Hurricane Mitch in 1998 triggered a major landslide on Volcano Casita in Nicaragua with at least 3,800 fatalities.

Debris avalanches and pyroclastic flows on islands and coastal volcanoes can cause tsunamis when they enter the sea (Siebert, 1984). Tsunamis can cause very large loss of life because of their scale, speed, and their potential for distant impact that can devastate coastal populations. In 1792 a debris avalanche from Mount Unzen, Japan, caused a tsunami with over 14,500 fatalities. Most of the 36,417 fatalities reported during the 1883 eruption of Krakatau in Indonesia were the result of lethal tsunamis generated from pyroclastic flows entering the sea (Mandeville et al., 1996). Moreover, volcanogenic tsunamis can be destructive tens to hundreds of kilometres from where they were generated. Oceanic islands with gentle slopes, such as Hawaii, have collapsed to form some of the largest known debris avalanches. Tsunamis associated with these collapses could have affected the coastlines of entire oceans, but these events are very rare.

2.4.6 Volcanic gases

Volcanic gases can directly cause fatalities, health impacts, and damage to vegetation, livestock, infrastructure and property [e.g. Chapters 10, 11 and 13]. Volcanic gases are dissolved in magma at depth in the subsurface but escape during reduction in pressure as the magma moves towards the surface. Whilst the main volcanic gas is water vapour (60-99%), there are many other volcanic gas types and associated aerosols. These may include: carbon dioxide (up to 10%), sulfur dioxide and other sulfur gases (up to 15%), halogens (including fluorine and chlorine, up to 5%), various metals such as mercury and lead (trace amounts) and trace amounts of carbon monoxide. The impact of volcanic gases varies widely and depends on the amount and type of gas emitted, the level at which it is injected into the atmosphere, local topography and the meteorological conditions at the time. Carbon dioxide (CO_2) is denser than air and will flow silently along the ground accumulating in depressions, including cellars. In 1986 a sudden release of CO_2 from Lake Nyos in Cameroon generated a gas flow that moved into surrounding villages with 1,800 fatalities as a result of asphyxiation (Kling et al., 1987). The volcanic crater lake had became saturated in CO_2 and became unstable so that the lake water overturned releasing several hundred thousand tons of gas. Sulfur gases, notably sulfur dioxide, are toxic in high concentrations and convert in the atmosphere to sulfate particles, a major cause of air pollution (Schmidt et al., 2011). Fluorine and chlorine-bearing gases can also be hazardous and may adhere to the surfaces of volcanic ash. People and animals can be affected by fluoride poisoning if they consume affected water, soil, vegetation or crops [Chapter 11].

2.4.7 Lava

Lava flows (Figure 2.8d) usually advance sufficiently slowly to allow people and animals to self-evacuate, but anything in the pathway of a lava flow will be damaged or destroyed, including buildings, vegetation and infrastructure. A few exceptional volcanoes produce lavas with unusual chemical compositions that can flow rapidly. For example, Nyiragongo in the Democratic Republic of Congo has a lake of very fluid lava at the summit and when the crater wall fractured in 1977 lava flowed downhill at speeds of more than 80 kilometres per hour, killing an estimated 282 people (Auker et al., 2013). In 2002 lava from Nyiragongo (Komorowski, 2003) caused great damage, many injuries and fatalities (Figure 2.12f) [Chapter 11].

2.4.8 Volcanic earthquakes.

Earthquakes at volcanoes are typically small in magnitude. The cumulative effects of repeated small volcanic earthquakes can include damage to man-made structures, as well as ground deformation and cracks. Larger earthquakes can be associated closely in time with volcanic eruptions. However, most volcanoes are in major earthquake zones so many of these larger earthquakes are likely to be a coincidence. There is also evidence that some large earthquakes can very occasionally trigger eruptions and that volcanic eruptions can trigger earthquakes. However, volcanoes tend to be in tectonically active areas, and so many earthquakes are caused by fault movement or hydrothermal fluids, rather than magma movement.

2.4.9 Lightning

Explosive volcanic eruptions are commonly accompanied by spectacular lightning due to the friction charges that build up in the eruption column. A few fatalities have been reported as a result of volcanic lightning (Auker et al., 2013). McNutt & Williams (2010) found that lightning has been reported in 10% of VEI 6 eruptions, with smaller eruptions less commonly associated with lightning.

2.4.10 Environmental and secondary effects on communities

The effects of volcanic eruptions on communities and the environment are many and varied, as already described above for individual hazards. These are further discussed throughout this chapter. While evacuation saves lives from hazardous volcanic eruptions, it can be very disruptive for communities. There is typically great uncertainty in how long a volcanic emergency will last and the impact of an eruption will be exacerbated if emergency accommodation or facilities are poor.

Volcanic phenomena can lead to damaging secondary hazards. For example the Nyiragongo eruption in 2002 [Chapter 11] illustrates the impact of an eruption on a major city in the middle of a humanitarian crisis. Some deaths were reportedly caused by an explosion at a petrol station due to inundation by lava. Economic impacts caused by evacuations or closing volcano access to tourists, as well as infrastructure failure, are other examples of secondary hazards on society. Crop failure and livestock losses can cause famine and epidemic disease outbreaks can occur. The environmental effects of the 1783 Laki eruption in Iceland and the 1815 eruption of Tambora, Indonesia, have already been described in Section 2.3.3. An open question is how the modern globalised world will cope with eruptions with a magnitude similar to, or greater than, Tambora (1815) or Laki (1783-4). There is about a 1 in 3 chance of an eruption of this magnitude occurring in the twenty-first century.

2.4.11 Fatalities

Historic fatalities provide a valuable source of information to assess the impact of volcanic eruptions and assess the relative importance of different volcanic hazards. There are several sources of data on volcanic fatalities. Here we use an integrated database, which is available from the Smithsonian institution. All fatality data in this chapter are derived from analysis of this database (Auker et al., 2013).

Volcanic eruptions have caused 278,880 known fatalities (Figure 2.13). This is modest compared to other major natural hazards. However, a small number of disasters dominate the record. Five eruptions have caused 58% of recorded fatalities, and just ten eruptions 70% of all fatalities. Such large loss of life in the one event is catastrophic for the communities affected.

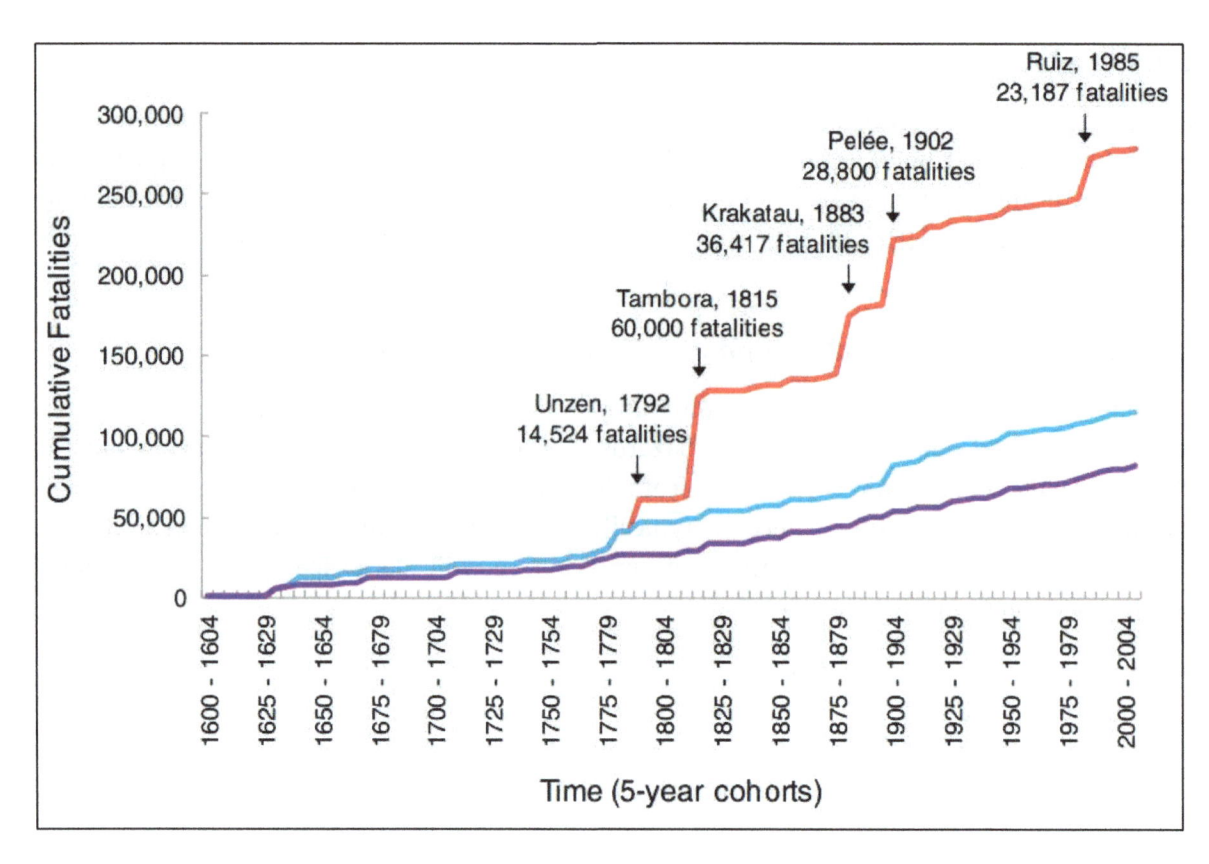

Figure 2.13 Cumulative number of fatalities directly resulting from volcanic eruptions (Auker et al., 2013). Shown using all 533 fatal volcanic incidents (red line), with the five largest disasters removed (blue line), and with the largest ten disasters removed (purple line). The largest five disasters are: Tambora, Indonesia in 1815 (60,000 fatalities); Krakatau, Indonesia in 1883 (36,417 fatalities); Pelée, Martinique in 1902 (28,800 fatalities); Nevado del Ruiz, Colombia in 1985 (23,187 fatalities); Unzen, Japan in 1792 (14,524 fatalities). The sixth to tenth largest disasters are: Grímsvötn, Iceland, in 1783 (9,350 fatalities); Santa María, Guatemala, in 1902 (8,700 fatalities); Kilauea, Hawaii, in 1790 (5,405 fatalities); Kelut, Indonesia, in 1919 (5,099 fatalities); Tungurahua, Ecuador, in 1640 (5,000 fatalities). Counts are calculated in five-year cohorts.

Statistical analysis of fatalities and eruption size from 4350 BC to 2011(Figure 2.14) shows that the most likely eruption size (mode) for number of fatal incidents is actually a modest VEI 3 and for number of fatalities is VEI 4. More than 80% of fatal incidents caused fewer than 100 deaths; however these amount to less than 10% of total fatalities. 85% of fatalities have occurred within the tropics, partially due to high rainfall contributing to extensive lahars and the tendency of populations to live at altitude, on the flanks of volcanoes. Indonesia, the Philippines, the West Indies and Mexico and Central America dominate the spatial distribution of fatalities. These regions have large populations and higher population densities proximal to volcanoes than other volcanically active zones.

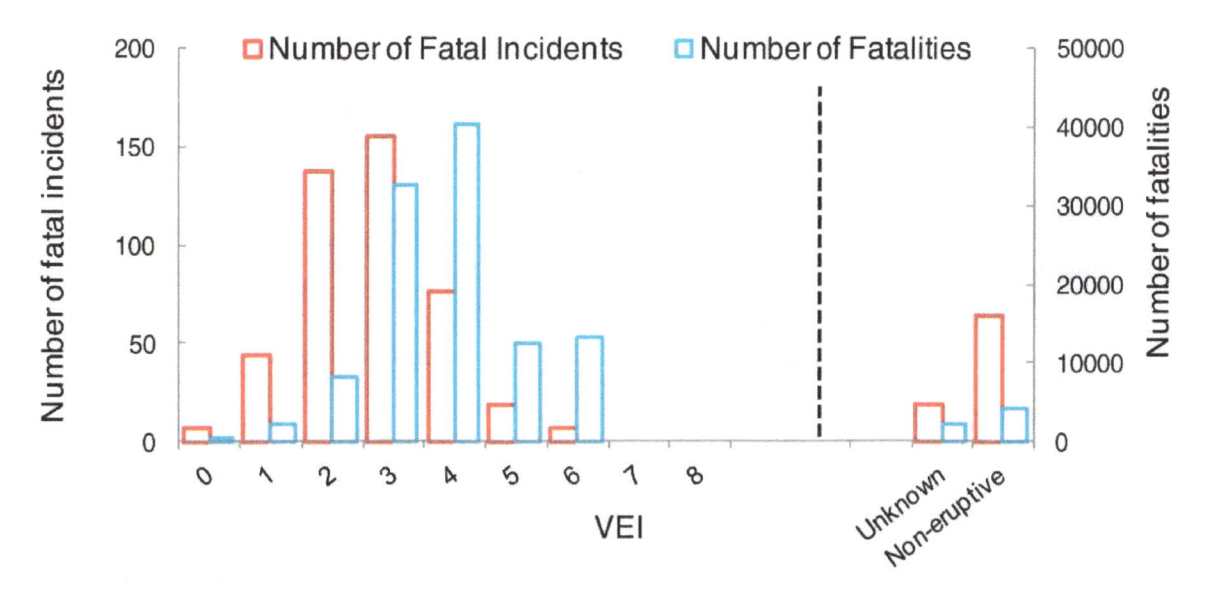

Figure 2.14 Distribution of fatalities and fatal incidents across VEI levels (not including the five largest disasters from statistical analysis of fatalities and eruption magnitude data). The five volcanic disasters with largest fatalities were discarded in order to investigate the relationship of fatalities with magnitude in an unskewed dataset, (Auker et al., 2013).

Despite exponential population growth, the number of fatalities per eruption has declined markedly in the last few decades, suggesting a reduction in societal vulnerability and exposure to volcanic hazards through an increase in monitoring and resulting improvements in hazard assessments, early warnings, evacuations, awareness and preparedness at specific volcanoes identified as posing a high risk (e.g. the UNISDR Decade Volcanoes). Volcano observatories have had a major role to play in this achievement. A conservative estimate is that at least 50,000 lives have been saved over the last century as a consequence of these improvements (Auker et al., 2013). Millions of people have been evacuated during volcanic emergencies over the 100 years and likely the number of lives saved is much higher. However, the potential for mass fatalities is still increasing as populations grow, causing an overall increase in exposure to volcanoes and their hazards. The margin of safety for recent mitigation successes has been alarmingly narrow (Newhall & Punongbayan, 1996b). Future fatalities may arise from eruptions at volcanoes where the risk is either not yet recognised, from large-magnitude eruptions, from unmonitored volcanoes, or from logistical, technical and management challenges in evacuating large numbers of people in time.

Figure 2.15 shows the distribution of fatalities between the different volcanic hazards and between direct (the hazards themselves) and indirect effects (e.g. famine and disease). Pyroclastic density currents emerge as the most significant hazard with lahars accounting for about half the number of fatalities attributed to pyroclastic density currents. Indirect effects are more pronounced in the greatest disasters. It is likely that fatalities resulting from secondary factors not directly related to primary volcanic hazards are under-represented in the fatalities. For example in the 1991 eruption of Pinatubo (Philippines) about 500 indigenous Aeta children died of measles in evacuation camps, because their parents distrusted Western-trained lowland doctors and refused help [Chapter 7].

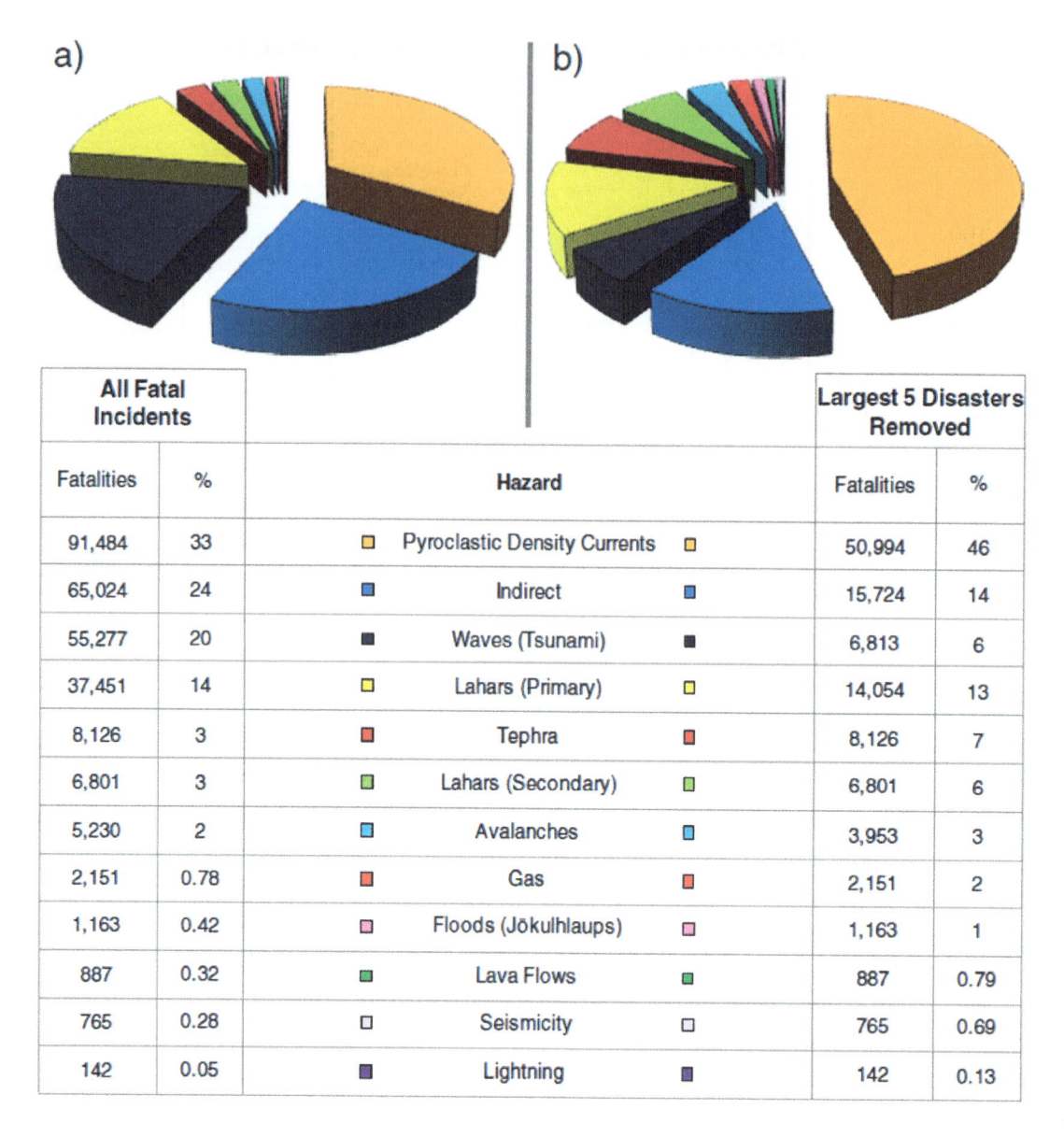

	All Fatal Incidents		Hazard		Largest 5 Disasters Removed	
Fatalities	%				Fatalities	%
91,484	33	☐	Pyroclastic Density Currents	☐	50,994	46
65,024	24	■	Indirect	■	15,724	14
55,277	20	■	Waves (Tsunami)	■	6,813	6
37,451	14	☐	Lahars (Primary)	☐	14,054	13
8,126	3	■	Tephra	■	8,126	7
6,801	3	■	Lahars (Secondary)	■	6,801	6
5,230	2	■	Avalanches	■	3,953	3
2,151	0.78	■	Gas	■	2,151	2
1,163	0.42	■	Floods (Jökulhlaups)	■	1,163	1
887	0.32	■	Lava Flows	■	887	0.79
765	0.28	☐	Seismicity	☐	765	0.69
142	0.05	■	Lightning	■	142	0.13

Figure 2.15 a) Distribution of fatalities across fatality causes, for all fatal incidents; b) Distribution of fatalities across fatality causes, with the largest five disasters removed; from Auker et al. (2013).

2.5 Monitoring and forecasting volcanic eruptions

Effective monitoring and integrated, effective warning systems are central to protecting citizens and assets affected by volcanic eruptions and increasing resilience in communities living with volcanoes. There are many benefits of volcano monitoring and it can be shown to be cost effective (Newhall et al., 1997). Costs of monitoring itself are typically between several tens of thousands USD per year at the lowest level to several million per year at the highest level. There are also costs of any evacuation (temporary housing, food, transport for evacuees), and indirect costs of evacuation (e.g. foregone business, increased health problems), and the costs and benefits of other measures taken outside an evacuation zone, e.g. to save crops or protect power grids downwind. The main benefit is the saving of lives that would be lost if the volcano erupts and those at risk were not evacuated, multiplied by a country's official Value of a Statistical Life, VSL (Mrozek & Taylor, 2002). Other benefits include avoided losses of equipment, goods and anything else that can be moved out of the way or otherwise protected, and matters such as business continuity, continuity of electric power supply that can be maintained if operators are alerted from the monitoring. Since the VSL in most countries is in the order of several hundred thousand to several million USD, saving a few lives alone justifies the monitoring from a cost-benefit perspective.

There are very few examples of in-depth assessment of economic losses from volcanic emergencies and evacuations. An evaluation of the economic impact of the eruption of the Soufrière Hills volcano, Montserrat in the first few years (Clay et al., 1999) estimated costs of dealing with the emergency to the UK government of order £160 million over the first 6 years and total losses to end of 1998 of order £1 billion, much of which relate to unrecoverable losses and uninsured assets. The costs of the Montserrat Volcano Observatory in this period were a few million Pounds sterling. The cost of the 6-month evacuation of around 73,000 people from Basse Terre, Guadeloupe, has been estimated at 60% of the total annual per capita Gross Domestic Product (GDP) in 1976 or about 342 million USD using 1976 currency rates (Hincks et al., 2014), but the estimate excludes the losses of uninsured and other personal assets.

Volcanic eruptions are typically preceded by days to months of precursory activity, unlike other natural hazards like earthquakes. For example 50% of stratovolcanoes erupted after about one month of reported unrest (Phillipson et al., 2013). Detecting and recognising warning signs provides the best means to anticipate, plan for, and mitigate against potential disasters. More than half of 288 studied eruptions reached their climax within a week of their onset and more than 40% peaked within the first 24 hours of the eruption (Sigurdsson et al., 2015).

The localised character and individuality of volcanoes gives a special importance to dedicated volcano observatories, which play a key role in monitoring, hazard assessments and early warning around the world [Chapter 15]. A volcano observatory may be an institute or group of institutes whose role it is to monitor active volcanoes and provide early warnings of future activity to the authorities and in most cases the public too. The responsibilities of a volcano observatory differ from country to country. In some nations, a volcano monitoring organisation may be responsible only for maintaining equipment and ensuring a steady flow of scientific data to an academic or civil protection institution, who then interpret the data or make decisions. In other jurisdictions, the volcano observatory may provide interpretations of those data and undertake cutting edge research on volcanic processes. In most cases a volcano observatory will provide volcanic hazards information such as setting Volcanic Alert Levels and issuing forecasts of future activity, and in some instances, a volcano observatory may even provide advice on when civil actions should take place such as the timing of evacuation. Effective volcanic risk reduction is a partnership between volcanologists, responding agencies and the affected communities.

Monitoring of volcanoes can be done at ground level, from the air and from space (including satellite observations). Changes in ground movement, thermal signatures, gas emissions, presence of ash and earthquakes provide clues about the movement of magma in the subsurface and detection of an eruption if it occurs. Where monitoring is in place, scientists infer the causative processes and likelihood that they will lead to eruption. Quality data enable more accurate eruption forecasting. Where there is no monitoring volcano observatories do not have the ability to analyse current or past activity, or forecast eruptions. It is important that baseline data are collected so that the normal behaviour of a volcano can be better understood. Periods of unrest can therefore be more easily recognised.

Most volcanic eruptions have a rapid onset, following periods of dormancy, which are commonly much longer than the duration of eruptions (Sigurdsson et al., 2015). There are, however, examples of persistently active volcanoes, which pose threats to surrounding communities much of the time. Analytical studies of volcanic samples, experimental investigations and theoretical modelling are providing insights into the dynamics of magmatic systems, giving a physical framework with which to interpret volcanic phenomena. Magmas undergo profound changes in physical properties as pressure and temperature vary during magma chamber evolution, magma ascent and eruption. Active volcanic systems also interact strongly with their surroundings, causing ground deformation, material failure and other effects such as disturbed groundwater systems and degassing. These processes and interactions lead to geophysical and phenomenological effects, which precede and accompany eruptions.

2.5.1 Monitoring

Instrumental monitoring is the basis of early warning, forecasting and scientific advice to decision-making authorities. Monitoring programmes at volcano observatories typically include: tracking the location and type of earthquake activity under a volcano; measuring the deformation of the ground surface as magma moves beneath a volcano; sampling and analysing gases and water being emitted from the summit and flanks of a volcano; observing volcanic activity using webcams and thermal imagery; measurements of other geophysical properties such as electrical conductivity, magnetism or gravity. Volcano observatories may have ground-based sensors measuring these data in real-time or they may have staff undertaking campaigns to collect data on a regular basis (e.g. weekly, monthly, annually) from sensors that are left in the field or deployed temporarily. Some volcano observatories also have the capability to collect and analyse satellite data.

The ability of a volcano observatory to make short-term forecasts effectively about the onset of a volcanic eruption or an increase in hazardous behaviour real-time is dependent on many factors including having functioning monitoring equipment and telemetry, real-time data acquisition and processing, a long baseline of data to take into account the variability of natural systems. However, it is fundamental to have good knowledge of the past behaviour of the volcano and a conceptual model for how the volcano works to interpret monitoring data, quantify uncertainty, and thus contribute to efficient risk management. Longer-term forecasts are based mainly upon geological and geochronological data if it has been collected.

The range and sophistication of the detection systems has increased dramatically in the last few decades (Sparks et al., 2012). Advanced models of volcanic processes are helping to interpret monitoring data. Spectacular advances in computing have led to improvements in power and speed, data transmission, data analysis and modelling techniques.

Monitoring volcanic earthquakes (seismicity) lies at the heart of every volcano observatory. Volcanic earthquakes are typically very low magnitude in comparison to tectonic earthquakes and maximum magnitudes rarely exceed 5. This means they may not be detected on regional and global seismic networks, requiring a dedicated network of seismometers near or on the volcanic edifice. There are several different causes of volcanic earthquakes, which can commonly be distinguished from the diagnostic characteristics of seismic signals (Chouet, 1996, McNutt, 2005). One very common kind of earthquake, known as a volcano-tectonic earthquake (VT), results from magma forcing its way towards the Earth's surface by breaking surrounding rocks. Identification of the location of VT earthquakes can help outline magma chambers (Figure 2.16) (Þorkelsson, 2012), and enable the pathways for magma ascent to be identified (Figure 2.17). In some cases VT earthquakes can track the migration of earthquakes towards the surface (Toda et al., 2002), enabling forecasts of the start of an eruption. A different kind of VT earthquake occurs up to tens of km away from the volcano, along pre-existing tectonic faults that are reactivated by increased pore water pressure as intruding magma compresses surrounding country rock.

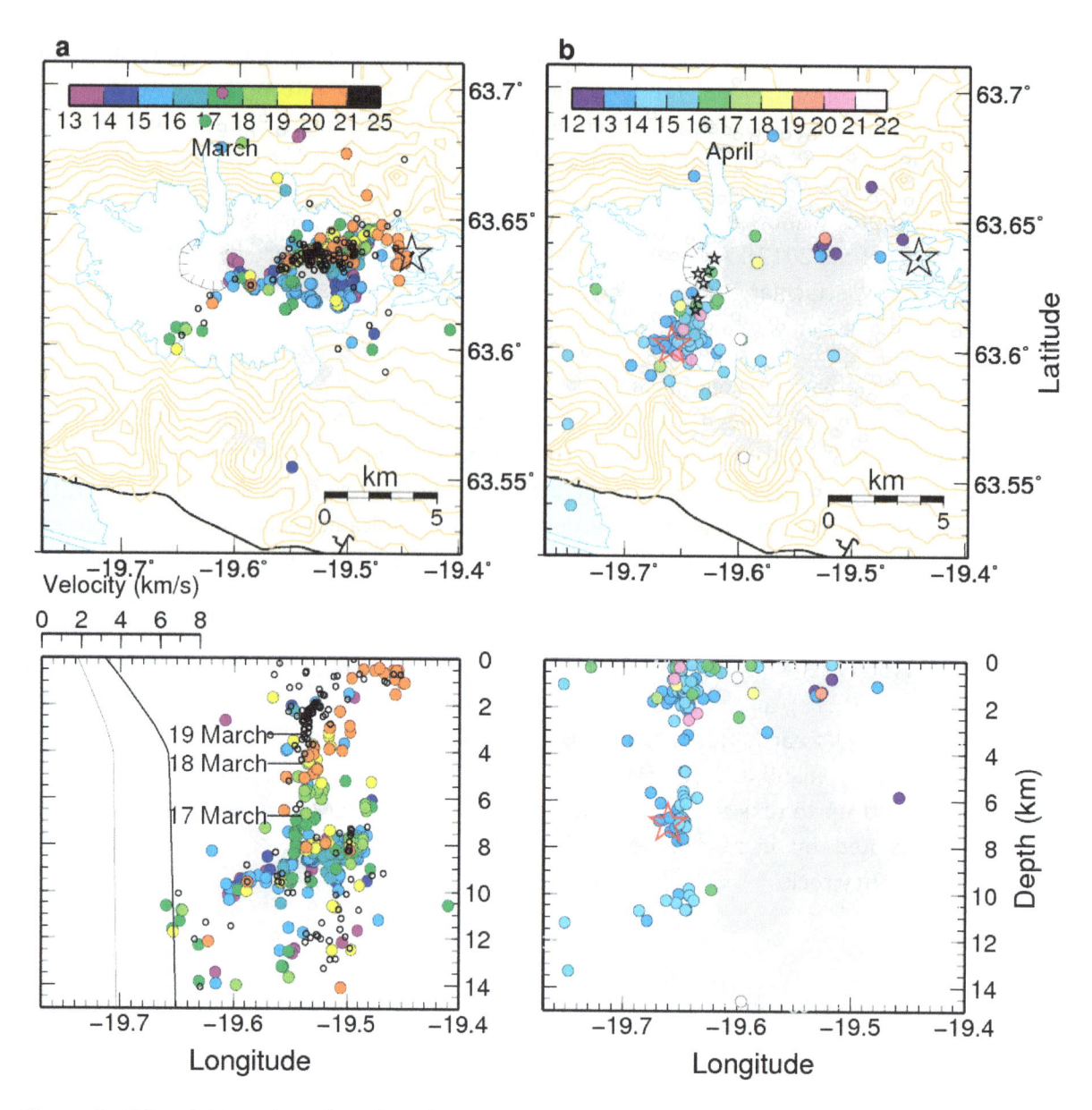

Figure 2.16 Spatial location of earthquakes before and during the eruption of Eyjafjallajökull volcano, Iceland in 2010. a) the flank eruption and b) the summit eruption. a) Earthquakes 4-12 March are gray and 13-24 March coloured or black. A flank eruption site is marked by a star. Crustal velocity model is shown below. b) Earthquakes 12-21 April coloured by date. Seismicity 2009 to March 2010 is grey. Red star shows location of the M 2.7 earthquake on 13 April. Further details can be found in Þorkelsson (2012). Diagram from Þorkelsson (2012) courtesy of Sigurlaug Hjaltadottir (Icelandic Meteorological Office).

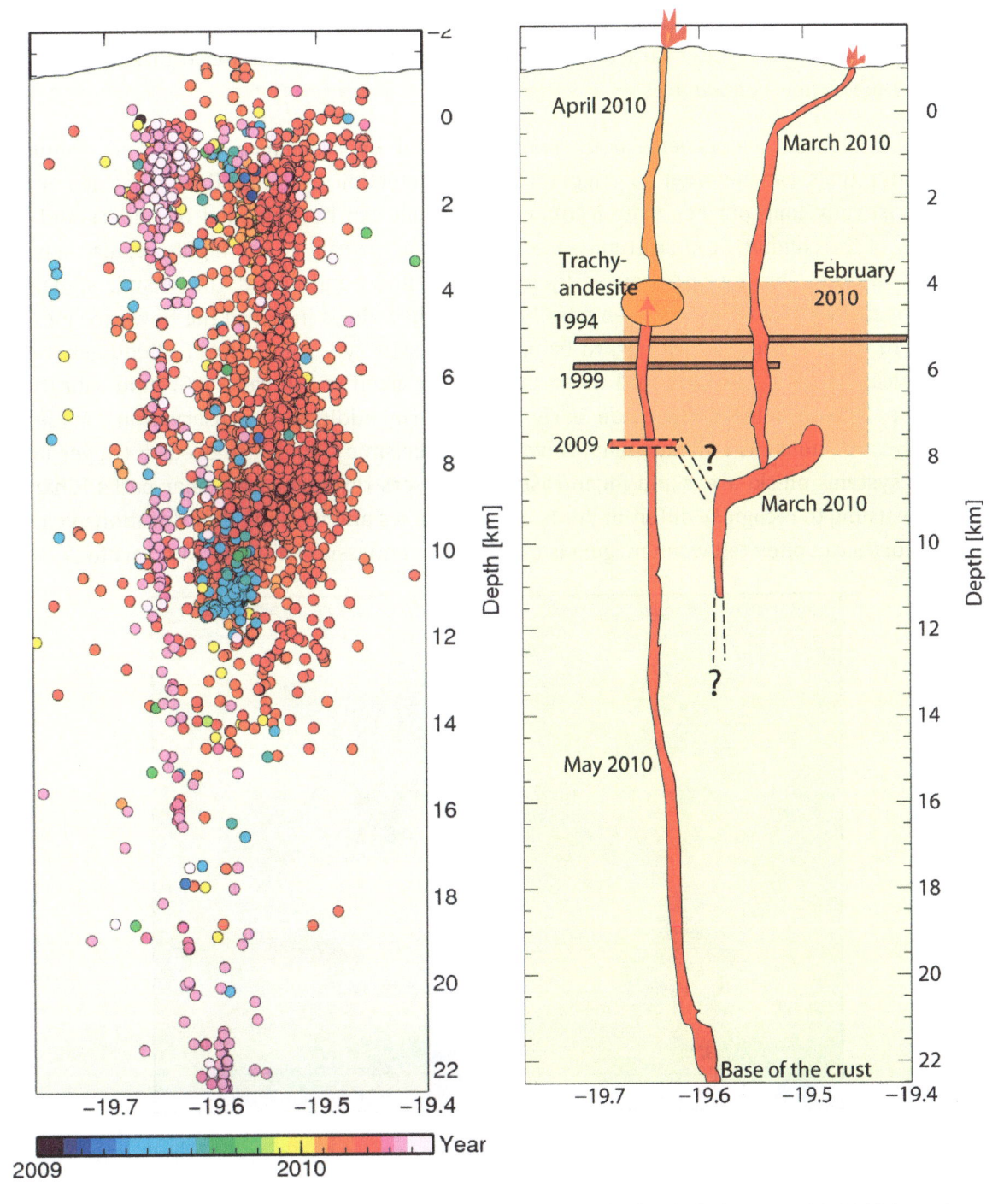

Figure 2.17 A schematic drawing of magma pathways in Eyjafjallajökull during 2009-2010 based on earthquake distribution (left) (Þorkelsson, 2012). Vertical scale is stretched. The red transparent box indicates roughly the extent of the February activity (intrusions), south-east of the main clusters. Diagram from Þorkelsson (2012) courtesy of Sigurlaug Hjaltadottir (Icelandic Meteorological Office).

The interaction of volcanic fluids with the surrounding rocks can lead to distinctive earthquakes with relatively low frequency content. These can provide insights into the internal dynamics of the volcanic system and can also provide a useful indicator of an impending eruption. Volcanic earthquakes are often highly repetitive as a result of similarities in both the location and mechanism of the earthquake. Swarms of many thousands of earthquakes can last for many hours and such swarms often occur in the immediate build-up to an eruption. Many volcanoes also display continuous seismic signals, know as volcanic tremor, as well as discrete earthquakes. Changes in the characteristics of these tremor signals can also provide valuable insights into volcanic behaviour.

Seismic signals are also generated by explosions and by resonance of volcanic conduits. Explosion signals can be used to trigger eruption detection systems. Resonance generates characteristically long periods (low frequencies), which can be modelled to understand the geometry of the conduit. Stop-start movement of the magma can lead to highly regular seismic patterns. Ground vibrations are caused by pyroclastic flows and lahars, the seismic signals of which are easily recognised. It is even possible to tell the valleys in which the flows are moving at night. In Iceland, the greatest hazard from volcanic activity is jökulhlaup (a flood generated when volcanoes melt glaciers) and often evacuations need to be called. Ground vibrations picked up by seismometers provide early warning (in addition to observations made by hydrological stations). On Mt Ruapehu, New Zealand, seismic signals are used to trigger lahar warning systems on ski areas and for infrastructure users threatened by lahar paths [Chapter 16]. By learning to recognise different kinds of earthquakes and tracking their locations volcano observatories can often tell when magma is on the move and issue forecasts and warnings.

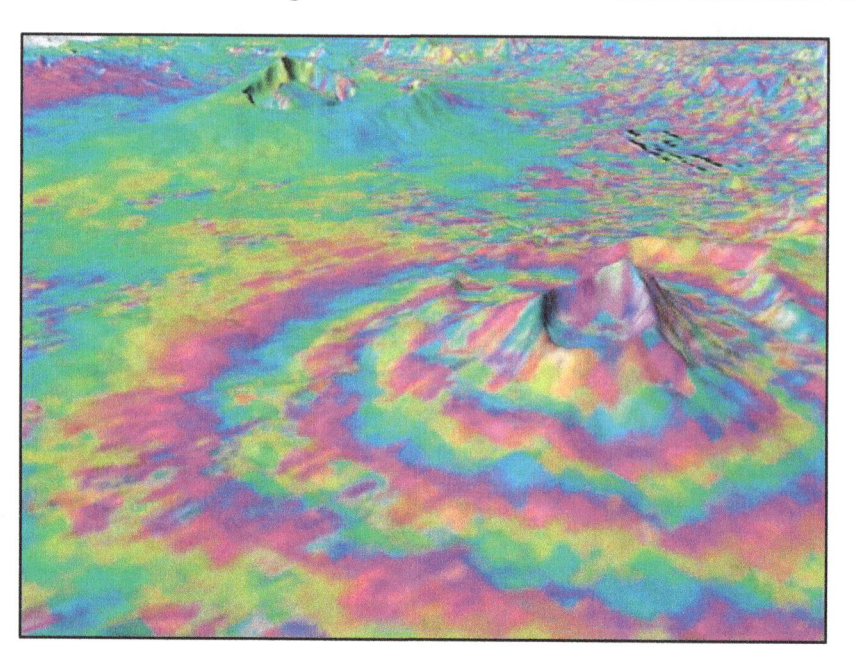

Figure 2.18 This InSAR image (Sparks et al., 2012) shows a pulse of uplift during 2004 to 2006 at Mount Longonot, Kenya, a volcano previously believed to be dormant. The image, from the ESA satellite Envisat, is draped over a digital elevation model from the Shuttle Radar Topography Mission. Each complete colour cycle (fringe) represents 2.8 cm of displacement towards the satellite. The distance between craters is ~35 km.

Underground magma movement and swelling of magma chambers, which can precede eruptions at many volcanoes, leads to surface ground movements, also known as *deformation* (Dzurisin, 2003). Movements of a few millimetres over a distance of a kilometre at rates of a few centimetres to a few decimetres per year are typical and can be easily detected by a variety of instruments. The value of deformation using Global Navigation Satellite System (GNSS) is exemplified by the 2007 eruption of Kilauea volcano, Hawaii (Larson et al., 2010). Nowadays electronic distance measurements using lasers and networks of GPS stations have largely replaced labour-intensive precise levelling and electronic distance surveys. The change of slope of the ground due to swelling or subsidence can be detected by electronic tiltmeters. Radar images from satellites [Chapter 17] have resulted in a major advance in the ability to detect ground movements (Figure 2.18). Two images taken at different times by synthetic aperture radar (InSAR) can be subtracted from one another to obtain an interference image, which shows the deformation. Unlike continuous GPS receivers, which measure deformation at specific locations, InSAR data indicate the entire deformation field over a wide area, and the two forms of geodetic measurements are quite complementary. InSAR has revolutionised the ability to monitor deformation at many of the world's volcanoes (Biggs et al., 2014), including those in remote places or those that are otherwise not monitored at all. The method also identifies volcanoes that are dormant but are showing current unrest (Figure 2.18; Chapter 17) (Biggs et al., 2009, Sparks et al., 2012). Currently, repeat times of radar satellites are too long to help scientists on the ground in short-term eruption forecasting, but these repeat times will soon be as short as a few days and InSAR will become an important tool for volcano observatories.

Volcanic gas monitoring has traditionally required volcanologists to visit high-risk locations to collect samples from fumaroles, hot springs and volcanic craters. The samples had to be returned to a laboratory for analysis with an unavoidable time delay. While such methods continue and are valuable both for monitoring and research, real-time volcanic gas monitoring nowadays is largely done remotely using ground-based and satellite-based instruments [Chapter 17] (Aiuppa et al., 2010, Nadeau et al., 2011). Different gases are released by magma rising through the Earth's crust, and hence the location of the magma can be inferred by measuring their concentrations through time. Although water and CO_2 are usually the major gases they are also abundant in the atmosphere, making it challenging to make accurate measurements. Therefore volcanic gases that have low background concentrations in the atmosphere are usually measured, in particular sulfur dioxide (SO_2), especially for satellite measurements. However, significant progress is currently being made on measuring CO_2 by both satellite and ground-based methods.

Measurements of volcanic gas composition and flux not only yield key insights into the subsurface volcanic processes but are also a vital tool for eruption forecasting. In the case of Pinatubo (Philippines) prior to its eruption in 1991, a ten-fold increase in SO_2 emission was attributed to magma ascent and a subsequent sudden drop in SO_2 suggested that magma was very near the surface and would erupt soon [Chapter 7] (Newhall & Punongbayan, 1996a). Once an eruption has commenced, observations of syn-eruptive degassing provide information on how an eruption is progressing and may give warnings about an imminent increase in explosivity. Gas measurements at Soufrière Hills, Montserrat, in 1998 revealed continuing activity of the volcano, even though seismic and geodetic signals had tapered off. The gas measurements were a vital diagnostic tool in hazard assessment in advance of a resumption of lava dome growth in 1999 (Wadge et al., 2014). In early 2008, Kilauea volcano (Hawaii) was emitting unusually high fluxes of sulfur dioxide from the otherwise inactive summit crater. There was no seismicity or ground deformation indicating an imminent eruption (in fact, the summit crater was in a stage of deflation), but the opposite was indicated by gas measurements. Kilauea summit crater subsequently erupted explosively three times within a month. Increased soil degassing of CO_2 was also the first sign of upwards migration of gas-rich magma preceding the 2008 eruption of Etna (Aiuppa et al., 2010).

Sulfur dioxide emissions from volcanoes are monitored in near real-time by the OMI satellite (http://so2.gsfc.nasa.gov/). The importance of SO$_2$ as a precursor to warn of an increase in explosivity was shown very clearly during the 2010 eruption of Eyjafjallajökull in Iceland. SO$_2$ emissions rose dramatically between 4 and 5 May (Figure 2.19), while ash output did not increase until several hours after the gas emissions (Figure 2.20). This ash-production phase caused widespread disturbance to European airspace during most of the following week.

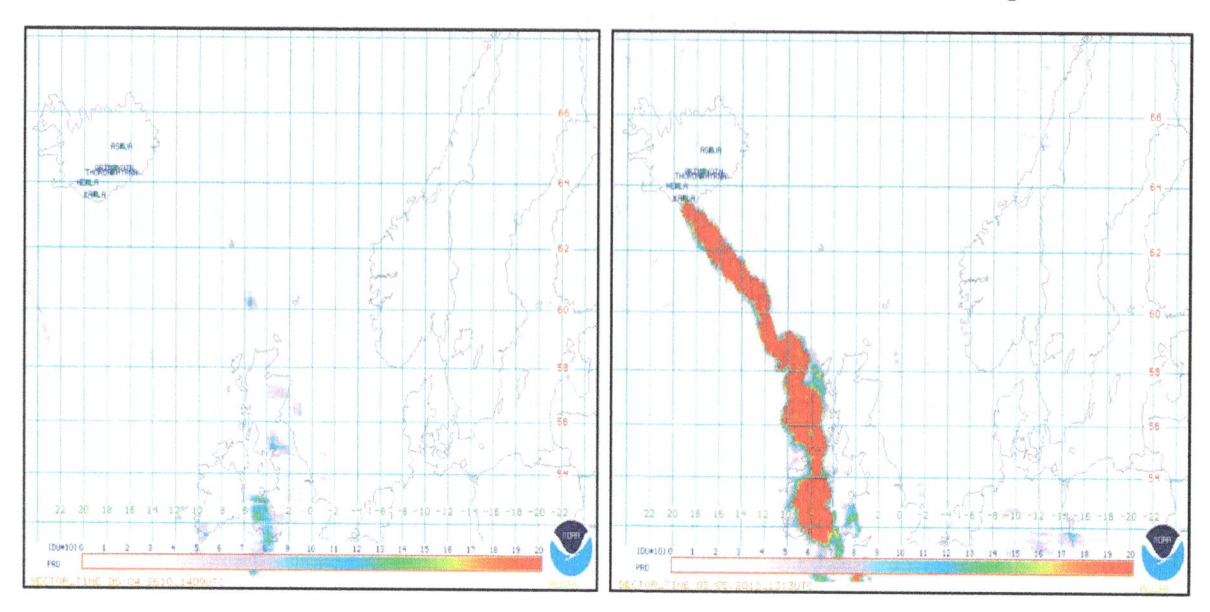

Figure 2.19 SO$_2$ emissions measured from satellites during the 2010 eruption of Eyjafjallajökull, Iceland. a): Low SO$_2$ emissions measured on 4 May 2010, 14:09 UTC (OMI satellite, NOAA). b) Drastic increase in SO$_2$ measured on 5 May 2010, 13:13 UTC (OMI satellite, NOAA).

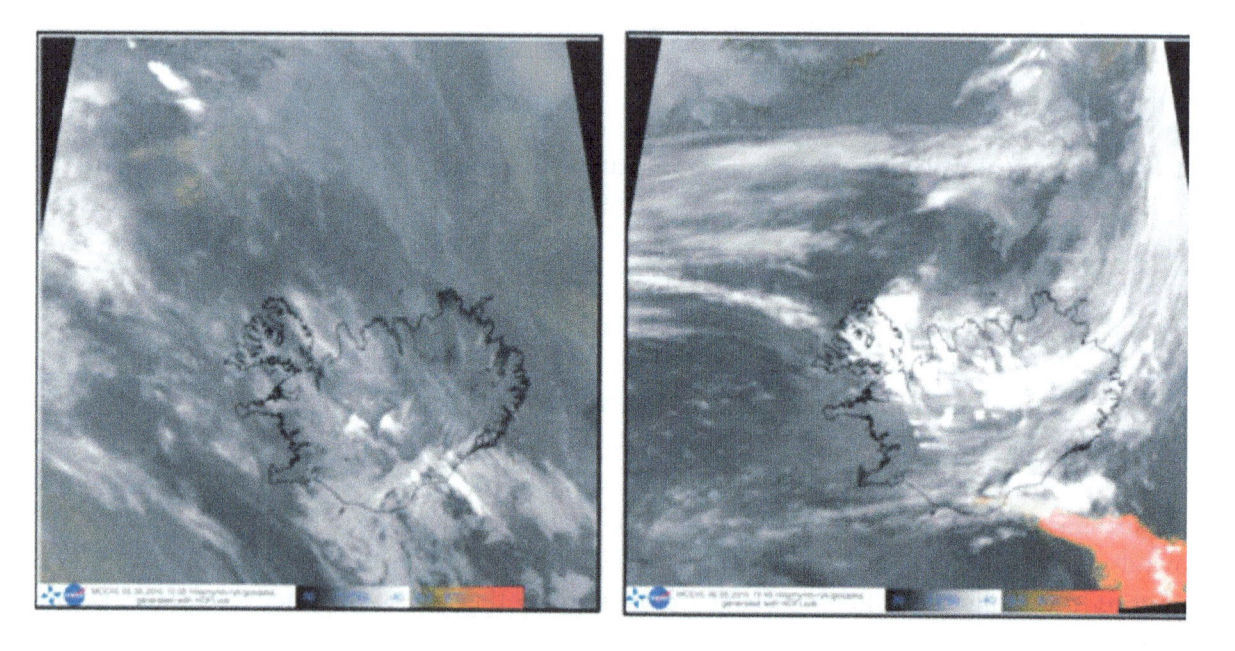

Figure 2.20 Ash output from Eyjafjallajökull volcano, Iceland in 2010. a) Low ash output on May 5th 13:05 UTC does not coincide with the high SO$_2$ gas flux (Figure 2.19b). b) Ash output has increased when measured on 6 May 13:45 UTC, several hours after the increase in SO$_2$ output. MODIS satellite, NOAA, processed by IMO.

Near real-time gas measurements (including satellite and ground-based monitoring) are augmented by studies of volcanic rocks, which trap gases prior to eruption (Edmonds, 2008), although this is a time-consuming technique currently unsuitable for real-time monitoring. The processing of satellite gas measurements needs to be improved to decrease the significant uncertainties and time delay. There also needs to be wider deployment of new ground-based UV and IR multi-gas sensors for gas measurements. To date SO_2 has been the main focus of gas monitoring because it is so easily detected in the atmosphere. The launch of NASA's OCO_2 and Spain's UVAS missions raises the possibility of detecting large CO_2 emissions and using these observations for early warning.

The three main stalwarts of volcano monitoring (earthquakes, deformation and gas) are augmented by many other kinds of data. These might include: visual observations of volcanic activity (sometimes from imaging networks); acoustics; thermal imaging; environmental measurements such as groundwater levels; other geophysical techniques such as gravity, electrical and magnetic measurements; rock geochemistry and petrology; measurements in rivers, crater lakes and hot springs; ground-based infra-sound (Johnson & Ripepe, 2011) and borehole strain meters (Roberts et al., 2011). Much of these data will be collected in real-time by automated sensors. Other data may be collected on a regular basis (e.g. weekly, monthly, annually). Research is ongoing to identify novel methodologies; for example there is current work on using muons (subatomic particles) to image volcanoes (Lesparre et al., 2012).

Factors that inhibit other methods being widely and routinely employed on monitoring and forecasting include expense, field logistics and expertise. For example borehole strainmeters may cost of order US$ 150K to build and install, and require significant processing to interpret. Samples sent back to a laboratory may take days or even weeks to process and require laboratory funding for a rapid response (e.g. petrological characterisation).

One kind of measurement is usually not sufficient for assessing the state of an active or potentially active volcano. Increasingly multiple strands of evidence are necessary for a confident diagnosis. Indeed volcano monitoring can be usefully compared with medical symptoms, one of which on their own might not identify a disease, but where symptoms taken together can. In the explosive eruption of Pinatubo in 1991 it was a combination of earthquake data, SO_2 measurements, minor precursory eruptions and a geologic record of huge eruptions that led to the advice to evacuate large areas a few days before the cataclysmic eruption of 15 June (Punongbayan et al., 1996). Pattern recognition is commonly a key element of using monitored data (Sparks, 2003, Segall, 2013). In 1997 the combination of earthquake swarms, deformation measured by tiltmeter, spurts of dome growth and explosions allowed the Montserrat Volcano Observatory to recognise regular patterns of activity and forewarn the public and authorities in advance when major escalation of activity was expected (Voight, 1999).

2.5.2 Volcanic unrest

Almost all eruptions are preceded by periods of *volcanic unrest* [Chapter 18], which can be defined as the deviation from the background behaviour of a volcano that might presage an eruption. Changes may occur in seismicity (the number, location or types of small earthquakes), surface deformation, gas emissions, geochemistry or fumarolic activity. However, there are many cases when such unrest does not lead to eruption. Due to the active tectonic settings of volcanoes, and often the presence of geothermal fields, unrest phenomena can occur without any magma movement. There are still major challenges when assessing whether unrest will actually lead to an eruption or wane with time.

A survey of reports of unrest at 228 volcanoes active in the period 2000-2011 (Phillipson et al., 2013) indicated that 47% of periods of reported unrest led to an eruption. There is therefore considerable uncertainty about whether an eruption will occur when unrest occurs. The same survey indicated that the duration of reported unrest episodes (both those that lead to eruption and those that don't) is typically an average of about 500 days and can vary from as short as a few days to several years. Unrest prior to eruption at volcanoes that have been long dormant tends to be shorter and averages about one month, but there is a wide range in the durations of precursory unrest. Long-dormant volcanoes are also more likely to have large eruptions. Intensification of the unrest sometimes makes it evident that an eruption is very likely, but this is typically only a few days or hours in advance. These traits mean that volcanic emergencies can begin with prolonged periods of unrest without eruption. To minimise the possibility of costly disruption and no eruption, evacuations are commonly only called at a late stage.

Unrest has been documented at hundreds of volcanoes globally in the last few decades (Dzurisin, 2003, Phillipson et al., 2013, Segall, 2013, Biggs et al., 2014) and is usually based on changes in seismic behaviour and more recently deformation. The duration of pre-eruptive unrest differs according to volcano type: roughly half of the stratovolcanoes studied erupted after about one month of reported unrest, whereas shield volcanoes had a significantly longer unrest period before the onset of eruption (~ 5 months). For stratovolcanoes there appears to be no link between the length of time between eruptions and the duration of pre-eruptive unrest, suggesting that apparently dormant volcanoes will not necessarily experience extended periods of unrest before their next eruption.

A major challenge in interpreting unrest is that the majority of volcanoes globally are presently not monitored effectively or at all. It is difficult to determine how many of these experience unrest. In 1996, it was estimated that ~200 volcanoes had some form of seismic monitoring, but far fewer are classed as well-monitored. Satellite studies of surface deformation enable us to perform systematic studies of large numbers of volcanoes and can be crucial in identifying the first signs of unrest at unmonitored volcanoes in remote and inaccessible locations. As of 1997, ground-based methods (e.g. traditional surveying, tiltmeters, or GPS) had observed surface deformation at 44 different volcanoes but by 2010, the use of satellite data had increased this number to 110 and the list currently stands at 210 (Biggs et al., 2014).

Systematic studies of ~200 volcanoes using satellite data show that after 18 years, 54 had experienced deformation and of these 25 had erupted (Biggs et al., 2014). In terms of assessing eruption potential, this dataset has strong evidential worth, with roughly half of the volcanoes that deformed over the past 18 years also erupting, compared with only 6% of the volcanoes at which no deformation was observed. Continuous, ground-based monitoring is still required for making short-term evacuation decisions and to validate satellite data, but complementary satellite observations may enable strategic deployment of additional ground-based monitoring systems as needed.

Progress in knowledge of volcanic unrest is hampered by historical isolation of observatories and their data, and, even today, a lack of systematic reporting. To improve the knowledge-base on volcanic unrest, the World Organisation of Volcano Observatories (WOVO) is building WOVOdat, a web-accessible, open archive of monitoring data from around the world (www.wovodat.org). Such data are important for the short-term forecasting of volcanic activity amid technological and scientific uncertainty and the inherent complexity of volcanic systems. The principal goal of WOVOdat, in contrast with databases at individual observatories, is to enable rapid comparisons of unrest at various volcanoes, rapid searches for particular patterns of unrest, and other operations on data from many volcanoes and episodes of unrest. WOVOdat will serve volcanology as epidemiological databases serve medicine. WOVOdat is an example of increasing collaboration between volcano observatories.

2.5.3 Forecasting and early warning

An ability to forecast the onset of an eruption and tracking the course of an eruption once it starts, are key components of an effective early warning system and support for emergency management. Recent eruptions, such as the 2010 eruptions of Eyjafjallajökull (Iceland) and Merapi (Indonesia) have benefitted from dense monitoring networks composed of seismometers and deformation instruments, integrated with satellite imagery. Investments in monitoring precursory activity, alongside hazard assessment, mitigation, civil protection, and preparedness has enabled a number of successful evacuations and resulted in a measurable reduction in society's vulnerability to volcanic hazards during the 20th century. One conservative analysis (Auker et al., 2013) suggests that at least 50,000 lives were saved during the twentieth century, and Voight et al. (2013) notes that a similar number have already been saved in the three decades that postdate the tragedy at Nevado del Ruiz.

Intensive monitoring of recent eruptions has generated integrated time series of data, which have resulted in several successful examples of warnings being issued on impending eruptions [Chapters 7 and 9] (Sparks, 2003, Sparks & Aspinall, 2004). Forecasting of hazardous volcanic phenomena is becoming more quantitative and based on understanding of the physics of the causative processes. Forecasting is evolving from empirical pattern recognition to forecasting based on models of the underlying dynamics. This has led to the development and use of models for forecasting volcanic ash-fall and pyroclastic flows, for example. However, volcanoes are complex systems where the coupling of highly non-linear and complex kinetic and dynamic processes leads to a rich range of behaviours. Due to intrinsic uncertainties and the complexity of non-linear systems, precise prediction is usually not achievable. These system attributes mean that probabilistic modelling is the most appropriate way to characterise the uncertainties associated with volcanic hazards and risks, so that forecasts of eruptions and hazards can be developed in a manner similar to weather forecasting (Marzocchi & Bebbington, 2012).

Despite wide-ranging technological advances in monitoring, a major challenge for volcano observatories continues to be determination of whether an episode of volcanic unrest will culminate in eruption. It is also almost impossible to predict the exact timing of eruption onset, even where very good monitoring systems are in place. When evacuations are called by authorities but then nothing happens, public trust may be undermined especially if large population and commercial centres are affected, whereas evacuations that are called too late or not at all can lead to tragedy. In Iceland, the warning period for the onset of eruption typically varies between less than an hour and up to several days. Onset of small eruptions may not even be instrumentally detected (e.g. the explosion of Ontake Volcano, Japan on 27 September 2014). Well-informed local populations may self-evacuate if they feel uncomfortable and have emergency housing options, and this may precede official calls.

2.5.4 Volcano observatories

A volcano observatory is an organisation (national institution, university or dedicated observatory) whose role it is to monitor active volcanoes and provide early warnings of anticipated volcanic activity to the authorities and usually the public too [Chapter 15]. The fact that most volcano observatories face similar problems and challenges has led to increasing collaboration between volcano observatories and the formation of the World Organisation of Volcano Observatories (WOVO; http://www.wovo.org/). There are over 100 members of WOVO. The exact constitution and responsibilities of a volcano observatory may differ country-by-country, but it is typically the source of authoritative short-term forecasts of volcanic activity and information on volcanic hazards.

There are a variety of models of volcano observatory. Observatories range from sophisticated scientific centres with dedicated buildings, multiple staff, and comprehensive state-of-the art instrumentation to simple observation posts. Some very active high-risk volcanoes may have a dedicated observatory; the Montserrat Volcano Observatory in the Eastern Caribbean operated by the Seismic Research Centre in Trinidad, and the Osservatorio Vesuviano operated by the Istituto Nazionale Geofisica and Vulcanologia, Italy, are examples. Some observatories, such as the Cascades Volcano Observatory (US Geological Survey) monitor several volcanoes in a region. Some national scientific institutions have a mandate for all the nation's volcanoes (e.g. the Philippines Institute of Volcanology and Seismology). The approach in Indonesia with 147 active Volcanoes (by Pusat Vulkanogi dar Mitigasi Bencana Geologi; CVGHM) is to have intense monitoring on a small number of high-risk and very active volcanoes, a permanently staffed observation post with seismometers on 70 volcanoes, and to deploy teams to typical long-dormant volcanoes with no dedicated monitoring should they become restless.

Volcano observatories play a critical role in supporting communities to reduce the adverse effects of eruptions through: hazard assessments for pre-emergency planning to protect populations and environments; providing early warning when volcanoes threaten to erupt; providing forecasts and scientific advice during volcanic emergencies; and supporting post-eruption recovery and remediation. Capacity to monitor volcanoes is thus a central component of disaster risk reduction for volcanism.

Volcano observatories are involved in all parts of the risk management cycle. In times between eruptions observatories may assess hazards as preparation for emergencies and for long-term land-use planning. They are often involved in outreach activities so that authorities and the communities can better understand the potential risk from their volcanoes. Outreach may also involve regular exercises with civil protection agencies and aviation authorities to generate and test planning for eruption responses. During the lead up to an eruption, volcano observatories may provide regular updates on activity which inform decisions on evacuations or mitigation actions to reduce risk to people or to critical infrastructure. For example, power transmission companies may choose to shut off high-voltage lines if there is a high probability of ash fall

(Durant et al., 2010). The extent of involvement of volcano observatories in decision-making varies greatly between different countries. A volcano observatory is usually responsible for raising the alert and communicating to the relevant authorities (e.g. civil protection and Volcanic Ash Advisory Centres -VAACs) when monitoring data and observations indicate that an eruption is probable in the short term, is imminent, or has commenced. During an eruption, volcano observatories provide up-to-date information about the progression of activity. For an explosive eruption, information might include the duration, the height that ash reaches in the atmosphere and areas being impacted on the ground. This can inform decisions such as search and rescue attempts or provide input to ash dispersion forecasts for aviation. After an eruption has ceased, volcano observatories can aid recovery through advice about ongoing hazards such as remobilisation of ash deposits due to high winds or heavy rainfall. In addition to scientific analysis of the eruptive activity and erupted products, they may also carry out retrospective analysis of emergencies to help improve future response from lessons learnt (Loughlin et al., 2002, Voight et al., 2013).

WOVO, IAVCEI and regional organisations have been active in organising workshops and meetings that promote knowledge sharing, best practice and interactions with disaster risk managers. Two Volcano Observatory Best Practice workshops, organised by WOVO with support from GVM and IAVCEI, were held in 2011 and 2013 at Erice, Italy. Cities on Volcanoes is a biennial meeting of IAVCEI with a strong focus on hazard and risk, disaster risk reduction and knowledge exchange between scientists, public officials and citizens.

2.5.5 Volcano observatories and aviation safety

Since the start of commercial airline travel in the 1950s, 247 volcanoes have been active, some with multiple eruptions. There were at least 129 encounters of volcanic ash by aviation from 1953-2009 (Guffanti et al., 2010), including a number of very near major catastrophic accidents. Two of the most significant encounters occurred in the 1980s which resulted in total engine shut-down, and 16 more encounters with the 1991 ash of Mt Pinatubo led the International Civil Aviation Organization (ICAO) to set up nine regional Volcanic Ash Advisory Centres or VAACs [Chapters 12 and 14]. They provide volcanic ash advisories to the aviation community for their own geographical area of responsibility (Figure 2.21).

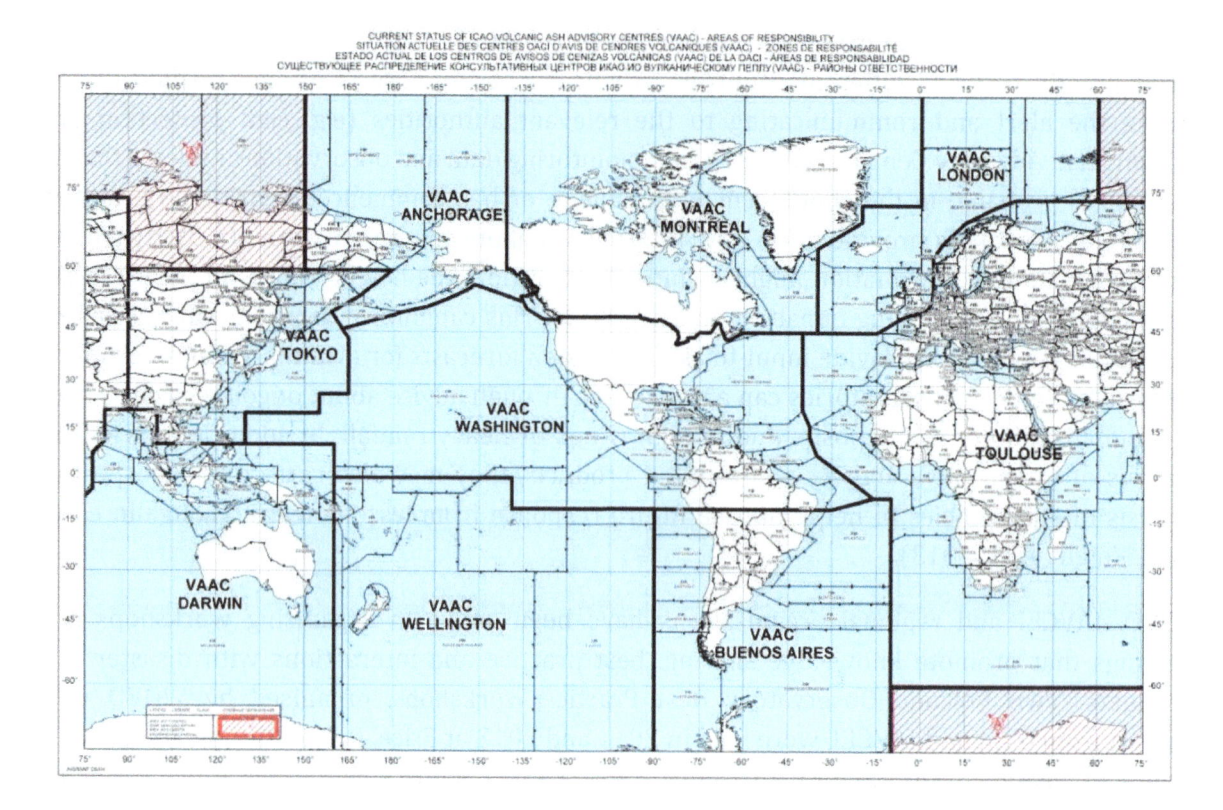

Figure 2.21 Map showing areas of responsibility of the nine Volcanic Ash Advisory Centres (ICAO). This figure can also be seen in Chapter 14, as Figure 14.1.

There are several different alerting systems used worldwide [Chapter 14], each with the aim to update those in local population centres close to the volcano and the aviation community (see Section 2.7.2). The United States Geological Survey (USGS) uses one notification (aviation colour code) specifically for the aviation sector (Neal & Guffanti, 2010), and another (volcano alert system) for hazards that might affect the surrounding population (Gardner & Guffanti, 2006). ICAO adapted its international aviation colour code system from that of the USGS.

Minimising aviation impact requires rapid notices from volcano observatories and pilots, real-time monitoring, detection and tracking of ash clouds using satellites, modelling to forecast ash movement and global communication. International working groups, task forces and meetings have been assembled to tackle the problems related to volcanic ash in the atmosphere. The World Meteorological Organization-International Union of Geology and Geophysics (WMO-IUGG) held workshops on ash dispersal forecast and civil aviation in 2010 (Bonadonna et al., 2012) and 2013. ICAO assembled the 2010-2012 International Volcanic Ash Task Force (IVATF) as a focal point and coordinating body of work related to volcanic ash at global and regional levels. ICAO's International Airways Volcano Watch Operations Group (IAVWOPSG) preceded and continues the work of the IVATF. Globally, there can be many erupting volcanoes that are potentially hazardous to aviation. Therefore, the VAACs and local volcano observatories must work closely together to provide the most effective advisory system and ensure the safety of all those on the ground and in the air.

Volcano observatories play a key role in the system of providing early warning of ash hazard to aviation. Early notification of eruptions is critical for air traffic controllers and airlines so that they can undertake appropriate mitigation of risk to aircraft, such as changing routes. Ideally, the volcano observatory and the regional VAAC should have regular communication both during and in between eruptive crises. A good example is the Icelandic Meteorological Office (volcano observatory) in Iceland that gives a weekly report on the volcano activity status to London VAAC during quiet periods; while during eruptions the reports are issued every three hours and these reports have also been found to have value for other sectors including civil protection. This observatory to VAAC communication channel was significantly improved based on the experience from the 2010 eruption, but it must be noted that not all volcano observatories worldwide have well-defined relationships with their regional VAAC. One of the main roles of WOVO has been to link more volcano observatories with VAACs to enhance communication between volcano observatories and the aviation sector. WOVO is also developing discussions on best practice, for example in short-term forecasting and communication of hazard and risk information.

2.5.6 Global monitoring capacity

Of 1,551 Holocene volcanoes, 596 have recorded historical activity (VOTW4.22). Monitoring activities are largely focussed on historically active volcanoes. A full catalogue called the Global Volcano Research and Monitoring Institutions Database (GLOVOREMID, see Chapter 19) is in development. GLOVOREMID will allow an understanding of global capabilities, equipment and expertise distribution and will highlight gaps. GLOVOREMID began as a study of monitoring in Latin America, comprising 314 Holocene volcanoes across Mexico, Central and South America [Chapter 19]. This database has been populated by the relevant Latin American monitoring institutions and observatories. Monitoring levels were assigned based on the use of seismic, deformation and gas analyses. The catalogue shows that 36% of Holocene volcanoes and 70% of historically active volcanoes in this region are monitored at a basic level or better. About 27% of historically active volcanoes in Latin America have no monitoring, about 17% have basic seismic monitoring and 57% have seismic networks in place coupled with additional deformation or gas monitoring.

Efforts to expand GLOVOREMID to a global dataset are ongoing, but it is not yet complete. A preliminary appraisal of global monitoring has been undertaken here. Determining the monitoring capacity globally is not an easy task. Many monitoring institutions were approached to aid this understanding and the existing Latin American subset of GLOVOREMID was also used to help populate this dataset, together providing monitoring details for about 50% of the historically active volcanoes. The remaining 50% were investigated through online resources provided by monitoring institutions. This is complicated by the availability of information, outdated information, reduced web presence for some areas and, sometimes, contradictory information. Our effort established the monitoring situation of about a further quarter of historically active volcanoes. The monitoring situation at the remaining volcanoes is unknown, but likely to be poor.

For this work, we estimated the numbers of volcanoes in three categories: volcanoes without dedicated monitoring systems; those with some monitoring; and those that have adequate monitoring for basic assessments of magma movements and some quantitative assessments of the probability of future volcanic events. The number of seismometers on a volcano is a relatively easy metric to establish and can be used to estimate the level of monitoring at different volcanoes. Although a single seismometer is of limited use in determining the location of earthquakes and for forecasts of volcanic activity, it can be used, often in combination with the larger regional seismic network, to alert the relevant authorities and commence the deployment of further monitoring systems. This may be particularly useful in countries where resources are prioritised at recently active or high hazard or risk volcanoes.

Ideally, a multi-station network of 4 or more seismometers is required to establish accurately the location and size of seismic events beneath a volcano, allowing for swarms of micro-quakes to be detected and for the establishment of the cause of earthquakes, e.g. volcano-magmatic, glacier movements, rockfalls and others. As such, the three levels of monitoring derived in this study are:

- No monitoring: No known dedicated volcano monitoring equipment. No dedicated seismometers. No dedicated volcano observatory.
- Some monitoring: 1 to 3 or fewer seismometers dedicated to volcano monitoring, and a volcano observatory or institution that is responsible for monitoring. Additional monitoring techniques such as deformation and gas analysis may also be in place.
- Adequate monitoring: 4 or more seismometers dedicated to volcano monitoring, and a volcano observatory or an institution that is responsible for monitoring and equipment maintenance. Additional monitoring techniques may also be in place.

Of the historically active volcanoes worldwide between 25% and 45% are unmonitored. This large uncertainty exists due to an absence of information for about a quarter of historical volcanoes. Further research needs to be undertaken to better constrain this detail. Some of these unmonitored volcanoes are located in densely populated areas and have histories including large magnitude eruptions. About 14% of historically active volcanoes are described with 1 to 3 seismometers and the majority of these volcanoes have this seismic monitoring alone. About 35% of historically active volcanoes have four or more seismometers within 20 km distance and most of these volcanoes also have GPS stations, tiltmeters, or other deformation instruments.

2.5.7 Low-cost systems for monitoring volcanoes in repose

With half of the world's historically active volcanoes having repose of more than 100 years before eruption it is not always practical or cost-effective to have permanent and extensive monitoring networks. Financial constraints are a major obstacle to maintaining monitoring networks at volcanoes in long repose. However, technological advances and international agreements are yielding opportunities for the low-cost monitoring of volcanoes in repose that do not have conventional, permanent ground networks.

Satellite-based Earth Observation (EO) provides the best means of bridging the currently existing volcano-monitoring gap [Chapter 17]. EO data are global in coverage and provide information on some of the most common eruption precursors, including ground deformation, thermal anomalies, and gas emissions. Ideally, EO data must be processed and appropriate products made available and accessible to volcano observatories (often in a nominated national institution) in a timely manner. In addition, training will ensure EO data products can be analysed and used effectively by volcano observatories. Such systematic global provision will come at a modest cost although it will be highly cost-effective. Scientists receiving EO information products about volcanic unrest can then potentially enhance ground monitoring networks and instigate additional mitigation measures with the authorities and populations at risk.

Once an eruption is in progress, continued tracking of these parameters, as well as ash emission and dispersal, is critical for modelling the temporal and spatial evolution of the hazards and the likely future course of the eruption. The need for volcano-monitoring EO data is demonstrated by a number of international projects [Chapter 17]. The costs of satellite monitoring, in particular, are declining through the supply of free data and Free and Open Source Software (FOSS) for image processing and map-making.

2.6 Assessing volcanic hazards and risk

Knowledge of volcanoes and the ability to anticipate their behaviour is improving. However, the great complexity of volcanoes means that in most circumstances we cannot give precise predictions of the onset and evolution of volcanic eruptions and their consequences. Precise prediction of the time and place of an eruption and its associated hazards is exceptional (Sparks, 2003). Likewise deterministic assessment of footprints for different kinds of volcanic hazard is not a realistic expectation. However, volcanologists are improving their ability to anticipate future volcanic events and their likely footprints. Forecasting the outcomes of volcanic unrest and ongoing eruptions is implicitly or explicitly probabilistic and forecasts are becoming increasingly quantitative (Sparks, 2003, Sparks et al., 2012). This trend reflects the fact that in natural systems and especially volcanoes, multiple eruptive outcomes and consequences are possible over any given time period. Every volcano, as well as the hazards and risks associated with it, is unique in some respects and requires dedicated investigation. This diversity has led to different methods being developed and applied in hazard and risk assessment in different places (Sparks & Aspinall, 2004, Marzocchi & Bebbington, 2012, Sparks et al., 2013). Some generic methodologies have proven successful for several eruptions, while for a few high-risk volcanoes significant research efforts have been undertaken and advances include development of novel techniques.

Like other natural phenomena, volcanic hazard and risk are linked to one another through exposure and vulnerability.

2.6.1 Hazards maps

At many volcanoes, hazards assessments take the form of maps, which may be qualitative, semi-quantitative or quantitative in nature [Chapter 20]. Most are based upon a geological and historical knowledge of past eruptions over a given period of time (Tilling, 1989). A typical study involves mapping young volcanic deposits to generate maps for each type of hazard, reflecting areas that have been affected by past volcanic events. An important limitation though, is that the distribution of previous events (even if known in their entirety), does not represent all possible future events. Increasingly such studies are augmented by computational modelling of the processes involved. Computer simulations are run under the range of conditions thought to be plausible for the particular volcano and commonly calibrated to observed deposit distributions.

The hazards of most widespread concern, as indicated by frequency of occurrence on hazards maps and fatality data (Auker et al., 2013), are: lahars, pyroclastic density currents and tephra fall. Currently, tephra hazards (which can have the widest distribution and far-reaching economical impacts) are the best quantified. Lahars and pyroclastic density currents both have more localised impacts, but account for far greater loss of life, infrastructure and livelihoods. These hazard types present greater challenges for modelling, and as a result quantitative hazard analysis for lahar and pyroclastic density currents lags behind that for tephra fall (see Chapter 3).

Hazard maps take many forms from geology-based maps reflecting the distribution of previous events, to circles of a given radius around a volcano or different zones likely to be impacted by different hazards to probabilistic maps based on stochastic modelling, to administrative maps constructed to aid in crisis management. Hazard maps can also be produced for a region with many volcanoes that consider cumulative hazard from all possible eruption types weighted to frequency and magnitude. Hazards maps can represent specific eruptive scenarios (e.g. dome eruption, explosive Plinian or subplinian eruption), or be based on a scenario from a specific historic eruption of a volcano that is thought to be representative of a likely future eruption or can be hazard specific (e.g. hazard from tephra fall, pyroclastic density currents or lahars). These different kinds of map and hazards information are commonly integrated together so that the area around a volcano is divided into zones of decreasing hazard. A common type of integrated ("bulls-eye") hazard map will have a red zone of high hazard, orange or yellow zones of intermediate hazard (often both), and a green zone of low hazard. Hazard maps used for most volcanoes worldwide today indicate these zones qualitatively or semi-quantitatively, whereas quantitative (fully probabilistic) maps are actually the exception.

The boundaries between zones on a hazard map are typically marked initially by lines on maps based on judgement by scientists about the levels of different hazards. The position of zone boundaries on hazard maps is implicitly probabilistic. Increasingly boundaries are explicitly based on fully quantitative probabilistic analysis. The precise boundary position may be modified to take account of administrative issues and practical matters, such as evacuation routes, as determined by civilian or political authorities. At this point these maps become directly relevant to the planning and decision-making process and more closely aligned to the analysis of risk. Recently, volcanologists are making greater efforts to integrate risk knowledge collaboratively into hazard zone maps. For example a new hazard map Mt. Tongariro was produced in 2012 by GNS Science New Zealand [Chapter 16] (Figure 2.22). The area impacted by the eruptions includes a section of the popular Tongariro Alpine Crossing walking track. Requirements of tourists, concessionaires and local residents were considered alongside scientific modelling and geological information to produce an effective communication product, which was tailored for use during that specific eruption.

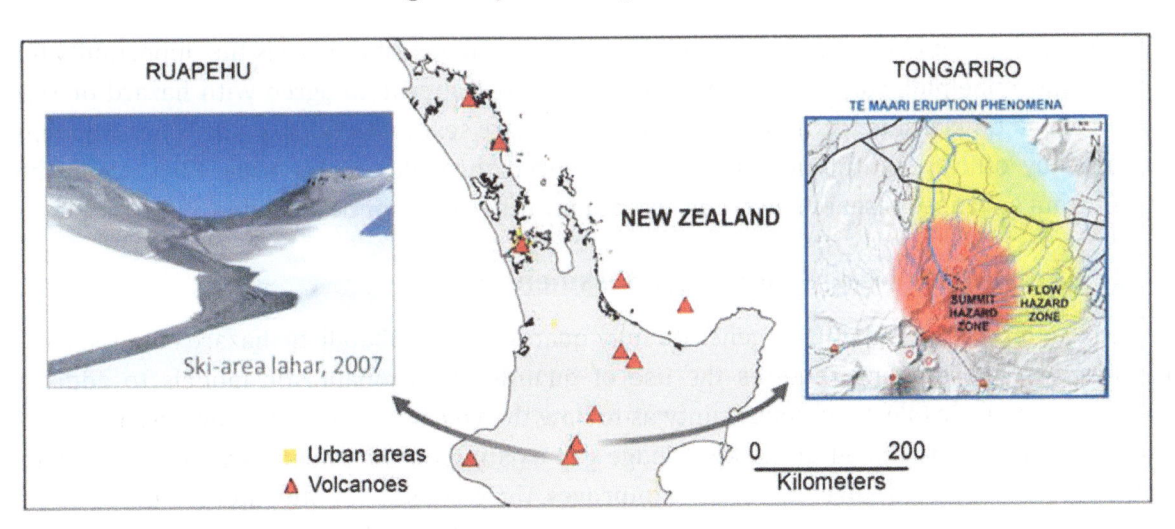

Figure 2.22 Risk management at two volcanoes in New Zealand. The comprehensive Tongariro hazard map can be found at www.gns.cri.nz/volcano. This figure can also be seen in Chapter 16, as Figure 16.1.

Drawing lines on maps to demark safe from unsafe zones sounds easy in principle, but is difficult in practice especially if the threshold that defines the line is itself hard to estimate and the uncertainties in these estimates are large. This problem was very well illustrated during the 2010 Icelandic ash emergency. Initially the operational guidelines for response of air traffic control involved ash avoidance, so computer simulations simply had to forecast where ash would go rather than how much ash there was. However, after a few days of almost complete shutdown of European air space, engine manufacturers effectively announced that engines would not be compromised if the ash concentrations were less than 2 milligrams of ash per cubic metre of air. Forecasting precise atmospheric concentrations, as well where concentrations are above a given threshold, is much more challenging and requires advances in scientific knowledge and modelling methods (Bonadonna et al., 2012).

In volcanic risk, hazards maps are used for multiple purposes such as raising awareness of hazards and likely impacts, for planning purposes and to help emergency managers mitigate risks. Hazard and derivative risk management maps of volcanoes are produced by volcano observatories (or their parent institutions such as geological surveys) and a variety of other organisations (e.g. private sector). Geological surveys or other government institutions typically have official responsibility for providing scientific information and advice to civilian or political or military authorities, who have the responsibility to make policy or decisions such as whether to evacuate. Efforts are underway to classify hazards maps, harmonise terminology and develop discussions around good practices such as how to account for uncertainties, what time interval is taken for the magnitude-frequency analysis of past eruptive behaviour, and what scale of events to consider. There is consensus that the basic foundation on which any hazard analysis should be undertaken is the establishment of an understanding about a volcano's evolution and previous eruptive behaviour through time, based on combined field geology and geochemical characterisation of the products. However, bringing together experts in modelling and statistical analysis, together with field scientists, is then key. Driven by the needs of today's stakeholders there is also a need for future research efforts to advance the science that will aid in the production of a new generation of robust, fully quantitative, accountable and defendable hazard maps.

Academic groups and insurance companies also generate maps, and there is the opportunity for serious, and unhelpful, contention if any of these do not appear to agree with hazard or risk maps from an official source. On the other hand, there are several examples where hazard maps produced by official institutions have been enhanced through collaboration. Such cases can benefit from advanced research methods, which may otherwise not be available.

2.6.2 Probabilistic hazard and risk assessment

There is an increasing impetus to generate fully quantitative probabilistic hazards assessments and forecasts. Forecasting requires the use of quantitative probabilistic models to address adequately intrinsic (aleatory) uncertainty as to how the volcanic system may evolve, as well as epistemic uncertainty linked to the knowledge gap existing on the phenomena or the volcano. As the extent and resolution of monitoring improves, the process of jointly interpreting multiple strands of indirect evidence becomes increasingly complex. The use of new probabilistic formalism for decision-making (e.g. Bayesian Belief Network analysis, Bayesian event decision trees) (Marzocchi & Bebbington, 2012, Sparks et al., 2013, Hincks et al., 2014), could significantly reduce scientific uncertainty and better assist public officials in making urgent evacuation decisions and policy choices when facing volcanic unrest, although these methods have yet to be applied widely. Selection of appropriate mitigation actions using probabilistic forecast models and properly addressing uncertainties is particularly critical for managing the evolution of a volcanic emergency at high-risk volcanoes, where mitigation actions require advance warning and incur considerable costs, including those of evacuation.

There are a variety of probabilistic maps that depend on the nature of the hazard. For volcanic flows (pyroclastic density currents, lahars and lavas) the map typically displays the spatial variation of inundation probability over some suitable time period or given that that the flow event takes place. For volcanic ashfall hazard the probabilistic analysis can be represented using the exceedance of some threshold of thickness or ground loading, volumetric concentration at some specific atmospheric level, or even particle size.

Recent developments mean that ashfall hazard has been considered far beyond individual volcanic or even country settings. In SE Asia, volcanic ashfall is the volcanic hazard most likely to have widespread impacts since a single location may receive ashfall at different times from different volcanoes [Chapter 12]. Probabilistic curves and maps of ashfall thickness can be calculated using volcanic histories and simulations of eruption characteristics, eruption column height, ash volume and wind directions at multiple levels in the atmosphere (Jenkins et al., 2012). In this example risk is expressed via the amount of ash deposited and its characteristics (hazard), as well as the numbers and distribution of people and assets (exposure), and the ability of people and assets to cope with ashfall impacts (vulnerability). By combining probabilistic hazard estimates with freely available exposure data and a proxy for human vulnerability (the UN Human Development Index for a country), each component contributes towards an overall 'risk' score (Figure 2.23).

This approach offers a synoptic insight into what drives a city's risk. When applied to the Asia-Pacific region, home to 25% of the world's volcanoes and over two billion inhabitants, Tokyo's risk is dominated by the high cumulative hazard (54 active volcanoes lie within 1,000 km), Jakarta's risk is dominated by population exposure, and Port Moresby's risk by the vulnerability. Chapter 3 presents a much more detailed assessment on ashfall hazard and risk.

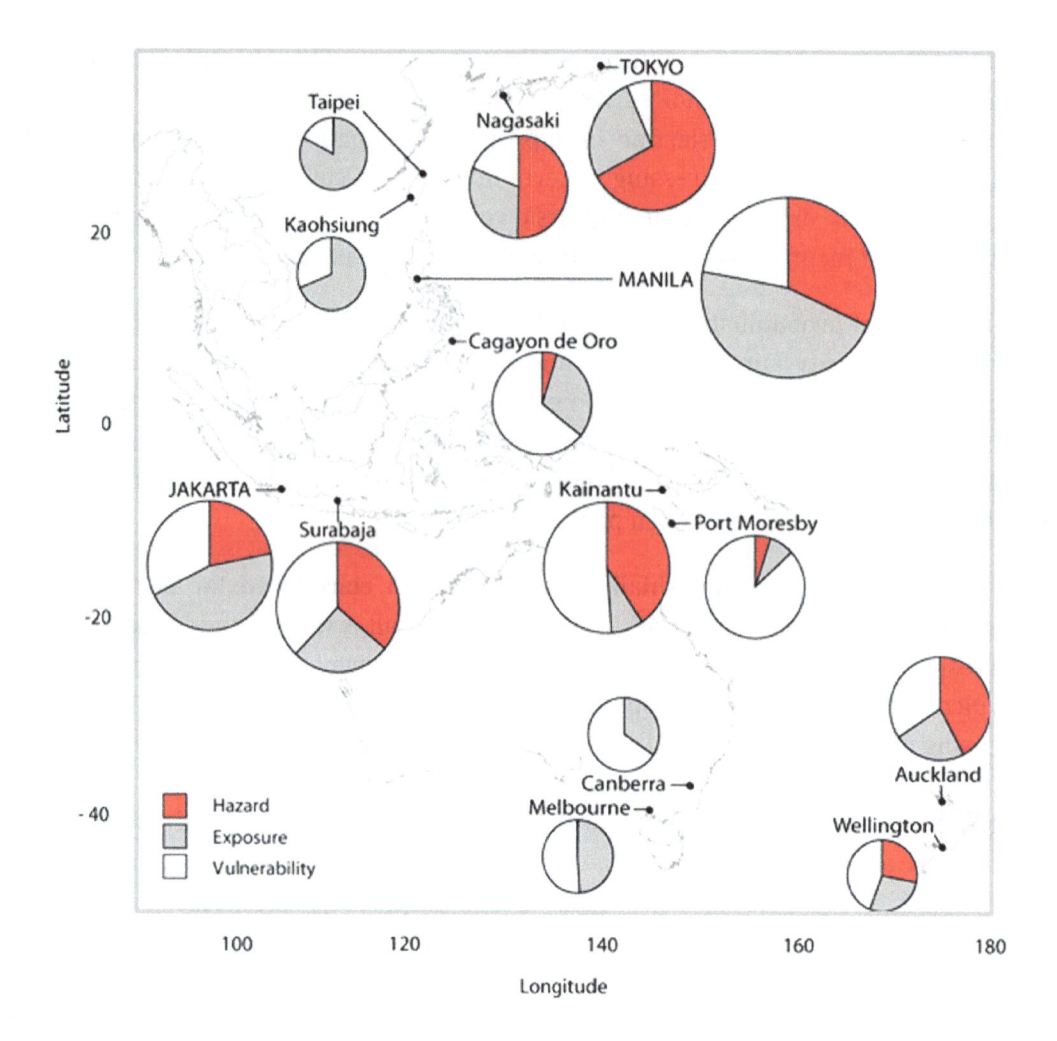

Figure 2.23 Relative contributions of the three factors comprising overall risk score for cities in the Asia-Pacific region (see Chapters 3 and 12). This figure can also be seen in Chapter 12, as Figure 12.3.

2.6.3 Event trees

Event trees are useful logical frameworks for discussing probabilities of possible outcomes at volcanoes showing unrest or already in eruption [Chapter 18] (Newhall & Hoblitt, 2002, Sparks & Aspinall, 2004). They can be valuable for discussion between scientific teams but also with authorities and the public. Each branch of an event tree leads from a necessary prior event to a more specific outcome and is allocated a conditional probability, e.g. given the occurrence of an earthquake swarm the probability of an eruption is estimated. The probability estimates might be based on past and current activity (empirical), expert elicitation (Aspinall, 2010), numerical simulations, or a combination of methods. The probabilities can be revised regularly as knowledge or methodologies improve or when the character of volcanic activity changes. Event trees have been successfully used at many eruptions worldwide since the 1980s, including the eruption of the Soufrière Hills Volcano, Montserrat (Sparks and Aspinall, 2004). Event trees are also commonly used in Volcano Disaster Assistance Program (VDAP) responses to volcanic emergencies [Chapter 25].

2.6.4 Exposure and vulnerability

There can be many different kinds of loss as a consequence of volcanic eruptions including: loss of life and livelihoods (Kelman & Mather, 2008, Usamah & Haynes, 2012), detrimental effects on health [Chapter 13], destruction or damage to assets (e.g. buildings, bridges, electrical lines and power stations, potable water systems, sewer systems, agricultural land) (Blong, 1984); economic losses; threats to natural resources including geothermal energy (Witter 2012); systemic vulnerability and social capital. Each of these will have its own specific characteristics in terms of exposure and vulnerability (Spence et al., 2005, Wilson et al., 2012b).

Thus moving from hazards to risk requires an assessment of exposed populations and assets, taking account of their vulnerabilities (both physical and social). Vulnerability is a key means by which the impact of volcanic hazards can be amplified or attenuated according to circumstance. Vulnerability is an attribute of individuals and their assets as well as institutions, critical services and cultural or political groupings. Like volcanic hazards, these attributes of vulnerability vary in both space and time (Wisner et al., 2004) and can be expected to affect the outcome for populations at risk in several ways (Figure 2.24). Risks also usually have to be placed within a suitable time-frame appropriate for decision-making and actions. Risk assessments might need to be carried out in near real-time at the height of a volcanic crisis, while long-term planning is normally undertaken over longer timeframes. Thus the nature of the loss and the time scale over which the loss is being considered are critical in characterising vulnerability and exposure. In volcanology there are several examples of analysis of individual facets of vulnerability, particularly health and assets, but rather less on individual and social vulnerability or the dynamics of these components under stress.

In the vicinity of volcanoes, direct loss of life and evacuation of people from high-risk areas have been a priority concern. Hazard footprints arising from hazard assessments are traditionally superimposed on census data to identify exposed populations for preliminary potential societal risk calculations. Here we use the metric of numbers of people within 100 km of a volcano. The greatest numbers of people living close to volcanoes are found in Indonesia, the Philippines and Japan [Chapter 4]. However, many countries such as Guatemala and Iceland have a higher proportion of their total population within 100 km of a volcano (with >90%) and some small island communities may have all their population within 100 km. Hazards footprints can be used to identify exposed assets such as buildings and infrastructure, agriculture, critical systems, supply chains, livelihoods and so on. The scale of assessments (local to global) brings in different uncertainties and assumptions due to availability of data. There is a need for harmonisation of methods and data sources.

Vulnerability is a major determinant of the impact of volcanic hazards and is a key concept for understanding the resilience of a community and its assets. There are both social and physical vulnerabilities, which are commonly linked. Physical aspects of vulnerability include, for example, building quality (Spence et al., 2005) and to what extent transport systems enable rapid evacuation.

Social vulnerability is defined (Adger, 2006) as "the propensity of a society to suffer from damages in the event of the occurrence of a given hazard". Assessment of social vulnerability is complex as the characteristics of communities and individuals, like volcanoes, vary in both space and time. It is widely acknowledged that marginalised communities, be that geographically, socially or politically, often suffer the most from natural hazards (Dibben & Chester, 1999, Gaillard, 2007, 2008, Lavigne et al., 2008). Tourists have also been recognised as a vulnerable group unlikely to be aware of evacuation procedures or how to receive emergency communications when volcanic activity escalates (Bird et al., 2010). Like hazards, vulnerabilities can only be assessed within the community and with a strong understanding of the local social, cultural and political landscape. Identifying the factors that lead to social vulnerability is challenging and is only just beginning to be applied for volcanic risk (Crosweller & Wilmshurst, 2013).

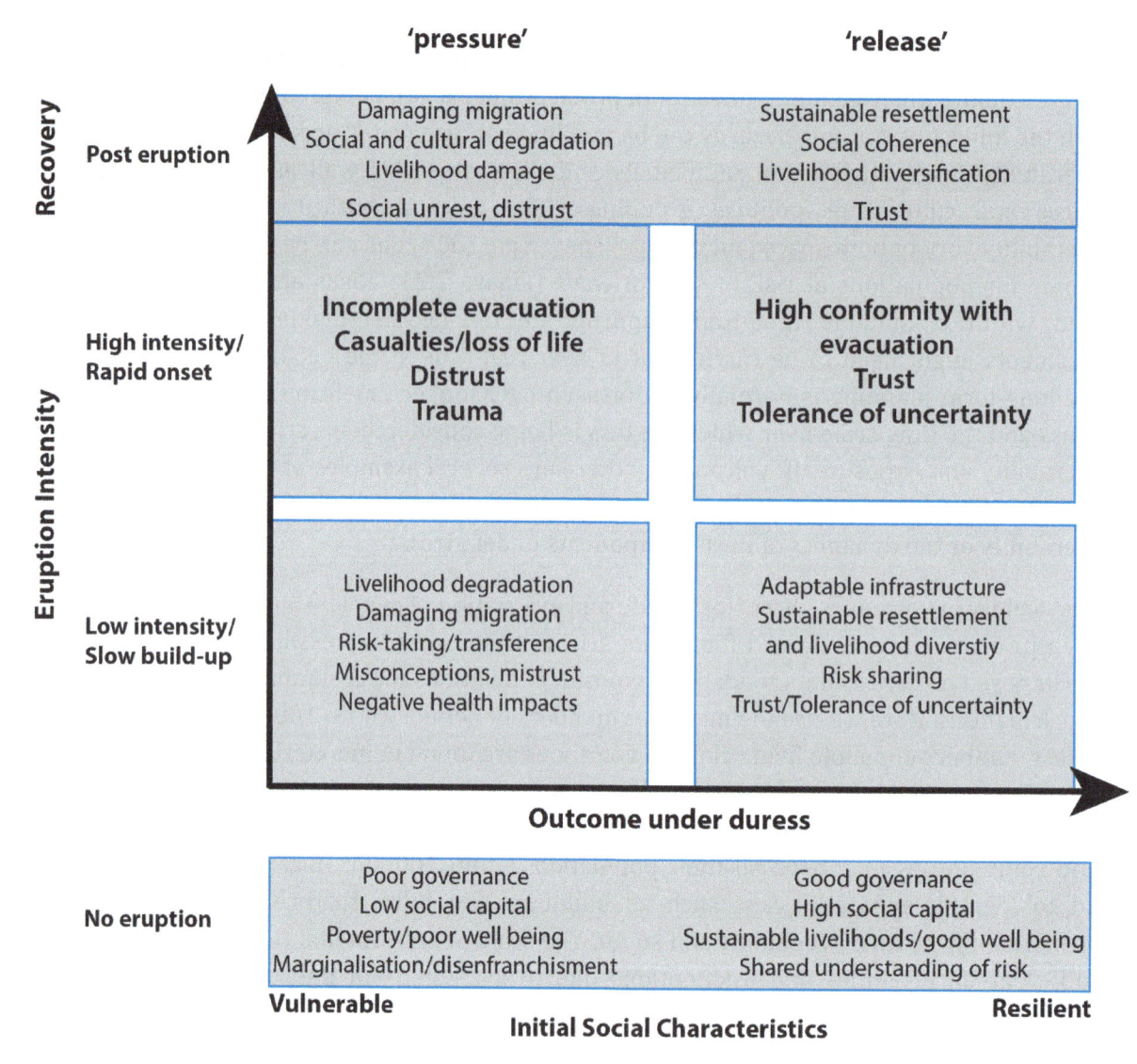

Figure 2.24 Summary of outcomes associated with the interaction of differing population types with volcanic hazards at differing stages of an eruptive cycle. The 'response' phase of volcanic crises has been sub-divided according to intensity and duration. Low-intensity, slow-building or long-duration activity (e.g. persistent small explosions or lahars) contains most characteristics of intensive risk for affected populations, whereas the higher intensity activity (e.g. Plinian explosions, sector collapse) has impacts more consistent with extensive risk. 'Pressure' and 'release' draws an analogy with the Pressure and Release using a model for vulnerability and disaster risk. Image (C) J.Barclay, modified from Barclay et al. (2015).

At volcanoes, it is recognised that livelihood is a key factor that plays a role in the vulnerability of societies and individuals. In particular, in equatorial settings large populations live and farm at elevation on volcanic slopes due to the combination of fertile soil and mild climate (Small & Naumann, 2001). Farmers, particularly those at subsistence level, need to maintain their crops and livestock in order to secure an income, so even short eruptions can be very damaging. If farmlands are evacuated, the longer the period of evacuation, the more likely it is that attempts will be made to return to evacuated land to care for crops and livestock in at-risk areas. This behaviour has been documented many times around volcanoes (e.g. Philippines, (Seitz, 2004); Ecuador (Lane et al., 2003); Indonesia (Laksono, 1988), Tonga (Lewis, 1999)).

If the conditions under which evacuees must live are poor, individuals are more likely to return to their homes in at-risk areas. For example, in Montserrat (Lesser Antilles) evacuated families were living in temporary shelters for months and ultimately years (Clay et al., 1999), and some sought peace and quiet at their homes in the evacuated zone or continued to farm, resulting in 19 unnecessary deaths in 1997 (Loughlin et al., 2002). Concerns about looting also cause people to delay evacuation and return frequently to at-risk areas.

A health and vulnerability study for the Goma, Democratic Republic of Congo, volcanic crisis in 2002 (Baxter et al., 2003) considered human, infrastructural, geo-environmental and political vulnerability following the spontaneous and temporary evacuation of 400,000 people at the onset of the eruption [Chapter 11]. The area was already in the grip of a humanitarian crisis and a chronic complex emergency involving armies and armed groups of at least six countries. The potential for cascading health impacts (e.g. a cholera epidemic) as a result of such a large displaced population was extremely high. The evacuation of large numbers of people into temporary accommodation for even short periods can lead to significant public and other risks.

The extent to which a population is willing or able to take the appropriate action in the face of a threat is a major factor in vulnerability. The complex pre-existing and dynamic political and cultural landscape is known to impact on likelihood to take action with many other messages competing with emergency and preparedness information. In the cases of the emergencies in 2010 at Merapi [Chapter 10] and in 2002 in the city of Goma [Chapter 11] many people were familiar with previous eruptions and this knowledge led to prompt life-saving actions and positive responses to official advice. In the case of La Soufrière (Guadeloupe) in 1976 [Chapter 8] publically debated disagreement on the future course of the eruption by scientists and officials, a highly disruptive massive evacuation largely perceived as an exaggerated and political application of the precautionary principle, and the non-occurrence of a significant eruption led to a loss of trust in science and public policy, making the population now more vulnerable to future eruptions. Volcanoes that have not erupted historically (e.g. Pinatubo; Chapter 7) or in living memory pose more problems in that volcanic activity and attendant hazards are outside the experience of the exposed population, crisis managers and policy decision-makers.

The forensic analysis of past volcanic disasters offers an opportunity to identify and investigate risk factors in different situations and also to identify evidence of good practice (Integrated Research on Disaster Risk, Forensic Investigations of Disasters: http://www.irdrinternational.org/projects/forin/). Long-lived eruptions such as Soufrière Hills volcano, Montserrat, and Tungurahua, Ecuador, also offer opportunities to assess adaptation to extensive risks, for example coping with the cascading impacts of repeated ashfall (Sword-Daniels, 2011) and developing new risk assessment methodologies.

Like natural hazards, understanding all the factors that contribute to social and physical vulnerability at any moment in time is challenging. Nevertheless, growing knowledge, improved methodologies and an increasing willingness to integrate information across disciplines should contribute to increased understanding of risk drivers. Increasing the opportunities to integrate knowledge and experience from scientists (of all disciplines), authorities and communities at risk should enable us collectively to increase resilience and reduce risk.

2.6.5 Quantification and representations of volcanic risk

In recent years volcanologists have started to make quantitative assessments of risk. Not all kinds of risk can be easily calculated so the focus to date has largely been on risk to life. The Soufrière Hills Volcano (SHV), Montserrat, has been erupting episodically since 1995, with life-threatening pyroclastic flows generated by lava dome collapse and explosive events. It has provided a testing ground for methods to calculate and track risk during a major volcanic emergency [Chapter 21] (Wadge & Aspinall, 2014). Volcanic activity is monitored by the Montserrat Volcano Observatory (MVO). With an international membership, the Scientific Advisory Committee on Montserrat Volcanic Activity (SAC), provides regular quantitative hazard and risk assessments. Advanced quantitative risk analysis techniques have been developed, forming an important basis for mitigation decisions.

Over 18 years, the SAC has used the following sources of information and methods: MVO data on activity at the SHV and other lava dome volcanoes; computer models of hazardous volcanic processes; formalised elicitations of probabilities of future hazards scenarios using structured expert judgement methods (Aspinall, 2010); probabilistic event trees; Bayesian belief networks; census data on population numbers and distribution and Monte Carlo modelling of risk levels faced by individuals, communities and the whole island population. The combined methods characterises uncertainty, which is regarded as an essential element for informed and effective decision-making.

Risk assessments are presented to the authorities and public via open reports in a manner that is understandable. Societal casualty risks and individual risk of death are both calculated. MVO and the SAC developed a means of representing risk, which follows methods used for industrial accidents. Societal risk is calculated in terms of the probability (F) of exceeding a given number of fatalities (N) in a specified time. F-N curves have been used successfully in the emergency management on Montserrat. The F-N plot from 2003 (Figure 2.25) shows the probability of N or more fatalities due to the volcano (red, with uncertainty), the reduced risk if the main at-risk area is evacuated (green) and comparative hurricane and earthquake risks. An individual risk ladder from 2011 is shown (Figure 2.26) with both residential zone risk levels and work-related risk levels plotted, with uncertainties. Central to the effectiveness of this approach is that comparative values from familiar circumstances are shown for reference.

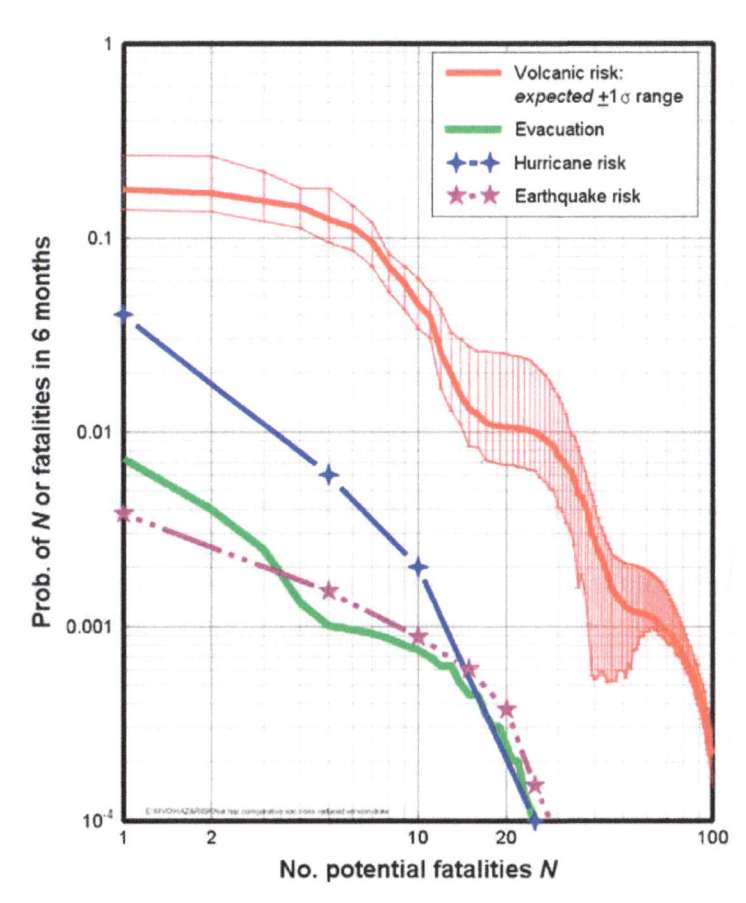

Figure 2.25 The F-N plot from 2003 for the Soufrière Hills Volcano, Montserrat shows the probability of N or more fatalities due to the volcano (red, with uncertainty), the reduced risk if the main at-risk area is evacuated (green) and comparative hurricane and earthquake risks. The curves are compared with societally accepted risks from regional earthquakes and hurricanes on Montserrat. Figure modified from Wadge & Aspinall (2014) see Figure 21.1 and Chapter 21 for further details.

These risk assessments have been used to inform critical decisions on Montserrat, including evacuation and re-occupation (Figure 2.25), and development of management controls on sand mining (Figure 2.26). In both cases the risks are compared to more familiar risks to aid communication with decision-makers. One very difficult issue is the assessment of the probability that an eruption has ended. This has major societal implications. Although this is the most uncertain area of volcanic risk assessment, end-of-eruption criteria have been proposed for the current eruption on Montserrat. They are systematically evaluated with probabilistic analysis as proxies of the internal behaviour of the volcanic system and to provide support for public decision-making. The Montserrat emergency has lasted over 18 years and so has given the opportunity to assess the hazards forecasts, which form a key component of the risk assessments. The performance of probabilistic event forecasts against actual outcomes has been measured using the Brier Skill Score: more than 80% of life-critical forecasts had positive scores indicating very reliable hazard forecast.

Evacuations may be short term but during some long-lived eruptions evacuations may become regular occurrences as populations continue to live and work alongside a sporadically active volcano (e.g. Tungurahua, Ecuador, Chapter 26) or become permanent large-scale movements of populations (e.g. Montserrat 1997). Once a permanent evacuation has occurred, risk assessments are needed to manage access into evacuated areas (e.g. White Island, New Zealand), and to manage access and land use in marginal zones (e.g. Montserrat), also to consider the potential for hazards of even greater impact.

Concern about the risk to human health from volcanic ash [Chapter 13] motivated an example of a fully quantitative probabilistic risk assessment on the exposure of population on Montserrat to very fine respirable ash (Hincks et al., 2005). Here volcanology, sedimentology, meteorology and epidemiology had to be combined together to assess the probability of exposure to ash of different population groups over a 20-year period. The study illustrates the multi-disciplinary character of risk assessments, where diverse experts are needed. Figure 2.27 illustrates some results that led to the conclusion that health risks for most people living on Montserrat was low but that risks were high enough to cause concern for certain more exposed occupations such as gardeners.

Vulnerability is commonly converted to indices to facilitate semi-quantitative approaches to risk. For example the structural vulnerability of roofs to collapse following ashfall (physical vulnerability) can be assessed using an index of different roof types and thresholds for collapse under different conditions (Spence et al., 2005). Although semi-quantitative approaches can be used to incorporate assessments of vulnerability into risk assessments, in order to be useful for near real-time emergency response such assessments need to be fine-grained, ideally at a household/building scale.

In most cases so far, despite the considerable potential and proven value of quantitative risk assessment approaches, volcanic risks have largely been managed without being measured. Small Island Developing States and cities at risk are examples of situations where a quantitative approach will support effective risk management.

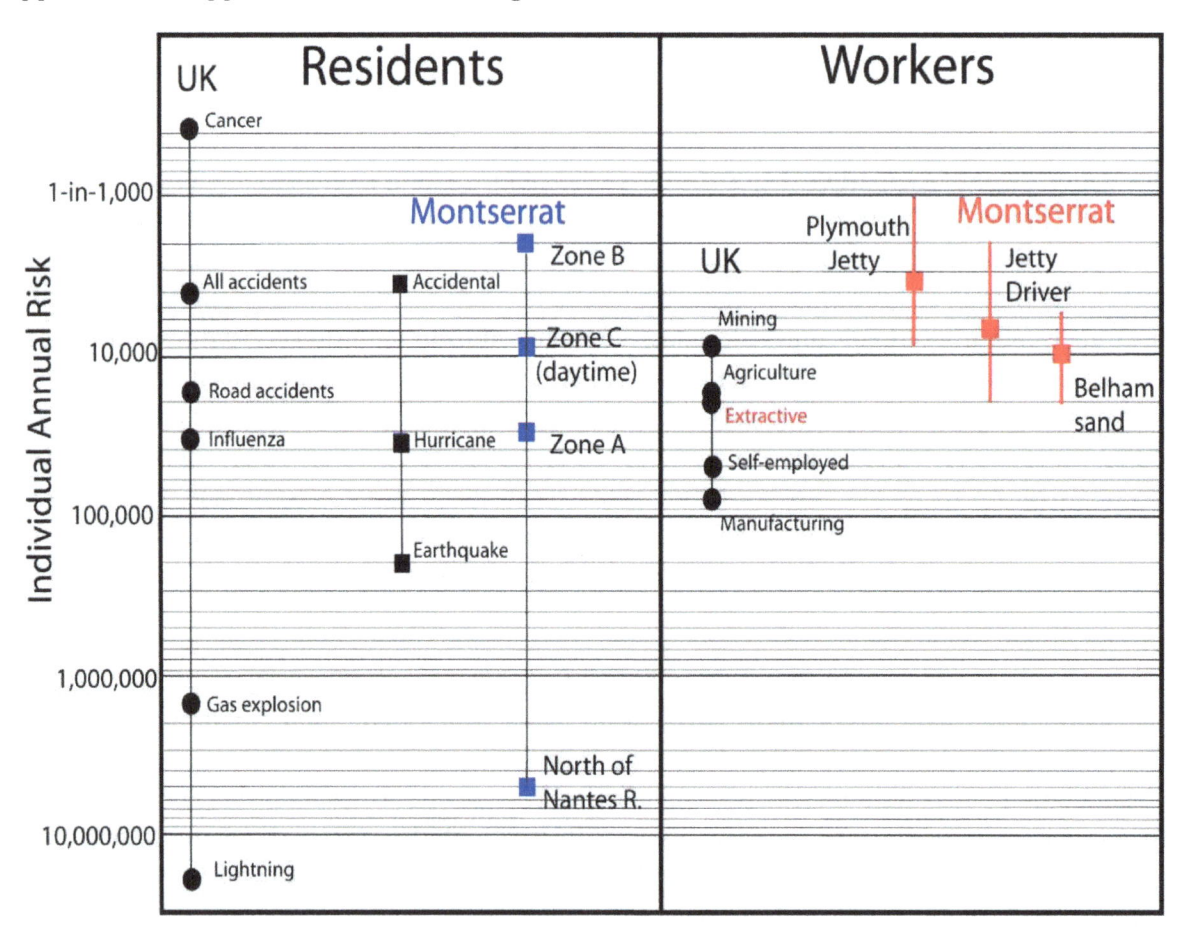

Figure 2.26 (previous page) An individual risk ladder from 2011 for the Soufrière Hills Volcano, Montserrat is shown with both residential zone risk levels and work-related risk levels plotted, with uncertainties. Comparative values from familiar circumstances are shown for reference. Figure modified from Wadge & Aspinall (2014): see Figure 21.1 and Chapter 21 for further details.

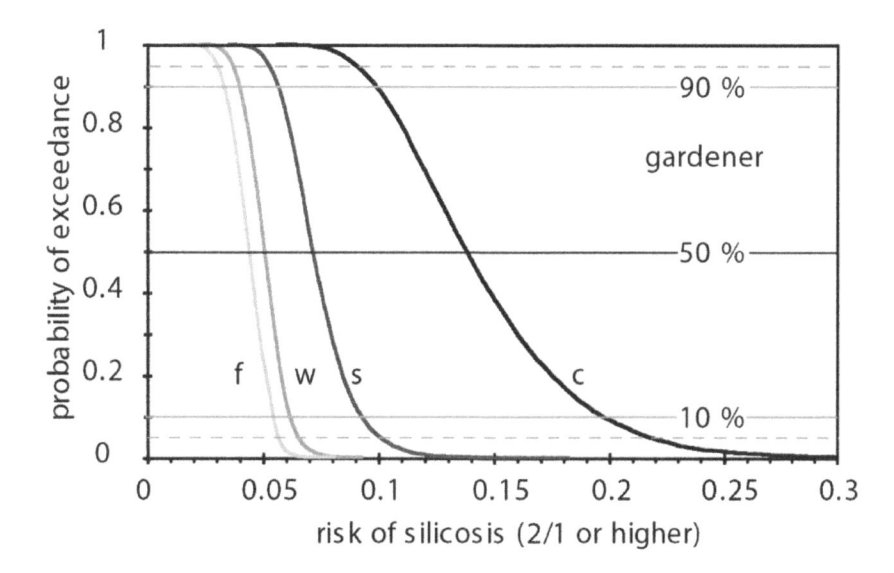

Figure 2.27 Probability of exceedance curve for risk of silicosis (classification ≥2/1) for gardeners, calculated from simulated cumulative exposures: see Hincks et al. (2005) for key to curves, which are for specific Montserrat locations. The study involved Monte Carlo sampling of probability distributions for key factors determining exposure to very fine respirable ash. Four curves are shown for different locations on Montserrat. In the location closest to the volcano the risk to gardeners exceeds air quality standards and risk of silicosis is at levels that cause concern to the authorities, resulting in precautionary measures. This figure can also be seen in Chapter 13 as Figure 13.2.

2.6.6 Global and regional assessment

There is also a need for more synoptic assessment of volcanic hazard and risk on global, regional and country scales. This scale of assessment provides a basis for identifying gaps in knowledge, prioritising resources on the highest risk volcanoes and assessing the overall volcanic risk in regions and countries.

Vulnerabilities to various volcanic hazards can be assessed in a wide variety of ways. The vulnerability of communities is sometimes based on indices such as the Human Development Index (HDI) – a composite measure of development and human well-being, the assumption being that higher levels of development enhance capacity to recover. The Human Vulnerability Index is also used (1-HDI). Vulnerabilities are diverse and complex and a continuing challenge to incorporate effectively into risk assessment and analysis.

A Volcano Hazard Index (VHI) has been developed to characterise the hazard level of volcanoes based on their recorded eruption frequency, modal and maximum recorded VEI levels and occurrence of pyroclastic density currents, lahars and lava flows. The full methodology is summarised in Chapter 22. The index builds on previous index approaches (Ewert et al., 2005, Aspinall et al., 2011). The supplementary online report (Appendix B) is a compendium of regional and country profiles, which use these indices to identify high-risk volcanoes.

The VHI is too coarse for local use, but is a useful indicator of regional and global threat. VHI can also help identify knowledge gaps. The VHI can be modified for volcanoes as more information becomes available and if there are new occurrences of either volcanic unrest, or eruptions, or both. There are 328 volcanoes with eruptive histories judged sufficiently comprehensive to calculate VHI and most of these volcanoes (305) have had historical eruptions since 1500 AD. There are 596 volcanoes with post-1500 AD eruptions, so the VHI can currently be applied to about half the world's recently active volcanoes. A meaningful VHI cannot be calculated for the remaining 1,223 volcanoes due to lack of information in the eruption record. The absence of thorough eruptive histories for most of the world's volcanoes makes hazard assessments at these sites particularly difficult, and this is a major knowledge gap that must be addressed. Research is ongoing investigating the use of the well known, classified volcanoes, to inform the hazard assessment at the poorly known, unclassified volcanoes, for example through the identification of analogous volcanoes. However, this would be associated with significant uncertainties and will not substitute for a thorough understanding of individual systems.

The Population Exposure Index (PEI) is an indicator based on populations within 10, 30 and 100 km of a volcano, which are then weighted according to evidence on historical distributions of fatalities with distance from volcanoes. It effectively amalgamates the Volcano Population Index values at fixed distances given in the VOTW4.0 and uses evidence from historic fatalities to derive a single value. The methodology [Chapter 4] extends previous concepts (Ewert & Harpel, 2004; Aspinall et al. 2011). Volcano population data derived from VOTW4.0 is used to calculate PEI, which is then divided into seven levels from sparsely to very densely populated areas to provide an ordinal ranking indicator. The PEI provides an indication of direct risk to life and can be used as a proxy for economic impact based on the distance from the volcano. However, this does not account for indirect fatalities caused by secondary impacts such as famine and disease or far-field economic losses to aviation and agriculture caused by the dispersion of volcanic ash, gas and aerosols.

Here VHI is combined with the PEI to provide an indicator of risk, which is described as Risk Levels I to III, with increasing risk. The essential aim of the scheme is to identify volcanoes that are high risk due to a combination of high hazard and population density. 156, 110 and 62 volcanoes classify at Risk Levels I, II and III respectively. In the country profiles (Appendix B) plots of VHI versus PEI provide a way of understanding volcanic risk. Indonesia and the Philippines are plotted as an example (Figure 2.28). Volcanoes with insufficient information to calculate VHI should be given serious attention and their relative threat can be assessed through PEI. Being unclassified does not mean a volcano is not hazardous. There are 288 unclassified volcanoes with a high PEI (PEI5-7), indicating volcanoes of potentially high risk.

The calculation of VHI and Risk Levels from PEI also enables knowledge gaps to be identified and provides a benchmark with which to measure progress in improving knowledge of the status of hazard knowledge on the world's volcanoes.

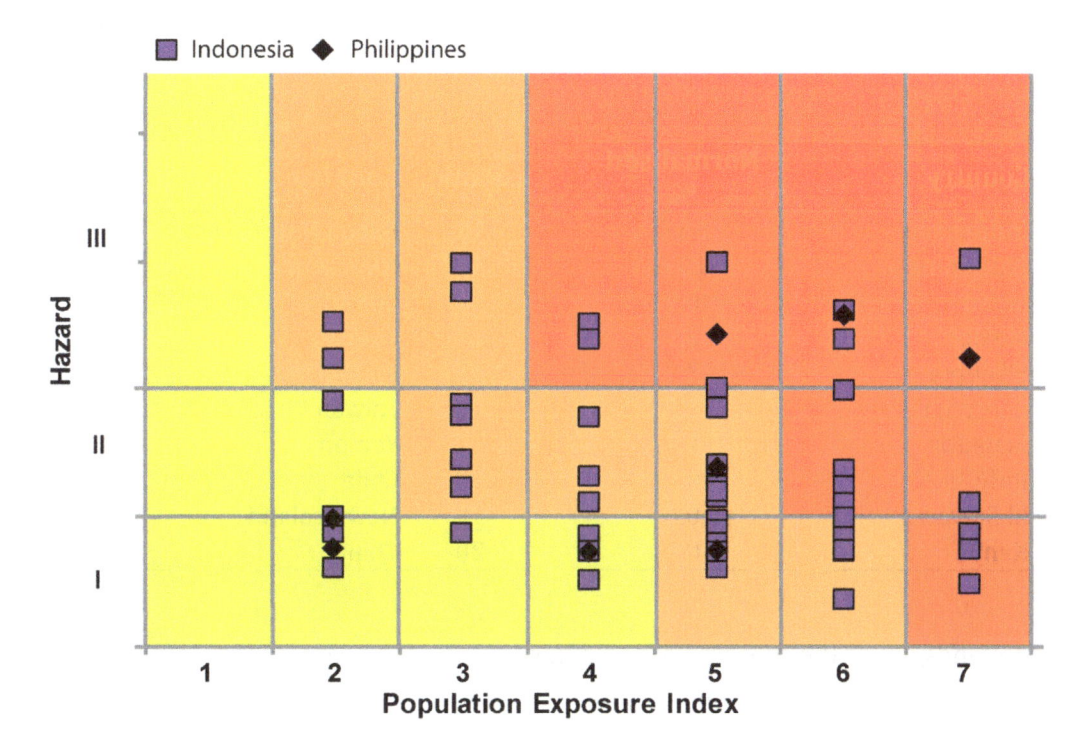

Figure 2.28 Plot of volcanic Hazard Index (VHI) and Population Exposure Index (PEI) for Indonesia and the Philippines, comprising only those volcanoes with adequate eruptive histories to calculate VHI. The warming of the background colours is representative of increasing risk through Risk Levels I to III.

2.6.7 Distribution of volcanic threat between countries

In this section the distribution of volcanic threat (potential loss of life) is investigated to help understand how volcanic threat is distributed and to identify countries where threat is high. 'Risk' requires assessment of vulnerability, which has not been assessed here, therefore the term 'threat' is used as a combination of hazard and exposure. Two measures have been developed, combining the number of volcanoes in the country, the size of the population living within 30 km of active volcanoes (Pop30) and the mean hazard index score (VHI). Population exposure is determined using LandScan (Bright et al., 2012) data to calculate the total population within a country living within 30 km of one or more volcanoes with known or suspected Holocene activity. Countries are ranked using the two measures. Each measure focuses on a different perspective of threat. The full methodology and results are presented in Chapter 23.

Measure 1 is of overall volcanic threat country by country based on the number of active volcanoes, an estimate of exposed population and average hazard index of the volcanoes.

Measure 1 = mean VHI x number of volcanoes x Pop30 (1)

The sum of Measure 1 for all countries is itself a simple measure of total global volcanic threat. The distribution of threat between countries can be evaluated and countries can be placed in rank order using a normalised version of Measure 1. Table 2.2 shows the distribution of Measure 1 between the 20 highest scoring countries.

Table 2.2 The top 20 countries with highest overall volcanic threat. The normalised percentage represents the country's threat as a percentage of the total global threat.

Rank	Country	Normalised %	Rank	Country	Normalised %
1	Indonesia	66.0	11	Papua New Guinea	0.4
2	Philippines	10.6	12	Nicaragua	0.4
3	Japan	6.9	13	Colombia	0.4
4	Mexico	3.9	14	Turkey	0.4
5	Ethiopia	3.9	15	Costa Rica	0.3
6	Guatemala	1.5	16	Taiwan	0.2
7	Ecuador	1.1	17	Yemen	0.2
8	Italy	0.9	18	Chile	0.2
9	El Salvador	0.8	19	New Zealand	0.2
10	Kenya	0.4	20	China	0.2

Indonesia stands out as the country with two thirds of the share of global volcanic threat due to the large number of active volcanoes and high population density. Table 2.3 shows the distribution of threat by region to provide a broader picture of global distribution of volcanic threat. The results are compared with the ranking of these regions based on known historical fatalities (Auker et al., 2013).

Table 2.3 Regional ranking using Measure 1 and known historical fatalities with the ten largest disasters removed. Following Auker et al. (2013) the regions used here comprise only the countries or territories named, allowing for comparison of ranks with the fatality data.

Measure 1 rank	Region* (Country)	Fatalities rank
1	Indonesia (Indonesia)	1 (=)
2	Philippines and China (Philippines, SE China)	3 (-1)
3	Japan (Japan)	6 (-3)
4	Mexico and Central America (Costa Rica, El Salvador, Guatemala, Mexico, Nicaragua)	4 (0)
5	Africa and Red Sea (Cameroon, DRC, Ethiopia, Tanzania)	9 (-4)
6	South America (Chile, Colombia, Ecuador, Peru)	7 (-1)
7	Mediterranean (Italy, Greece, Turkey)	5 (+2)
8	Melanesia (Papua New Guinea, Solomon Islands, Vanuatu)	2 (+6)
9	New Zealand to Fiji (New Zealand, Tonga)	11 (-2)
10	North America (Alaska, Canada, USA-contiguous states)	12 (-2)
11	Atlantic Ocean (Azores, Canary Islands, Cape Verde)	10 (+1)
12	Kuril Islands and Kamchatka (Russia)	14 (-2)
13	Indian Ocean (Comoros, French territories)	15 (-2)
14	Iceland (Iceland)	16 (-2)
15	West Indies (Martinique and Guadeloupe, Montserrat, St. Vincent and the Grenadines)	8 (+7)
16	Hawaii (Hawaii)	13 (+3)

Measure 1 is an overall measure of threat distribution and may be misleading because individual countries may vary considerably in the proportion of their population that is exposed to volcanic threat as nation states vary greatly in size and in their populations from, for example, China with 1.3 billion people (<1% exposed) to St. Kitts and Nevis in the Caribbean with only 54,000 people (100% exposed). Thus we need a measure of threat that reflects how important volcanic threat is to each country. Volcanic threat is very much higher in small island nations with active volcanoes than in larger countries. Measure 2 was developed to rank the importance of threat in each country that is independent of the country's size, so numbers of volcanoes and exposed population numbers are not included in the calculation. Measure 2 is defined:

$$Measure\ 2 = \frac{Pop30}{TPop} \times Mean\ VHI \qquad (2)$$

The top 20 countries according to this measure are listed in Table 2.4.

Table 2.4 The top 20 countries or territories ranked by proportional threat: the product of the proportion of the population exposed per country and the mean VHI.

Rank	Country	Rank	Country
1	UK-Montserrat	11	Guatemala
2	St. Vincent & the Grenadines	12	Sao Tome & Principe
3	France – West Indies	13	Spain – Canary Islands
4	St. Kitts & Nevis	14	Grenada
5	Dominica	15	Vanuatu
6	Portugal – Azores	16	Nicaragua
7	St. Lucia	17	Samoa
8	UK – Atlantic	18	USA – American Samoa
9	El Salvador	19	Armenia
10	Costa Rica	20	Philippines

Here the countries identified are those that have very high overall vulnerability to volcanic hazards and are completely different to the rankings using Measure 1. They are a collection of small island states and small countries. Ranking on a broader regional basis using Measure 2 is shown in Table 2.5.

Table 2.5 Ranking by region using Measure 2. Note the Kuril Islands region is not included.

Relative risk rank	Region	Relative risk rank	Region
1	West Indies	10	Philippines & SE Asia
2	Mexico & Central America	11	Indonesia
3	Atlantic Ocean	12	Japan, Taiwan, Marianas
4	Africa & Red Sea	13	Iceland & Arctic
5	New Zealand to Fiji	14	Alaska
6	Melanesia & Australia	15	Hawaii & Pacific
7	Mediterranean & West Asia	16	Kamchatka & Mainland Asia
8	Middle East & Indian Ocean	17	Canada & Western USA
9	South America	18	Antarctica

Again the ordering of regions is completely different with the West Indies at the top using Measure 2 and near the bottom using Measure 1.

There is no suggestion which of these different country and regional rankings should be preferred. They are simply providing contexts and answers to different perspectives and questions. If the issue is to identify where most volcanic threat is concentrated then SE Asia and East Asian countries like Indonesia, the Philippines and Japan have a large share of the total global volcanic threat. If the question is which countries and regions, irrespective of size, are most vulnerable to volcanic hazards then the West Indies and small nation states are indicated, where the potential losses could be most significant in the context of the country's size.

2.7 Volcanic emergencies and disaster risk reduction

Volcanic eruptions vary in type, frequency and magnitude, and occur over quite variable periods of time (days to years). Compared to other natural hazards such as the passage of a hurricane, volcanic emergencies can be prolonged with potentially a series of impacts caused by different primary and secondary volcanic hazards. Importantly though, volcanic unrest may precede an eruption giving some early warning if a monitoring capability and responding institutions are in place. Likewise, monitoring data can be used to forecast imminent increases in hazardous activity once an eruption has begun. Volcanic eruptions do not respect national borders and frequently impact several different countries in different ways; for example eruptions at Chilean volcanoes frequently affect neighbouring Argentina, primarily through ash dispersal (Appendix B; Viramonte et al. 2001). Scientists, regulators and emergency managers need to coordinate their activities with other nations in such situations, adding another layer of complexity to emergency activities. Establishing that an eruption is over can be challenging and the end of an eruption does not necessarily imply a lack of hazard. Some secondary hazards (e.g. lahars) may continue for years post-eruption thus requiring continued mitigation and response efforts (e.g. post-eruptive lahars of Pinatubo [Chapter 7]), and some volcanoes are persistently active so the risks arising from primary and secondary volcanic hazards have to be continually managed. In the majority of cases, volcanic emergencies have to be managed in the face of considerable uncertainty.

The official responsibility for volcanic risk management and risk reduction at a societal level usually lies with government agencies, but to be effective also relies on the engagement of communities (including the private sector and NGOs) and individuals. Evacuations are called by these authorities following short-term forecasts and early warning from scientists. During volcanic eruptions evacuations may be short-lived or prolonged, both affecting lives and livelihoods. In some cases towns, villages, land and infrastructure may be completely destroyed requiring resettlement and resulting inevitably in a long and protracted period of recovery. Based on priorities and capacities, each individual has a different tolerance to risk and this may in some cases differ substantially from the tolerances considered acceptable for society by civil authorities. For example, some may self-evacuate long before official calls to do so, others may resist evacuation in order to maintain assets or care for livestock.

2.7.1 Role of scientists

The primary role of scientists (volcanologists) in risk management and risk reduction is to provide timely and impartial information, volcanic hazards assessments, and both long- and short-term eruption forecasting to the civil authorities so they can make effective risk-based decisions, for example, about evacuation. In practice, especially if a volcano has been dormant for a long period and accounting for staff turnover among the responding authorities, the scientists may need to provide basic knowledge about the potential hazards and impacts of volcanic eruptions based on lessons learnt at previous eruptions. Good scientific decision-making, effective hazards assessments and forecasts depend not only on good leadership, but also on the capabilities of scientists, the availability of reliable data and often on supportive national and international scientific networks [Chapters 7 and 10]. To turn good science into effective disaster risk reduction requires good communication, strong long-term relationships, cooperation and coordination amongst stakeholders, effective public engagement and ultimately on the capacity of communities to respond.

Volcano observatories are by necessity connected to a place since volcanoes don't generally move and likely areas of impact are known in advance if appropriate geological studies have been carried out. This means scientists and technicians are active within the communities they serve, enhancing the potential for long-term relationship building, knowledge exchange, good communication and joint activities in resilience building and risk reduction.

The 1976 volcanic emergency at La Soufrière, Guadeloupe was a pivotal event in highlighting the challenges of effective communication and the issue of trust in scientists, especially under conditions of uncertainty brought about by limited monitoring [Chapter 8] (Hincks et al., 2014). Here a publically expressed lack of consensus by scientists led to a loss of trust. There was no comprehensive monitoring network prior to the 1976 crisis, limited knowledge of the eruptive history of the volcano, large uncertainties in the interpretation of available scientific data, and awareness of devastating Caribbean eruptions in the past. Following the controversial management of the 1976 eruption (a large-scale evacuation of the capital city with no subsequent major eruption), a major effort in disaster risk reduction began in the area around La Soufrière. A dedicated volcano observatory was established and new methods in hazard and risk assessment are being developed alongside cost-benefit analysis in support of pragmatic long-term development and risk mitigation.

Since this episode, other high-profile volcanic eruptions showed that some of the issues experienced in Guadeloupe tended to recur, but there were also examples of very good scientific practice. The International Association of Volcanology and Chemistry of the Earth's Interior (IAVCEI), the professional body for volcanologists, published a protocol on the behaviour of scientists working in volcanic emergencies (Newhall et al., 1999). The protocol highlights some key issues and offers guidelines as to how international scientists could support rather than hinder in-country scientists before, during and after eruptions. One of the key principles of this 'IAVCEI protocol' is that there should be a 'single message' established by the official in-country scientific institution (e.g. volcano observatory), often in consultation with others. This might take the form of notices or reports compiled by multiple scientists or agencies. Any disagreement should be handled among scientists themselves and be incorporated in the advice (e.g. within measures of uncertainty) if appropriate. Whenever potentially conflicting material is produced, with or without the knowledge of an official institution (e.g. volcano observatory), it undermines in-country scientists and their relationships with authorities and the public. This can result in ineffective risk reduction measures by the authorities and hence puts lives at stake.

Litigation is emerging as a major concern for scientists providing advice on volcanic risk, following the conviction of seven scientists in the L'Aquila earthquake trial in Italy. Governments will likely need to provide re-assurance to scientists who are willing to serve the community in emergencies. The ongoing debate on the implications of the L'Aquila trial within the scientific community may well lead to new protocols and guidelines for scientists working on volcanic emergencies. We note that the conviction of six of the scientists was overturned in appeal in 2014.

2.7.2 Alert levels

One of the main functions of a volcano observatory is to provide early warning to communities and authorities. Early warnings are needed both for hazards on the ground and airborne hazards. Volcanic Alert Levels (Fearnley, 2013, Potter et al., 2014) are commonly used as a means to rapidly communicate the status of volcanic activity to those who need it around the volcano, in a simple and understandable manner. They differ in the number of levels, the types of labels used (e.g., using colours, numbers, words or symbols), and their emphasis on unrest vs. eruptions. Some systems incorporate or are closely associated with response actions (e.g. evacuation), depending on the roles and responsibilities between scientists and authorities in each country. There is variation in the amount of forecasting language included in alert level systems. Alert level systems need to be effective for local communities and emergency responders, as well as for the scientists who typically set the levels (Chapter 16). There is also an established International Aviation Colour Code system which, although optional, is specifically intended to aid communication between volcano observatories and Volcanic Ash Advisory Centres.

The use of an alert level system is exemplified by the 2010 eruption of Merapi [Chapter 9]. Merapi is one of the most active volcanoes in Indonesia. The 2010 Merapi eruption affected two provinces and four regencies and led to evacuation of about 400,000 people (Surono et al., 2012). Indonesia applies four levels of warnings for volcano activity (Figure 2.29). From the lowest to highest: at Level I (Normal), the volcano shows a normal (background) state of activity; at Level II (Advisory) visual and seismic data show significant activity that is above normal levels; at Level III (Watch) the volcano shows a trend of increasing activity that is likely to lead to eruption; and at Level IV (Warning) there are obvious changes that indicate an imminent and hazardous eruption, or a small eruption has already started and may lead to a larger and more hazardous eruption. At Level III people must be prepared for evacuation and at Level IV evacuations are required.

	ALERT LEVEL	DATES	RADIUS	ERUPTION
DECREASING	NORMAL	15-9-2011		
	ADVISORY	30-12-2010		
	WATCH	3-12-2010		
		4-11-2010	20 KM (11:00 UTC)	4 Nov. 17:05 UTC (16,5 km)
INCREASING		3-11-2010	15 KM (08:05 UTC)	3Nov. 08:30 UTC (9 km)
	WARNING	25-10-2010	10 KM (11:00 UTC)	26-10-2010 (10:02 UTC)
	WATCH	21-10-2010		
	ADVISORY	20-9-2010		
	NORMAL	17-9-2010		

Figure 2.29 (Previous page) The chronology of warnings and radius of evacuations used during the 2010 eruption of Merapi, Indonesia (time increases from the bottom of the diagram upwards (see Chapter 10). The four alert levels are indicated by colour: green (I); yellow (II); orange (III); and red (IV).

During the crisis, there was rapid escalation of seismicity, deformation and high rates of initial lava extrusion [Chapter 9] (Surono et al., 2012). All monitoring parameters exceeded levels observed before and during previous eruptions of the late twentieth century. This raised concerns of an impending much larger hazardous event. Satellite monitoring (radar) provided an additional and valuable tool to establish the rapidly changing topography at the summit of the volcano. Consequently, a Level IV warning was issued and evacuations were carried out within 10 km of Merapi's summit. The exclusion zone was then extended to 15 and then to 20 km as the eruptive activity escalated. Each evacuation was followed within hours by devastating pyroclastic density currents that travelled to increasing distances. This very effective anticipation of hazard and risk was possible due to the combination of effective real-time monitoring, effective in-country institutions with strong relationships, communications and a well-practiced response system, good interaction with communities and good international scientific support networks. About 300 people died because they either would not or could not evacuate, but between 10,000 and 20,000 were saved by warnings and evacuations.

2.7.3 Effective communication and relationships

Communication is a critically important aspect of volcanic risk reduction [Chapter 24]. It has been shown repeatedly that networks of responding institutions with recognised roles and responsibilities (in the form of response protocols) must be established well before volcanic unrest and eruption if there is to be an effective response. Ideally, regular activities and exercises between a volcano observatory, the authorities and communities at risk are needed to maintain relationships, trust, knowledge and to develop a common language. Volcanic eruptions are complex and may have multiple outcomes so all potentially affected sectors (from industry to households) need to develop some understanding of the implications for their area of responsibility in advance. Volcano observatories often have a significant educational role in terms of discussing hazards and their potential impacts with authorities and the public, ideally this is targeted to different audiences. Scientific data products and knowledge need to be provided in formats of value and of use to stakeholders, so often both scientists and non-scientists need to think in different ways in order to identify what is needed at different times and in different situations. For example, authorities may appreciate maps but the public may prefer 3D imagery. The development of user-friendly scientific products is an iterative process requiring long-term dialogue and mutual understanding.

During an emergency situation decisions must be made quickly and under pressure. This is too late to be learning about hazards and risks and to get to know key stakeholders, their roles, responsibilities and needs. Ideally, a nation will establish an 'emergency committee' capable of handling a range of risks but with specific experts identified for volcanic hazards and risks and this committee will meet regularly during a crisis situation to facilitate communication across sectors.

To be effective the authorities responding to an emergency must be confident in evaluating specialist advice. However, most decision-makers are unfamiliar with the scientific limits of forecasting volcanic behaviour and with the scale of disruption and damage that eruptions can produce. At the same time, scientists may be unfamiliar with how the demand for, and use of, scientific information is shaped by the needs of the user and by the political and institutional contexts in which decisions are made (Haynes et al., 2008b, Solana et al., 2008, Barclay et al., 2014). So, an ongoing dialogue between scientists and officials is essential in order to maintain mutual understanding. Commonly there is rapid turn-over of officials and decision-makers due to changes of government or structures of governance so this need for raising awareness is necessarily permanent.

The media and in particular social media are playing an increasing role in informing and updating populations on any event that takes place. The media can be an effective risk management tool, but again interaction with the media requires planning and management, for example through press offices or even dedicated media centres and the distribution of materials specifically for media uptake. Some information in the media will be erroneous and so the media needs to be handled proactively to reduce misinformation. Engagement with the media can take up considerable amounts of time during an emergency for both scientists and authorities, particularly if an eruption has cross-border impacts and needs to be planned in advance. During the 2010 eruption of Eyjafjallajökull volcano, the demand for information from scientists and the authorities was so extreme that two media centres were established in Iceland and regular press briefings and releases were organised and issued. Communications through the media and internet, and with local communities (through local media, community groups and public meetings) were ultimately very effective, but it was recognised that tourists are one group that is challenging to reach (Bird et al., 2010), especially if they do not actively seek information (e.g. from tourist information centres). A recent initiative in Iceland calls on tourists to register their mobile phone numbers so they can receive SMS texts in case of volcanic unrest.

2.7.4 International collaboration

There has been a long history of international collaboration around volcanoes and volcanic eruptions. In part this is because experience and lessons learnt from one emergency can often be applied to some extent at the next and sharing of this experience is considered critical to effective global progress in volcanic disaster risk reduction. The internet, together with open access journals and reports, are now facilitating this sharing of experience. International partners can also provide equipment, specialist expertise and extra hands during a crisis. There are several examples of international scientific collaboration that have led to effective disaster risk management sometimes through the application of novel techniques (e.g. Chapters 7, 10 and 11). Opportunities for scientists to engage across regions and internationally in collaborative and coordinated cross-disciplinary research have helped to support progress and to ensure research is integrated into operations [e.g. Chapter 19]. Regional scientific collaborations in the Pacific region, Europe and Latin America to support science and Disaster Risk Reduction in volcanic hazards are proving highly effective and productive initiatives.

The Volcano Disaster Assistance Program (VDAP) of the US Geological Survey has, for almost three decades, supported scientists and institutions in many countries during volcanic emergencies [Chapter 25]. VDAP adopts a strict policy of only coming in when invited through formal government mechanisms. VDAP has assisted at some of the major eruptions of the last few decades, including those of Pinatubo, Philippines in 1991, Soufrière Hills Volcano, Montserrat in 1995, and Merapi, Indonesia in 2010. Consortia of international scientists have also supported some volcanic emergencies [Chapter 11].

Experience has also shown that it is essential that an identified scientific group within an affected country lead the scientific response and are not undermined or contradicted by external scientists. The IAVCEI protocol (Newhall et al., 1999) went some way to addressing this concern (Section 2.7.1), but it remains possible for a scientific institution dealing with the demanding operational aspects of an eruption response to be overlooked by researchers keen to take advantage of a unique and time-limited research opportunity. Both opportunistic research and an effective operational response are needed but research before, during and after a crisis must be carried out with the knowledge and engagement of the official scientific institution. This ensures that there is no duplication, enhances the potential impact of the research and may enable negotiated access to additional datasets and research networks. For a volcano observatory, engagement with researchers can facilitate access to novel methods, provide more data to aid operational activities or conceptual understanding, and may facilitate the timely communication of scientific progress.

The global network of VAACs operating under ICAO guidelines [Chapter 14] provides a strong framework within which communications with volcano observatories can be practised and standardised reports can be produced. There is as yet no other formal standardised reporting system of global extent for volcanic unrest and eruption. The notices issued to the aviation sector are not available at the global scale, although there is now a clearly recognised demand for a global resource identifying the status of the world's volcanoes and potential threats to aviation and other sectors. The accessibility of such knowledge and enhanced awareness of volcanic unrest and activity is essential to improve preparedness for and mitigation of risks and to reduce losses. The Smithsonian Institution currently compiles weekly and monthly reports from volcano observatories and intends to increase this frequency to daily reports. Crucial to this effort of daily reporting will be voluntary contributions of observatory reports and forecasts of activity, and later validation of data and information by observatory staff.

Volcanic eruptions do not respect national boundaries and can thus impact adjacent nations equally, but the presence of different regulatory frameworks, either for civil aviation or development, can lead to inconsistent response and planning. The 2010 eruption of the Eyjafjallajökull volcano in Iceland, provided significant challenges across Europe. For example, the UK government had no planning in place for volcanic eruptions but the existing relationships (and regular exercises) between the London VAAC and the Icelandic Meteorological Office (volcano observatory) and between the British Geological Survey and the University of Iceland enabled a rapid ad-hoc response at government level. As the eruption progressed, a Memorandum of Understanding was signed between responding scientific organisations in the UK and Iceland to ensure the open sharing of data in support of emergency response and to support capacity building in both countries. This underpins ongoing collaborative activities.

2.7.5 Training and capacity

Some volcano observatories are fortunate enough to be linked to national institutions or a university that can provide training opportunities and career progression, but a common challenge at volcano observatories is to ensure that scientists and technicians are not overwhelmed by operational and technical demands and have sufficient time to develop their research and skills. Training and capacity building is one area where external support can be very useful, but of course it should respond to a needs-analysis led by the observatory itself.

Collaborative research projects can be extremely valuable for volcanologists and may lead to the application of new methodologies, ideas and techniques at volcano observatories. However, they can also be a great drain on observatory resources and so should include provision for the engagement of observatory scientists in the research as partners and the potential need for technician time and expertise should not be forgotten. Turning any research into a long-term operational capacity at a volcano observatory may be challenging given that observatory resources are already stretched and such activities also require funding. Training in the use of new equipment or technology, or in new methodologies (e.g. social science approaches) may be very welcome and can also be included in research project budgets. Research projects may leave monitoring equipment but the need for continued maintenance in the field, data acquisition and real-time interpretation can for example be a prohibitive long-term cost.

Volcano observatories typically take part in regular exercises to test operational responses with VAACs, civil authorities and communities. Such exercises can simply test existing communication lines or can test responses to more complex situations (e.g. scenario planning). The lessons learnt from such exercises are critical to building capacity, increasing resilience and continually improving preparedness and response.

Merapi [Chapter 10] is well known for a capacity building programme named *wajib latih* (mandatory training) required for people living near the volcano. The aim of this activity is to improve hazard knowledge, awareness and skill to protect self, family and community. In addition to the *wajib latih*, people also learn from direct experience with volcanic hazards, which at Merapi occur frequently. However, the 2010 Merapi eruption showed that well-trained and experienced people must also be supported by good management, and that training and mitigation programmes must consider not only "normal" but also unusually large eruptions (Mei et al., 2013). Another example of building capacity [Chapter 16] is provided by improvements to lahar warnings and hazard information for visitors to the ski areas on Mt Ruapehu, New Zealand (Figure 2.22). The assessment of multiple simulated events indicated potential actions aimed at improving future responses, such as increasing ski area staff training and improving hazard signage (Leonard et al., 2008). The communication tools were improved by repeating these activities annually and tracking perceptions of visitors through time in response to real events (Potter et al., 2014).

Cost-benefit analysis (Woo, 2014) is proving an increasingly valuable tool to support constructive dialogue between scientists and civil authorities on the merits of volcano monitoring and the management and mitigation of volcanic risks where it is currently minimal or lacking. Effective monitoring may, for example, help avoid unnecessary and costly evacuations and may support good risk management practice in the private sector. In some cases, existing monitoring capabilities can be harnessed to help monitor volcanoes (e.g. earthquake monitoring, groundwater monitoring, air quality monitoring).

2.7.6 Risk management

Risk management is usually led and implemented by relevant government authorities at different levels (national to local) with active response to disasters led by civil protection or civil defence and partners across sectors. Risk mitigation requires recognition of risk and allocation of budgets for mitigation measures, again at national to local scales. Effective management and mitigation of risk includes establishment and practice of early warning systems, the maintenance of effective political, legal and administrative frameworks, land use planning and efforts to influence the behaviour of populations at risk.

There are a variety of different disaster risk management options open to authorities. Attempts to reduce the hazard are rare, reflecting that this is in most cases not possible, but there have been some examples of engineering measures such as lava flow diversion and lahar barriers that have had some effect. Exposure can be reduced directly in the short-term through evacuation of people and can be reduced long-term by development of new assets in geographic areas of lower risk and by transferring existing assets to areas of lower risk. Vulnerability of individuals and communities is complex and diverse, reflecting in part ability to cope with disruption to lives and livelihoods and to understand and respond effectively to warnings. Improving the resilience of communities and society as a whole may include increasing awareness of hazards, effective planning at individual to national scale and enhancing capacity to adapt in the face of risks. Effective early warning and forecasts from volcanologists can, in combination with effective emergency management, good communication and participatory approaches, contribute to timely evacuation and hence reduction in exposure and vulnerability. In long-lived eruptions and at frequently erupting volcanoes there is a need to adapt and live alongside volcanoes, requiring careful identification of evolving risks alongside effective risk management. The need for different sectors to work together requires long-term relationships, trust and a common language to ensure effective communication (Haynes et al., 2008a).

During a volcanic crisis, civil authorities and scientists are under immense pressure and must make decisions in short time-frames and often with limited information. Decisions about evacuation, for example, may be based on pre-defined thresholds and probabilities. The example of the 1976 eruption in Guadeloupe [Chapter 8] in Section 2.7.1 shows what can go wrong when an evacuation is called with minimal preparation or planning. One effective way for civil authorities and scientists to work together to support risk management is to combine hazards and risk assessments with cost-benefit analysis. There are trade-offs involved in taking mitigating action in the interests of public safety that can be analysed within the economic decision framework of cost-benefit analysis (Leonard et al., 2008, Marzocchi & Woo, 2009) and this has been applied at Naples (Chapter 6). Another example of an analysis of the costs and benefits of evacuation has been carried out at Auckland, New Zealand (Chapter 5). Cost-benefit analysis of evacuation does raise some difficult issues, such as the value of human life, but can be used to support any aspect of decision-making not just evacuation, such as land use planning and the establishment of monitoring capability. Importantly it can be done before any crisis develops and allows difficult topics to be considered outside an emergency situation.

Experience and lessons learnt inevitably improves risk management and this is particularly evident around volcanoes with long-lived eruptions. The Tungurahua volcano in Ecuador erupted in 1999 after decades of inactivity. Thousands were forcibly evacuated for three months leading to acrimony between the authorities, the community and scientists. A more collaborative approach to risk management was quickly adopted in which community volunteers (now 25 of them) became key players in the official communication network. They use VHF radios to share volcano observations with the volcano observatory on a daily basis, and also manage sirens and facilitate local evacuations (with support from Civil Defence). The network has been sustained largely on a voluntary basis thanks to commitment from all parties for over 14 years and has resulted in several effective evacuations since 2000 (Chapter 26). The communities themselves take account of the most vulnerable individuals in their evacuation planning and many have alternative agricultural land and homes (allocated as part of the risk management procedures) to which they can retreat during periods of elevated volcanic activity.

Threat to livelihoods has been identified as a critical risk driver in many cases and during volcanic eruptions there have been many cases of farmers returning to evacuated land to care for livestock and harvest crops (Loughlin et al., 2002), or business owners returning to retrieve capital assets. Safeguarding livelihoods by providing alternative land or opportunities can significantly contribute to disaster risk reduction. This is linked to planning and development both of which need to be closely connected to disaster risk reduction activities.

Long-lived eruptions also demonstrate the critical difference between intensive and extensive risk. Intensive risk may be extreme and short-lived leading to evacuation and disaster risk management activities. In long-lived eruptions, however, populations may be exposed to low levels of semi-continuous ash fall for weeks or months, which do not result in specific risk management or risk reduction measures. Volcanic ash fall can have many impacts on infrastructure, agriculture, environment, water, transport and also on psychological well-being but these are not characterised holistically and thus may not lead to an identifiable economic impact.

In practice, in many places, individuals have their own risk tolerance thresholds and will tend to act upon that. One significant challenge for individuals relying on such an approach in close proximity to an active volcano is a tendency to require visual or audible signals before action, even though it is well known that dangerous hazards such as pyroclastic density currents and lahars may give no audible signal and may go unnoticed if one is not watching the volcano (Loughlin et al., 2002). Even a large explosion may go unheard on the flanks of a volcano but will be heard clearly much farther away.

More participation of communities in risk assessment, risk management and risk reduction can have considerable benefits to the community (Kelman & Mather, 2008, Usamah & Haynes, 2012) and can influence the psychological and sociological aspects of risk, it can also benefit scientists (increase in trust) and civil authorities (more efficient and timely response). For example there is evidence that uncertainties may be better understood and there is more acceptance of risk reduction actions taken in the face of uncertainty.

2.7.7 Planning and preparedness

A milestone in international collaboration for natural disaster risk reduction was the approval of the "Hyogo Framework for Action 2005-2015: Building the Resilience of Nations and Communities to Disasters" (International Strategy for Disaster Reduction, 2005: Hyogo Framework for Action 2005-2015). This document, which was approved by 164 UN countries during the World Conference on Disaster Reduction in Kobe, January 2005, clarified international working modes, responsibilities and priority actions for the following 10 years. Here we discuss volcanic risk through the prism of these five HFA priorities for action:

1. Ensure that disaster risk reduction is a national and local priority with a strong institutional basis for implementation.
2. Identify, assess and monitor disaster risks and enhance early warning.
3. Use knowledge, innovation and education to build a culture of safety and resilience at all levels.
4. Reduce the underlying risk factors.
5. Strengthen disaster preparedness for effective response at all levels.

Priority for Action 1 states that each nation has the prime responsibility for preventive measures to reduce disaster risk, and is expected to take concrete actions. This principle can be interpreted as the need to establish an institution or mandate an existing institution to monitor active volcanoes within each country. Ideally every active volcano should have a dedicated monitoring system, but this ideal is unrealistic due to the large resources implied, especially for countries with many active volcanoes. Thus some prioritisation is needed where high-risk volcanoes are identified. Our assessment of monitoring capacity indicates significant improvements around the world in the last 10 years, but some high-risk volcanoes remain unmonitored or poorly monitored and major geological knowledge gaps exist. There is still a lack of a comprehensive data on monitoring capacity, which prevents documenting progress and identifying gaps. This principle also encourages community participation in all aspects of disaster risk reduction.

Priority for Action 2 states that emphasis should be put on making and regularly updating risk assessments at local to national scale, establishing and/or enhancing early warning systems, capacity building and anticipation of regional and emerging risks. There is also emphasis on data sharing, information systems, space-based earth observation and dissemination. There would be great benefits if more effort were put into these areas. Volcanic risk organisations such as WOVO, IAVCEI and GVM provide platforms for assuring quality of information, identifying best practice and protocols in data collection, standardisation of terminology and harmonised approaches to database construction and hazard mapping for example. The consideration of risk at different scales from local to global is of key importance and there are implications for approaches used at different scales. This principle also encourages people-centred early warning and we have presented several examples of good practice in this area between volcano observatories and communities at risk.

Priority for Action 3 focuses on using knowledge, innovation and education to enhance disaster risk reduction at all levels and across disciplines and regions. To a large extent international cooperation to assist in volcanic emergencies is working well in volcanology. The US Geological Survey's VDAP has been exemplary in supporting countries that need assistance and there are several examples of bi-lateral and multi-lateral assistance responses. There are also examples of workshops, short courses and technical training activities organised within the volcanological and Observatory communities at national, regional and global levels that promote technical training, knowledge transfer and sharing of good practise. However, there is a case that international cooperation should be defined in broader terms than this principle suggests. A model of high-income jurisdictions helping low-income countries does not fully reflect the need to support integrated efforts for disaster risk reduction. IAVCEI and GVM provide mechanisms for grassroots actions and coordinated international collaboration. Some volcanic emergencies cross borders so there is a need to support co-operation between countries. Support is needed for the international community to work together to create and maintain databases, collate information, work towards agreed standards and international authenticated methodologies and procedures, and share best practices.

Priority for Action 4 addresses the need to reduce identified risk factors encouraging partnerships, integration across sectors, risk sharing, land-use planning and development among other things. Identification of key risk factors in Action 2 will greatly facilitate the ability to reduce risk. There is certainly great potential for volcanologists to have a greater input into risk reduction activities across sectors, but especially in land use planning and development. Long-term risk assessments can be carried out over large geographic areas if there has been investment in geological and geochronological studies and hazard modelling.

Priority for Action 5 states that greater preparedness will lead to better response at all levels. There is strong evidence that good planning, well-prepared emergency services and a well-informed population with trusted advice from volcanologists and decisions made by the authorities based on this advice will greatly reduce the impact of a volcanic emergency. In densely populated areas around a volcano there is a need for regular review of hazard mitigation strategy, including spatial planning, mandatory disaster training, contingency planning and for regular evacuation drills.

Like many other countries, the UK and Iceland are actively responding to the HFA guidelines in their planning. In Iceland with many active volcanoes there are mature plans to raise awareness and prepare for future eruptions [see Supplementary Case Study 1 in the appendix to Chapter 1]. The eruptions of Eyjafjallajökull in 2010 and Grimsvötn in 2011 drew attention to how an eruption in one country with active volcanoes could affect many other countries. The UK government department handling civil protection in the UK, the Civil Contingencies Secretariat (CCS) of the Cabinet Office, introduced volcanic risks into the National Risk Register (NRR) for the first time [see Supplementary Case Study 2 in the appendix to Chapter 1]. In order to enhance UK preparedness for most types of eruption in Iceland and their distal impacts, two scenarios were included in the NRR based on past events: a small to moderate explosive eruption of several weeks duration (the Eyjafjallajökull eruption) and a large fissure eruption of several months duration (the 'Laki' eruption of Grimsvötn volcano). Hazard assessments based on these scenarios inform contingency planning across government (e.g. transport, health, environment) and at local authority level so the UK should be more resilient to future eruptions in Iceland.

Traditionally, consideration of volcanic risk has focused on loss of life. However, other potential losses such as livelihoods, critical infrastructure, buildings, health, agriculture and environment all benefit from rigorous hazard and risk assessment approaches.

2.8 The way forward

Here we have highlighted the wide range of hazards posed by volcanoes and described their diverse impacts on communities. As with other hazards an approach based on science and technology has developed to anticipate these hazards, increase societal resilience and reduce risk. Many volcanic hazards are localised around a particular volcano and each volcano is to some extent unique. Thus dedicated volcano observatories are at the frontline of emergency management and disaster risk reduction. Observatories and their linked scientific institutions can help communities prepare for future eruptions, can provide early warning and forecasts when a volcano threatens to erupt, and will be at the centre of emergency management during an eruption. There is also increasing recognition that, although science and technological monitoring are vital components of disaster risk reduction for volcanoes, scientists can also contribute to risk management and mitigation, and support the decision-making of individuals and authorities. Building resilience and living with an active volcano requires good communication between scientists, decision-makers, emergency services and the public, effective planning and exercise of emergency responses, development of trust, and understanding of cultural factors that affect community responses. Our study also highlights gaps in knowledge, best practices and shortfalls in capacity.

The benefits of preventive measures are increasingly recognised, locally, at the level of national governments and among international donors and the Hyogo Framework for Action 2005-15 provides an excellent blueprint for disaster risk reduction. Here, we identify three key pillars for the reduction of risks associated with volcanic hazards worldwide and list recommended actions with the underlying principle that volcanologists based in a specific country are the best to lead any national needs-analysis:

Pillar 1: Identify areas and assets at risk, and quantify the hazard and the risk

Without knowledge and characteristics of hazard and risk, it would not be meaningful to plan and implement mitigation measures. Many of the world's active volcanoes have only rudimentary records at best. Also, many of the data that do exist are not in a standardised form and lack quality control. These knowledge gaps can only be closed by systematic geological, geochronological and historical studies and support for international collaborative activities attempting to address issues around data collection, analysis methods and databases.

Action 1.1 The hazard level of many volcanoes is highly uncertain, mostly reflecting the paucity of geological knowledge and in many cases a low frequency or absence of historical eruptions. Those volcanoes with a combination of high uncertainty level and high population exposure index (in this study) should be prioritised for geological studies that document recent volcanic history for a hazard assessment context. Recommended studies include stratigraphy, geochronology, petrology, geochemistry and physical volcanology. Such studies greatly enhance the ability of volcanologists to interpret volcanic unrest and respond effectively when activity begins. In some cases, findings are likely to increase the currently known risk. This work requires government funding to resource geological surveys and research institutions as primary funds are not likely to come from the private sector. However, where there are commercial activities associated with active volcanoes, such as geothermal energy, tourism or insurance potential it would be reasonable to ask for contributions to this base-line work.

Action 1.2 Probabilistic assessment of hazard and risk that fully characterises uncertainty is becoming mandatory to inform robust decision-making. Deterministic approaches cannot fully characterise either hazard or risk, are limited and further can be highly misleading. Assessments and forecasts are typically combinations of interpreting geological and monitoring data and various kinds of modelling. Probabilistic event trees and hazard maps for individual volcanoes are best made by local or national scientists, and we recommend that these be made in advance for high-risk volcanoes. However, some data from beyond the specific volcano in question are also needed for these trees and maps, especially if the volcano in question is poorly known.

Action 1.3 Global databases can serve as references for local scientists, providing analogue data and distributions of likely eruption parameters. Creation and maintenance of global databases on volcanoes, volcanic unrest and volcanic hazards, and quality assurance on data, hazard assessment methods, forecast models, and monitoring capacity are best done through international co-operation. Funding compilation of such databases does not fit easily into national and regional research funding and needs stronger international support.

Action 1.4 Forensic assessments of volcanic hazards, their impact and risk drivers are needed during and after eruptions. Such studies are essential to improve knowledge of hazards and vulnerability in particular and to improve and test methodologies such as forecast modelling based on real observational data. National Governments should be encouraged to support their institutions to include timeline-based analysis of their actions and subsequent impacts, and to report successes and shortcomings of crisis responses. A great deal of valuable information about volcanic disasters is in the form of unpublished and often anecdotal information, so formal publication of post-hoc assessments of emergency responses should be encouraged. Evaluations of "lessons learnt" from past disasters are likewise important to improve future responses and avoid repetition of mistakes.

Action 1.5 Risks from volcanic ashfall associated with a particular volcano or region can be characterised by detailed probabilistic modelling, taking into account the range of physical processes (atmospheric and volcanic) and associated uncertainties. There is also a need to better understand the impacts of volcanic ash, and define thresholds of atmospheric concentration and deposit thickness for various levels of damage to different sectors. We recommend that further analysis be performed for all high-risk volcanoes, to enable more conclusive statements to be made about expected losses and disruption and to support resilience and future adaptation measures.

Pillar 2: Strengthen local to national coping capacity and implement risk mitigation measures

Mitigation means implementing activities that prevent or reduce the adverse effects of extreme natural events. In a broad perspective, mitigation includes: volcano monitoring, reliable and effective early warning systems, active engineering measures, effective political, legal and administrative frameworks. Mitigation also includes land-use planning, careful siting of key infrastructure in low risk areas, and efforts to influence the behaviour of at-risk populations in order to increase resilience. Good communication, education and community participation are critical ingredients to successful strategies. All these measures can help minimise losses, increase societal resilience and to assure long-term success.

Action 2.1 Many active volcanoes are either not monitored at all, or have only rudimentary monitoring. Some of these volcanoes are classified in this study as at high risk. A major advance for hazard mitigation would be if all active volcanoes had at least one volcano-dedicated seismic station with continuous telemetry to a nominated responsible institution (volcano observatory) combined with a plan for use of satellite services. This matches a strategy from space agencies to monitor all Holocene volcanoes and make data available (http://www.congrexprojects.com/2012-events/12m03/memorandum). For volcanoes in repose there are two suggested responses, namely implementation of low-cost systems for monitoring and raising awareness of volcanic hazards and risk among vulnerable populations. Provision of funding to purchase equipment must be complemented by support for scientific monitoring, training and development of staff and long-term equipment maintenance. We recommend this action as a high priority to address volcanic risk.

Action 2.2 Volcanoes identified as high risk should ideally be monitored by a combination of complementary multi-parameter techniques, including volcano-seismic networks, ground deformation, gas measurements and near real-time satellite remote sensing services and products (e.g. satellite-based geophysical change detection systems). We recommend that all high-risk volcanoes should have basic operational monitoring from all four domains. This should be maintained, interpreted and responded to by a nominated institution (volcano observatory). Donations of equipment and knowledge transfer schemes need to be sustainable long-term with respect to equipment maintenance and consumables. Supporting monitoring institutions and sustaining local expertise is essential.

Action 2.3 Technological innovation should strive towards reducing costs of instrumentation and making application of state-of-the-art science as easy as possible so more volcanoes can be monitored effectively. For example, satellite observation offers a new and promising approach to monitoring the world's volcanoes in isolated or remote locations as well as providing additional information to augment ground-based observatory monitoring systems. However, lower costs, easier access, technological training, and better and more timely sharing of data are needed to realise the potential. Many of the new models derived from research of volcanic processes and hazardous phenomena for forecasting can be made into accessible and easy to apply operational tools to support observatory work and decision-making. More such tools are needed to aid decision-making in general. There is also a lack of model comparison and validation to standards that might ensure robust application. More resources need to be put into converting research into effective tools.

Action 2.4 Volcanic hazards, monitoring capacity, early warning capability and communication by volcanologists are key risk factors. The behaviour, attitudes and perceptions of scientists, decision-makers and communities also influence risk. Reducing risk is thus possible with better assessment and awareness of the hazards, effective communication by scientific institutions and authorities, well-practiced response protocols, participatory activities with communities and a greater awareness by all of key risk factors and how they can be managed/reduced. We recommend open, transparent interaction and communication with effective exchange of knowledge. In addition, well-thought-out plans for emergencies are essential in all sectors of society.

Pillar 3: Strengthen national and international coping capacity

Efforts should be made to increase coping capacity to address a wide range of hazards, especially relatively infrequent events like major volcanic eruptions. Many countries are enhancing their own disaster preparedness as suggested in the Hyogo Framework for Action. New resources have been made available. In addition, a number of countries have over the last decade also assisted developing countries where the risk associated with natural hazards is high. A key challenge with all projects from donor countries is to be assured that they are needs-based, sustainable and well anchored in the host countries' own development plans. Another challenge is coordination, which often has proven to be difficult because the agencies generally have different policies and the implementation periods of various projects do not overlap. A growing number of (recipient) countries want 100% ownership of their DRR activities.

On the other hand some volcanic emergencies cross borders and there may be hazards and attendant risks at regional or global scales. The threat to aviation from airborne volcanic ash is an example, another may be threats to health and agriculture from ashfall. A volcanic summit may lie in one country but valleys at risk from lahars and pyroclastic flows may lie across a border. Co-ordinated planning, mitigation and response from two or more countries are needed in these situations.

Action 3.1 Exchange visits, workshops, Summer Schools, and international research collaboration are good ways to share experience and expertise in volcano monitoring, appraisal of unrest, assessment of hazard and risk, and communication. Topics could include hazard mapping, physical volcanology, real-time interpretation of multi-parameter data, process modelling especially with respect to practical hazards assessment and forecasting tools, remote sensing and risk assessment. Cross-disciplinary training is particularly useful in earth science, remote sensing, social science, atmospheric science and technology. The value of interdisciplinary science is becoming more evident and an understanding of methodologies available in other disciplines can greatly strengthen effective collaboration. Volcanoes often have cross-border impacts so collaborative regional networks of countries can work together to build capacity, carry out research, carry out coordinated monitoring and planning, and make effective use of leveraged resources.

Action 3.2 There needs to be much more effort to integrate volcanic hazard and risk assessments with development and land use planning activities, preferably before eruptions occur so issues around livelihood, evacuation and potential resettlement are considered as part of resilience building and risk reduction activities.

Action 3.3 Free and easy access to the most advanced science and data will greatly enhance the ability to manage and reduce volcanic risk. Access to knowledge is globally very uneven between the developed and developing nations. For volcanic hazards, easy access to high-resolution digital elevation data and remote sensed data, together with appropriate training would significantly improve the scientific capacity of many countries. We encourage ISDR to promote open access of scientific knowledge to all and support the deployment of advanced technologies and information wherever it is needed. Equally important, ground-based data need to be shared among volcano observatories and with the EO community (for validation purposes). Progress toward this goal has been slow but steady and some volcano observatories already make their data freely available. Great effort and expense goes into both ground-based and satellite-based data, but the volcano community is still far from full utilisation of those data. As spatial, temporal, and spectral resolution of satellite data continues to improve, satellite and ground-based data simply must reach and be integrated by observatories in near real-time. For applications beyond minute-to-minute monitoring, WOVOdat, the GVM Task Force for Volcano Deformation, and several other initiatives are developing searchable archives with useful derivatives from raw ground- and satellite data monitoring data

Action 3.4 Index-based methods to characterise hazard, exposure, risk and monitoring capacity used in this study are straightforward, intended to provide a basic broad overview of volcanic hazard and risk across the world. The Volcanic Hazards Index and Population Exposure Index should not be used to assess or portray hazard and risk in detail at individual volcanoes, which is the responsibility of national institutions and volcano observatories. Nonetheless, combinations of the two at many volcanoes will enable improved and more robust global and regional assessments and identification of knowledge gaps.

References

Adger, W. N. 2006. Vulnerability. *Global Environmental Change,* 16, 268-281.

Aiuppa, A., Burton, M., Caltabiano, T., Giudice, G., Guerrieri, S., Liuzzo, M., Mur, F. & Salerno, G. 2010. Unusually large magmatic CO_2 gas emissions prior to a basaltic paroxysm. *Geophysical Research Letters,* 37, L17303.

Aspinall, W. 2010. A route to more tractable expert advice. *Nature,* 463, 294-295.

Aspinall, W., Auker, M., Hincks, T., Mahony, S., Nadim, F., Pooley, J., Sparks, R. & Syre, E. 2011. Volcano hazard and exposure in GFDRR priority countries and risk mitigation measures-GFDRR Volcano Risk Study. *Bristol: Bristol University Cabot Institute and NGI Norway for the World Bank: NGI Report,* 20100806, 3.

Auker, M., Sparks, R., Siebert, L., Crosweller, H. & Ewert, J. 2013. A statistical analysis of the global historical volcanic fatalities record. *Journal of Applied Volcanology,* 2:2, 1-24.

Barberi, F., Corrado, G., Innocenti, F. & Luongo, G. 1984. Phlegraean Fields 1982–1984: Brief chronicle of a volcano emergency in a densely populated area. *Bulletin Volcanologique,* 47, 175-185.

Barclay, J., Haynes, K., Houghton, B. & Johnston, D. M. 2015. Social processes and volcanic risk reduction. *In:* Sigurdsson, H., Houghton, B., McNutt, S., Rymer, H. & Stix, J. *Encyclopedia of Volcanoes: 2nd Edn.* Academic Press.

Barclay, J., Haynes, K., Mitchell, T. O. M., Solana, C., Teeuw, R., Darnell, A., Crosweller, H. S., Cole, P., Pyle, D., Lowe, C. & Fearnley, C. 2014. Framing volcanic risk communication within disaster risk reduction : finding ways for the social and physical sciences to work together. Geological Society, London, Special Publications 305.1, 163-177.

Baxter, P., Allard, P., Halbwachs, M., Komorowski, J., Andrew, W. & Ancia, A. 2003. Human health and vulnerability in the Nyiragongo volcano eruption and humanitarian crisis at Goma, Democratic Republic of Congo. *Acta Vulcanologica,* 14, 109.

Biggs, J., Anthony, E. Y. & Ebinger, C. J. 2009. Multiple inflation and deflation events at Kenyan volcanoes, East African Rift. *Geology,* 37, 979-982.

Biggs, J., Ebmeier, S. K., Aspinall, W. P., Lu, Z., Pritchard, M. E., Sparks, R. S. J. & Mather, T. a. 2014. Global link between deformation and volcanic eruption quantified by satellite imagery. *Nature Communications,* 5, 3471-3471.

Bird, D. K., Gisladottir, G. & Dominey-Howes, D. 2010. Volcanic risk and tourism in southern Iceland: Implications for hazard, risk and emergency response education and training. *Journal of Volcanology and Geothermal Research,* 189, 33-48.

Blong, R. J. 1984. *Volcanic Hazards. A Sourcebook on the Effects of Eruptions,* Australia, Academic Press.

Badan Nasional Penaggulangan Bencana (BNPB) (Indonesian National Board for Disaster Management). 2011. Ketangguhan Bangsa Dalam Menghadapi Bencana: dari Wasior, Mentawai, hingga Merapi. *GEMA BNPB,* 2, 48-48.

Bonadonna, C., Folch, A., Loughlin, S. & Puempel, H. 2012. Future developments in modelling and monitoring of volcanic ash clouds: Outcomes from the first IAVCEI-WMO workshop on Ash Dispersal Forecast and Civil Aviation. *Bulletin of Volcanology*, 74, 1-10.

Bright, E. A., Coleman, P. R., Rose, A. N. & Urban, M. L. 2012. *LandScan 2011. Oak Ridge National Laboratory, Oak Ridge, TN, USA, digital raster data.* Available: http://web.ornl.gov/sci/landscan/

Brown, S. K., Crosweller, H. S., Sparks, R. S. J., Cottrell, E., Deligne, N. I., Ortiz Guerrero, N., Hobbs, L., Kiyosugi, K., Loughlin, S. C., Siebert, L. & Takarada, S. 2014. Characterisation of the Quaternary eruption record: analysis of the Large Magnitude Explosive Volcanic Eruptions (LaMEVE) database. *Journal of Applied Volcanology*, 3:5, pp.22.

Carey, S., Sigurdsson, H., Mandeville, C. & Bronto, S. 1996. Pyroclastic flows and surges over water: an example from the 1883 Krakatau eruption. *Bulletin of Volcanology*, 57, 493-511.

Carlsen, H. K., Hauksdottir, A., Valdimarsdottir, U. A., Gíslason, T., Einarsdottir, G., Runolfsson, H., Briem, H., Finnbjornsdottir, R. G., Gudmundsson, S. & Kolbeinsson, T. B. 2012. Health effects following the Eyjafjallajökull volcanic eruption: a cohort study. *BMJ Open*, 2, e001851.

Cashman, K. V., Stephen, R. & Sparks, J. 2013. How volcanoes work: A 25 year perspective. *Bulletin of the Geological Society of America*, 125, 664-690.

Charbonnier, S. J., Germa, A., Connor, C. B., Gertisser, R., Preece, K., Komorowski, J. C., Lavigne, F., Dixon, T. & Connor, L. 2013. Evaluation of the impact of the 2010 pyroclastic density currents at Merapi volcano from high-resolution satellite imagery, field investigations and numerical simulations. *Journal of Volcanology and Geothermal Research*, 261, 295-315.

Chouet, B. A. 1996. Long-period volcano seismicity: its source and use in eruption forecasting. *Nature*, 380, 309-316.

Clay, E., Barrow, C., Benson, C., Dempster, J., Kokelaar, P., Pillai, N. & Seaman, J. 1999. An evaluation of HMG's response to the Montserrat Volcanic Emergency. Part 1. DFID (UK Government) Evaluation Report EV5635.

Cottrell, E. 2014. Global Distribution of Active Volcanoes. *In:* Papale, P. (ed.)*Volcanic Hazards, Risks and Disasters.* Academic Press.

Cronin, S. J. & Sharp, D. S. 2002. Environmental impacts on health from continuous volcanic activity at Yasur (Tanna) and Ambrym, Vanuatu. *International Journal of Environmental Health Research*, 12, 109-123.

Crosweller, H. S., Arora, B., Brown, S. K., Cottrell, E., Deligne, N. I., Guerrero, N. O., Hobbs, L., Kiyosugi, K., Loughlin, S. C. & Lowndes, J. 2012. Global database on large magnitude explosive volcanic eruptions (LaMEVE). *Journal of Applied Volcanology*, 1:4, pp.13.

Crosweller, H. S. & Wilmshurst, J. 2013. Natural hazards and risk: the human perspective. *In:* Rougier, J., Sparks, R. S. J. & Hill, L. Cambridge: Cambridge University Press. pp. 548-569

Decker, R. & Decker, B. 2006. *Volcanoes*, W. H. Freeman.

Deligne, N. I., Coles, S. G. & Sparks, R. S. J. 2010. Recurrence rates of large explosive volcanic eruptions. *Journal of Geophysical Research: Solid Earth,* 115, B06203.

Dibben, C. & Chester, D. K. 1999. Human vulnerability in volcanic environments: The case of Furnas, Sao Miguel, Azores. *Journal of Volcanology and Geothermal Research,* 92, 133-150.

Druitt, T. H., Costa, F., Deloule, E., Dungan, M. & Scaillet, B. 2012. Decadal to monthly timescales of magma transfer and reservoir growth at a caldera volcano. *Nature* 482 (7383), 77-80

Durant, A. J., Bonadonna, C. & Horwell, C. J. 2010. Atmospheric and environmental impacts of volcanic particulates. *Elements,* 6, 235-240.

Dzurisin, D. 2003. A comprehensive approach to monitoring volcano deformation as a window on the eruption cycle. *Reviews of Geophysics,* 41.

Edmonds, M. 2008. New geochemical insights into volcanic degassing. *Philosophical transactions. Series A, Mathematical, Physical, and Engineering Sciences,* 366, 4559-4579.

Ewert, J. W., Guffanti, M. & Murray, T. L. 2005. An assessment of volcanic threat and monitoring capabilities in the United States: Framework for a National Volcano Early Warning System (NVEWS). *US Geological Survey Open-File Report,* 1-62.

Ewert, J. W. & Harpel, C. J. 2004. In harm's way: population and volcanic risk. *Geotimes,* 49, 14-17.

Fearnley, C. J. 2013. Assigning a volcano alert level: negotiating uncertainty, risk, and complexity in decision-making processes. *Environment and Planning A,* 45, 1891-1911.

Furlan, C. 2010. Extreme value methods for modelling historical series of large volcanic magnitudes. *Statistical Modelling,* 10, 113-132.

Gaillard, J.-C. 2007. Resilience of traditional societies in facing natural hazards. *Disaster Prevention and Management,* 16, 522-544.

Gaillard, J. C. 2008. Alternative paradigms of volcanic risk perception: The case of Mt. Pinatubo in the Philippines. *Journal of Volcanology and Geothermal Research,* 172, 315-328.

Gardner, C.A. & Guffanti, M.C. 2006 U.S. Geological Survey's alert notification system for volcanic activity: U.S. Geological Survey Fact Sheet 2006-3139.

Guffanti, M., Casadevall, T. J. & Budding, K. 2010. Encounters of aircraft with volcanic ash clouds: a compilation of known incidents, 1953-2009. *US Geological Survey Data Series,* 545, 12-12.

Haynes, K., Barclay, J. & Pidgeon, N. 2008a. The issue of trust and its influence on risk communication during a volcanic crisis. *Bulletin of Volcanology,* 70, 605-621.

Haynes, K., Barclay, J. & Pidgeon, N. 2008b. Whose reality counts? Factors affecting the perception of volcanic risk. *Journal of Volcanology and Geothermal Research,* 172, 259-272.

Hincks, T. K., Aspinall, W. P., Baxter, P. J., Searl, a., Sparks, R. S. J. & Woo, G. 2005. Long term exposure to respirable volcanic ash on Montserrat: a time series simulation. *Bulletin of Volcanology,* 68, 266-284.

Hincks, T. K., Komorowski, J.-C., Sparks, S. R. & Aspinall, W. P. 2014. Retrospective analysis of uncertain eruption precursors at La Soufrière volcano, Guadeloupe, 1975–77: volcanic hazard assessment using a Bayesian Belief Network approach. *Journal of Applied Volcanology,* 3:3, pp.26.

Horwell, C. & Baxter, P. J. 2006. The respiratory health hazards of volcanic ash: a review for volcanic risk mitigation. *Bulletin of Volcanology,* 69, 1-24.

Jenkins, S., Komorowski, J. C., Baxter, P. J., Spence, R., Picquout, A., Lavigne, F. & Surono 2013. The Merapi 2010 eruption: An interdisciplinary impact assessment methodology for studying pyroclastic density current dynamics. *Journal of Volcanology and Geothermal Research,* 261, 316-329.

Jenkins, S., Magill, C., McAneney, J. & Blong, R. 2012. Regional ash fall hazard I: a probabilistic assessment methodology. *Bulletin of Volcanology,* 74, 1699-1712.

Johnson, J. B. & Ripepe, M. 2011. Volcano infrasound: A review. *Journal of Volcanology and Geothermal Research,* 206, 61-69.

Kar-Purkayastha, I., Horwell, C. & Murray, V. 2012. Review of evidence on the potential health impacts of volcanic ash on the population of the UK and ROI. Centre for Radiation, Chemical and Environmental Hazards, Health Protection Agency, United Kingdom. Available: www.hpa.org.uk

Kelman, I. & Mather, T. A. 2008. Living with volcanoes: The sustainable livelihoods approach for volcano-related opportunities. *Journal of Volcanology and Geothermal Research,* 172, 189-198.

Kling, G. W., Clark, M. A., Wagner, G. N., Compton, H. R., Humphrey, A. M., Devine, J. D., Evans, W. C., Lockwood, J. P., Tuttle, M. L. & Koenigsberg, E. J. 1987. The 1986 Lake Nyos gas disaster in Cameroon, West Africa. *Science (New York, N.Y.),* 236, 169-175.

Komorowski, J.-C., Jenkins, S., Baxter, P. J., Picquout, A., Lavigne, F., Charbonnier, S., Gertisser, R., Preece, K., Cholik, N. & Budi-Santoso, A. 2013. Paroxysmal dome explosion during the Merapi 2010 eruption: Processes and facies relationships of associated high-energy pyroclastic density currents. *Journal of Volcanology and Geothermal Research,* 261, 260-294.

Komorowski, J. 2003. The January 2002 flank eruption of Nyiragongo volcano (Democratic Republic of Congo): Chronology, evidence for a tectonic rift trigger, and impact of lava flows on the city of Goma. *Acta vulcanologica,* 14, 27.

Komorowski, J. C., Boudon, G., Semet, M., Beauducel, F., Anténor-Habazac, C., Bazin, S. & Hammouya, G. 2005. Guadeloupe. *In:* Lindsay, J. M., Robertson, R. E. A., Shepherd, J. B. & Ali, S. *Volcanic Atlas of the Lesser Antilles.* Seismic Research Unit, University of the West Indies.

Laksono, P. M. 1988. Perception of volcanic hazards: villagers versus government officials in Central Java. *The Real and Imagined Role of Culture in Development: Case Studies from Indonesia.* Honolulu: University of Hawaii Press.

Lane, L. R., Tobin, G. a. & Whiteford, L. M. 2003. Volcanic hazard or economic destitution: hard choices in Baños, Ecuador. *Environmental Hazards,* 5, 23-34.

Lara, L. E., Moreno, R., Amigo, Á., Hoblitt, R. P. & Pierson, T. C. 2013. Late Holocene history of Chaitén Volcano: New evidence for a 17th century eruption. *Andean Geology,* 40, 249-261.

Larson, K. M., Poland, M. & Miklius, a. 2010. Volcano monitoring using {GPS:} Developing data analysis strategies based on the June 2007 Kilauea Volcano intrusion and eruption. *Journal of Geophysical Research-Solid Earth,* 115, B07406-B07406.

Lavigne, F., De Coster, B., Juvin, N., Flohic, F., Gaillard, J. C., Texier, P., Morin, J. & Sartohadi, J. 2008. People's behaviour in the face of volcanic hazards: Perspectives from Javanese communities, Indonesia. *Journal of Volcanology and Geothermal Research,* 172, 273-287.

Lavigne, F., Degeai, J.-P., Komorowski, J.-C., Guillet, S., Robert, V., Lahitte, P., Oppenheimer, C., Stoffel, M., Vidal, C. M., Surono, Pratomo, I., Wassmer, P., Hajdas, I., Hadmoko, D. S. & de Belizal, E. 2013. Source of the great A.D. 1257 mystery eruption unveiled, Samalas volcano, Rinjani Volcanic Complex, Indonesia. *Proceedings of the National Academy of Sciences,* 110, 16742-16747.

Le Maitre, R. W., Streckeisen, A., Zanettin, B., Bas, J. L., Bonin, B. & Bateman, P. 2002. *Igneous Rocks: A Classification and Glossary of Terms: Recommendations of the International Union of Geological Sciences Subcommission on the Systematics of Igneous Rocks*, 2nd Edition. Cambridge University Press.

Leonard, G. S., Johnston, D. M., Paton, D., Christianson, A., Becker, J. & Keys, H. 2008. Developing effective warning systems: ongoing research at Ruapehu volcano, New Zealand. *Journal of Volcanology and Geothermal Research,* 172, 199-215.

Lesparre, N., Gibert, D., Marteau, J., Komorowski, J.-C., Nicollin, F. & Coutant, O. 2012. Density muon radiography of La Soufrière of Guadeloupe volcano: comparison with geological, electrical resistivity and gravity data. *Geophysical Journal International,* 190, 1008-1019.

Lewis, J. 1999. *Development in Disaster-prone Places: Studies of Vulnerability,* London, Intermediate Technology Publications. pp.174.

Lindsay, J. M. Volcanoes in the big smoke: a review of hazard and risk in the Auckland Volcanic Field. Geologically Active. Delegate Papers of the 11th Congress of the International Association for Engineering Geology and the Environment (IAEG), 2010.

Lockwood, J. P. & Hazlett, R. W. 2013. *Volcanoes: Global Perspectives*, John Wiley & Sons. pp.552.

Loughlin, S., Baxter, P., Aspinall, W., Darroux, B., Harford, C. & Miller, A. 2002. Eyewitness accounts of the 25 June 1997 pyroclastic flows and surges at Soufrière Hills Volcano, Montserrat, and implications for disaster mitigation. *Geological Society, London, Memoirs,* 21, 211-230.

Luhr, J. F., Simkin, T. & Cuasay, M. 1993. *Paricutin: A Volcano Born in a Mexican Cornfield,* Phoenix, Arizona, Geoscience Press. pp. 427.

Mandeville, C. W., Carey, S. & Sigurdsson, H. 1996. Sedimentology of the Krakatau 1883 submarine pyroclastic deposits. *Bulletin of Volcanology,* 57, 512-529.

Marzocchi, W. & Bebbington, M. S. 2012. Probabilistic eruption forecasting at short and long time scales. *Bulletin of Volcanology,* 74, 1777-1805.

Marzocchi, W. & Woo, G. 2009. Principles of volcanic risk metrics: Theory and the case study of Mount Vesuvius and Campi Flegrei, Italy. *Journal of Geophysical Research,* 114, B03213.

Mastin, L. G., Guffanti, M., Servranckx, R., Webley, P., Barsotti, S., Dean, K., Durant, A., Ewert, J. W., Neri, A., Rose, W. I., Schneider, D., Siebert, L., Stunder, B., Swanson, G., Tupper, A., Volentik, A. & Waythomas, C. F. 2009. A multidisciplinary effort to assign realistic source parameters to models of volcanic ash-cloud transport and dispersion during eruptions. *Journal of Volcanology and Geothermal Research,* 186, 10-21.

McNutt, S. & Williams, E. R. 2010. Volcanic lightning: global observations and constraints on source mechanisms. *Bulletin of Volcanology,* 72, 1153-1167.

McNutt, S. R. 2005. Volcanic Seismology. *Annual Review of Earth and Planetary Sciences,* 33, 461-491.

Mei, E. T. W., Lavigne, F., Picquout, A., de Bélizal, E., Brunstein, D., Grancher, D., Sartohadi, J., Cholik, N. & Vidal, C. 2013. Lessons learned from the 2010 evacuations at Merapi volcano. *Journal of Volcanology and Geothermal Research,* 261, 348-365.

Mrozek, J. R. & Taylor, L. O. 2002. What determines the value of life? A meta-analysis. *Journal of Policy Analysis and Management,* 21, 253-270.

Nadeau, P. A., Palma, J. L. & Waite, G. P. 2011. Linking volcanic tremor, degassing, and eruption dynamics via SO$_2$ imaging. *Geophysical Research Letters,* 38, L01304.

Neal, C.A. & Guffanti, M.C. 2010. Airborne volcanic ash; a global threat to aviation: U.S. Geological Survey Fact Sheet 2010-3116.

Newhall, C., Aramaki, S., Barberi, F., Blong, R., Calvache, M., Cheminee, J.-L., Punongbayan, R., Siebe, C., Simkin, T., Sparks, R. S. J. & Tjetjep, W. 1999. IAVCEI Subcommittee for Crisis Protocols: Professional conduct of scientists during volcanic crises. *Bulletin of Volcanology,* 60, 323-334.

Newhall, C., Hendley II, J. W. & Stauffer, P. H. 1997. *Benefits of volcano monitoring far outweigh the costs - the case of Mount Pinatubo. U.S. Geological Survey Fact Sheet 115-97.* [Online]. Available: http://pubs.usgs.gov/fs/1997/fs115-97/.

Newhall, C. & Hoblitt, R. 2002. Constructing event trees for volcanic crises. *Bulletin of Volcanology,* 64, 3-20.

Newhall, C. G. 2007. Volcanology 101 for Seismologists. *In:* Schubert, G. & Kanamori, H. (ed.) Treatise on Geophysics, 4 ed. Elsevier B.V., 351-388.

Newhall, C. G. & Punongbayan, R. 1996a. *Fire and Mud: Eruptions and Lahars of Mount Pinatubo, Philippines*, Philippine Institute of Volcanology and Seismology Quezon City.

Newhall, C. G. & Punongbayan, R. S. 1996b. The narrow margin of successful volcanic-risk mitigation. *In:* Scarpa, R. & Tilling, R. I. *Monitoring and Mitigation of Volcano Hazards.* Berlin: Springer-Verlag.

Newhall, C. G. & Self, S. 1982. The volcanic explosivity index (VEI) an estimate of explosive magnitude for historical volcanism. *Journal of Geophysical Research,* 87, 1231-1231.

Papale, P. 2014. *Volcanic Hazards, Risks and Disasters*, Academic Press, pp. 532.

Phillipson, G., Sobradelo, R. & Gottsmann, J. 2013. Global volcanic unrest in the 21st century: an analysis of the first decade. *Journal of Volcanology and Geothermal Research,* 264, 183-196.

Potter, S. H., Jolly, G. E., Neall, V. E., Johnston, D. M. & Scott, B. J. 2014. Communicating the status of volcanic activity: revising New Zealand's volcanic alert level system. *Journal of Applied Volcanology,* 3, 1-16.

Potter, S. H., Scott, B. J. & Jolly, G. E. 2012. Caldera Unrest Management Sourcebook. *GNS Science Report.* GNS Science.

Punongbayan, R. S., Newhall, C. G., Bautista, M. L. P., Garcia, D., Harlow, D. H., Hoblitt, R. P., Sabit, J. P. & Solidum, R. U. 1996. Eruption hazard assessments and warnings. *Fire and Mud: Eruptions and Lahars of Mount Pinatubo, Philippines,* 415-433.

Ragona, M., Hannstein, F. & Mazzocchi, M. 2011. The impact of volcanic ash crisis on the European Airline industry. *In:* Alemanno, A. (ed.)*Governing Disasters: The Challenges of Emergency Risk regulations.* Edward Elgar Publishing.

Roberts, M. R., Linde, A. T., Vogfjord, K. S. & Sacks, S. Forecasting Eruptions of Hekla Volcano, Iceland, using Borehole Strain Observations, EGU2011-14208. 2011.

Robock, A. 2000. Volcanic eruptions and climate. *Reviews of Geophysics,* 38, 191-219.

Schmidt, A., Ostro, B., Carslaw, K. S., Wilson, M., Thordarson, T., Mann, G. W. & Simmons, A. J. 2011. Excess mortality in Europe following a future Laki-style Icelandic eruption. *Proceedings of the National Academy of Sciences,* 108, 15710-15715.

Schmincke, H. U. 2004. *Volcanism,* Berlin Heidelberg, Springer-Verlag, pp. 324.

Segall, P. 2013. Volcano deformation and eruption forecasting. *Geological Society, London, Special Publications,* 380, 85-106.

Seitz, S. 2004. *The Aeta at the Mount Pinatubo, Philippines: A Minority Group Coping with Disaster,* Quezon City, New Day Publishers.

Self, S. & Blake, S. 2008. Supervolcanoes: Consequences of explosive supereruptions. *Elements,* 4, 41-46.

Siebert, L. 1984. Large volcanic debris avalanches: characteristics of source areas, deposits, and associated eruptions. *Journal of Volcanology and Geothermal Research,* 22, 163-197.

Siebert, L., Simkin, T. & Kimberley, P. 2010. *Volcanoes of the World, 3rd edn,* Berkeley, University of California Press.

Sigurdsson, H., Houghton, B., Rymer, H., Stix, J. & McNutt, S. 2015. *Encyclopedia of Volcanoes,* Academic Press.

Simkin, T. 1993. Terrestrial volcanism in space and time. *Annual Review of Earth and Planetary Sciences,* 21, 427-452.

Small, C. & Naumann, T. 2001. The global distribution of human population and recent volcanism. *Global Environmental Change Part B: Environmental Hazards,* 3, 93-109.

Smithsonian. 2013. *Volcanoes of the World 4.0* [Online]. Washington D.C. Available: http://www.volcano.si.edu.

Solana, M. C., Kilburn, C. R. J. & Rolandi, G. 2008. Communicating eruption and hazard forecasts on Vesuvius, Southern Italy. *Journal of Volcanology and Geothermal Research,* 172, 308-314.

Sparks, R.S.J., Biggs, J. & Neuberg, J. 2012. Monitoring volcanoes. *Science,* 335, 1310-1311.

Sparks, R. S. J. 2003. Forecasting volcanic eruptions. *Earth and Planetary Science Letters,* 210, 1-15.

Sparks, R. S. J. & Aspinall, W. P. 2004. Volcanic activity: frontiers and challenges in forecasting, prediction and risk assessment. In: Sparks, R.S.J. & Hawkesworth, C.J. *The State of the Planet: Frontiers and Challenges in Geophysics.* American Geophysical Union, Washington D.C.

Sparks, R. S. J., Aspinall, W. P., Crosweller, H. S. & Hincks, T. K. 2013. Risk and uncertainty assessment of volcanic hazards. *In:* Rougier, J., Sparks, R. S. J. & Hill, L. *Risk and Uncertainty Assessment for Natural Hazards.* Cambridge: Cambridge University Press, 364-397.

Spence, R., Kelman, I., Baxter, P., Zuccaro, G. & Petrazzuoli, S. 2005. Residential building and occupant vulnerability to tephra fall. *Natural Hazards and Earth System Science,* 5, 477-494.

Surono, Jousset, P., Pallister, J., Boichu, M., Buongiorno, M. F., Budisantoso, A., Costa, F., Andreastuti, S., Prata, F., Schneider, D., Clarisse, L., Humaida, H., Sumarti, S., Bignami, C., Griswold, J., Carn, S., Oppenheimer, C. & Lavigne, F. 2012. The 2010 explosive eruption of Java's Merapi volcano-A '100-year' event. *Journal of Volcanology and Geothermal Research,* 241-242, 121-135.

Sword-Daniels, V. 2011. Living with volcanic risk: The consequences of, and response to, ongoing volcanic ashfall from a social infrastructure systems perspective on Montserrat. *New Zealand Journal of Psychology,* 40, 131-138.

Thordarson, T. & Self, S. 2003. Atmospheric and environmental effects of the 1783–1784 Laki eruption: a review and reassessment. *Journal of Geophysical Research: Atmospheres (1984–2012),* 108, AAC-7.

Þorkelsson, B. 2012. The 2010 Eyjafjallajokull Eruption, Iceland: Report to ICAO. Icelandic Meteorological Office.

Tilling, I. 1989. Volcanic hazards and their mitigation: Progress and problems. *Reviews of Geophysics,* 27:2, 237-269.

Toda, S., Stein, R. S. & Sagiya, T. 2002. Evidence from the AD 2000 Izu islands earthquake swarm that stressing rate governs seismicity. *Nature,* 419, 58-61.

Usamah, M. & Haynes, K. 2012. An examination of the resettlement program at Mayon Volcano: What can we learn for sustainable volcanic risk reduction? *Bulletin of Volcanology,* 74, 839-859.

US Geological Survey (2015). *Volcano Hazards Program.* [Online] Available: http://volcanoes.usgs.gov/

Viramonte, J.G., Peralta, C.M., Garrido, D. and Felpeto, A. 2001. Uso de sensores remotos para la mitigación de efectos causados por erupciones volcánicas : Elaboración de mapas de Riesgo volcánico y alertas para la aeronavegación: Un caso de estudio. [Online] Available:www.conae.gov.ar/WEB_Emergencias/Links_del_Cuerpo_Principal/Volcanes/Informe%20Riesgo%20Volcanico.htm

Voight, B. 1990. The 1985 Nevado del Ruiz volcano catastrophe: anatomy and retrospection. *Journal of Volcanology and Geothermal Research,* 42, 151-188.

Voight, B., Sparks, R.S.J., Miller, A.D., Stewart, R.C., Hoblitt, R.P., Clarke, A., Ewart, J., Aspinall, W.P., Baptie, B., Calder, E.S., Cole, P., Druitt, T.H., Hartford, C., Herd, R.A., Jacksomn, P., Lejeune, A.M., Lockhart, A.B., Loughlin, S.C., Luckett, R., Lynch, L., Norton, G.E., Robertson, R., Watson, I.M., Watts, R. & Young, S.R. 1999. Magma Flow Instability and Cyclic Activity at Soufriere Hills Volcano, Montserrat, British West Indies. Science, 283.5405, 1138-1142

Voight, B. 2000. Structural stability of andesite volcanoes and lava domes. *Philosophical Transactions of the Royal Society A: Mathematical, Physical and Engineering Sciences,* 358, 1663-1703.

Voight, B., Calvache, M. L., Hall, M. L. & Mosalve, M. L. 2013. Nevado del Ruiz volcano, Colombia 1985. *In:* Bobrowsky, P. T. (ed.). Berlin Heidelberg London: Springer, 732-738.

Wadge, G. & Aspinall, W. 2014. A review of volcanic hazard and risk assessments at the Soufrière Hills Volcano, Montserrat from 1997 to 2011. *In:* Wadge, G., Robertson, R. & Voight, B. *The Eruption of Soufriere Hills Volcano, Montserrat, from 2000 to 2010: Geological Society Memoirs, Vol. 39.* Geological Society of London.

Wadge, G., Voight, B., Sparks, R. S. J., Cole, P. D., Loughlin, S. C. & Robertson, R. E. A. 2014. An overview of the eruption of Soufriere Hills Volcano, Montserrat from 2000 to 2010. *Geological Society, London, Memoirs,* 39, 1-40.

Wilson, T. M., Stewart, C., Bickerton, H., Baxter, P. J., Outes, V., Villarosa, G. & Rovere, E. 2012a. Impacts of the June 2011 Puyehue Cordón-Caulle volcanic complex eruption on urban infrastructure, agriculture and public health. GNS Science Report.

Wilson, T. M., Stewart, C., Sword-Daniels, V., Leonard, G. S., Johnston, D. M., Cole, J. W., Wardman, J., Wilson, G. & Barnard, S. T. 2012b. Volcanic ash impacts on critical infrastructure. *Physics and Chemistry of the Earth, Parts A/B/C,* 45, 5-23.

Wisner, B., Blaikie, P., Cannon, T. & Davis, I. 2004. *At Risk: Natural Hazards, People's Vulnerability and Disasters*, Routledge.

Woo, G. 2014. Cost-benefit analysis in volcanic risk. *In:* Papale, P. (ed.)*Volcanic Hazards, Risks and Disasters.* Academic Press, 289-300.

Woodhouse, M. J., Hogg, A. J., Phillips, J. C. & Sparks, R. S. J. 2013. Interaction between volcanic plumes and wind during the 2010 Eyjafjallajökull eruption, Iceland. *Journal of Geophysical Research: Solid Earth,* 118, 92-109.

World Organisation of Volcano Observatories (WOVO) (2015) Members' Observatories Directory. [Online] Available: http://www.wovo.org/observatories/

World Organisation of Volcano Observatories Database of Volcanic Unrest (WOVOdat) (2015) Available: http://www.wovodat.org/

Chapter 3

Volcanic ash fall hazard and risk

S.F. Jenkins, T. Wilson, C. Magill, V. Miller, C. Stewart, R. Blong, W. Marzocchi, M. Boulton, C. Bonadonna, A. Costa

Executive summary

All explosive volcanic eruptions generate volcanic ash, fragments of rock that are produced when magma or vent material is explosively disintegrated. Volcanic ash is then convected upwards within the eruption column and carried downwind, falling out of suspension and potentially affecting communities across hundreds, or even thousands, of square kilometres. Ash is the most frequent, and often widespread, volcanic hazard and is produced by all explosive volcanic eruptions. Although ash falls rarely endanger human life directly, threats to public health and disruption to critical infrastructure services, aviation and primary production can lead to potentially substantial societal impacts and costs, even at thicknesses of only a few millimetres. Communities exposed to any magnitude of ash fall commonly report anxiety about the health impacts of inhaling or ingesting ash (as well as impacts to animals and property damage), which may lead to temporary socio-economic disruption (e.g. evacuation, school and business closures, cancellations). The impacts of any ash fall can therefore be experienced across large areas and can also be long-lived, both because eruptions can last weeks, months or even years and because ash may be remobilised and re-deposited by wind, traffic or human activities.

Given the potentially large geographic dispersal of volcanic ash, and the substantial impacts that even thin (a few mm in thickness) deposits can have for society, this chapter elaborates upon the ash component of the overviews provided in Chapters 1 and 2. We focus on the hazard and associated impacts of *ash falls;* however, the areas affected by volcanic ash are potentially much larger than those affected by ash falling to the ground, as fine particles can remain aloft for extended periods of time. For example, large portions of European airspace were closed for up to five weeks during the eruption of Eyjafjallajökull, Iceland, in 2010 because of airborne ash (with negligible associated ash falls outside of Iceland). The distance and area over which volcanic ash is dispersed is strongly controlled by wind conditions with distance and altitude from the vent, but also by the size, shape and density of the ash particles, and the style and magnitude of the eruption. These factors mean that ash falls are typically deposited in the direction of prevailing winds during the eruption and thin with distance. Forecasting ash dispersion and the deposition 'footprint' is typically achieved through numerical simulation.

Jenkins, S.F., Wilson, T., Magill, C., Miller, V., Stewart, C., Blong, R., Marzocchi, W., Boulton, M., Bonadonna, C. & Costa, A. (2015) Volcanic ash fall hazard and risk. In: S.C. Loughlin, R.S.J. Sparks, S.K. Brown, S.F. Jenkins & C. Vye-Brown (eds) *Global Volcanic Hazards and Risk,* Cambridge: Cambridge University Press.

In this chapter, we discuss volcanic ash fall hazard modelling that has been implemented at the global and local (Neapolitan area, Italy) scales (Section 3.2). These models are probabilistic, i.e. they account for uncertainty in the input parameters to produce a large number of possible outcomes. Outputs are in the form of hazard maps and curves that show the probabilities associated with exceeding key hazard thresholds at given locations. As with any natural hazard, these results are subject to uncertainty and the local case study describes how ongoing research is working to better quantify this uncertainty through Bayesian methods and models. Further to the ash fall hazard assessments, we discuss the key components required to carry these hazard estimates forward to risk: namely identification of likely impacts and the response (vulnerability) of key sectors of society to ash fall impact. The varied characteristics of volcanic ash, e.g. deposit thickness and density, particle size and surface composition, the context, e.g. timing and duration of ash fall, and resilience of exposed people and assets can all influence the type and magnitude of impacts that may occur. We draw from data collected during and following past eruptions and experimental and theoretical studies to highlight likely impacts for key sectors of society, such as health, infrastructure and the economy (Section 3.3). In many parts of the world, the failure, disruption or reduced functionality of infrastructure or societal activities, e.g. ability to work or go to school, is likely to have a larger impact on livelihoods and the local economy than direct damage to buildings. Broad relationships between ash thickness (assuming a fixed deposit density) and key levels of damage is also outlined (Section 3.4); however, vulnerability estimates are typically the weakest part of a risk model and detailed local studies of exposed assets and their vulnerability should ideally be carried out before a detailed risk assessment is undertaken.

Greater knowledge of ash fall hazard and associated impacts supports mitigation actions, crisis planning and emergency management activities, and is an essential step towards building resilience for individuals and communities. This chapter concludes with a discussion on where some of the important advances in ash fall hazard and risk assessment may be achieved, providing a roadmap for future research objectives.

Glossary

Ash (tephra) fall: Volcanic ash and lapilli dispersed by winds away from the volcano, falling out of suspension to form a deposit. Tephra falls are commonly referred to colloquially as ash falls.

Exposure: Societal assets (e.g. people, property, infrastructure networks) present in hazard zones that are thereby subject to potential losses.

Hazard: Potential threat arising from the physical phenomenon.

Hazard intensity: A parameter describing the severity of a hazard at a location, e.g. ash fall thickness.

Impact: Function of the hazard and vulnerability on the exposed asset.

Lapilli: Fragments of rock, between 2 and 64 mm in diameter, produced explosively during a volcanic eruption.

Resilience: The ability of a system, community or society exposed to hazards to resist, absorb, accommodate to and recover from the effects of a hazard in a timely and efficient manner, including through the preservation and restoration of its essential basic structures and functions.

Risk: Function of both the characteristics of the hazardous event and the consequences for the exposed assets.

Tephra: Fragments of rock, regardless of size, that become airborne during a volcanic eruption.

Volcanic ash: Tephra less than 2 mm in diameter, produced explosively during a volcanic eruption.

Vulnerability: The degree to which characteristics and the circumstances of an individual, community, system, network or asset, and any interdependencies, makes it susceptible to the damaging effects of a hazard.

3.1 Introduction

Volcanic eruptions can produce a number of hazards that vary widely in their spatial distribution and in the impacts they can have for society. Localised phenomena, such as pyroclastic density currents and lahars, are the most destructive and dangerous (Auker et al., 2013); however, the most frequent, and often widespread, volcanic hazard is tephra, occurring in more than 90% of all eruptions (Newhall & Hoblitt, 2002). Tephra comprises fragments of rock produced when magma or vent material is explosively disintegrated during an eruption. Larger tephra ('blocks' or 'bombs') follow ballistic trajectories and pose a local threat (usually < 5 km). Smaller diameter tephra (Figure 3.1) - comprising lapilli (2 to 64 mm) and ash (< 2 mm) – are convected upwards within the eruption column or convective plume and dispersed away from the volcano by winds or buoyancy forces. Ash particles can be carried hundreds or even thousands of kilometres away from source falling out of suspension over large multi-national areas to form 'ash falls'. For example, the 1991 eruption of Hudson volcano in Chile covered an estimated 150,000 km² of land with ash, most notably in Argentina (Scasso et al., 1994). Ash falls are the volcanic hazard with the greatest potential to directly or indirectly affect the largest number of people worldwide (Simkin et al., 2001, Witham, 2005).

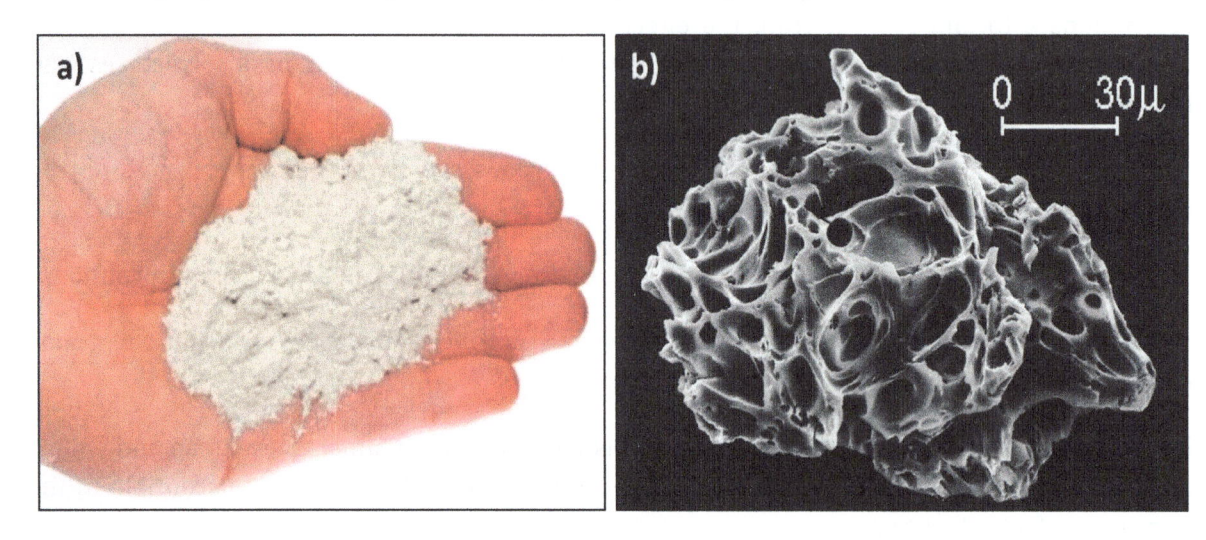

Figure 3.1 a) Rhyolitic ash produced by the VEI 4 eruption of Chaitén volcano, Chile, in 2008 (median particle size: 0.02 mm); b) Scanning electron microscope image of an ash particle from the eruption of Mount St Helens, USA, in 1980. Vesicles, formed as gas expanded within solidifying magma, are clearly visible and increase the particle surface area; typically, ash is highly abrasive because of its irregular shape and hardness. Photos: a) G. Wilson, University of Canterbury; b) A.M. Sarna-Wojcicki, USGS. This figure is reproduced in Chapter 12, as Figure 12.1.

Ash falls usually do not endanger human life directly, except where very thick falls cause structural damage (e.g. roof collapse) or where casualties are indirectly sustained, for example during ash clean-up operations or in traffic accidents. However, ash falls can threaten public health, with the short-term health effects of ash exposure typically being irritation of the eyes and upper airways and exacerbation of pre-existing asthma. Ash falls can also disrupt critical infrastructure services (e.g. electricity and water supply, aviation and other transport routes), damage buildings, and damage or disrupt agricultural production and other economic activities (Wilson et al., 2012b). Even relatively thin ash falls of a few millimetres can cause significant societal impacts, through widespread damage, disruption and economic loss. Impacts can be long-lived, either because eruptions may be long-duration or because ash falls may be remobilised by wind, water, traffic or human activities.

This chapter elaborates upon ash fall hazard and risk as discussed in Chapters 1 and 2. We provide information on how the volcanological community is addressing three major components of ash fall risk assessment: 1) quantifying the hazard; 2) identifying likely impacts; and 3) estimating vulnerability for key sectors of society. Understanding the breadth and severity of hazard and risk for ash falls of different thicknesses and characteristics is key to reducing the potential impact on livelihoods and socio-economic activities during future events.

In this technical background paper, we focus on the hazard and associated impacts of *ash falls*; however, volcanic ash poses problems for society both while being dispersed in the atmosphere and on the ground once deposited. In general, areas affected by airborne ash are much larger than those affected by ash falls because fine particles can remain aloft for extended periods of time creating a hazard for aviation and health but not producing discernible ash falls. For example, the direct impacts of ash fall from the relatively small (VEI 4 on a logarithmic scale from 0 to 8[1]) volcanic eruption of Eyjafjallajökull, Iceland, in 2010 were limited to the farming communities of southern Iceland. By contrast, large portions of European airspace were closed for up to five weeks because of airborne volcanic ash (with negligible associated ash falls). This airspace closure caused major disruption to the aviation industry and affected the transport of goods worldwide (IATA, 2010). In terms of the potential for multiple casualties and large economic losses, interaction of volcanic ash with aircraft represents the most important hazard from airborne ash, although respiratory health impacts are also of concern. Airborne volcanic ash is an important societal hazard but beyond the scope of this paper; it is covered in some detail in the technical overview (Chapter 2) and associated case studies (Chapter 4 - 26).

3.2 Quantifying ash fall hazard

The distance and area over which volcanic ash is dispersed is strongly controlled by wind direction and speed, and atmospheric conditions (e.g. the presence of water) with distance from the vent, which can vary hourly, daily and seasonally, and with altitude above the vent, affecting particles as they fall through the atmosphere. Winds at the surface are usually different in direction and speed than winds higher in the atmosphere and winds throughout the atmospheric column typically vary in strength and direction according to the time of year. Thus the height of the eruption column, and the time of year, strongly influence where ash is dispersed to. Figure 3.2 shows the distribution and speeds of winds with height above a selection of cities in active volcanic areas; generally, wind speeds increase and directions become more consistent with increasing height because of global circulation patterns. While wind conditions have a strong control on ash dispersal, dispersal is also affected by the size, shape and density (and therefore fall velocities) of the ash particles, and the eruption style and magnitude.

---

[1] Volcanic Explosivity Index (VEI) – An estimate of explosivity magnitude for volcanic eruptions. VEI ranges from 0 to 8 on a logarithmic scale so that, for example, a VEI 5 eruption has ten times the erupted volume of a VEI 4 eruption (Newhall and Self, 1982).

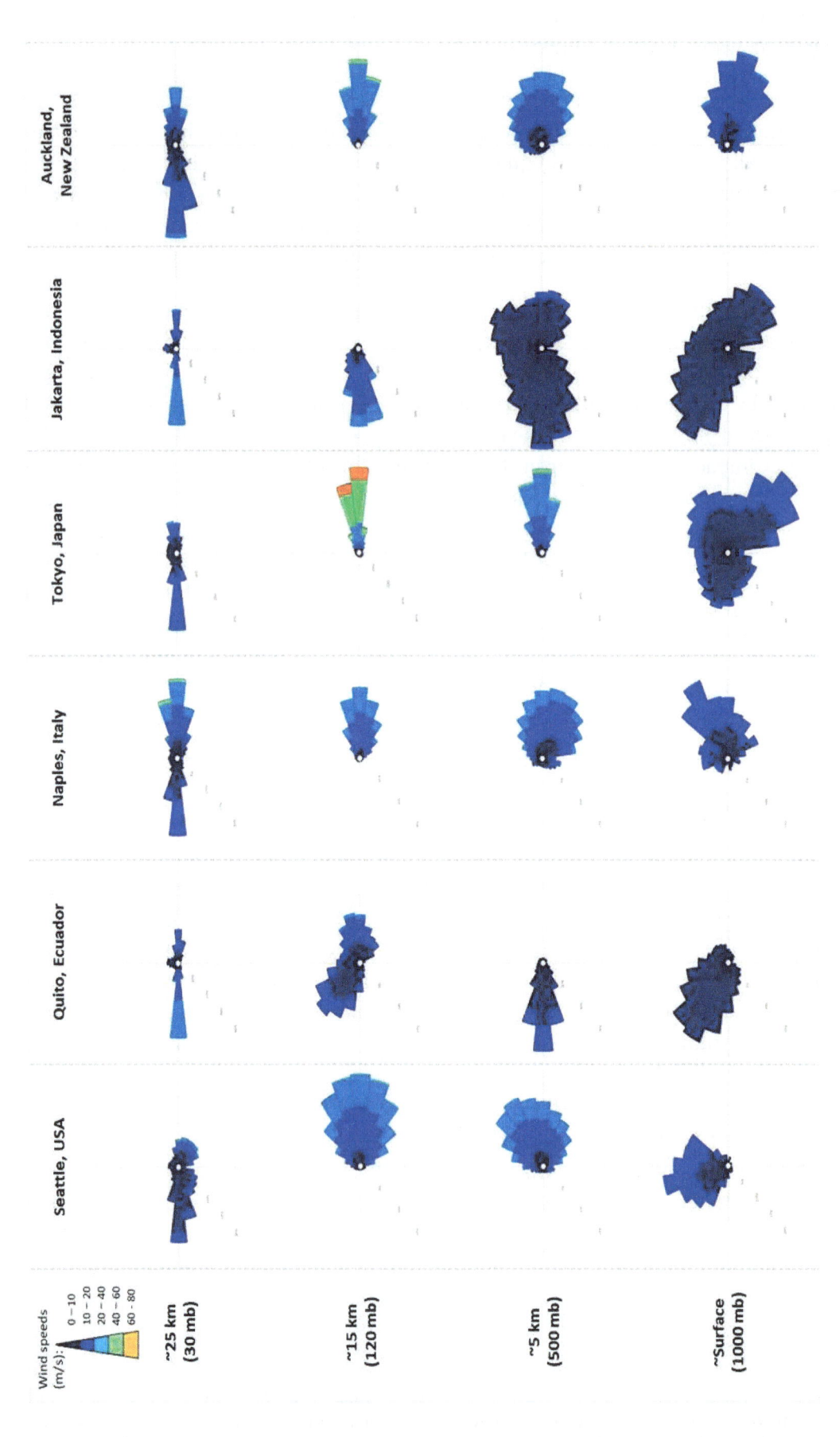

Figure 3.2 Wind roses with height above sea level for a selection of six cities in active volcanic areas. The wind roses show the frequency of winds, and the range of wind speeds, blowing towards particular directions during the period 2000 to 2009 (NCEP/NCAR 6 hourly reanalysis wind data). Standard compass directions are used with north at the top and east to the right of the wind roses.

Larger, heavier fragments are typically deposited closer to source and smaller, lighter fragments dispersed farther downwind. The resulting deposit accumulates in a radial pattern around the volcano only where wind is absent; more typically, it is dispersed in the direction of the prevailing winds throughout the eruption (Figure 3.3). As a general rule, deposit thickness and particle size distribution reduce with distance from the volcano: decreases that can be well fitted by a Weibull distribution (Bonadonna & Costa, 2013). Aggregation of particles can lead to secondary maxima in deposit thickness and particle sizes at varying distances from the volcano (Figure 3.3). Quantifying the hazard from future ash falls can therefore be problematic, in part because of data limitations that make eruption characteristics uncertain, but also because, given an eruption, the distribution of ash is then controlled by time and altitude-varying wind and weather conditions.

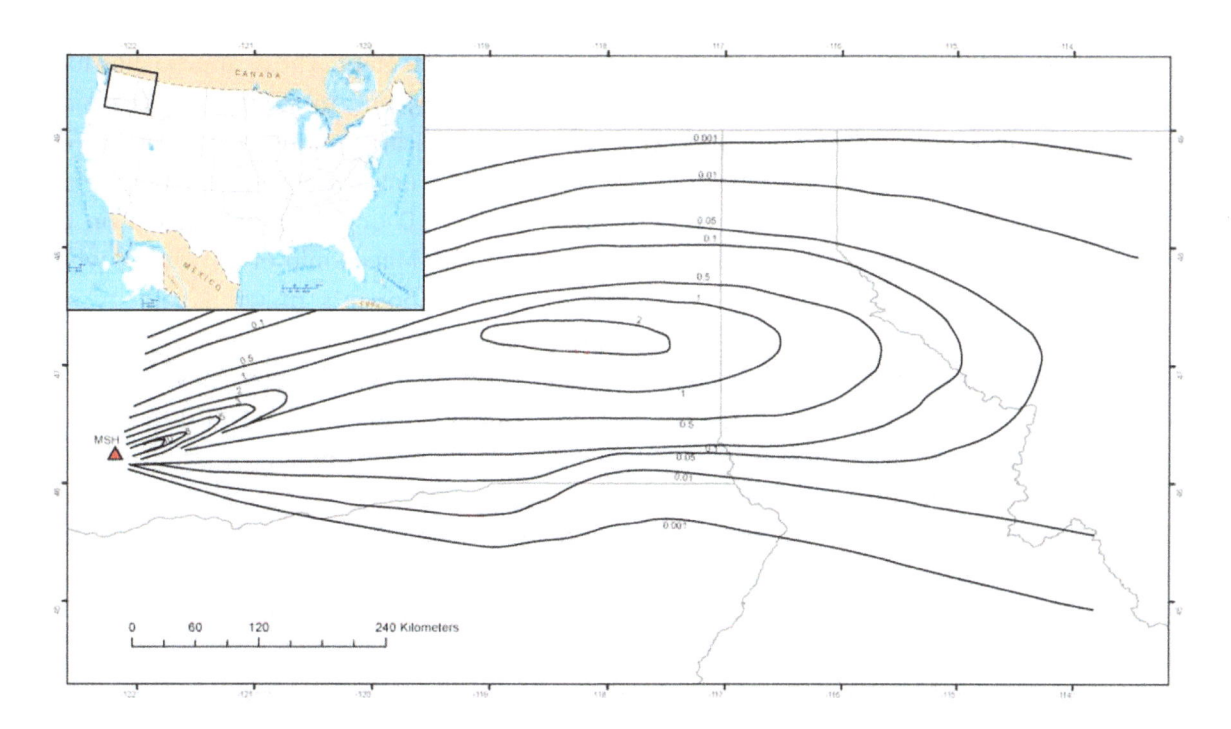

Figure 3.3 Isopachs (lines of equal ash fall thicknesses, here for cm of un-compacted ash) for ash fall produced during the 18 May 1980 eruption of Mount St Helens, USA (modified from Durant et al., 2009 and originally reproduced from, Sarna-Wojcicki et al., 1981). The secondary maximum in deposit thickness approximately 400 km east of the volcano is likely attributed to aggregation of particles. Winds were predominantly westerly and ash was dispersed across multiple US states, affecting transport, health, agriculture and infrastructure (BGVN, 1980). Clearly, these impacts would have been markedly different had they been erupted into an easterly wind, with most ash being dispersed to sea.

Quantitative ash fall hazard assessments have commonly been deterministic, in that they carry out single scenario simulations of historical or reference eruptions. Simulating historical scenarios can be useful for model calibration and for assessing what could happen if a well-known eruption took place in a modern day setting ('what if' scenarios). Reference scenarios build on this by considering that a particular volcano's eruptive history offers an indication of the range of future activity so that 'most likely' or 'maximum considered' eruption scenarios can be defined. These scenarios have typically been used for emergency management and response planning (e.g. Vesuvius, Italy: Zuccaro et al., 2008). However, as with many other natural hazard events, deterministic studies are increasingly being combined with, or subsumed within, probabilistic frameworks to allow understanding of the full range of plausible outcomes. Probabilistic methods are particularly important for determining volcanic ash fall hazard because the range of potential magnitudes, styles and external factors, such as wind conditions, may be large. A full hazard assessment should ideally account for the uncertainties associated with forecasting how much ash will be erupted and where it will be dispersed and deposited, and should be updateable as new information becomes available (e.g. Bonadonna, 2006).

Due to the complexities involved in forecasting ash dispersion and deposition, algorithms and numerical models are employed. Traditionally, numerical modelling of ash dispersal and deposition has been carried out for three purposes: 1) to better understand and quantify plume dynamics and particle sedimentation mechanisms, 2) to assess the hazard associated with ash fallout at ground level and 3) to assess the hazard from ash within the atmosphere. We focus here on the use of models and methods for ash fall hazard assessment. These approaches have been comprehensively reviewed by Bonadonna & Costa (2013) but, to summarise, there are three main approaches of differing complexity: 1) 1D theoretical thinning relationships (e.g. Bonadonna & Costa, 2012, Pyle, 1989), which typically utilise empirical data to estimate ash sedimentation with distance from the volcano; 2) analytical 1D or 2D models based on the advection-diffusion-sedimentation equation (e.g. Suzuki, 1983, Macedonio et al., 1988), which assume effective horizontal diffusion, negligible vertical components of wind velocity and diffusion and uniform atmospheric conditions with distance from the volcano; and 3) numerical 3D time-dependent models, based on Eulerian, Lagrangian or hybrid approaches to particle transport (e.g. Heffter & Stunder, 1993, Barsotti et al., 2008, Ryall & Maryon, 1998) that can be used to assess airborne ash concentrations. All ash dispersal models are subject to uncertainty because of inherent natural variability in volcanic and climatic systems and processes and because of an incomplete understanding of the processes by which ash from different eruption magnitudes and styles is dispersed. In theory, this latter uncertainty is being reduced, or at least better quantified, through continued model development and calibration, application of probabilistic approaches and sensitivity testing. The output from all ash dispersal models is a spatial appreciation of ash intensity (thickness, load or atmospheric concentration).

In the following sections, we present probabilistic ash fall hazard modelling approaches at the global and local scales, as a demonstration of the range of methods available for this type of analysis. Alternative methods are also available, as discussed above. When considering the appropriate model to use for ash hazard analysis, one must consider the purpose of the assessment, its end-users and assessment period, the required resolution, available data and any logistical constraints, such as computational resources or time. Careful consideration of the methodology is necessary to ensure that the results are valid and fit for purpose. For example, models at global or regional scales are typically only appropriate to long-term (years to decades) hazard estimation because of high-level data requirements and analyses that typically prohibit near real-time assessment. Locally applied models can incorporate time-dependent knowledge such as monitoring data to allow short-term hazard estimation, i.e. hourly, daily or weekly, during times of unrest.

3.2.1 Global assessment

Few global studies of volcanic hazard or risk have been attempted: Yokoyama et al. (1984), Small & Naumann (2001) and Dilley et al. (2005) all undertook global volcanic hazard analyses but, for the only study in which volcanic hazard was defined spatially (Small & Naumann, 2001), a constant level of hazard was assumed within concentric circles of 200 km radii extending from the source volcano. These types of analyses, while providing valuable early estimations, were erroneous and misleading with respect to ash fall hazard because they did not account for the varying dispersion of ash due to time- and altitude-varying wind conditions and did not account for potentially different eruption magnitudes or styles from the volcano in question.

In an attempt to deal with these uncertainties, Jenkins et al. (2012a, 2012b) developed a quantitative framework for assessing ash fall hazard on a regional scale. This methodology employed stochastic simulation techniques, utilising statistical analyses of eruption frequencies and magnitudes, a numerical ash dispersal model and multi-decadal time-series of wind conditions at each volcano in the Asia-Pacific Region. Typically, ash fall hazard assessments consider the hazard from a single volcano (e.g. Macedonio et al., 2008, Costa et al., 2009). However, such studies may not fully capture the ash fall hazard in volcanically active areas where one location may potentially be affected by ash falls from multiple volcanoes (e.g. modelling suggests that at least 55 volcanoes are capable of depositing ash in central Tokyo: Jenkins et al., 2012b). The approach suggested by Jenkins et al. (2012a) deliberately accumulates ash fall hazard on a cell-by-cell basis so that all volcanoes capable of impacting each cell are considered. For the UNISDR 2015 Global Assessment Report (Jenkins et al., 2014b), and as part of the Global Volcano Model, we expanded the existing methodology of Jenkins et al. (2012a) to carry out a first global probabilistic ash fall hazard assessment.

Methodology

- Identifying volcanoes for analysis

Data regarding eruptive behaviour for volcanoes around the world were sourced from the Smithsonian Institution's newly revised 2013 Global Volcanism Program catalogue of Holocene (approximately the last 10,000 years) events (www.volcano.si.edu). To be considered in our analysis, volcanoes had to have exhibited at least one recorded eruption in the Holocene, and not be classified as submarine, hydrothermal, fumarolic or of unknown type, giving a total of 720 volcanoes (Figure 3.4).

- Deriving unique eruption frequency-magnitude relationships

A statistical analysis of the global eruption record was carried out to determine unique eruption frequency-magnitude relationships for each of the 720 volcanoes based upon the eruption history of each volcano and the averaged eruptive behaviour of analogous volcanoes globally, following the methodology of Jenkins et al. (2012a) (Appendix A). To determine which portion of a volcano's eruption record could be considered complete, and thus useful for data mining, step changes in the cumulative number of eruptions with time (representing completeness dates) were identified for different geographic regions and eruption magnitudes (results shown in Mead & Magill, 2014). While we calculate average recurrence intervals for all possible eruption magnitudes, we only simulate large magnitude eruptions (VEI ≥ 4) due to their potential for generating thicker and more widely dispersed ash falls. Ash falls from smaller magnitude eruptions (VEI ≤ 3) will not extend as far from the source volcano or produce such large thicknesses because of the volume of ash produced and the column height to which it is erupted. However, they will occur more frequently: a VEI 3 eruption is around twice as likely as a VEI 4 eruption, but about an order of magnitude smaller (Newhall & Self, 1982).

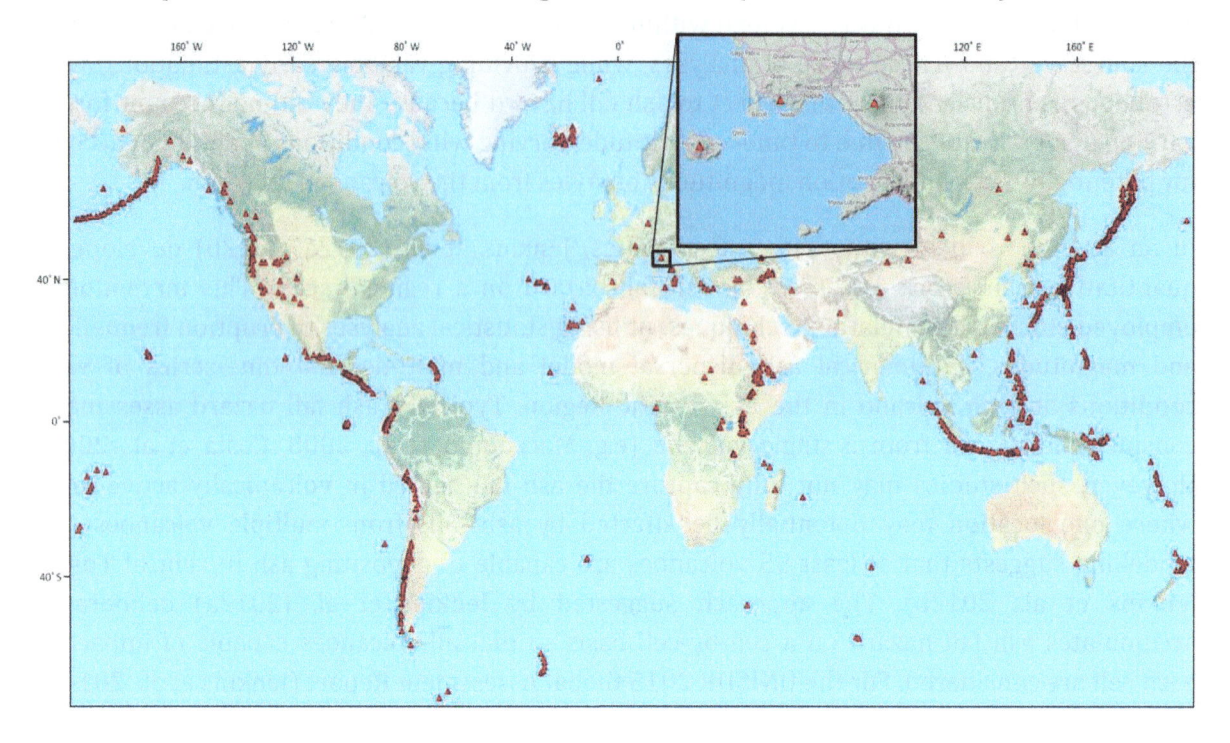

Figure 3.4 Map showing volcanoes considered in the global assessment of ash fall hazard (n=720). Local case study volcanoes (Section 3.2.2) are shown in the inset box. Base map provided by OpenStreetMap.

- Ash dispersal modelling

The ash dispersal model ASHFALL (Hurst, 1994) was used to simulate a large range (up to 19,200) of possible ash footprints for each volcano, on a 10 x 10 km grid that extended 1,000 km from the volcano in question. Analytical 2D models like ASHFALL are well suited to probabilistic assessments because a large number of simulations can be carried out efficiently using available computing resources. However, local use of these types of models should be limited to relatively small computational domains for which the assumption of uniform wind is likely to be valid (e.g. a few hundreds kms). Variations in possible erupted ash volume, column height, wind conditions and parameters, such as settling velocity (derived from particle size distribution and density), were accounted for by randomly sampling from predefined probability distributions (Table 3.1). The probability distributions dictated the magnitude and allowable range for each variable, and their relative likelihoods within this range.

Table 3.1 ASHFALL model inputs, derivations, ranges and references used for the global ash hazard assessment. The horizontal diffusion coefficient and Suzuki constant were not sampled and remained constant for all simulations. For further details the reader is referred to Jenkins et al. (2012a).

Input	Derivation	Min.	Max.	Source
VEI	An equal number of simulations (n=4800) were carried out for each VEI class to allow for possible variations in wind conditions	VEI 4	VEI 7	Newhall and Self, 1982
Eruption volume	Sampled from a power law distribution within the determined VEI classification	0.1 km^3 (low VEI 4)	1000 km^3 (high VEI 7)	
Column height (Z)	Related to eruption volume (V) by [2]: $Z = 8.67 \log_{10}(V) + 20.2$	11.5 km	46 km	Carey and Sigurdsson, 1989; Jenkins et al., 2007
Settling velocities	Means and standard deviations varied for each simulation to give a truncated lognormal distribution	0.3 m/s	9.5 m/s	Walker, 1981; Woods and Bursik, 1991; Sparks et al., 1992
Wind conditions	Uniform sampling of wind speed and direction at 35 height levels from a 40 year ERA-Interim record closest to the volcano (1.5 degree grid).			European Centre for Medium range Weather Forecasting global re-analysis project (www.ecmwf.int)
Horizontal diffusion coefficient		6,000 m^2/s		Hurst, 1994
Suzuki constant		5		Macedonio et al., 1988; Hurst, 1994

Results

For each modelled grid cell, the ash fall hazard was identified by considering all simulations from each volcano that impacts the grid cell and summing the corresponding annual eruption probabilities over all eruptions, potentially from a number of volcanoes. For each cell, the annual probability (or inversely the average recurrence interval) for ash thickness exceeding certain accumulation thresholds can then be calculated – for a given volcano, a selection or all volcanoes, and for different seasons or eruption magnitudes. Here we show the aggregated results (Figures 3.5 - 3.7), incorporating the full range of possible eruption scenarios and environmental conditions. The influence of tectonic setting and prevailing wind conditions (e.g. predominantly to the east in the Americas) on the aggregated ash hazard is clear, with greater hazard being calculated along the major subduction zones where higher eruption frequencies and magnitudes may be expected.

---

[2] This relationship is valid for the large eruptions simulated here (VEI ≥ 4) for which duration is typically a few hours. For long-duration or smaller magnitude eruptions, a stronger empirical relationship exists between column height and mass eruption rate (Wilson and Walker, 1987).

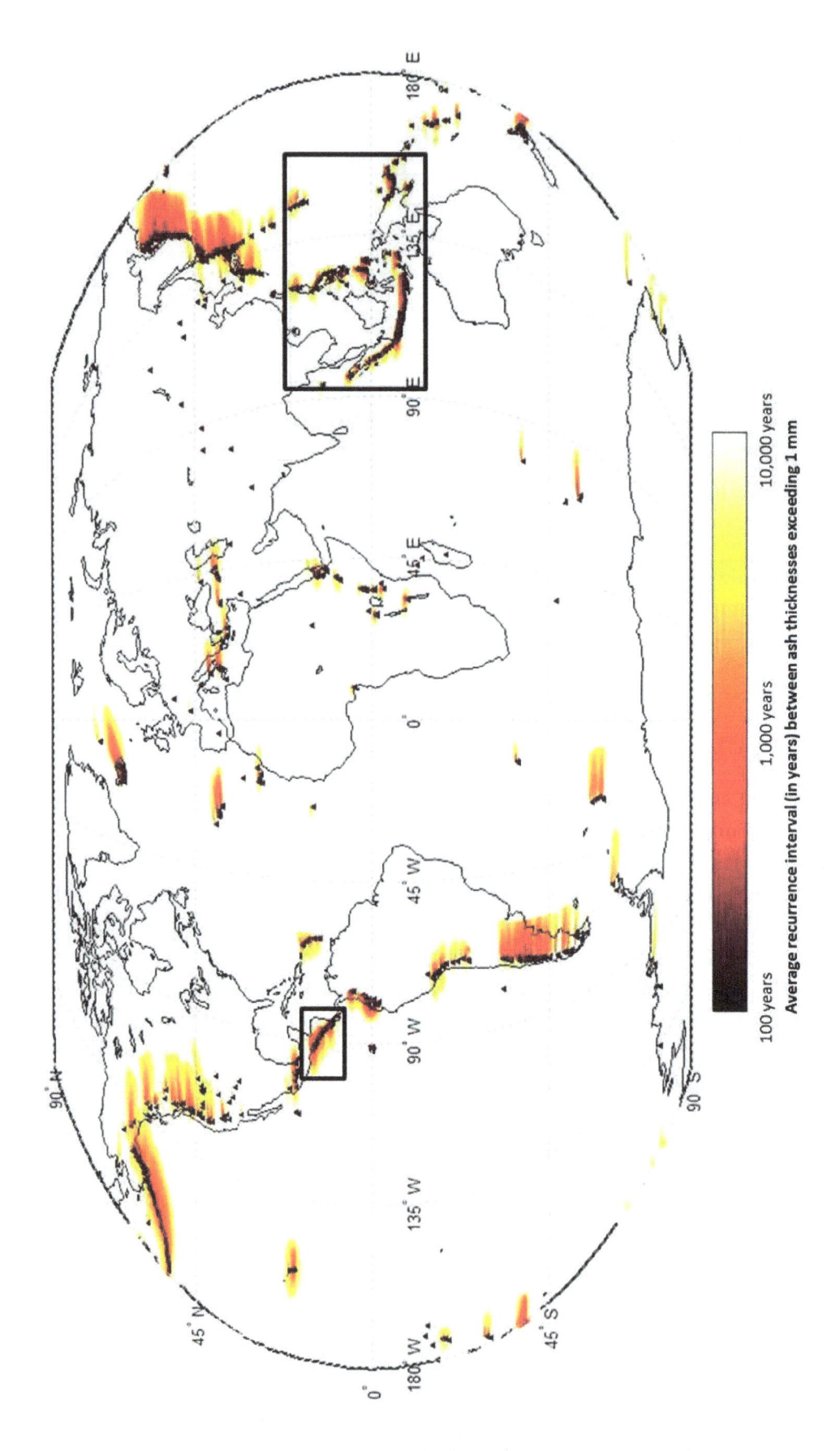

Figure 3.5 Global map of probabilistic ash fall hazard, displayed here as the average recurrence interval between ash fall thicknesses exceeding 1 mm: a threshold that may cause concern for the aviation industry and critical infrastructure. Ash concentration in the atmosphere (and thus the cause of flight disruptions during the 2010 Eyjafjallajökull eruption in Iceland) are not output by 2D modelling and so disruption can occur, and has occurred, in areas where ash fall thicknesses are significantly less than 1 mm (e.g. Biass et al., 2014, Folch & Sulpizio, 2010). Enlarged maps are shown for the inset boxes in Figure 3.6.

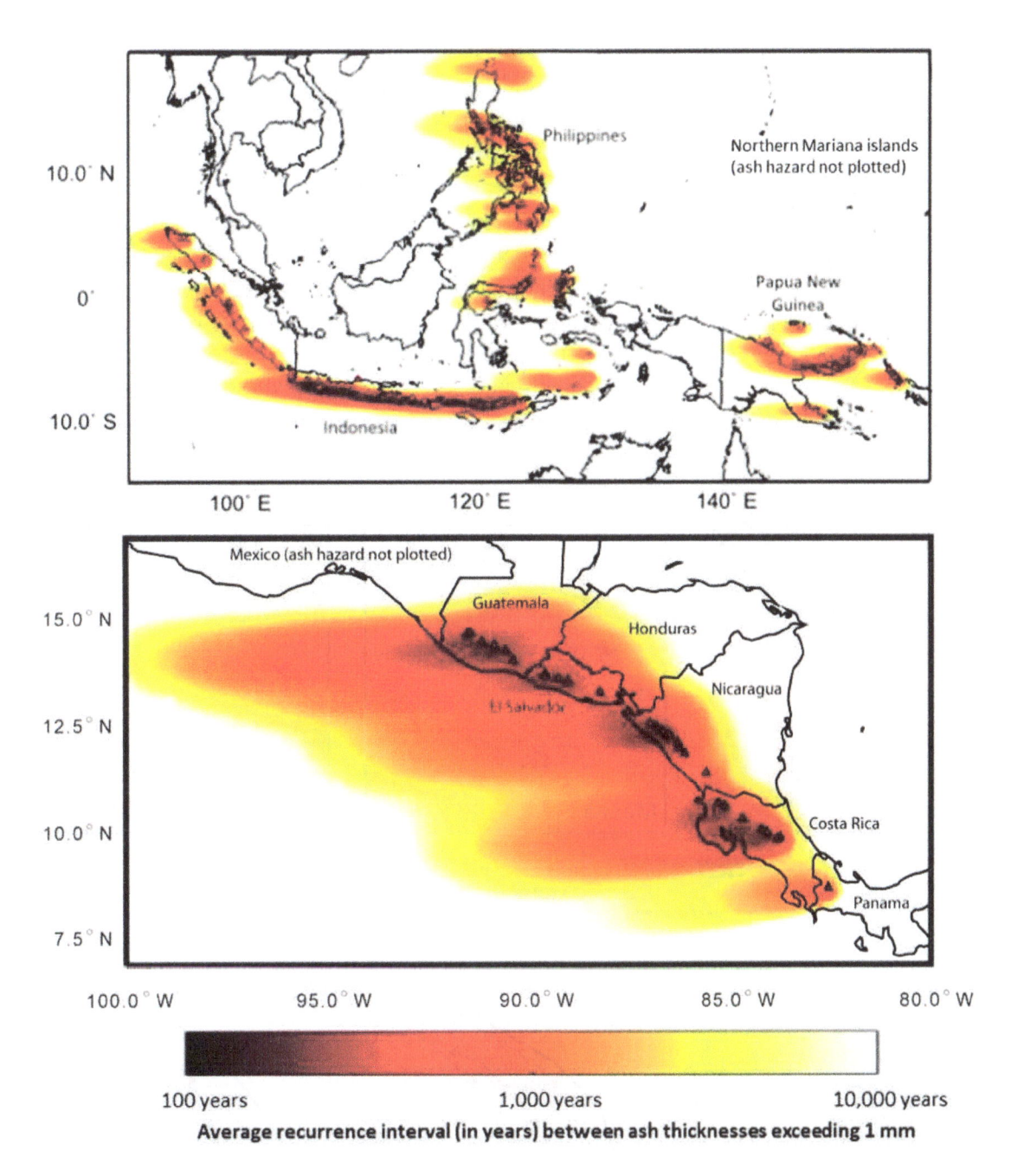

Figure 3.6 Regional maps of probabilistic ash fall hazard, displayed here as the average recurrence interval between ash fall thicknesses exceeding 1 mm. A few mm of ash can disrupt transport, power and water systems and, at the wrong time of year, can affect crop pollination and productivity.

The advantage of an aggregated multi-volcano probabilistic approach is that the ash fall hazard at any grid cell can be estimated for all ash fall thicknesses and all simulated average recurrence intervals, allowing a hazard curve to be built. Hazard curves compare hazard intensity (in this case thickness) to the expected annual probability of exceedance or average recurrence intervals and are useful for comparing the hazard at important locations, e.g. cities, airports or proposed infrastructure sites. They can act as fore-runners to more detailed local study, particularly by identifying which volcano or volcanoes may contribute most to the overall hazard (Figure 3.7). Tokyo, Manila and Jakarta are megacities of more than 10 million people located in volcanically active regions where ash falls of 10 mm or more may be expected at least every 1,000 years. Within 1,000 km of Tokyo, Jakarta and Manila, respectively, 57, 35 and 13 potentially active volcanoes have been identified. However, ash fall hazard in each city is not a direct function of volcano proximity and will be dictated by likely eruption frequencies, magnitudes and wind conditions. Figure 3.7 shows that Tokyo is estimated to have the highest relative ash fall hazard for thicknesses exceeding 1 mm, which may be expected given the large number of surrounding volcanoes. Despite having the lowest number of surrounding volcanoes of the three cities, the frequency of ash falls in Manila is comparable to that of Jakarta. This is a result of the ash fall hazard in Manila being dominated by few local sources upwind from the city, which are capable of producing relatively large thicknesses in Manila. By contrast, ash fall hazard in Jakarta is expected to be influenced more by numerous distal sources resulting in a greater number of thinner and less frequent falls (hence the more shallow hazard curve). Readers are reminded that these values represent a minimum hazard as small magnitude and long-quiescent volcanoes are not considered.

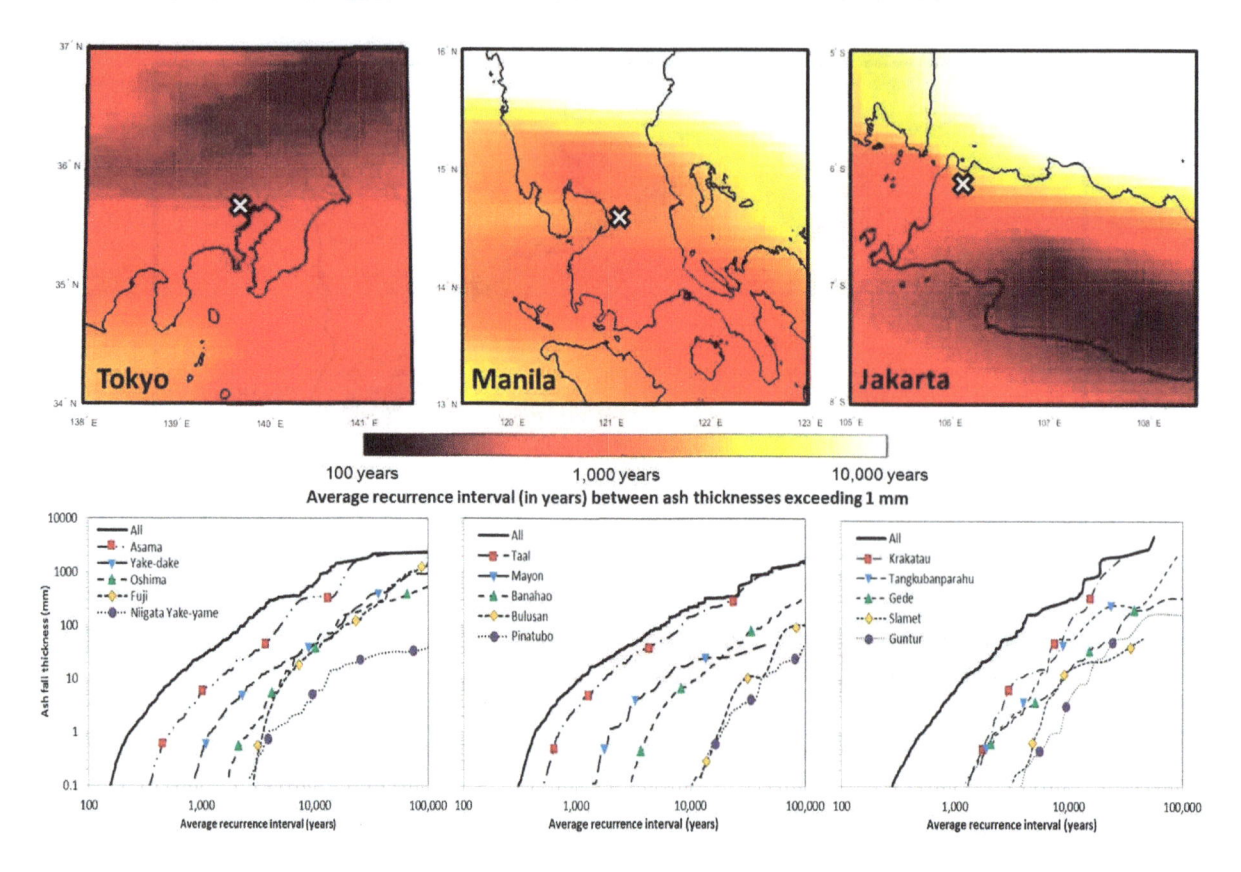

Figure 3.7 Ash fall hazard maps (average recurrence intervals for thicknesses of 1 mm or more) and curves across a) Tokyo, Japan, b) Manila, Philippines, and c) Jakarta, Indonesia. The cross marks the centre of each city. Ash hazard curves show the average recurrence intervals between exceeding the range of ash thicknesses, aggregated across all volcanoes (solid left most curve) and disaggregated to show hazard by volcano (curves to right). These draw from the global hazard modelling and are therefore useful for highlighting the relative hazard from large magnitude eruptions. They should not be used in place of more detailed local studies and represent a minimum hazard as discussed in the text.

Discussion

A global probabilistic ash fall hazard assessment has been produced for the first time using numerical simulation. Probabilistic methods were applied to capture the range of possible eruption scenarios (VEI ≥ 4) at each active volcano around the world (n=720) and, critically, to incorporate likely variability in wind speeds and directions, with altitude at each volcano (12.9 million scenarios total globally). As with any numerical representation of a complex and variable natural phenomenon, there are limitations to this approach. These are outlined in Jenkins et al. (2012a), but include the following:

- Uncertainty in estimating the completeness of available eruption data affects the calculation of eruption probabilities, particularly in areas where breakpoints between complete and incomplete subsets of data are difficult to identify. The uncertainty in identifying breakpoints has been reduced in this study by applying the statistical techniques shown in Mead & Magill (2014).
- Volcanoes that have no recorded eruptions in the Holocene (approximately the last 10,000 years) - either because there have been no eruptions, few or no geological studies or observations at the volcano or inaccurate dating of the deposits - have not been included in our assessment. The probability-weighted contribution to the hazard from volcanoes with long quiescent intervals in the order of 10,000 years or more that have not erupted in the Holocene will be minimal. However, calculated eruption frequencies at volcanoes with inadequate eruption records are subject to greater uncertainty until further studies are carried out. For example, prior to the VEI 4 eruption of Chaitén volcano in Chile in 2008, the eruption record contained one previous eruption, approximately 9,400 years earlier. Geological investigations in the years since 2008 have identified a further two large magnitude eruptions approximately 5,100 and 370 years before present.
- By drawing data from volcano analogues to supplement eruption frequency-magnitude relationships, we average behaviour across multiple volcanoes. A more detailed volcano-by-volcano assessment is not possible for many areas because few geological or historical records exist.
- To limit data and computing requirements, and thus allow simulation of a large number of eruption scenarios, our model assumes wind fields remain uniform with time and distance from the source volcano: this means that in areas with markedly variable wind fields model results are potentially less reliable with distance.
- Only large magnitude explosive eruptions (VEI 4 to 7) are considered, with ash fall simulated to a maximum distance of 1,000 km from the volcano due to the rarity and low intensity of ash falls beyond that extent. Therefore, in proximal areas, the ash hazard contribution from smaller explosive eruptions is missing and in very distal areas (>1,000 km) the contribution from very large explosive eruptions is not considered. VEI 8 eruptions are known to have occurred in the past, e.g. at Yellowstone volcano, USA, 640,000 years ago; however, none have been produced in the Holocene and are thus absent from this assessment.

The use of this approach is justified here in providing a broad, state-of-the-art assessment of ash fall hazard for all active terrestrial volcanoes across the globe, allowing us to identify areas of higher, and lower, hazard and volcanoes and regions which may benefit from further, more detailed, study. A comprehensive volcanic hazard assessment must also consider the spectrum of additional volcanic hazards, such as pyroclastic density currents, lahars and lava flows.

3.2.2 Local assessment

The Neapolitan area in Italy represents one of the highest volcanic risk areas in the world, both because of the presence of three potentially explosive and active volcanoes (Vesuvius, Campi Flegrei and Ischia), and because of the extremely high exposure (over a million people located in a very large and important metropolitan area). Even though pyroclastic density currents are likely to be more destructive, ash falls pose a serious threat to the area because they impact over much larger areas and may cause major disruption to the European economy (Folch & Sulpizio, 2010).

Previous studies for the Neapolitan area have combined field data of ash fall deposits and numerical simulations of ash dispersal (often considering tens of thousands of wind profiles to account for wind variability) to assess ash fall hazard from Neapolitan volcanoes (Figure 3.8) (e.g. Barberi et al., 1990, Cioni et al., 2003, Macedonio et al., 2008, Costa et al., 2009). However, these studies have produced conditional ash load probability maps (i.e. not accounting for the probability of eruption) for a specific scenario (e.g. fixing magnitude, intensity and vent position) or for a few reference scenarios at Mount Vesuvius and Campi Flegrei. This could be considered a semi-probabilistic approach because variability in wind conditions is accounted for, but variability in eruption style, frequency, magnitude and location, and their associated probabilities of occurrence, is not. These approaches are still largely used in volcanology, but have severe limitations in representing the real volcanic hazard. Selva et al. (2010b) and Marzocchi et al. (2010) have attempted to overcome some of these limitations by building on the Campi Flegrei ash fall hazard assessment of Costa et al. (2009). In particular, this work has developed the Bayesian Event Tree Volcanic Hazard tool (BET_VH: Marzocchi et al., 2010; vhub.org/resources/betvh), which accounts for the most important sources of uncertainty and natural variability in eruptive processes, statistically combining the contribution to the final probabilistic volcanic hazard analysis from all possible scenarios, making use of the law of total probability.

Methodology

BET-VH is an open source tool, based on Bayesian inference modelling, that properly accounts for the aleatoric (intrinsic natural variability) and epistemic (linked to our limited knowledge of the eruptive process) uncertainties, propagating these through to a probabilistic ash hazard assessment. This method, including accounting for the related uncertainties, can be described by a probability density function denoted as $[\theta]$. The analysis is based on an Event Tree (see Figure 3.9), and all uncertainties are assessed at each level, namely on the eruption occurrence ($[\theta_{1\text{-}2\text{-}3}]$), on vent position ($[\theta_4]$), on the eruptive scale ($[\theta_5]$), on the production of ash ($[\theta_6]$) and on its transport, dispersal and deposition by the wind ($[\theta_7]$ and $[\theta_8]$), according to the following equation (from Marzocchi et al., 2010):

$$[\theta^{(k)}] = [\theta_{1-2-3}] \sum_{i=1}^{N_v} \left\{ [\theta_4^{(i)}] \sum_{j=1}^{N_s} [\theta_{5,i}^{(j)}][\theta_{6,j}][\theta_{7,i,j}^{(k)}][\theta_{8,i,j}^{(k)}] \right\}$$

(1)

where *k* indicates a specific point in the target domain, *i* and *j* indicate respectively a given vent position (out of the possible N_v) and size class (out of the possible N_s, see Marzocchi et al., 2010 for more details). The BET_VH tool has been used in other volcanic areas to produce a full probabilistic hazard assessment for ash fall and other volcanic hazardous phenomena (Sandri et al., 2012, Sandri et al., 2014). An ongoing improvement of the method aims at performing the production of fully probabilistic hazard curves that include uncertainty bounds, based on the method proposed for seismic hazard by Selva & Sandri (2013). Site-specific hazard curves represent the most complete information about the hazard allowing volcanologists to produce hazard maps at different levels of probability and/or different levels of hazard intensity (SSHAC, 1997).

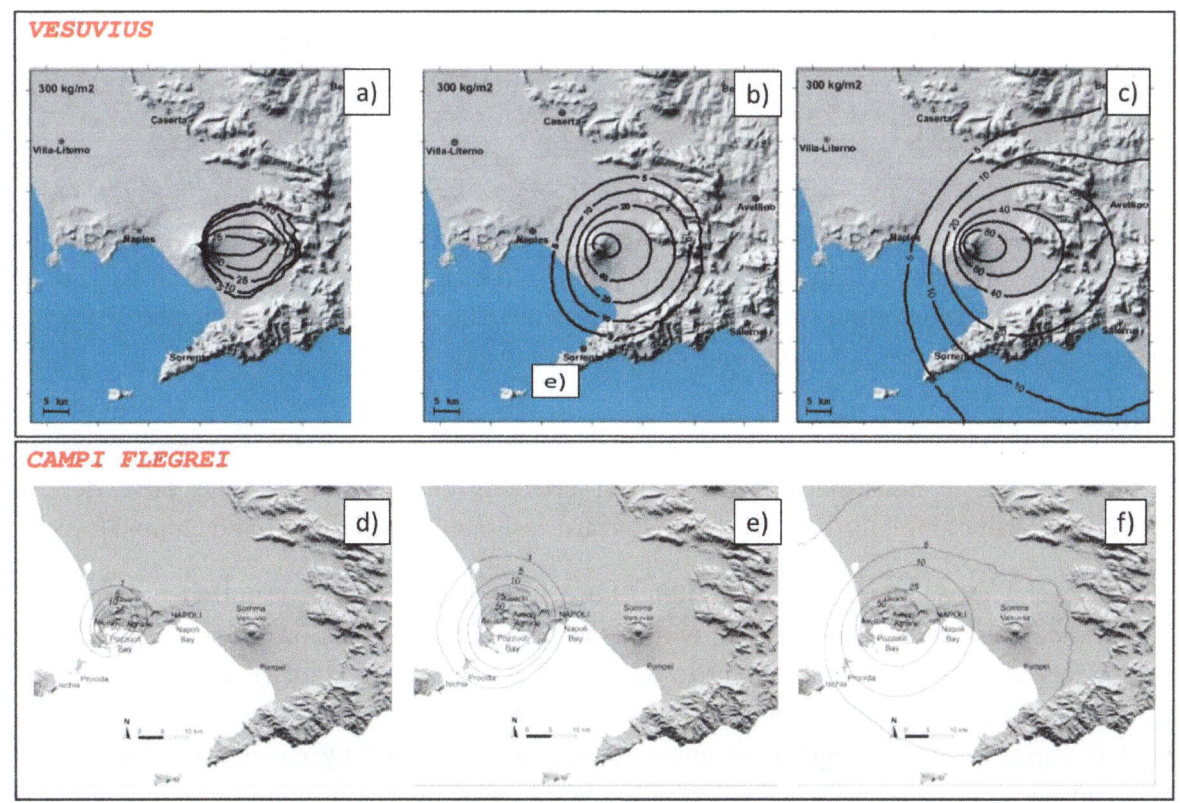

Figure 3.8 Maps showing the probability of ash fall loads exceeding 300kg/m² given the occurrence of an eruption of size a) Violent Strombolian, b) Subplinian and c) Plinian at Vesuvius (Macedonio et al., 2008), and of size d) low, e) medium and f) high explosive, from the eastern vent (Averno-Monte Nuovo) at Campi Flegrei (Costa et al., 2009). A loading of 300 kg/m² is expected to result in significant (>80%) of weak flat wooden, reinforced concrete and steel roofs (Costa et al., 2009, Zuccaro et al., 2008). This figure is reproduced in Chapter 6.

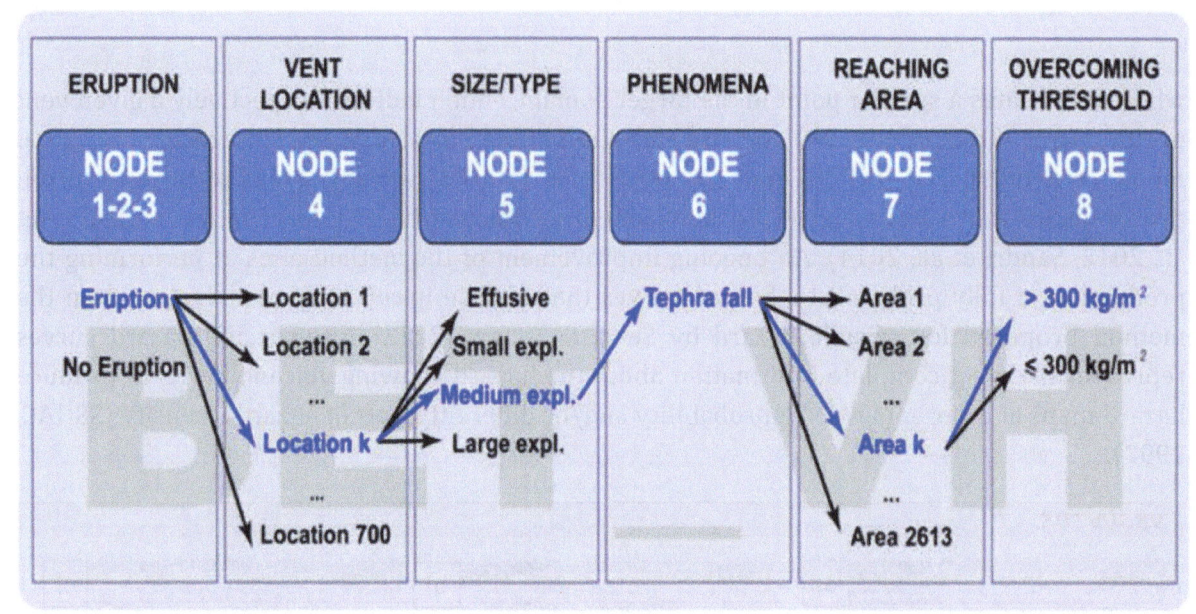

Figure 3.9 Event tree of the model BET_VH for a specific volcano to evaluate the PVHA for ash fall loads above 300 kg/m². This figure is reproduced in Chapter 6, as Figure 6.1.

Large-scale global or regional approaches to probabilistic ash hazard assessment (Section 3.2.1) are currently only appropriate for long-term (years to decades) assessment because the current unrest state of each volcano is not taken into account. The advantage of a tool like BET_VH is that it can be used at both the long- and short- (hours to days) timescales.

Results

For the factors in Equation 1, a number of previous and ongoing studies have derived appropriate values for long- and short-term hazard assessment at Vesuvius and Campi Flegrei:

- For Vesuvius, Marzocchi et al. (2008, 2004) have estimated the eruption probabilities, locations and style (nodes 1 - 5), while Macedonio et al. (2008) have used ash dispersal models to estimate the probabilities associated with spatial extent (node 7) and critical thresholds (node 8: 300 kg/m² is shown here as example).
- For Campi Flegrei, a significant improvement was achieved by considering all possible eruptive sizes and vents, conditional on the occurrence of an eruption (Selva et al., 2010a). This was computed by accounting for the uncertainty in the vent location (node 4: Selva et al., 2012), the eruption style (node 5: Orsi et al., 2009), and the probability weighted dispersion of the ash over impacted areas (nodes 6 - 8: see Figure 3.10). This kind of approach is particularly useful for large and potentially very explosive calderas, such as Campi Flegrei, for which the position of the vent is critical and can impose a large uncertainty on the final hazard assessment.

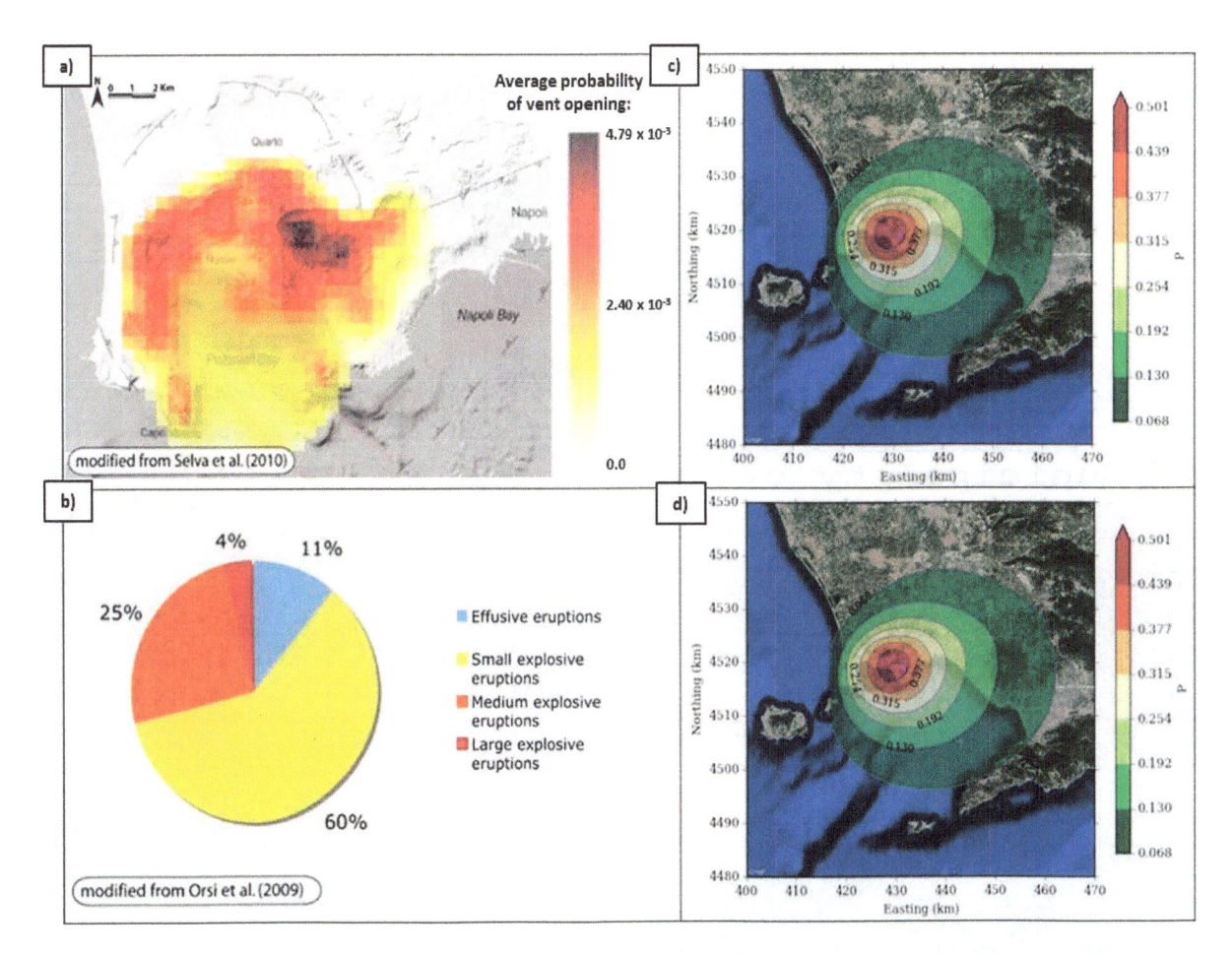

Figure 3.10 Panel a) and b) show the mean probability of vent opening and possible eruptive sizes at Campi Flegrei, respectively. Panels c) and d) show the mean probability of an ash fall load and a load larger than 300 kg/m², respectively, conditional on the occurrence of an eruption at Campi Flegrei from any possible vent location and of any possible size, properly combined according to the probabilities shown in panels a) and b). A loading of 300 kg/m² is shown here and is expected to result in significant (>80%) of weak flat wooden, reinforced concrete and steel roofs (Costa et al., 2009, Zuccaro et al., 2008). This figure is reproduced in Chapter 6.

Discussion

Despite the recent significant steps towards achieving a full and comprehensive probabilistic ash hazard assessment for the Neapolitan area, much work has still to be done.

- Long-term ash fall hazard assessment

Two ongoing Italian projects (ByMuR, 2010-2014, DPC-V1, 2012-2013) have been working towards providing further improvements in long-term ash hazard assessment for Vesuvius, Campi Flegrei and Ischia. A preliminary merging of the full probabilistic ash fall hazard assessments at Vesuvius and Campi Flegrei for the municipality of Naples is shown in Figure 3.11. The variability of eruptive parameters within each size class must also be modelled, to evaluate its importance and impact on the final assessment. The production of hazard curves, as mentioned above, is a necessary step if results are to be included into quantitative risk assessment procedures. The assessment of the epistemic uncertainty on the hazard curves represents the most complete results that we aim to achieve. An updated probabilistic volcanic hazard assessment for the municipality of Naples is planned to be ready for the end of the project ByMuR (late 2014) and will consist of hazard curves for a set of confidence levels (regarding epistemic uncertainties), for each cell of the grid covering the municipality of Naples (Figure 3.11).

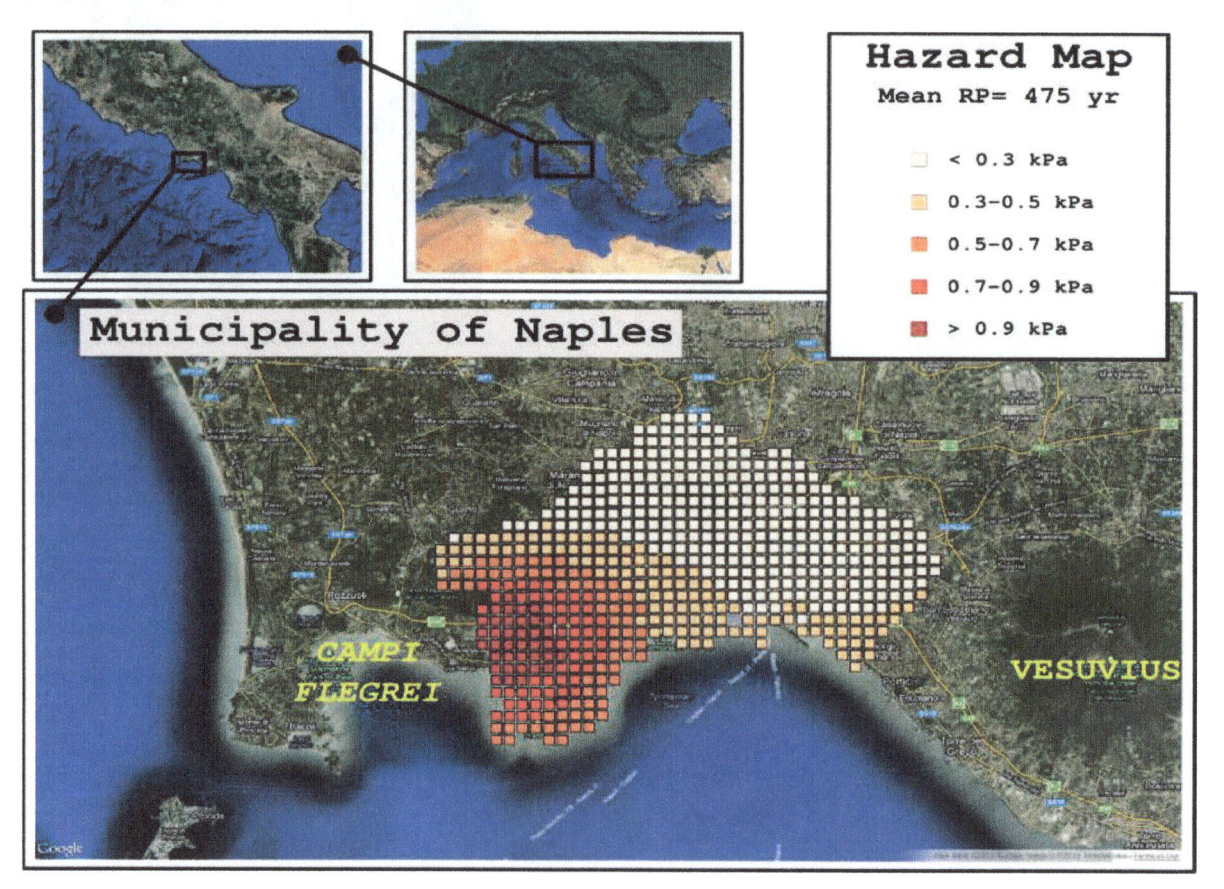

Figure 3.11 Hazard map (mean) for ash fall loading with an average recurrence interval of 475 years (exceedance probability threshold equal to 0.1 in 50 years), considering both Vesuvius and Campi Flegrei on the municipality of Naples (area shown by pixels). This figure is reproduced in Chapter 6.

- Short-term ash fall hazard assessment

Two research projects (the Italian DPC-V2, 2012-2014, and the EC MEDSUV, 2013-2015) aim at providing quantitative improvements for short-term ash fall hazard assessment at Vesuvius and Campi Flegrei in order to allow such hazard assessments to be operational for emergency management. This approach can provide a tool of primary importance during potential volcanic unrest episodes and for ongoing eruptions. The tool could be updated frequently (hourly or daily) to account for a rapidly evolving crisis situation, providing crucial information for crisis management. Theoretically, short-term assessments should be based on sound modelling procedures stemming from frequently updated meteorological forecasts and information about the crisis evolution. In addition, the relevance of epistemic uncertainties arising from the forecast of the future eruption dynamics and wind conditions, and from the ash dispersal model, should be estimated.

3.3 Ash fall impacts

Understanding how ash falls can impact communities, built environments and economies, and the consequences with varying hazard intensities, is critical to understanding ash fall risk. In this section, we draw from data collected during and following past eruptions, as well as experimental and theoretical studies, to highlight the impacts an ash fall may have for key sectors of society such as health, infrastructure and the economy. We also discuss how the varied characteristics of volcanic ash and the context, e.g. the timing and duration of an ash fall or the environmental conditions at the time, can influence impacts. Section 3.4 then builds upon these observations and experiments by identifying approximate thresholds for damage and reduced functionality. Although we mostly consider impacts associated with discrete ash falls, we note that remobilisation of ash by weather (e.g. wind, rain, snow) or human actions (e.g. clean-up, movement of vehicles) can lead to repeated impacts over weeks, months or even years after an eruption.

3.3.1 Hazard intensity

Historically, thickness of ash fall deposits has been the most common parameter used to assess the severity of an ash fall at a site (the hazard intensity) in relation to damage because it is relatively easy to measure in the field or interpret from eyewitness accounts and because most published data use this intensity measure (Wilson et al., 2012b). Most recently, ash loading (in mass per unit area, typically kg/m^2) has become a preferred measure of intensity as it is more informative when considering impacts to structures and agriculture and is a common output of numerical ash fall models. Measurements or estimates of load also avoid complications with measuring ash thickness due to ash compaction or rain following deposition. A water-saturated ash deposit can be up to twice the load of a similar thickness of dry ash (Macedonio & Costa, 2012).

3.3.2 Ash characteristics

The quantity of ash (thickness or loading) is not the only mechanism by which ash can cause damage or disruption. In practice, the impacts sustained by societal assets, such as critical infrastructure networks and primary production, are determined by a complex interplay of factors. These include other ash fall characteristics such as ash surface composition and particle size, site specific factors, such as climate, and factors related to the affected assets, such as system design and pre-existing vulnerability (Wilson et al., 2012b). The physical and chemical properties of ash are largely controlled by eruptive dynamics and magma composition, although dispersal conditions, e.g. wind and rain, also play a role. Ash properties can therefore vary among different eruptions and within the same eruption.

Condensation of strong mineral acids (such as HCl, HF and H_2SO_4) in a cooling plume can lead to a strongly acidic surface coating on freshly-fallen ash (Delmelle et al., 2007). Complex interactions between primary gases and mineral surfaces in fragmenting magma lead to the presence of readily-soluble salts on ash surfaces (Delmelle et al., 2007) which are released upon contact with water or other body fluids (Stewart et al., 2013). The presence of soluble salts may also make freshly fallen ash highly conductive to electrical currents in certain weather conditions, which leads to a high risk of ash deposition causing disruption to electrical transmission and distribution systems via the mechanism of insulator flashover.

The density and mechanical strength (friability) of individual ash particles affects how difficult it is to remove deposits, for example if ash enters drainage networks and also how easily deposits can be ground down and re-suspended by wind or rain. Smaller particles are carried farther downwind from the vent, are more easily eroded and remobilised by wind, and can penetrate smaller openings than larger grains. Penetration of ash particles into the respiratory tract is dependent on particle size. Larger particles (>10 μm diameter) lodge in the upper airways, while those in the 4-10 μm size range deposit in the trachea and bronchial tubes. Very fine (< 4 μm diameter) particles may penetrate deeper into the lungs (Horwell & Baxter, 2006). Small particles can also adhere more readily to vegetation and infrastructure components. Magnetic minerals in some volcanic ashes may also influence adhesion, although there are very few studies on this phenomenon.

3.3.3 Context

As well as the intensity and characteristics of ash falls, their duration and timing within an eruption also strongly determines what impacts may occur, particularly for the agriculture production cycle. In addition to the eruption duration and timing with respect to season, environmental factors such as wind and rain on ash dispersal and remobilisation, will also dictate the level of impact a community experiences. For example, dry, windy environments facilitate remobilisation of ash deposits near the ground which can 'ash-blast' vegetation and cause challenging living conditions for decades after the eruption (e.g. Hudson 1991 eruption: Wilson et al., 2011). Conversely, high precipitation can saturate ash deposits and increase the load causing greater structural and agricultural damage as well as large, potentially damaging mudflows (e.g. Pinatubo eruption 1991: Newhall & Punongbayan, 1996). Deposits of ash have been observed to form surface crusts which may serve to inhibit water infiltration and soil-gas exchange and increase surface runoff (Blong, 1984). Ash falls onto glaciated or snow-covered areas can act to increase or reduce run-off, depending upon the ash fall thickness, with associated consequences for surrounding populated areas (preliminary assessment provided by a Supraglacial Ash fall Likelihood Index: Supplementary Case Study 3).

3.3.4 Assessment of ash fall impacts

The assessment of ash fall impacts is a comparatively underdeveloped field compared to, for example, seismic impact assessment. This is in part because large explosive eruptions impacting populated areas are relatively infrequent and because few comprehensive quantitative impact studies have been carried out. Impacted zones can be dangerous and sometimes inaccessible for long periods. Broadly speaking, three main zones of ash fall impact may be expected, each requiring a different approach to impact management and planning: 1) Destructive and potentially life-threatening (Zone I); 2) Damaging and/or disruptive (Zone II); 3) Mildly disruptive and/or a nuisance (Zone III). These zones are summarised as a cartoon in Figure 3.12 where physical ash impacts to selected societal assets are depicted against ash deposit thickness – which generally decreases with distance from the volcano source. The more severe impacts depicted in Zones I and II may not eventuate in smaller magnitude eruptions because insufficient ash fall occurs. A further impact zone (Zone IV) could be applied to areas that extend beyond those receiving ash fall, where aviation may be disrupted by airborne ash. For example the Puyehue Cordón-Caulle eruption in Chile in 2011 deposited ash over 75,000 km² (Buteler et al., 2011), primarily in Argentina, but the airborne ash was detected in air sampling equipment in Porto Alegre, Brazil, 2,000 km away (De Lima et al., 2012) and caused flight cancellations around the southern hemisphere, e.g. in south east Australia 10,500 km from the volcano (Lahey, 2012).

Quantitative collection of ash fall impact data began in earnest with the explosive eruption (VEI 5) of Mount St Helens in the USA in 1980. Ash was erupted 23 km into the atmosphere and dispersed across 11 states (Figure 3.3: BGVN, 1980)) affecting critical infrastructure, health and socio-economic activities (Blong, 1984). More recent collection of quantitative impact data following subsequent eruptions of varying styles and magnitudes in a range of climates, communities, built environments and socio-economic settings now allows more robust impact estimation for future ash falls. Drawing on this available literature, we have identified six primary sectors that are important with reference to ash fall impacts in each of the zones identified in Figure 3.12 and for which impacts can be approximately related to ash fall thickness. A qualitative overview of ash impacts as a function of thickness is provided for each of these sectors in Tables 3.2 - 3.5. Each category is generic and specific impacts will depend strongly on the network or system design characteristics (typology), ash fall volume and characteristics, and the effectiveness of mitigation strategies that are applied. Impacts are considered with reference to ash fall thickness because it is the parameter commonly used as a metric of ash fall magnitude, being most easily measured or estimated in the field. However, as discussed earlier, other ash characteristics may influence or control the impact, and these have been highlighted. Below we provide some recommended eruption case studies and selected review articles as supplements to Tables 3.2 - 3.5:

1) *Health:* Horwell & Baxter (2006); VEI 4 eruptions of Eyjafjallajökull and Grímsvötn, Iceland, in 2010 and 2011 (Carlsen et al., 2012b, Carlsen et al., 2012a); VEI 5 eruption from Mount St Helens, USA, in 1980 (Baxter et al., 1981).
2) *Critical infrastructure:* Wilson et al. (2012b); Mount St Helens eruption, USA, in 1980 (Blong, 1984); VEI 4 eruption from Puyehue Cordón-Caulle, Chile, in 2011 (Wilson et al., 2012a).
3) *Buildings*: VEI 4 eruption from Rabaul, Papua New Guinea, in 1994 (Blong, 2003a); VEI 6 eruption from Pinatubo, Philippines, in 1991 (Spence et al., 1996).
4) *Agriculture:* VEI 3 eruption from Ruapehu, New Zealand, in 1995-1996: Cronin et al. (1998); VEI 5 eruption from Mt Hudson, Chile, in 1991 (Wilson et al., 2011); Mount St Helens eruption, USA, in 1980 (Cook et al., 1981).
5) *Clean-up:* VEI 2 eruption from Shinmoedake, Japan, in 2011 (Magill et al., 2013); Mount St Helens eruption, USA, in 1980 (Blong, 1984).
6) *Economic activities:* VEI 3 eruption of Etna, Italy, 2002 to 2003 (Munich-Re, 2007); Pinatubo eruption, Philippines, in 1991 (Newhall & Punongbayan, 1996).

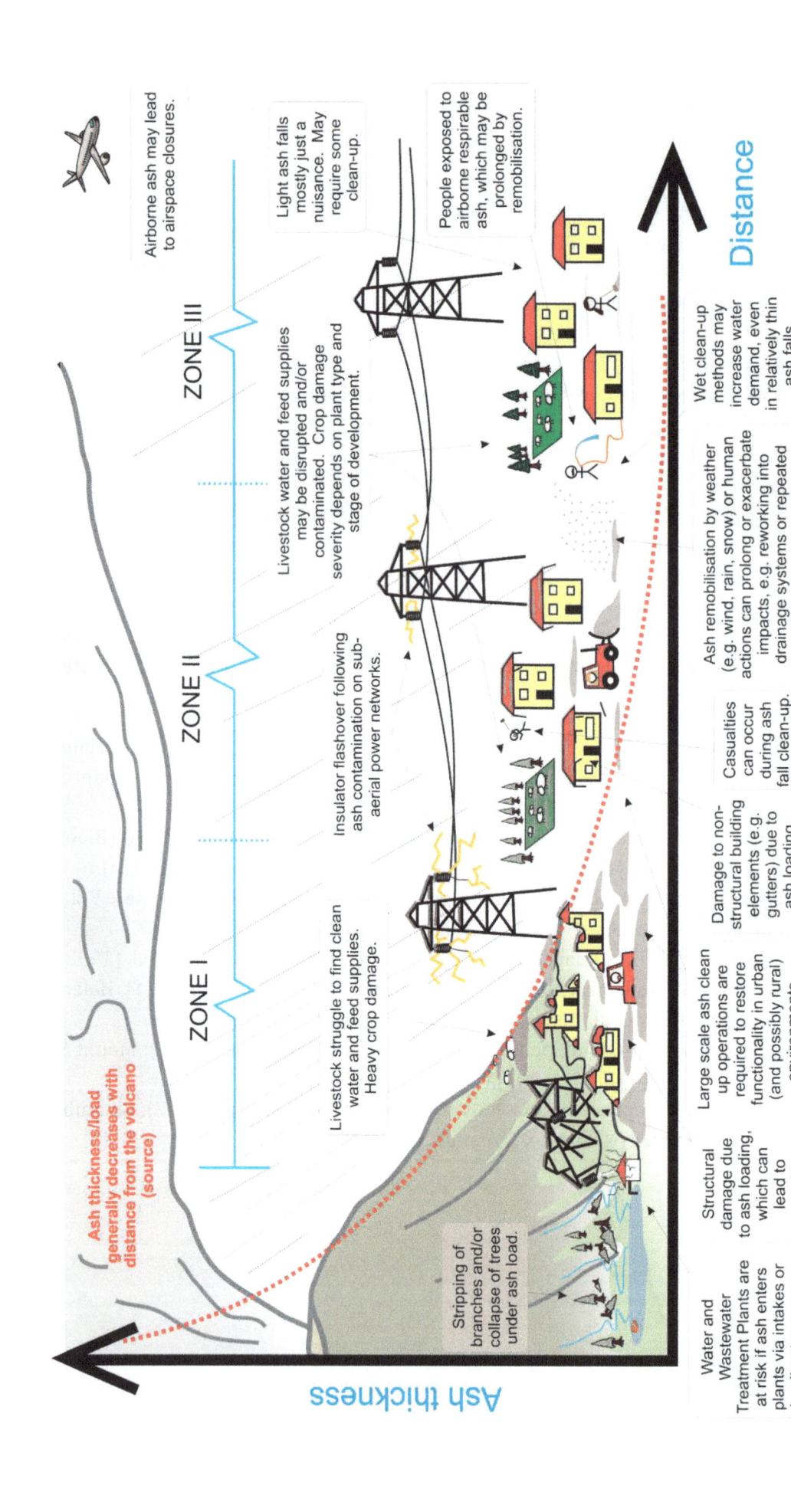

Figure 3.12 Schematic of some ash fall impacts with distance from a volcano. This assumes a large explosive eruption with significant ash fall thicknesses in the proximal zone and is intended to be illustrative rather than prescriptive. Three main zones of ash fall impact are defined: 1) Destructive and immediately life-threatening (Zone I); 2) Damaging and/or disruptive (Zone II); 3) Disruptive and/or a nuisance (Zone III).

Table 3.2 Potential impacts from thin (~ 1 to 10 mm) ash falls, categorised by sector and including key ash characteristics that determine the level of damage or disruption. Note that overloading of a system or network through high-demand or damage can lead to closure, exhaustion or restrictions at critical times, e.g. the exhaustion of water supplies during clean-up activities and these impacts will also exist where thicknesses are greater, potentially with higher severity.

	Potential consequences	Relevant ash characteristics
Public health	Casualties from fall deposits may occur, e.g. due to people falling from roofs during clean-up. However, the most commonly reported public health effects of ash exposure are irritation of the eyes and upper airways and exacerbation of pre-existing respiratory conditions, such as asthma. Serious health problems (requiring hospitalisation) are rare. A proportion of the population is likely to experience increased levels of psychological distress due to factors such as increased workload during the eruption and uncertainty about effects on health and livelihoods. People with pre-existing health vulnerability are more at risk of psychological distress. Individuals who may be exposed to ashy conditions will require protective clothing and masks.	Particle size; Mineral composition; Surface area; Morphology; Soluble salt burden; Thickness
Infrastructure	Most road markings obscured; traction and visibility problems; Airports often closed and requiring clean-up. Increased wear of engine and brakes; possible signal failure on railway lines.	Thickness; Particle size; Mechanical strength
	Possible clogging of air-and water-handling and filtration systems, mechanical and electrical equipment and abrasion damage by waterborne ash to pump impellers and turbines. Minor short-term increases, particularly in streams and small reservoirs, in elements leached from ash. Blockage of water intake structures, particularly in streams. Suspended ash in water intakes and sewer lines a possible threat to water/wastewater treatment plants. 'Open' systems (e.g. with open air sand filters) are more vulnerable.	Soluble salt burden; Conductivity; Abrasiveness; Particle size
	Potential flashover of power lines and transformers (particularly in light wet weather conditions). Corrosion and/or abrasion of, e.g. paintwork, windscreens, metallic elements, some air- and water-handling, mechanical, electrical equipment or engines. Possible damage to external telecommunication components or power cables.	
Buildings	No structural damage. Possible infiltration and internal contamination and corrosion of metallic components. Roofing materials may be abraded or damaged by human actions during ash removal.	Loading; Soluble salt burden; Abrasiveness

Continued overleaf

Table 3.2 continued...

	Potential consequences	Relevant ash characteristics
Agriculture	Effects on livestock expected to be minor but may include irritation of eyes and skin. Animals may ingest ash along with feed but small quantities are unlikely to cause harm. Cases of fluorosis (if ash contains moderate to high levels of soluble F) are rare but have been reported for these depths of ash fall. Ash coverage of crops, may lead to aesthetic discolouration and/or acid damage to leaves and fruits and abrasion during clean-up or harvest. Ash fall may have beneficial effects on soil if ash contains high levels of available plant growth nutrients such as sulfur, and may provide a beneficial mulching effect. Ash may also encourage or discourage pests and/or pollinating insects.	Soluble salt burden; Particle size; Abrasiveness
Clean-up	Minor clean-up required: sweeping of roads, paved areas and roofs/gutters usually sufficient. Ash falls of only a few mm depth will generate large volumes of ash for collection and disposal and clean-up is a time-consuming, costly and resource-intensive operation. Water demand may remain high for months afterwards if wind-remobilised ash requires dampening.	Thickness; Density; Abrasiveness
Economy	Some economic activities may increase, e.g. volcano tourism, but most will be disrupted. Disruption to work and travel. Clean-up cost. Increased maintenance costs (e.g. at water supply plants where sand filter beds may need to be cleaned more frequently). Increased labour and health and safety requirements.	Thickness; Presence

Table 3.3 Potential impacts from moderate (~ 10 to 100 mm) ash falls, categorised by sector and including the key ash characteristics that determine the level of damage or disruption. These impacts are in addition to those experienced at lower thicknesses, i.e. Table 3.2.

	Potential consequences	Relevant ash characteristics
Public health	Short-term effects as for Table 3.2 but with greater proportions of the population affected. Ash deposit remobilisation may prolong effects of ash exposure. Increased likelihood of long-term physical and psychosocial effects in vulnerable individuals. Self or official evacuations more likely.	Thickness; Particle size; Mineral composition; Surface area; Soluble salt burden
Infrastructure	Ground transport networks disrupted by visibility and traction problems. Airports closed.	Thickness; Particle size
	Ash accumulation may overload lines, weak poles and light structures, and cause additional tree-fall onto lines.	
	Suspended ash in water intakes and sewer lines a threat to water/wastewater treatment plants. Changes to surface water composition from elements leached from ash still likely to be minor. Possible blockage of storm drains, ditches and/or sewers, leading to surface flooding (depending on catchment size and system or network design).	Thickness; Particle size; Abrasiveness
Buildings	In rare instances, non-engineered and long span roofs may be vulnerable to damage, particularly when ash falls wet or is subsequently wetted.	Loading
	Non-structural elements such as gutters and overhangs may suffer damage. Some infiltration of dry ash into interiors.	
Agriculture	Effects on livestock as for Table 3.2 but likely to be more pronounced. Thicker ash falls may smother feed and restrict access to drinking water, leading to starvation and/or dehydration. Ingestion of larger quantities of ash along with feed may cause intestinal blockages. Livestock likely require supplementary feed due to ash coverage of pasture. Cases of fluorosis (if ash contains moderate to high levels of soluble F) more probable.	Thickness; Mechanical strength; Soluble salt burden
	Ash coverage of plants may lead to loading damage and shading from light. Windblown ash may strip foliage. Acidic ash leachate may damage fruit or foliage.	
	Ash coverage of soils will have increasingly detrimental impacts with increasing thickness (with some dependency on grain sizes).	

Continued overleaf

Table 3.3 continued...

	Potential consequences	Relevant ash characteristics
Clean-up	All roads and paved areas on public and private properties require cleaning. Private properties usually require assistance with clean-up/disposal. Need for co-ordination of clean-up. Ash dump(s) established and maintained to reduce ash mobilisation.	Thickness; Abrasiveness
Economy	Considerable production losses possible from impact to crops and livestock. Loss of assets possible, e.g. farm land disruption requiring at least temporary land-use change. Repeated ash falls or remobilisation of deposits drives health concerns that can disrupt business activities, e.g. employee anxiety, and the tourism industry.	Thickness; Soluble salt burden
Example/s	*Acidic ash leachate damage to chillis during the Merapi 2006 eruption, Indonesia © G. Kaye*	*20 - 30 mm of ash covering aeroplanes during the 2011 Puyehue-Cordón Caulle eruption, Chile. © Bariloche Airport*

Table 3.4 Potential impacts from thick (100 to 1,000 mm) ash falls, categorised by sector and including the key ash characteristics that determine the level of damage or disruption. These impacts are in addition to those experienced at lower thicknesses, i.e. Tables 3.2 and 3.3.

	Potential consequences	Relevant ash characteristics
Public health	Casualties (blunt trauma injuries from impact or falling debris, suffocation/inhalation injuries) possible from collapse of buildings or trees. Prolonged or permanent relocation possible due to concerns about health and disruption to critical infrastructure and economic activities. Prolonged exposure to ash leading to high prevalence of effects described in Table 3.2. Increased levels of stress may manifest as psychological morbidity (e.g. post-traumatic stress syndrome or depression). Prolonged exposure to ash may increase risk of chronic silicosis or inflammatory responses.	Particle size; Mineral composition; Surface area; Morphology; Soluble salt burden; Loading
Infrastructure	Transport networks severely affected by reduced traction. Very thick ash falls may create extra loading on bridges, especially when wet. Substantial and repeated clean-up of sites required; dry, windy conditions exacerbate remobilisation and drifting.	Thickness; Density; Particle size
	Most generation plants likely to be disrupted.	
	Major and ongoing damage to water supply intake structures located in streams; blockages of canals, intake structures and pipes. Blockages of underground drainage leading to surface flooding. High risk of severe damage to water- and wastewater- treatment plants if ash enters systems through intakes, or by direct ash fall.	Thickness; Particle size; Mechanical strength
Buildings	Non-engineered and long span or low pitched roofs more vulnerable to collapse.	Loading
	Non-structural elements (such as gutters) likely to be damaged.	
Agriculture	Burial of pasture and most crops; loading damage to tree crops and shade trees likely. Soil rehabilitation, e.g. through mixing or removal of the ash, is typically required for agricultural production to be restored. The buried soil's fertility may wane over time as key soil processes are broken or inhibited by ash blanket (i.e. water, oxygen and nitrogen cycles). Damage/burial of farm assets, including machinery, irrigation systems, access-ways, etc.	Thickness; Loading; Density; Particle size
	Tree damage possible in production forests, particularly for those not adapted to snow.	

Continued overleaf

Table 3.4 continued...

	Potential consequences	Relevant ash characteristics
Clean-up	As for thinner ash falls, but much larger volumes will require greater resources, time, repeated and frequent clean-up. Vegetated areas (e.g. parks and gardens) may require cleaning.	Loading; Thickness
Economy	Losses associated with business interruption, supply chain failures, reconstruction and relocation costs and loss of land. Prolonged disruption to economic activities may drive evacuations and associated direct and indirect economic losses.	Loading; Thickness
Example/s	*Collapse of a weak metal sheet roof under 150 - 300 mm of ash during the 1991 Pinatubo eruption, Philippines ©T.J. Casadevall, USGS.*	*A tractor mixes volcanic ash with the underlying soil following the 1991 eruption of Pinatubo in the Philippines. © C. Newhall, USGS.*

Table 3.5 Potential impacts from very thick (>1,000 mm) ash falls, categorised by sector and including the key ash characteristics that determine the level of damage or disruption. These impacts are in addition to those experienced at lower thicknesses, i.e. Tables 3.2, 3.3 and 3.4.

	Potential consequences	**Relevant ash characteristics**
Public health	Casualties possible from collapse of buildings or trees (see Table 3.4). Rescue operations are likely to be required. Prolonged or permanent relocation possible due to land-use change from very thick ash deposits.	Thickness; Particle size; Mineral composition; Surface area; Morphology; Soluble salt burden; Loading
Infrastructure	Structural damage to all structures possible. Most infrastructure systems will be damaged and disrupted for extended periods. Ground transport networks rendered impassable to most vehicles. Substantial and repeated clean-up of sites required; fluvial remobilisation of unconsolidated deposits is likely for years to decades; dry, windy conditions exacerbate remobilisation and drifting.	Thickness; Density; Particle size; Loading
Buildings	Structural damage to most buildings likely.	Loading
Agriculture	Total burial of crops/pastures, likely requiring land-use change without rehabilitation. The buried soil's fertility will decrease over time as key soil processes are sealed off. Tree damage likely in production forests.	Thickness; Loading; Density; Particle size
Clean-up	As for thinner falls, but much larger volumes will require greater resources, time and repeated and frequent clean-up. It may not be economic to undertake a clean-up operation.	Loading; Thickness
Economy	Losses associated with business interruption, supply chain failures, reconstruction and relocation costs and loss of land. Prolonged disruption to economic activities may drive evacuations and associated direct and indirect economic losses.	Loading; Thickness

Thick ash falls (Zone I, Tables 3.4 and 3.5) can cause structural damage to buildings and infrastructure, with associated casualties. They can also bury crops and create major clean-up demands, but are typically confined to within tens of kilometres of the vent. By contrast, relatively thin falls of a few millimetres (Zone III: Table 3.2) can occur over very large areas disrupting transport and causing significant disruption to everyday activities. For example, the 1995/96 eruptions of Ruapehu volcano, New Zealand, were small (VEI 3), but still covered a large proportion of the North Island with thin ash falls of <1-3 mm. Effects included significant disruption and damage to aviation, a hydro-electric power generation facility, electricity transmission lines, water supply networks, wastewater treatment plants, agriculture and the tourism industry. The total cost of the eruption was estimated to be approximately NZ$ 130 million (US$ 91 million), in 1996 value (Johnston et al., 2000). Even the threat of ash fall can be sufficient to generate public concern with losses associated with precautionary mitigation strategies, e.g. procurement and distribution of face masks.

Social impacts

In addition to the physical and health impacts associated with ash falls, affected communities can experience considerable direct and indirect social impacts. These impacts are difficult to relate to any one hazard intensity measure and often highly context- and circumstance-specific. For example, one manifestation of the economic hardship and anxiety associated with the 1995-1996 eruptions of Ruapehu volcano was that communities at the base of the volcano experienced increased levels of post-traumatic stress in school-aged children (Ronan, 1997). Socio-economic impacts such as these, and particularly those that are indirect, can develop over a long period of time and can be difficult to predict, plan for and assess; they are likely to be more pronounced in long-duration crises with continuous or intermittent ash falls or extended remobilisation episodes (e.g. Sword-Daniels et al., 2014). As with other natural hazards, socio-economic impacts are often felt disproportionately within and between communities due to a range of pre-existing and inter-related socio-cultural, political and economic vulnerabilities (e.g. gender, age, ethnicity, physical ability, social capital, economic status, political representation, health and education) (Wisner et al., 2012). For example, groups marginalised due to ethnic or economic status may live in higher risk areas in unsuitable housing (Gaillard, 2006), e.g. rural dwellings of weak construction, and often experience considerable financial hardship due to hazard impacts. Communities exposed to any magnitude of ash fall commonly report anxiety about the health impacts of inhaling or ingesting ash (and impacts on animals and property), which may lead to temporary (and sometimes unnecessary) socio-economic disruption (e.g. evacuation, school closures, cancellations).

Communities affected by one or few large ash falls within a short duration - often rare, low probability, events –*may* have limited experience, tools or resources to react leading to exacerbation of the original impacts and a comparatively poor capacity to cope. Conversely, chronic ash impacts (through repeated eruptions or ash fall remobilisation) *can* drive adaptation and mitigation activities. For example, Kagoshima city in Japan has been frequently affected by ash from Sakura-jima volcano over the last 50 years but structural adaptations and mitigation measures have successfully reduced the disruption of ongoing ash falls (Durand et al., 2001). The city has modified critical services (such as covering waste-water treatment plants and jacketing electrical insulators to reduce the likelihood of flashover), changed building designs to prevent ash ingress and sedimentation, and invested in ash clean-up equipment and development of plans for clean-up operations. As such, the city copes relatively well with the frequent ash falls and experiences little disruption (Durand et al., 2001). In summary, forecasting how a community will respond to ash falls is complex and will depend upon many factors including individual and collective experiences, socio-economic setting, the vulnerability of community assets to damage and disruption, the level of planning, leadership, and social capital.

3.4 Estimating vulnerability

Empirical evidence for the impacts of volcanic eruptions is relatively limited for the reasons given in Section 3.3. Experimental data (e.g. Oze et al., 2014) are also limited, and, where available, refer exclusively to materials and the building codes or standards of the tested component or system. In the absence of empirical or experimental data, theoretical calculations of the response of components, systems or networks to hazard parameters can add to the availability of quantitative vulnerability information (e.g. Spence et al., 2005), but may be subject to various uncertainties. As a result, vulnerability estimates for volcanic impacts are typically less robust than those for earthquake or wind, for example, because the research field is relatively new and because volcanoes exhibit relatively low recurrence intervals. However, volcanic eruptions have the potential to cause more widespread high-consequence impacts. Reducing ash fall risk requires an understanding of the breadth and severity of potential impacts and understanding the vulnerability of key community assets to ash fall is fundamental in assessing potential impacts.

A commonly used approach is to categorise impacts by sector (see Section 3.3) and consider the relationship between hazard intensity (in this case ash thickness) and damage, or disruption state, for each sector. In Tables 3.6 - 3.8, we present some broad guides that relate dry ash thickness (as the ash fall characteristic most easily measured and observed in the field) to critical infrastructure, buildings and agriculture. We consider these to be the sectors most relevant to how communities cope with ash fall and for which there is empirical evidence to inform assignment of vulnerability values. We note that wet ash has a substantially higher density than dry ash, which influences loading: saturated ash can as much as double the load the deposit imparts, such that the damage states reported in Table 3.7 may occur under half the thicknesses of dry ash. Brief descriptions of likely disruption (impact on functionality) and damage are outlined for each state (D0 - D5), and ordered by increasing dry ash fall thicknesses for severe damage or disruption (D4). We outline approximate costs (as a proportion of replacement cost) for roofs damaged by ash fall (following Blong, 2003a and unpublished studies around Vesuvius, Italy). For damage states with no damage, there may still be associated costs. For example, closure of airports because of airborne trace ash falls can result in significant loss of revenue. Losses may also result from precautionary risk management activities, e.g. covering water supplies, precautionary evacuations or loss of business activity, associated with the threat of ash fall. At higher hazard intensities, where losses may be expected due to physical damage, there may also be a significant cost associated with management and assessment. For example, of the approximately NZ$ 130 million (US$ 91 million in 1996) losses associated with the 1995-1996 eruptions of Ruapehu volcano in New Zealand, around NZ$ 6 million (US$ 4.2 million) were attributed to management of the crisis and subsequent damage assessment (Johnston et al., 2000).

The sectors shown in Tables 3.6 - 3.8 are those for which ash fall thickness is an appropriate measure of hazard intensity relative to damage and functional state. This list is therefore not exhaustive, but highlights some key sectors that are expected to suffer damage or disruption under ash falls. We deliberately excluded some sectors that are vulnerable to ash impacts, such as water treatment plants, because damage is not proportionally related to ash fall thickness due to complex and variable system and network design.

Minimal structural damage may be expected below about 100 mm of ash fall, although cleaning and repair of critical infrastructure and building interiors and fittings such as air-conditioning units, in addition to the interruptions associated with their reduced functionality, could be costly and disruptive (Tables 3.6 and 3.7). Depending upon when an ash fall occurs (the worst times being during the juvenile growing stage and just before and during harvest), thicknesses of around 100 mm may be associated with total loss of some crops and very little disruption for more robust agricultural types such as forestry (Table 3.8). Machinery used for harvest and transportation of horticultural and forestry crops will be adversely affected. Above around 300 mm of ash fall, we may expect to see changes in land use and potentially indefinite closure of assets becoming an issue.

In reality, vulnerability to ash fall impacts is dependent on building or infrastructure condition, construction, maintenance, network design, ash volume and particle characteristics, as well as the effectiveness of applied mitigation strategies. The values provided in Tables 3.6 - 3.8 are intended as broad indicators only and are subject to significant uncertainty; few empirical data are available and, by necessity, vulnerability studies are typically supplemented by expert judgement. A number of assumptions have been made in these impact estimates: namely, that assets were subject to one discrete ash fall event; ash was not removed during deposition (either by wind, water or human actions); and no mitigation actions had been taken prior to impact (e.g. internal strengthening of the building). The hazard-damage/disruption relationships shown here are based on few data and assume generic building types, systems and networks of average condition, design and construction. Vulnerability estimates are thus subject to significant uncertainty and intended as broad guidelines based on the evidence available, with more detailed assessment suggested for individual cases and conditions; the generalisations in Tables 3.2 - 3.5 and 3.6 - 3.8 should be considered with this caveat in mind. The approach is simplistic and does not capture system interdependence, i.e. the relationships and dependencies between affected sectors (Meheux et al., 2007). An understanding of how each component or impact-area interfaces and is connected with others is critical for addressing vulnerability. The loss of one component or node in a system may create cascading effects to other sectors, causing dynamic vulnerability of the system and its environment. For example, ash-blocked storm-water drains can cause secondary flooding that may recur for weeks or months following an eruption, until the system is cleared of ash. The secondary flooding can cause road access problems, and can prevent access to impacted infrastructure for maintenance or repair (Wilson et al., 2012b). In this way, the ash fall impacts on the water infrastructure system can have a knock-on effect to the functionality of transport systems. The approach used in Tables 3.6 - 3.8 also does not recognise the (sometimes complex) multi-hazard environment which may emerge following an ash fall, such as where a house roof survives a thick ash fall loading, but ash remobilised as lahars impact the house making it unsafe for habitation (Blong, 2003b).

Table 3.6 Approximate median (and interdecile) hazard intensities (using dry ash thickness as a proxy) that relate to key damage and functionality states for a range of critical infrastructure. Water and telecommunication networks are not included here because damage states are difficult to relate them to a single hazard intensity, i.e. thickness. The response of a system or network to ash fall thicknesses will depend upon the system/network design and type, its components and the characteristics of the ash fall.

Code:		D0	D1	D2	D3	D4	D5
	Description:	No damage	Cleaning required		Repair required		Beyond economic repair
Airports	Function	Fully functional		Closure of runway			Indefinite closure
	Damage	No damage (but loss of revenue costs)	Possible runway surface degradation	Possible runway surface degradation	Collapse of critical buildings; possible runway surface degradation [3]		Complete burial
	Thickness	0 mm		>0 mm			>500 mm
Power	Function	Fully functional	Temporary disruption, e.g. flashover of insulators		Disruption requiring repair		Permanent disruption
	Damage	No damage	No damage to components		Damage to critical components; long delays in receiving replacement components.		Structural damage
	Thickness	0 (0-20) mm	5 (1-20) mm		20 (2-100) mm		>500 mm (100-1000 mm)
Railways	Function	Fully functional	Reduced visibility and traction	Signals disrupted	Loss of traction making operation unsafe; Possible derailing through ash accumulation		Impassable
	Damage	No damage		Possible abrasion and/or corrosion of signal components and track			Complete burial
	Thickness	0 (0-5) mm	0.5 (0.1-10) mm	1 (0.1-20) mm	30 (2-100) mm		100 (50-200) mm
Roads	Function	Fully functional	Reduced visibility and traction	Road markings obscured	2WD vehicles obstructed	4WD vehicles obstructed	Impassable
	Damage	No damage		Possible road surface and marking abrasion	Road surface and marking abrasion		Complete burial
	Thickness	0 (0-5) mm	0.5 (0.1-10) mm	2 (1-20) mm	50 (10-100) mm	150 (50-300) mm	n/a [4]

[3] Rarely observed
[4] Usually, thick ash deposits can be compacted and a road reinstated.

Table 3.7 Approximate median (and interdecile) hazard intensities that relate to key damage and functionality states for a range of generic roof types (following Spence et al., 2005 and Jenkins et al., 2014). Approximate cost ratios are estimated, following the work of Blong (2003b) and unpublished studies around Vesuvius, Italy 5. Dry ash fall thickness (in mm) is used as a proxy for hazard intensity and a load density of 1000 kg/m² is assumed. A saturated ash deposit would result in the damage states identified at as little as half the suggested thicknesses.

Code:	D0	D1	D2	D3	D4	D5
Description:	No damage	Minor/basic repair required	Moderate repair required	Major/specialist repair required	Major/specialist repair required	Beyond economic repair
Function:	Functional	Repeated clean-up required; Some loss of functionality for some contents and fittings	Ash infiltration or threat of roof and/or wall collapse may prohibit habitation	Ash infiltration or threat of roof and/or wall collapse may prohibit habitation	Ash infiltration or threat of roof and/or wall collapse may prohibit habitation	Retired
Cost (% of replacement cost):	0 – 1	1 – 5	5 – 20	20 – 60	20 – 60	>60
Structural damage:	No damage	No damage	No damage to principal roofing supports	Partial or complete failure of the supporting structure, e.g. battens or trusses; Partial or moderate damage to the vertical structure	Partial or complete failure of the supporting structure, e.g. battens or trusses; Partial or moderate damage to the vertical structure	Collapse of roof and supporting structure over 50% of roof area; External walls may be destabilised
Non-structural damage:	No damage	Minor damage to roof coverings, e.g. abrasion and corrosion of metallic roofs.	Potential damage to gutters and roof covering, e.g. excessive bending, and overhangs	Severe damage or partial collapse of roof overhangs; Collapse or partial collapse of roof covering	Severe damage or partial collapse of roof overhangs; Collapse or partial collapse of roof covering	Partition wall/s destroyed in some cases
Contents and fittings:	Some infiltration of ash possible	Ash infiltration and potential damage to fittings, e.g. air-con, and appliances	Variable levels of contamination and damage	Variable levels of contamination and damage	Variable levels of contamination and damage	Damage to most contents and fittings is irreversible, or salvage is uneconomical
GENERIC ROOF TYPES: Timber board on weak timber supports	1 mm?	10 mm?			200 mm (100 – 400 mm) [6]	
Tiles on timber supports	1 mm?	10 mm?			300 mm (150 – 600 mm) [6]	
Modest sheeting on timber supports	1 mm?	10 mm?			300 mm (150 – 600 mm) [6]	
Domestic reinforced concrete	1 mm?	10 mm?			700 mm (400 - 1400 mm) [6]	

[5] The few available studies of roof damage from ash fall loading consider roof failure (D5) as either 1) collapse of the roof, roof support and supporting walls (e.g. Blong, 2003b), such studies are typically aimed at economic loss estimation; or 2) collapse of the roof and roof support (e.g. Spence et al., 2005), these studies are mostly concerned with occupant safety. Logically, damage cost ratios will differ between the two forms of failure; however, for the two available studies of damage cost ratio, values are comparable. Further studies are required.

[6] Estimates are based upon few data and strongly dependent upon building-specific aspects such as build quality, roof span, pitch, condition, fixings and design code (e.g. for wind or snow). These are likely to vary significantly between and within countries and will be larger when considering a more comprehensive form of failure, i.e. collapse of the entire building.

Table 3.8 Approximate median (and interdecile) hazard intensities (using dry ash thickness as a proxy) that relate to key levels for loss of production for a range of agriculture types. These hazard-loss of productivity relationships are based on expert judgement and very few empirical and experimental data. Impacts assume that the crops are in the growing stage (a worst-case impact); season modifiers will allow impacts during other periods of the growing cycle to be accounted for. These estimates are thus intended as broad guidelines and should be refined for individual cases and conditions.

Code:		D0	D1	D2	D3	D4	D5
Description:		No damage	Disruption to harvest operations and livestock grazing of exposed feed	Minor productivity loss: less than 50 %/crop	Major productivity loss: more than 50 %/crop; Remediation required	Total crop loss; Substantial remediation required	Major rehabilitation required/ Retirement of land[7]
AGRICULTURE TYPE:	Horticulture & Arable — Ground Crops & Arable	0 mm (0-20 mm)	1 mm (0.1-50 mm)	5 mm (1-50 mm)	50 mm (1-100 mm)	100 mm (25-200 mm)	300 mm (100- 500 mm)
	Horticulture & Arable — Tree Crops	0 mm (0-20 mm)	1 mm (0.1-50 mm)	5 mm (1-50 mm)	50 mm (1-100 mm)	200 mm (5-500 mm)	300 mm (200- 500 mm)
	Pastoral	0 mm (0-20 mm)	3 mm (0.1-50 mm)	25 mm (1-70 mm)	60 mm (20-150 mm)	100 mm (30-200 mm)	300 mm (100- 500 mm)
	Paddies	0 mm (0-50 mm)	1 mm (0.1-50 mm)	30 mm (1-75 mm)	75 mm (20 - 300 mm)	150 mm (75 – 300 mm)	300 mm (100- 750 mm)
	Forestry	0 mm (0-75 mm)	5 mm (0.1-75 mm)	200 mm (20-300 mm)	1000 mm (100-2000 mm)	1500 mm (100->2000 mm)	?

---

[7] Land will likely only be retired if there is an alternative commodity source. Where there are no alternatives, farmers need to change production and manage poorer and fewer produce.

Vulnerability estimates are typically the weakest part of a risk model and, wherever possible, detailed local studies of exposed assets and their vulnerability (e.g. Galderisi et al., 2013, Zuccaro et al., 2008) should be carried out before a full risk assessment is undertaken. In many parts of the world the failure, disruption or reduced functionality of infrastructure is likely to have a larger impact on livelihoods and the local economy than direct damage to buildings and infrastructure. In some cases impacts can exacerbate existing economic, social or agronomic decline (e.g. Wilson et al., 2011). We therefore recommend that future physical vulnerability assessments for volcanic areas always include surveys of exposed buildings *and* infrastructure, and any interdependencies, as well as considering direct and indirect socio-economic impacts.

3.5 Looking ahead

In this chapter we have concentrated on volcanic ash, as the volcanic hazard that most frequently affects large populations and assets. This work has been reviewed by the Global Volcano Model network (GVM) and the International Association of Volcanology and Chemistry of the Earth's Interior (IAVCEI), who wish to highlight the importance of ash as a far-reaching volcanic hazard that can disrupt human activities even at very small intensities. Volcanic hazard and risk research is a rapidly growing and evolving field and ongoing research is working towards better quantification and communication of the uncertainties associated with current approaches. The broad overviews and examples illustrated in this paper are only some of the many strands of volcanic hazard and risk research being undertaken by the volcanology community. For example, the CAPRA risk modelling platform (ecapra.org) has been used to provide preliminary estimates of potential building damage around active volcanoes in the Asia-Pacific Region using simplified volcanic hazard outputs from a statistical emulator (see Case Study box: Unravelling volcanic ash fall hazard for risk-managers). A more local example is the KazanRisk loss model (riskfrontiers.com/kazanrisk.htm), which uses numerical dispersal modelling of ash fall in Greater Tokyo to estimate potential losses associated with building damage, clean-up and reductions in agricultural productivity.

Volcanic risk assessment is still in its infancy compared to other natural hazards, such as earthquake and tropical cyclone. Our knowledge regarding the impacts of ash fall are constrained to anecdotal evidence and a relatively small number of post-event investigations. Over the rest of this section we highlight some of the key advancements that may be expected in the coming years, including improved knowledge of exposure relevant to volcanic hazards, better understanding the vulnerability of these exposed communities and assets, and advances in modelling hazard dynamics and intensity.

Volcanic risk assessment is still in its infancy compared to other natural hazards, such as earthquake and tropical cyclone. Our knowledge regarding the impacts of ash fall are constrained to anecdotal evidence and a relatively small number of post-event investigations. Over the rest of this section we highlight some of the key advancements that may be expected in the coming years, including improved knowledge of exposure relevant to volcanic hazards, better understanding the vulnerability of these exposed communities and assets, and advances in modelling hazard dynamics and intensity.

Unravelling volcanic ash fall hazard for risk-managers
A.N. Bear-Crozier, V. Miller, V. Newey, N. Horspool, and R. Weber

With the objective of examining volcanic ash fall hazard and components of the associated risk on a regional-to-global scale, the CAPRA platform is utilised in GAR15 (www.ecapra.org; CIMNE et al., 2013). Preliminary volcanic ash fall hazard information is combined with regional vulnerability curves for structural damage to buildings (Maqsood et al., 2014) and exposure information (De Bono, 2013), in requisite formats for the CAPRA platform (ash fall-load in kg/m² at 5 km spacing for each average return period). Up-scaling local ash dispersal modelling methods to the regional-to-global scale is both time consuming and computationally intensive. The Probabilistic Volcanic Ash Hazard Analysis (PVAHA) methodology was developed and implemented at Geoscience Australia (Bear-Crozier et al., 2014) for determining ash fall hazard, consistent with the objectives and constraints outlined above. PVAHA considers a multitude of volcanic eruption occurrences using associated ash fall-load attenuation relationships (determined for a range of parameters including eruption magnitude, column height, duration, particle size distribution, wind velocity) the results from each applicable event are integrated to derive a preliminary annual exceedance probability for each site of interest.

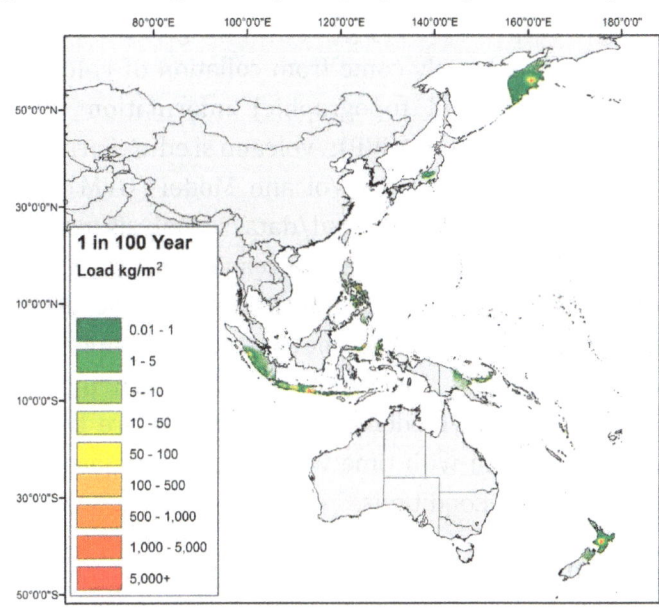

Estimated maximum ash fall-load for the Asia-Pacific Region at the 100 year average return period.

In the Asia-Pacific Region 276 volcanoes were identified and statistically characterised in order to generate an emulated catalogue of volcanic eruptions. Sites were identified using a 5 km grid extending to a 500 km distance from each volcano, a total of 178,935 sites covering ~4,600,000 km² of the region (Bear-Crozier et al., 2014). The influence of wind direction on the distribution of the ash fall-load was greatly simplified, with each direction assumed to have an equal likelihood; therefore hazard and risk will be underestimated in downwind areas and overestimated upwind of the volcano. The preliminary hazard curves of annual probability of exceedance, and (the inverse) average return periods, versus ash fall-load (kg/m²) were calculated at each site, with 12 return periods provided for integration into the CAPRA model (see above for 100 year return period results). Calculating hazard curves of annual probability of exceedance versus ash fall load at individual sites allows for disaggregation of the hazard results by, e.g., magnitude, distance or source at each site, which is a useful tool for analysing risk at a site of interest e.g. a densely populated city or piece of critical infrastructure (see left for a preliminary example showing disaggregation of hazard by volcanic source in Jakarta, Indonesia), or for identifying priority areas for more detailed hazard modelling at the local-scale.

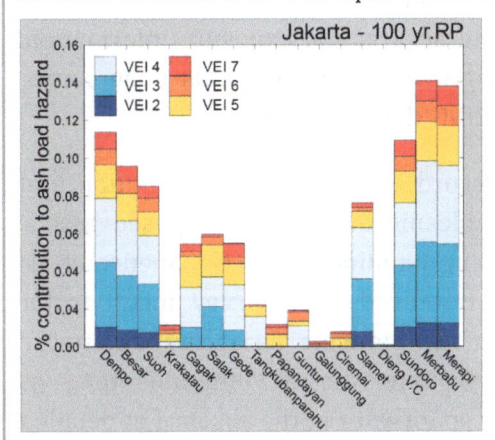

Disaggregating the percentage contribution to ash fall-load in Jakarta from volcanoes around Indonesia for the 100 year average return period

This hazard modelling and associated risk results must be considered preliminary; probabilistic consideration of wind conditions is required before incorporating results into hazard or risk decision making.

Source: Geoscience Australia

3.5.1 Better understanding of hazard

When considering natural hazards in general, the success of hazard and risk assessment rests on the availability, resolution and accuracy of relevant data. In terms of ash fall hazard, geological and historical information describing past events is critical. A comprehensive eruption history allows time-dependent probabilities for various eruptive styles and magnitudes to be most accurately determined. While geological studies mean that it is possible to improve the understanding of eruptive activity for individual volcanoes, it is difficult to improve eruption history greatly at the global level. Improvements in global hazard assessment will therefore largely come from collation of volcano-specific eruption and environmental (i.e. meteorological and topographic) information into comprehensive databases (e.g. Global Volcanism Program (GVP): volcano.si.edu; Large Magnitude Explosive Volcanic Eruptions: bgs.ac.uk/vogripa; Global Volcano Model (GVM): globalvolcanomodel.org; NCEP/NCAR wind Reanalysis: esrl.noaa.gov/psd/data/reanalysis/reanalysis.shtml; ERA-Interim wind Reanalysis: ecmwf.int/en/research/climate-reanalysis/era-interim), in-depth statistical analysis of these databases, development of increasingly sophisticated modelling techniques and more advanced computing resources. Each new explosive eruption provides a calibration opportunity for ash dispersal models and, with an understanding of the underlying physics, can lead to improvements in model accuracy. Models are thus tested against an increasing set of input scenarios, which with time will make them more applicable to a range of eruption styles and environmental conditions.

3.5.2 Increased exposure information

For ash fall risk assessment, relevant exposure information includes the spatial distributions of population, demographics, building types, infrastructure and land usage. To best assess potential impacts, the location and characteristics of exposed elements should be included at the same scale as the modelled hazard. Improvements in satellite and ground-based acquisition and improved availability of local geo-referenced databases will mean that we will see continued improvement in this area, which can be carried through into future risk assessments. In particular, the development of automated image recognition tools and rapid and remote data collection will likely expand the availability, coverage and accuracy of exposure information. Global exposure datasets are improving rapidly with a growing trend towards open-source information (e.g. OpenStreetMap: openstreetmap.org; UNEP-GRID: grid.unep.ch).

At the local level, inventories can be developed that record the numbers, distribution and characteristics (with respect to their vulnerability to volcanic hazards) of exposed assets. Standardised field survey methodologies exist for assessing building stock (e.g. Spence et al., 1996, Marti et al., 2008, Jenkins et al., 2014a), but structured field survey methods for infrastructure are more difficult due to security concerns and the wide variability in network design, type and dependencies. However, assessing the characteristics of vulnerable items, as well as their dependency upon other networks, will be a step forward in the absence of detailed exposure information.

3.5.3 Improved understanding of impacts and associated vulnerability

Recent empirical, experimental and theoretical studies have greatly improved our understanding of the physical response of elements to ash fall. However, systematic and comprehensive documentation of observed impacts, including cascading impacts, under a wide range of ash fall conditions is lacking. For example, studies have typically not focused on impacts associated with thin (< 2 mm) ash falls, but on large eruptions and impacts associated with ash falls of more than 10 mm thickness. Thin, distal ash falls of a few mm or less are therefore a source of significant uncertainty for emergency management planning and loss assessment models, despite being the volcanic hazard most likely to be experienced by exposed critical infrastructure (Wilson et al., 2012b). In future damaging ash fall events, quantitative impact studies will be critical in supplementing the sparse existing empirical data set and for calibrating existing vulnerability estimates obtained through experimental and theoretical studies. To enable damaging events to be compared, standard data collection guidelines should be followed, for example the ash collection and analysis protocols promoted by the International Volcanic Health Hazard Network (ivhhn.org) and the building survey guidelines discussed in Section 3.4.2.

While post-eruption impact assessment studies will continue to strengthen and diversify our understanding of ash fall impacts, this reactive research model rarely allows for detailed analysis to determine with any certainty how and why observed impacts occurred. There is a lack of empirical knowledge regarding how ash properties such as abrasiveness and corrosiveness affect critical infrastructure components over varying timescales. Understanding how the geotechnical properties and surface chemistry of ash influences impacts to societal elements is a current gap in understanding, although there are exciting developments in the human health field. However, it is difficult to explore impact thresholds in the field due to difficulties in obtaining timely access, limited field time, variations in environmental conditions and incomplete ranges of ash characteristics such as thickness, grain size and surface coating composition. One solution is empirical performance testing of critical components using varying intensities, types and compositions of ash under laboratory conditions. This allows performance thresholds to be assessed for different configurations of equipment and different characteristics of ash fall. Crucially, it also provides a setting for proactive experimentation of potential mitigation methods.

In addition to empirical data sourced from impact assessments and experimental testing of critical components, developments in numerical impact and risk modelling (e.g. Magill et al., 2006, Spence et al., 2008, Zuccaro et al., 2008, Scaini et al., 2014) will allow us to test the effects of mitigation options and explore uncertainties in risk rapidly and in the absence of an actual eruption. Expert judgement focus groups (e.g. Aspinall & Cooke, 2013) can be useful in better quantifying components of risk where large uncertainty is present, for example the probability of an explosive eruption occurring.

3.6 Final words

Ash fall is the most widespread and frequent of the hazards posed by volcanic eruptions. A comprehensive volcanic hazard assessment must include ash fall in addition to more localised hazards such as pyroclastic density currents. However, the impacts of ash fall are arguably more complex and multi-faceted than for any of the other volcanic hazards and for that reason the GVM/IAVCEI contribution to GAR-15 includes a technical background paper dedicated solely to the hazard and risk posed by ash fall. Within this paper, a summary review of some approaches to volcanic ash fall hazard and risk assessment has been provided, elaborating on two ash fall hazard assessments that have been undertaken at the global and local scale (Section 3.2). In discussing how these hazard studies may be translated to risk, the key characteristics and likely impacts of ash fall for society have been highlighted (Section 3.3). Elements of society known to be impacted by volcanic ash include human populations, dwellings, infrastructure, aviation, agriculture and/or other economic activities. In this paper we focussed on the vulnerability of these sectors can influence the impacts experienced; we also considered socio-economic impacts and issues surrounding clean-up (Section 3.4). The ability to adequately quantify risk is clearly dictated by the availability and quality of each set of input data (hazard, exposed elements, vulnerability of these elements). The scarcity of detailed empirical data for ash fall impacts limits our ability to provide definitive assessment of future impacts but ongoing studies for ash fall, and other volcanic hazards, continue to improve our understanding and forecasting ability.

Acknowledgements

This chapter was originally drafted as a technical background paper for the UNISDR GAR-15 Report on Disaster Risk Reduction. We are grateful for the advice and support of UNISDR throughout the process. The global ash fall hazard assessment (Section 3.2.1) benefited from the input of Stuart Mead; the local ash fall hazard assessment (Section 3.2.2) benefits from the collaboration of co-authors Jacopo Selva and Laura Sandri, and invaluable discussions with Giovanni Macedonio and Roberto Tonini. The pilot volcanic ash fall hazard study for use in the CAPRA risk model (boxed text: Section 3.5) was undertaken with the collaboration of co-authors Adele Bear-Crozier, Vanessa Newey, Nick Horspool and Rikki Weber. This technical paper also benefitted from the support of Katharine Haynes and Claire Horwell for the social and health aspects of ash falls, respectively.

The authors would like to thank separate funding sources that have allowed them to contribute to this effort: The AXA Research Fund and VOLDIES ERC contract 228064 (SFJ), the New Zealand MBIE Natural Hazard Research Platform Subcontract C05X0804 (TW, CS) and Department of Foreign Affairs and Trade funding as part of the Australian Aid program (VM). This work was carried out under the auspices of the Global Volcano Model (GVM) and the International Association of Volcanology and Chemistry of the Earth's Interior (IAVCEI). The GVM Volcanic ash hazard working group and the IAVCEI Commission on Tephra hazard modelling supported the development of this work.

References

Aspinall, W. P. & Cooke, R. 2013. Expert Elicitation and Judgement. *In:* Rougier, J. C., Sparks, R. S. J. & Hill, L. *Risk and Uncertainty Assessment in Natural Hazards.* Cambridge: Cambridge University Press.

Auker, M., Sparks, R., Siebert, L., Crosweller, H. & Ewert, J. 2013. A statistical analysis of the global historical volcanic fatalities record. *Journal of Applied Volcanology,* 2:2, 1-24.

Barberi, F., Macedonio, G., Pareschi, M. T. & Santacroce, R. 1990. Mapping the tephra fallout risk: an example from Vesuvius, Italy. *Nature,* 344, 142-144.

Barsotti, S., Neri, A. & Scire, J. S. 2008. The VOL-CALPUFF model for atmospheric ash dispersal: I. Approach and physical formulation. *Journal of Geophysical Research,* 113, 85-96.

Baxter, P. J., Ing, R., Falk, H., French, J., Stein, G. F., Bernstein, R. S., Merchant, J. A. & Allard, J. 1981. Mount St Helens eruptions, May 18 to June 12, 1980: An overview of the acute health impact. *The Journal of the American Medical Association,* 246, 2585-2589.

Bear-Crozier, A. N., Miller, V., Newey, V., Horspool, N. & Weber, R. 2014. Emulating volcanic ash fall for multi-scale analysis - Development of the VAPAHR tool and application to the Asia-Pacific region. http://dx.doi.org/10.11636/Record.2014.036 Geoscience Australia Record 2014/36.

BGVN 1980. Bulletin of the Global Volcanism Network: 05/1980 (SEAN 05:05), Major eruption sends cloud to 23 km, destroys summit, and devastates region. . *In:* Smithsonian Institution (ed.).

Biass, S., Scaini, C., Bonadonna, C., Folch, A., Smith, K. & Höskuldsson, A. 2014. A multi-scale risk assessment for tephra fallout and airborne concentration from multiple Icelandic volcanoes – Part 1: Hazard assessment. *Natural Hazards and Earth System Sciences,* 2, 2463-2529.

Blong, R. 2003a. Building damage in Rabaul, Papua New Guinea, 1994. *Bulletin of Volcanology,* 65, 43-54.

Blong, R. J. 1984. *Volcanic hazards: a sourcebook on the effects of eruptions.*, Academic Press Australia.

Blong, R. J. 2003b. A new damage index. *Natural Hazards,* 30, 1-23.

Bonadonna, C. 2006. Probabilistic modelling of tephra dispersal. *In:* Mader, H., Cole, S. & Connor, C. B. *Statistics in Volcanology.* IAVCEI Series Volume 1: Geological Society of London.

Bonadonna, C. & Costa, A. 2012. Estimating the volume of tephra deposits: a new simple strategy. *Geology,* 40, 415–418.

Bonadonna, C. & Costa, A. 2013. Modeling of ash sedimentation from volcanic plumes. *In:* Fagents, S. A., Gregg, T. K. P. & Lopes, R. M. C. *Modeling Volcanic Processes: The Physics and Mathematics of Volcanism.* Cambridge University Press.

Buteler, M., Stadler, T., Lopez Garcia, G. P., Lassa, M. S., Trombotto Liaudat, D., D'Adamo, P. & Fernandez-Arhex, V. 2011. Propiedades insecticidas de la ceniza del complejo volcánico Puyehue-Cordón Caulle y su possible impacto ambiental. *Revista de la Sociedad Entomológica Argentina,* 70, 149-156.

ByMuR 2010-2014. Bayesian Multi-risk Assessment: a case study for the natural risks in the city of Naples. http://bymur.bo.ingv.it/.

Carey, S. & Sigurdsson, H. 1989. The intensity of plinian eruptions. *Bulletin of Volcanology*, 51, 28-40.

Carlsen, H. K., Gislason, T., Benediktsdottir, B., Kolbeinsson, T. B., Hauksdottir, A., Thorsteinsson, T. & Briem, H. 2012a. A survey of early health effects of the Eyjafjallajökull 2010 eruption in Iceland: a population-based study. *BMJ Open*, 2.

Carlsen, H. K., Hauksdottir, A., Valdimarsdottir, U. A., Gislason, T., Einarsdottir, G., Runolfsson, H., Briem, H., Finnbjornsdottir, R. G., Gudmundsson, S., Kolbeinsson, T. B., Thorsteinsson, T. & Pétursdóttir, G. 2012b. Health effects following the Eyjafjallajökull volcanic eruption: a cohort study. *BMJ Open*, 2.

CIMNE, ITEC SAS, INGENIAR LTDA & EAI SA 2013. Probabilistic Modelling on Natural Risks at the Global Level: Global Risk Model 2013. UNISDR: Geneva, Switzerland.

Cioni, R., Longo, A., Macedonio, G., Santacroce, R., Sbrana, A., Sulpizio, R. & Andronico, D. 2003. Assessing pyroclastic fall hazard through field data and numerical simulations: Example from Vesuvius. *Journal of Geophysical Research (Solid Earth)*, 108.

Cook, R. J., Barron, J. C., Papendick, R. I. & Williams, G. J. 1981. Impacts on agriculture of Mount St Helens eruption. *Science*, 211, 16-22.

Costa, A., Dell'Erba, F., Di Vito, M. A., Isaia, R., Macedonio, G., Orsi, G. & Pfeiffer, T. 2009. Tephra fallout hazard assessment at the Campi Flegrei caldera (Italy). *Bulletin of Volcanology*, 71, 259-273.

Cronin, S. J., Hedley, M. J., Neall, V. E. & Smith, R. G. 1998. Agronomic impact of ash fallout from the 1995 and 1996 Ruapehu Volcano eruptions, New Zealand. *Environmental Geology*, 34, 21-30.

De Bono, A. 2013. The Global Exposure Database for GAR 2013. UNISDR: Geneva, Switzerland.

De Lima, E. F., Sommer, C. A., Cordeiro Silva, I. M., Netta, A. P., LIndenberg, M. & Marques Alves, R. d. C. 2012. Morfologia e quimica de cinzas do vulcão Puyehue depositadas na região metropolitana de Porto Alegre em junho de 2011. *Revista Brasiliera de Geociências*, 42, 265-280.

Delmelle, P., Lambert, M., Dufrêne, Y., Gerin, P. & Óskarsson, N. 2007. Gas/aerosol-ash interaction in volcanic plumes: New insights from surface analyses of fine ash particles. *Earth and Planetary Science Letters*, 259, 159-170.

Dilley, M., Chen, R. S., Deichmann, U., Lerner-Lam, A. L. & Arnold, M. 2005. Natural disaster hotspots: a global risk analysis. *Disaster Risk Management 5*. The World Bank Hazard Management Unit.

DPC-V1 2012-2013. Valutazione della pericolosità vulcanica in termini probabilistici. http://istituto.ingv.it/l-ingv/progetti/progetti-finanziati-dal-dipartimento-di-protezione-civile-1/progetti-vulcanologici-2012.

DPC-V2 2012-2014. Precursori di eruzioni. http://istituto.ingv.it/l-ingv/progetti/progetti-finanziati-dal-dipartimento-di-protezione-civile-1/progetti-vulcanologici-2012.

Durand, M., Gordon, K., Johnston, D., Lorden, R., Poirot, T., Scott, J. & Shephard, B. 2001. Impacts of, and responses to ashfall in Kagoshima from Sakurajima volcano - lessons for New Zealand. *Institute of Geological & Nuclear Sciences science report.* Institute of Geological & Nuclear Sciences Limited.

Durant, A. J., Rose, W. I., Sarna-Wojcicki, A. M., Carey, S. & Volentik, A. C. M. 2009. Hydrometeor-enhanced tephra sedimentation: Constraints from the 18 May 1980 eruption of Mount St. Helens. *Journal of Geophysical Research, 114,* 114, 2.

Folch, A. & Sulpizio, R. 2010. Evaluating long-range volcanic ash hazard using supercomputing facilities: application to Somma-Vesuvius (Italy), and consequences for civil aviation over the Central Mediterranean Area. *Bulletin of Volcanology,* 72, 1039-1059.

Gaillard, J. C. 2006. Traditional societies in the face of natural hazards: the 1991 Mt.Pinatubo eruption and the Aetas of the Philippines. *International Journal of Mass Emergencies and Disasters,* 24, 5-43.

Galderisi, A., Bonadonna, C., Delmonaco, G., Ferrara, F. F., Menoni, S., Ceudech, A., Biass, S., Frischkcecht, C., Manzella, I., Minucci, G. & Gregg, C. E. 2013. Vulnerability assessment and risk mitigation: The case of Vulcano Island, Italy. *In:* Margottini, C., Canuti, P. & Sassa, K. *Landslide Science and Practice.* Springer-Verlag.

Heffter, J. L. & Stunder, B. J. B. 1993. Volcanic Ash Forecast Transport And Dispersion (VAFTAD) Model. *Weather and Forecasting,* 8, 533-541.

Horwell, C. & Baxter, P. J. 2006. The respiratory health hazards of volcanic ash: a review for volcanic risk mitigation. *Bulletin of Volcanology,* 69, 1-24.

Hurst, A. W. 1994. ASHFALL, a computer program for estimating volcanic ash fallout: report and users guide. *Institute of Geological & Nuclear Sciences Science Report 94.* GNS Science.

IATA 2010. Volcano Crisis Cost Airlines $1.7 Billion in Revenue - IATA Urges Measures to Mitigate Impact. *Press Release No. 15.* 21 April 2010.

Jenkins, S., Magill, C., McAneney, J. & Blong, R. 2012a. Regional ash fall hazard I: A probabilistic assessment methodology. *Bulletin of Volcanology,* 74, 1699-1712.

Jenkins, S., McAneney, J., Magill, C. & Blong, R. 2012b. Regional ash fall hazard II: Asia-Pacific modelling results and implications. *Bulletin of Volcanology,* 74, 1713-1727.

Jenkins, S. F., Magill, C. R. & McAneney, K. J. 2007. Multi-stage volcanic events: A statistical investigation. *Journal of Volcanology and Geothermal Research,* 161, 275-288.

Jenkins, S. F., Spence, R. J. S., Fonseca, J. F. B. D., Solidum, R. U. & Wilson, T. M. 2014a. Volcanic risk assessment: Quantifying physical vulnerability in the built environment. *Journal of Volcanology and Geothermal Research,* 276, 105-120.

Jenkins, S. F., Wilson, T. M., Magill, C. R., Miller, V., Stewart, C., Marzocchi, W. & Boulton, M. 2014b. Volcanic ash fall hazard and risk: Technical Background Paper for the UNISDR 2015 Global Assessment Report on Disaster Risk Reduction. www.preventionweb.net/english/hyogo/gar: Global Volcano Model and IAVCEI.

Johnston, D. M., Houghton, B. F., Neall, V. E., Ronan, K. R. & Paton, D. 2000. Impacts of the 1945 and 1995–1996 Ruapehu eruptions, New Zealand: an example of increasing societal vulnerability. *Geological Society of America Bulletin,* 112, 720-726.

Lahey, K. 2012. No volcanic ash clouds to rain on Australia's parade. *The Australian Journal of Emergency Management,* 27, 16-19.

Macedonio, G. & Costa, A. 2012. Brief Communication "Rain effect on the load of tephra deposits". *Natural Hazards Earth System Sciences,* 12, 1229-1233.

Macedonio, G., Costa, A. & Folch, A. 2008. Ash fallout scenarios at Vesuvius: Numerical simulations and implications for hazard assessment. *Journal of Volcanology and Geothermal Research,* 178, 366-377.

Macedonio, G., Pareschi, M. T. & Santacroce, R. 1988. A numerical simulation of the Plinian Fall Phase of 79 A.D. eruption of Vesuvius. *Journal of Geophysical Research,* 93, 14817-14827.

Magill, C., Blong, R. & McAneney, J. 2006. VolcaNZ--A volcanic loss model for Auckland, New Zealand. *Journal of Volcanology and Geothermal Research,* 149, 329-345.

Magill, C. R., Wilson, T. M. & Okada, T. 2013. Observations of tephra fall impacts from the 2011 Shinmoedake eruption, Japan. *Earth, Planets and Space,* 65, 677-698.

Maqsood, T., Wehner, M., Ryu, H., Edwards, M., Dale, K. & Miller, V. 2014. GAR15 Vulnerability Functions. Geoscience Australia Record 2014/38. http://dx.doi.org/10.11636/Record.2014.038.

Marti, J., Spence, R., Calogero, E., Ordoñez, A., Felpeto, A. & Baxter, P. 2008. Estimating building exposure and impact to volcanic hazards in Icod de los Vinos, Tenerife (Canary Islands). *Journal of Volcanology and Geothermal Research,* 178, 553-561.

Marzocchi, W., Sandri, L., Gasparini, P., Newhall, C. & Boschi, E. 2004. Quantifying probabilities of volcanic events: The example of volcanic hazard at Mount Vesuvius. *Journal of Geophysical Research: Solid Earth,* 109, B11201.

Marzocchi, W., Sandri, L. & Selva, J. 2008. BET_EF: a probabilistic tool for long- and short-term eruption forecasting. *Bulletin of Volcanology,* 70, 623-632.

Marzocchi, W., Sandri, L. & Selva, J. 2010. BET_VH: a probabilistic tool for long-term volcanic hazard assessment. *Bulletin of Volcanology,* 72, 705-716.

Mead, S. & Magill, C. 2014. Determining change points in data completeness for the Holocene eruption record. *Bulletin of Volcanology,* 76, 1-14.

MEDSUV 2013-2015. MEDiterranean SUpersite Volcanoes. http://ec.europa.eu/research/environment/geo/pdf/supersites/medsuv-puglisi.pdf.

Meheux, K., Dominey-Howes, D. & Lloyd, K. 2007. Natural hazard impacts in small island developing states: a review of current knowledge and future research needs. *Natural Hazards,* 40, 429-446.

Munich-Re 2007. Volcanism – Recent findings on the risk of volcanic eruptions. Schadenspiegel.

Newhall, C. G. & Hoblitt, R. P. 2002. Constructing event trees for volcanic crises. *Bulletin of Volcanology,* 64, 3-20.

Newhall, C. G. & Punongbayan, R. S. 1996. *Fire and mud: Eruptions and lahars of Mount Pinatubo, Philippines,* Quezon City, Seattle and London, Philippine Institute of Volcanology and Seismology and University of Washington Press.

Newhall, C. G. & Self, S. 1982. The volcanic explosivity index (VEI) - An estimate of explosive magnitude for historical volcanism. *Journal of Geophysical Research,* 87, 1231-1238.

Orsi, G., di Vito, M. A., Selva, J. & Marzocchi, W. 2009. Long-term forecast of eruption style and size at Campi Flegrei caldera (Italy). *Earth and Planetary Science Letters,* 287, 265-276.

Oze, C., Cole, J. W., Scott, A., Wilson, T. M., Wilson, G., Gaw, S., Hampton, S. J., Doyle, C. & Li, Z. 2014. Corrosion of metal roof materials related to volcanic ash interactions. *Natural Hazards,* 71, 785-802.

Pyle, D. M. 1989. The thickness, volume and grainsize of tephra fall deposits. *Bulletin of Volcanology,* 51, 1-15.

Ronan, K. R. 1997. The effects of a "benign" disaster: Symptoms of post-traumatic stress in children following a series of volcanic eruptions. *Australasian Journal of Disaster and Trauma Studies,* 1.

Ryall, D. B. & Maryon, R. H. 1998. Validation of the UK Met. Office's NAME model against the ETEX dataset. *Atmostpheric Environment,* 32, 4265-4276.

Sandri, L., Jolly, G., Lindsay, J., Howe, T. & Marzocchi, W. 2012. Combining long- and short-term probabilistic volcanic hazard assessment with cost-benefit analysis to support decision making in a volcanic crisis from the Auckland Volcanic Field, New Zealand. *Bulletin of Volcanology,* 74, 705-723.

Sandri, L., Thouret, J.-C., Constantinescu, R., Biass, S. & Tonini, R. 2014. Long-term multi-hazard assessment for El Misti volcano (Peru). *Bulletin of volcanology,* 76, 1-26.

Sarna-Wojcicki, A. M., Shipley, S., Waitt Jr., R. B., Dzurisin, D., Wood, S. H., Lipman, P. W. & Mullineaux, D. R. 1981. Areal distribution, thickness, mass, volume, and grain size of airfall ash from the six major eruptions of 1980. *In:* Lipman, P. W. & Mullineaux, D. R. *The 1980 Eruptions of Mount St. Helens, Washington.* U.S. Geological Survey Professional Paper.

Scaini, C., Felpeto, A., Marti, J. & Carniel, R. 2014. A GIS-based methodology for the estimation of potential volcanic damage and its application to Tenerife Island, Spain. *Journal of Volcanology and Geothermal Research,* 278, 40-58.

Scasso, R., Corbella, H. & Tiberi, P. 1994. Sedimentological analysis of the tephra from the 12–15 August 1991 eruption of Hudson volcano. *Bulletin of Volcanology,* 56, 121-132.

Selva, J., Costa, A., Marzocchi, W. & Sandri, L. 2010a. BET_VH: exploring the influence of natural uncertainties on long-term hazard from tephra fallout at Campi Flegrei (Italy). *Bulletin of Volcanology.* Springer Berlin / Heidelberg.

Selva, J., Costa, A., Marzocchi, W. & Sandri, L. 2010b. BET_VH: exploring the influence of natural uncertainties on long-term hazard from tephra fallout at Campi Flegrei (Italy). *Bulletin of Volcanology,* 72, 717-733.

Selva, J., Orsi, G., Di Vito, M. A., Marzocchi, W. & Sandri, L. 2012. Probability hazard map for future vent opening at the Campi Flegrei caldera, Italy. *Bulletin of Volcanology,* 74, 497-510.

Selva, J. & Sandri, L. 2013. Probabilistic Seismic Hazard Assessment: Combining Cornell-like approaches and data at sites through Bayesian inference. *Bulletin of the Seismological Society of America,* 103, 1709-1722.

Simkin, T., Siebert, L. & Blong, R. 2001. DISASTERS: Volcano fatalities- lessons from the historical record. *Science,* 291.

Small, C. & Naumann, T. 2001. The global distribution of human population and recent volcanism. *Global Environmental Change Part B: Environmental Hazards,* 3, 93-109.

Sparks, R. S. J., Bursik, M. I., Ablay, G. J., Thomas, R. M. E. & Carey, S. N. 1992. Sedimentation of tephra by volcanic plumes. Part 2: controls on thickness and grain-size variations of tephra fall deposits. *Bulletin of Volcanology,* 54, 685-695.

Spence, R., Kelman, I., Baxter, P., Zuccaro, G. & Petrazzuoli, S. 2005. Residential building and occupant vulnerability to tephra fall. *Natural Hazards and Earth Systems Science,* 5, 477-494.

Spence, R., Komorowski, J.-C., Saito, K., Brown, A., Pomonis, A., Toyos, G. & Baxter, P. 2008. Modelling the impact of a hypothetical sub-Plinian eruption at La Soufrière of Guadeloupe (Lesser Antilles). *Journal of Volcanology and Geothermal Research,* 178, 516-528.

Spence, R., Pomonis, A., Baxter, P., Coburn, A., White, M., Dayrit, M. & Field Epidemiology Training Program Team 1996. Building damage caused by the Mount Pinatubo eruption of June 15, 1991. *In:* Newhall, C. G. & Punongbayan, R. S. *Fire and Mud: Eruptions and Lahars of Mount Pinatubo, Philippines.* Quezon City & Seattle: Philippines Institute of Volcanology and Seismology & University of Washington Press.

SSHAC 1997. Senior Seismic Hazard Analysis Committee: Recommendations for probabilistic seismic hazard analysis: Guidance on uncertainty and use of experts. Technical Report, NUREG/CR-6372, U.S. Nuclear Regulatory Commission, Washington, D. C.

Stewart, C., Horwell, C., Plumlee, G., Cronin, S. J., Delmelle, P., Baxter, P., Calkins, J., Damby, D., Morman, S. & Oppenheimer, C. 2013. Protocol for analysis of volcanic ash samples for assessment of hazards from leachable elements Available at: www.ivhhn.org: IVHHN Report ratified by IAVCEI, Cities and Volcanoes Commission, USGS and GNS Science.

Suzuki, T. 1983. A theoretical model for dispersion of tephra. *In:* Shimozuru, D. & Yokohama, I. *Arc volcanism: physics and tectonics.* Tokyo: Terra Scientific Publishing.

Sword-Daniels, V., Wilson, T. M., Sargeant, S., Rossetto, T., Twigg, J., Johnston, D. M., Loughlin, S. C. & Cole, P. D. 2014. Consequences of long-term volcanic activity for essential services in Montserrat: challenges, adaptations and resilience. *Geological Society, London, Memoirs.*

Walker, G. 1981. Plinian eruptions and their products. *Bulletin of Volcanology,* 44, 223-240.

Wilson, T., Cole, J., Stewart, C., Cronin, S. & Johnston, D. 2011. Ash storms: impacts of wind-remobilised volcanic ash on rural communities and agriculture following the 1991 Hudson eruption, southern Patagonia, Chile. *Bulletin of Volcanology,* 73, 223-239.

Wilson, T. M., Stewart, C., Bickerton, H., Baxter, P. J., Outes, V., Villarosa, G. & Rovere, E. 2012a. Impacts of the June 2011 Puyehue Cordón-Caulle volcanic complex eruption on urban infrastructure, agriculture and public health. GNS Science.

Wilson, T. M., Stewart, C., Sword-Daniels, V., Leonard, G. S., Johnston, D. M., Cole, J. W., Wardman, J., Wilson, G. & Barnard, S. T. 2012b. Volcanic ash impacts on critical infrastructure. *Physics and Chemistry of the Earth, Parts A/B/C,* 45-46, 5-23.

Wisner, B., Gaillard, J. & Kelman, I. (eds.) 2012. *The Handbook of Hazards and Disaster Risk Reduction,* Oxon: Routledge.

Witham, C. S. 2005. Volcanic disasters and incidents: A new database. *Journal of Volcanology and Geothermal Research,* 148, 191-233.

Woods, A. W. & Bursik, M. I. 1991. Particle fallout, thermal disequilibrium and volcanic plumes. *Bulletin of Volcanology,* 53, 559-570.

Yokoyama, I., Tilling, R. I. & Scarpa, R. 1984. International mobile early-warning system(s) for volcanic eruptions and related seismic activities. *In:* UNESCO (ed.) *FP/2106-92-01 (2286).* Paris.

Zuccaro, G., Cacace, F., Spence, R. J. S. & Baxter, P. J. 2008. Impact of explosive eruption scenarios at Vesuvius. *Journal of Volcanology and Geothermal Research,* 178, 416-453.

Chapter 4

Populations around Holocene volcanoes and development of a Population Exposure Index

S.K. Brown, M.R. Auker and R.S.J. Sparks

A way of ranking the risk to life from volcanoes is to establish how many people live in their vicinity. In addition to being an indicator of lives under threat, population exposure is a proxy for threat to livelihoods, infrastructure, economic assets and social capital. This report uses two indicators of population density around volcanoes to assess the current global exposure and as a risk indicator for individual volcanoes, and discusses this in combination with the Human Development Index (HDI) as a proxy for vulnerability.

4.1 Background

Ewert & Harpel (2004) introduced the Volcano Population Index (VPI), which estimates the number of people living within 5 and 10 km radii of volcanoes (VPI5 and VPI10). These population statistics, and VPI30 and VPI100 (population within 30 and 100 km of Holocene volcanoes) are reported in the VOTW4.0 (2013) database (www.volcano.si.edu; Siebert et al. (2010)). The Population Exposure Index, (PEI), was developed by Aspinall et al. (2011) for a study of volcanic risk in the World Bank's Global Facility for Disaster Reduction and Recovery (GFDRR) priority countries. Here, populations within 10 and 30 km radii were estimated and combined using weightings that reflect how historic fatalities vary with distance from volcanoes.

The VPI was developed on the basis that most eruptions are small to moderate in size (VEI ≤3), with footprints of less than 10 km. The VPI therefore represents the population exposures for most eruptions. Indeed, eruptions of VEI2 occur at a rate of approximately one every few weeks, and VEI3 several times a year (Siebert et al., 2010). Eruptions of larger magnitudes (VEI≥4) are less frequent, but often cause fatalities at distances well beyond 10 km (Auker et al., 2013). Hazard footprints from such eruptions commonly extend to tens of kilometres. The PEI thus complements the VPI, accounting for the high threat from large eruptions and potentially distal hazard types. An advantage of PEI is that only a single indicator parameter captures the exposure of populations around each volcano with the various VPI populations all contributing to the index and weighted according to historical evidence on the distribution of fatalities with distance. Here we develop and apply an amended version of the PEI, which correlates quite well with VPI_{10}.

Brown, S.K., Auker, M.R. & Sparks, R.S.J. (2015) Populations around Holocene volcanoes and development of a Population Exposure Index. In: S.C. Loughlin, R.S.J. Sparks, S.K. Brown, S.F. Jenkins & C. Vye-Brown (eds) *Global Volcanic Hazards and Risk,* Cambridge: Cambridge University Press.

4.2 Population

The location and total population within circles of radius 10, 30 and 100 km of each volcano is derived from the VOTW4.0 (2013) database. Due to overlapping radii from multiple volcanoes, these population figures cannot simply be summed, therefore country-level data counting populations in the vicinity of multiple volcanoes only once were calculated by the Norwegian Geotechnical Institute (NGI) using the Oak Ridge National Laboratory LandScan 2011 dataset of Bright et al. (2012).

Holocene volcanoes are located in 86 countries. The total population within 100 km of these volcanoes is over 800 million (Table 4.1). With 142 volcanoes Indonesia has the greatest total population located within all distance categories (>8.6 million at 10 km, >68 million at 30 km and >179 million at 100 km). After Indonesia, the Philippines and El Salvador have the largest populations living within 10 km, both at >2 million.

Table 4.1 The total global population living within given radii of volcanoes, derived using volcano location data from VOTW4.0 and 2011 LandScan data. Populations within each country were calculated and summed: no population was counted twice.

Total population within 10 km	Total population within 30 km	Total population within 100 km
29,294,942	226,267,790	801,833,245

Indonesia, the Philippines and Japan have the greatest numbers of people living with 100 km of their volcanoes (Figure 4.1; left). The populations of small volcanic island nations, such as Tonga and Samoa, are almost all resident within 100 km. The percentage of the population living within 100 km of volcanoes is therefore calculated for those countries with an area of more than a circle of 100 km radius (Figure 4.1; right). These populations may be affected by volcanoes in bordering countries.

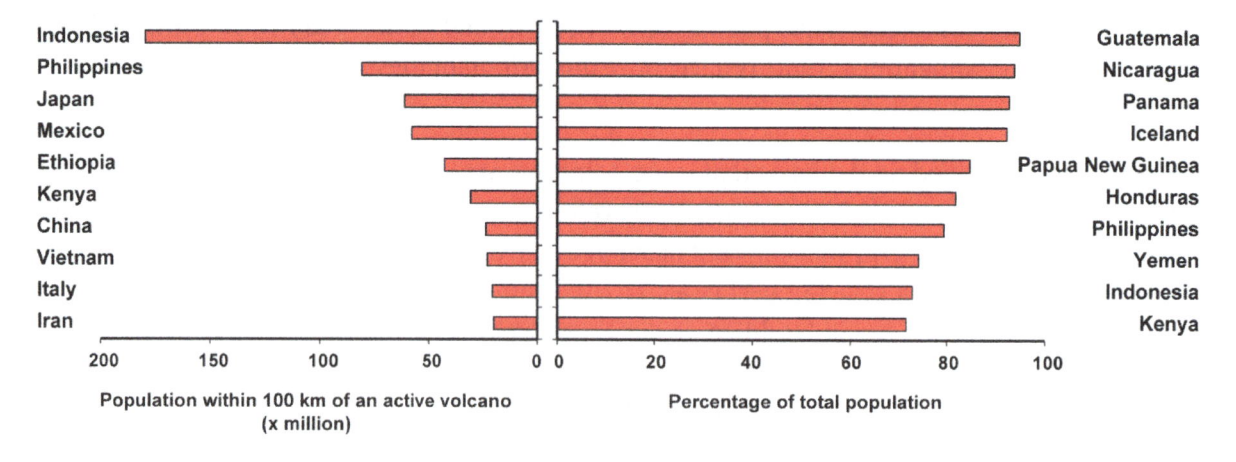

Figure 4.1 The top 10 countries for population within 100 km of a volcano (left) and the top 10 countries (area over 31,415 km²) for percentage of the total population (right).

Capital cities are frequently located close to volcanoes. The capitals of American Samoa, Wallis and Futuna islands, Montserrat, Dominica, Guadeloupe, Saint Kitts and Nevis and Nicaragua all lie within 10 km of volcanoes. A further 20 capitals lie between 10 and 30 km; 32 lie between 30 and 100 km.

Countries without Holocene volcanoes within their borders may also have populations within 100 km of a volcano (Table 4.2).

Table 4.2 Countries with populations living within 100 km of volcanoes beyond their borders. No populations in these countries live within 10 km of volcanoes.

Country	Population within 30 km	Population within 100 km
Jordan	48,278	5,690,340
Israel	1,056	1,884,367
Lebanon	0	3,141,870
Laos	0	26,512
Cambodia	0	1,409

The largest populations within 10 km of volcanoes are those around Michoacán-Guanajuato, Mexico (>5.7 million), Tatun Group, Taiwan (>5 million) and Leizhou Bandao, China (>3.2 million) (Table 4.3). These and many other volcanoes with high populations in their vicinity have poorly understood eruptive histories; the age or magnitude, in some cases both, of their last eruption are often unknown. Poorly constrained eruptive histories make hazard assessment at these volcanoes difficult.

The volcanoes with largest populations within 100 km differ considerably from those with largest populations within 10 km, illustrating variation in population distribution (Table 4.3). Indeed, Small & Naumann (2001) explored regional trends in population density around volcanoes, showing a globally averaged decrease in population density with increasing distance from volcanoes, dominated by tropical areas such as SE Asia and Central America. The opposite relationship is present in Japan and Chile. Eight of the ten most populous volcanoes at 100 km are located in Indonesia and Mexico; of these, only Chichinautzin and Popocatépetl have eruptions of VEI≥4 recorded in the Holocene.

The eruptive histories of almost half of the most populated volcanoes are poorly constrained (Table 4.3). The Holocene records for three volcanoes: Salak; Perbakti-Gagak; Tangkubanparahu (all Indonesia), solely contain VEI ≤2 eruptions, suggesting there is a low probability of significant effects beyond 10 km. However, large eruptions of M or VEI ≥4 often have long recurrence intervals, for which the Holocene may not be statistically representative. The volcanoes with largest proximal populations that have produced VEI≥4 eruptions in the Holocene are shown in Table 4.4.

Table 4.3 *The top ten volcanoes by population size within the given radii and details of their last recorded eruptions. As magnitude affects the extent of the hazard footprint, the maximum recorded Holocene VEI and model VEI are given, and the occurrence of a Pleistocene record (in the LaMEVE database) of M≥4 eruptions is shown.*

Radius	Volcano, Country	Population within given radius	Year of last eruption (VEI of last eruption)	Maximum recorded Holocene VEI and modal Holocene VEI	Pleistocene record of M≥4 events
10 km	Michoacán-Guanajuato, Mexico	5,783,287	1943 (VEI 4)	VEI 4. Modal 3.	
	Tatun Group, Taiwan	5,084,149	4100 BC (VEI 1)	VEI 1.	
	Leizhou Bandao, China	3,230,167	? (VEI ?)	Unknown	
	Kars Plateau, Turkey	3,067,709	1959? (VEI 2?)	VEI 2?	
	Malang Plain, Indonesia	2,397,210	? (VEI ?)	Unknown	
	Campi Flegrei, Italy	2,234,109	1538 (VEI 3)	VEI 5. Modal 4	Yes
	Ilopango, El Salvador	2,049,583	1879 (VEI 3)	VEI 6.	Yes
	Hainan Dao, China	1,731,229	1933 (VEI ?)	Unknown	
	Jabal ad Druze, Syria	1,487,860	? (VEI ?)	Unknown	
	San Pablo Volcanic Field, Philippines	1,349,742	1350 (VEI ?)	Unknown	
100 km	Gede, Indonesia	40,640,105	1957 (VEI 2)	VEI 3. Modal 2.	
	Salak, Indonesia	38,154,252	1938 (VEI 2)	VEI 2. Modal 2.	
	Perbakti-Gagak, Indonesia	36,630,568	1939 (VEI 1)	VEI 1. Modal 1.	
	Tangkubanparahu, Indonesia	32,855,731	2013 (VEI 2)	VEI 2. Modal 1.	Yes
	Hakone, Japan	30,282,197	1170 (VEI ?)	VEI 3. Most unknown.	Yes
	Papayo, Mexico	28,677,002	? (VEI ?)	Unknown	
	Chichinautzin, Mexico	28,030,794	400 (VEI 3)	VEI 4. Modal 3.	Yes
	Iztaccíhuatl, Mexico	27,276,280	? (VEI ?)	Unknown	
	Arayat, Philippines	27,216,491	? (VEI ?)	Unknown	
	Popocatépetl, Mexico	26,509,510	2013 (VEI 2)	VEI 5. Modal 2.	Yes

Table 4.4 *The top ten volcanoes by population size with a Holocene record of VEI ≥4 eruptions.*

Rank	Volcano, Country	Rank	Volcano, Country
1	Popocatépetl, Mexico	6	Merapi, Indonesia
2	Chichinautzin, Mexico	7	Galunggung, Indonesia
3	Fuji, Japan	8	Tengger Caldera, Indonesia
4	Kelut, Indonesia	9	Pinatubo, Philippines
5	Taal, Philippines	10	Izu-Toba, Japan

4.3 Population Exposure Index

A multitude of factors determine a population's vulnerability to volcanic hazards. Most volcanoes have the potential for a spectrum of eruption magnitudes and styles, with consequently varied footprints. In most cases, hazard and threat decreases with distance from the volcano. This is particularly true of pyroclastic density currents and lahars (the cause of almost 70% of all directly caused fatalities (Auker et al., 2013)), which are commonly confined to valleys. A maximum distance for consideration of population exposure of 100 km likely captures the majority of these hazards. However, the effects of the largest eruptions may extend beyond this distance.

The location and population within 10, 30 and 100 km of each volcano is derived from VOTW4.0. These populations are weighted on the basis of the area of each ring, and the number of fatal events recorded since 1600 AD within each distance category, using data from VOTW4.0 and the Smithsonian fatalities database described and analysed in Auker et al. (2013). Fatal incidents attributed to direct hazards (e.g. pyroclastic density currents, lahars, lava flows) are included, whilst indirect fatalities (e.g. famine) are excluded.

The fatality weighting is calculated based on the numbers of fatal incidents from direct hazards at each extent. The distance of fatalities from the volcano is only available for 27 of the 533 fatal incidents listed in Auker et al. (2013). These numbers differ from those used in Aspinall et al. (2011) due to the slight change in selection criteria. A total of 17 fatal incidents are recorded at 0-9 km, six fatal incidents at 10-29 km, and four fatal incidents at 30-100 km, giving proportional weightings of 0.63, 0.22 and 0.19 respectively.

The increase in area of each circle moving away from the volcano decreases the population density for any given population size. The area of the 10 km radius circle is nine times smaller than that of the 30 km circle, and 100 times smaller than the 100 km circle, giving weightings of 0.91, 0.08 and 0.01, respectively. These two sets of weights are combined and scaled, yielding a weighting of 0.967 for the 10 km ring, 0.03 for the 30 km ring, and 0.003 for the 100 km ring.

For each volcano, the population within each distance category is multiplied by the appropriate weighting, and the three figures are summed. These final weighted populations are then assigned one of seven index scores, from 1 to 7 (Table 4.5; amended from the scale of 0 to 3 of Aspinall et al. (2011). We refer to these seven index scores as the Population Exposure Index (PEI).

Table 4.5 Population Exposure Index (PEI), amended after Aspinall et al. (2011).

Weighted summed population	Population Exposure Index
0	1
<3,000	2
3,000 – 9,999	3
10,000 – 29,999	4
30,000 – 99,999	5
100,000 – 300,000	6
>300,000	7

There are inherent uncertainties in the population statistics and in the accuracy of volcano locations. The fatality weighting may be refined through further study of the historic record and improvement in the evidence on distances of fatalities.

4.3.1 Global PEI

The PEI is calculated for all volcanoes in VOTW4.0. Over 40% of volcanoes have a PEI of 2; with the exception of PEI 7 there is an approximately even spread across all other PEI classes. The division of the total global population within 100 km (Table 4.1) across PEI classes is also examined (Table 4.6). Analysis shows that 60% of the population living within 100 km are located around just 4% of volcanoes: the PEI 7 volcanoes. About a quarter of volcanoes are PEI ≥5, yet 96% of the total population are located here. The data indicate that most exposure (>95%) to volcanic hazards is distributed in volcanoes with PEI values of 5 to 7.

Table 4.6 The number of volcanoes in each PEI category globally and the percentage of the total number of volcanoes. The percentage of the total weighted population is also provided.

Population Exposure Index	Number of volcanoes (%)	Percentage of total weighted population
1	197 (12.7%)	0%
2	642 (41.4%)	0.4%
3	157 (10.1%)	1.0%
4	178 (11.5%)	3.5%
5	188 (12.1%)	11.4%
6	128 (8.3%)	23.8%
7	61 (3.9%)	59.9%

There are 61 PEI 7 volcanoes, of which 16 are in Indonesia. Africa and the Red Sea and Mexico and Central America have 11 PEI 7 volcanoes each, with all other regions having <10. Indonesia, Mexico and Central America and Africa and the Red Sea have the greatest numbers of PEI ≥5 volcanoes. The regions with the greatest proportions of PEI ≥5 volcanoes are Philippines and SE Asia (70%), Mexico and Central America (64%), and Indonesia (55%). Alaska, Antarctica, and the Kuril Islands have no PEI ≥5 volcanoes (Figure 4.2).

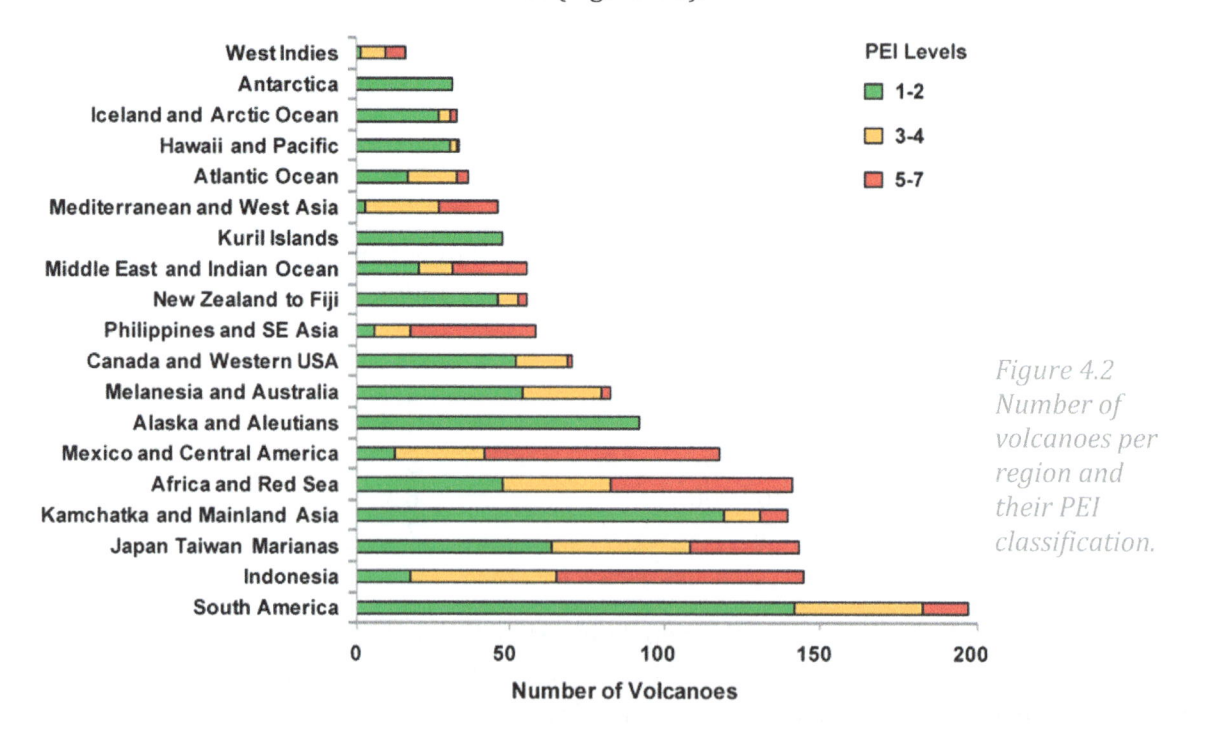

Figure 4.2 Number of volcanoes per region and their PEI classification.

4.3.2 PEI and VEI

Eruptions of M or VEI ≥4 are less frequent than smaller events, but have the potential for greater losses over larger areas. VOTW4.0 includes 864 volcanoes which have no recorded eruptions of known VEI. Of the remaining 687, 438 have no eruptions of VEI >3; 249 volcanoes have one or more eruptions of VEI ≥4, of which 61 have a PEI of 5-7. There is clearly the potential for far-reaching hazards to affect large populations.

4.3.3 PEI and HDI

The Human Development Index (HDI) combines details of life expectancy, education and income to provide a measure of social and economic development, calculated by the United Nations Development Programme (United Nations Development Programme (UNDP), 2013) and categorised as Low, Medium, High or Very High. HDI is available for most, though not all countries; notable exceptions are overseas territories and island volcanoes. HDI and other metrics such as Gross Domestic Product (GDP) provide an indication of the wealth and development of a country. A low HDI does not always reflect the resources dedicated to disaster preparedness and response; however, there is a general relationship between the wealth of a country and the losses sustained in disasters (Toya & Skidmore, 2007). Toya & Skidmore (2007) and references therein explained that populations of wealthier nations have greater expectations regarding safety and are therefore more likely to use expensive precautionary measures to improve safety. They found that disaster losses and the underlying socio-economic fabric within a country are also correlated.

There are 530 volcanoes located in countries of Very High HDI (dominantly in Japan, Chile and the USA, which account for 391 of these volcanoes; Figure 4.3). Countries of Low HDI have fewer volcanoes (218), though there is a broad negative correlation between HDI and PEI. Significant examples include Ngozi in Tanzania (Low HDI), which has a PEI of 7 and a Holocene eruption of VEI 5; and Masaya in Nicaragua (Medium HDI) which also classifies at PEI 7 and has a Holocene record of eruptions of VEI 5 and 6 and frequent eruptions of VEI 1 and 2. The 142 volcanoes of Indonesia dominate the distribution of PEIs amongst the 355 volcanoes in medium HDI countries.

Fewer than 20% of volcanoes in High and Very High HDI countries are PEI≥5, and over 60% are classed as PEI 1 or 2. Notable examples of PEI 7 volcanoes in Very High HDI countries are Vesuvius and Campi Flegrei. Both border Naples in Italy and have Holocene records of VEI 5 eruptions and the potential to generate far-reaching hazards. The Auckland Field in New Zealand (Very High HDI) is also PEI 7, due to its situation under the city of Auckland. The low relief of this volcanic field makes large explosive eruptions with volcanic flows that extend to tens of kilometres unlikely.

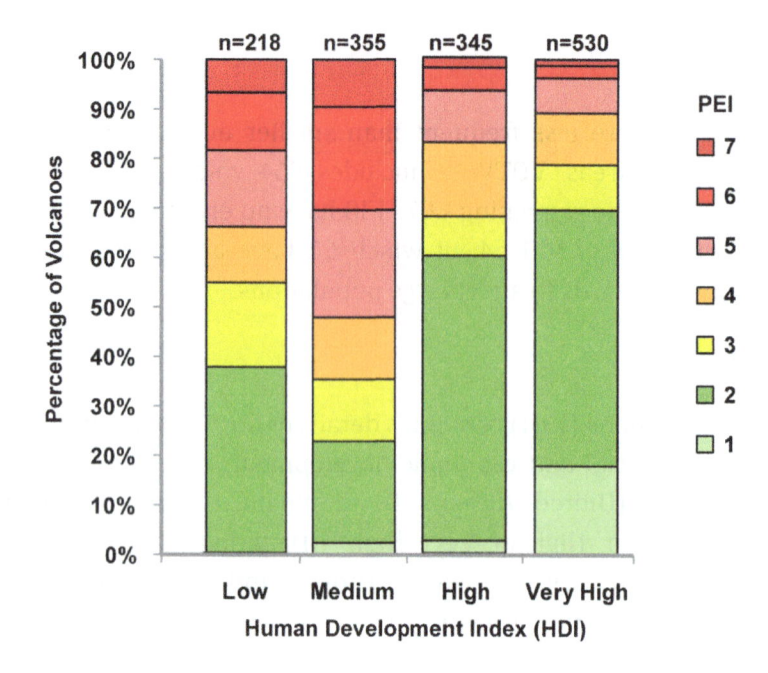

Figure 4.3 The four categories of the HDI and the proportion of volcanoes in each PEI band.

Combining PEI and HDI provides an indicator of the number of people in harm's way and societal and economic capacity to handle disasters. Volcanoes with large proximal populations in relatively low HDI countries may be more vulnerable and suffer greater relative losses. However, factors such as local prioritisation of resources or experience with natural hazards may counter this, and PEI and HDI do not account for differences in eruption styles or recurrence rates. Three main regions have high HDI x PEI rankings: Africa, SE Asia (dominated by Indonesia and the Philippines) and Central America (Figure 4.4).

4.4.4 Implications and use of PEI

The VPI and PEI enable volcanoes close to large populations to be identified. The PEI's weighting of populations at different extents aims to capture the factors that control the number of people exposed.

We show that the majority of people exposed to volcanic hazards live around the 61 PEI 7 volcanoes. Indonesia has the greatest number of PEI 7 volcanoes (16), and subsequently the greatest total number of people living within 100 km of volcanoes. PEI and HDI are generally negatively correlated, suggesting larger populations are situated close to volcanoes in developing countries, compared to developed. There are 61 volcanoes with recorded VEI ≥4 Holocene eruptions and a high PEI (PEI 5-7). Similarly large eruptions from these volcanoes have the potential to cause significant disruption and loss. Many other high PEI volcanoes may produce VEI ≥4 eruptions over longer time scales.

The PEI may be used as a basis for disaster risk reduction resource management decisions. However, population exposure is not the only component of volcanic threat, and use of a hazard index for volcanoes provides a fuller picture. The PEI is also not a substitute for in depth assessments of exposure and vulnerability at specific volcanoes. Indeed it is certain that volcanoes with the same index value may have very different exposed populations. Topographic factors in particular will have a large role in determining exposure. For example, many volcanoes have craters or flank collapse scars open in one direction that will channel flows and produce directed hazards that threaten populations on one side of the volcano to a far greater degree than the other. Also, populations close to river valleys and on flood plains are very vulnerable to lahars and pyroclastic flows, and populations in the dominant downwind direction are more exposed to ash fall hazards. Full assessment based on local factors may lead to different conclusions about priorities.

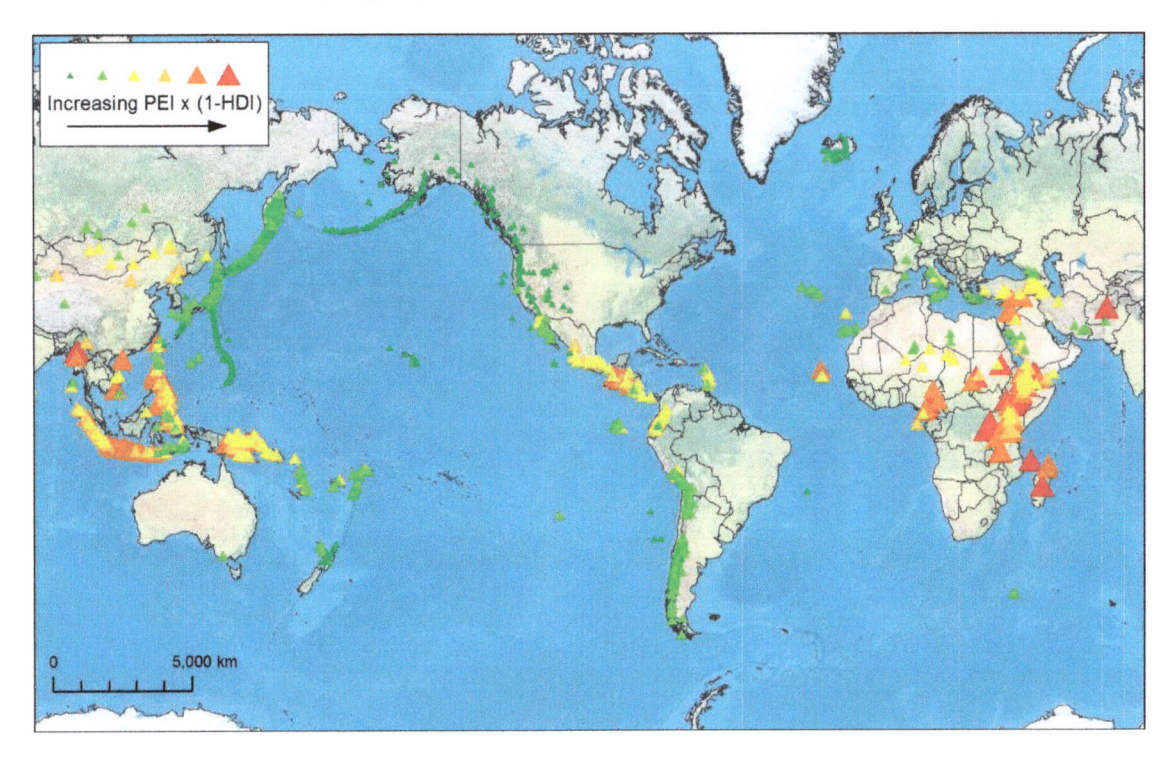

Figure 4.4 Global distribution of volcanoes coloured and scaled by PEI x (1-HDI), illustrating the locations of volcanoes with high proximal populations and lower HDI scores.

References

Aspinall, W., Auker, M., Hincks, T., Mahony, S., Nadim, F., Pooley, J., Sparks, R. & Syre, E. 2011. Volcano hazard and exposure in GFDRR priority countries and risk mitigation measures- GFDRR Volcano Risk Study. *Bristol: Bristol University Cabot Institute and NGI Norway for the World Bank: NGI Report,* 20100806, 3.

Auker, M. R., Sparks, R. S. J., Siebert, L., Crosweller, H. S. & Ewert, J. 2013. A statistical analysis of the global historical volcanic fatalities record. *Journal of Applied Volcanology,* 2, 1-24.

Bright, E. A., Coleman, P. R., Rose, A. N. & Urban, M. L. 2012. *LandScan 2011* [Online]. Oak Ridge, TN, USA. Available: http://www.ornl.gov/landscan/

Ewert, J. W. & Harpel, C. J. 2004. In harm's way: population and volcanic risk. *Geotimes,* 49, 14-17.

Siebert, L., Simkin, T. & Kimberley, P. 2010. *Volcanoes of the World, 3rd edn,* Berkeley, University of California Press.

Small, C. & Naumann, T. 2001. The global distribution of human population and recent volcanism. *Global Environmental Change Part B: Environmental Hazards,* 3, 93-109.

Smithsonian. 2013. *Volcanoes of the World 4.0* [Online]. Washington D.C. Available: http://www.volcano.si.edu.

Toya, H. & Skidmore, M. 2007. Economic development and the impacts of natural disasters. *Economics Letters,* 94, 20-25.

United Nations Development Programme (UNDP). 2013. *Human Development Report 2013: The Rise of the South: Human Progress in a Diverse World* [Online]. Available: www.hdr.undp.org/en/data

Chapter 5

An integrated approach to Determining Volcanic Risk in Auckland, New Zealand: the multi-disciplinary DEVORA project

N.I. Deligne, J.M. Lindsay and E. Smid

5.1 Background

Auckland, New Zealand, is home to 1.4 million people, over a third of New Zealand's population, and accounts for ~35% of New Zealand's GDP (Statistics New Zealand, 2014). The city is built on top of the Auckland Volcanic Field (AVF), which covers 360 km², has over 50 eruptive centres (vents), and has erupted over 55 times in the past 250,000 years, producing a cumulative volume of ~2 km³ of tephra, lava and other volcanic deposits[1] (see Figure 5.1). The field is likely to erupt again: the most recent eruption, Rangitoto, was only 550 years ago. Most AVF vents are monogenetic, i.e. they only erupt once. This means that it is very likely that the next vent will erupt in a new location within the field. Despite considerable scientific efforts, no spatial (where) or temporal (when) patterns have been identified; indeed, the oldest (Pupuke volcano) and the youngest (Rangitoto) vents are located next to each other. As such, it is wholly unknown where or when the next eruption will be. The size of the next eruption is also difficult to address, as the last eruption, Rangitoto, accounts for nearly half of the erupted volume of the field, and it is unclear whether this eruption is an anomaly or signals a change in the eruptive behaviour of the field. These difficulties of assessing location, time and size of next eruption pose a considerable problem for emergency and risk managers. The main challenges facing Auckland and other populated areas coinciding with volcanic fields include:

- uncertainty of where and when the next eruption will take place;
- communicating to the public how an eruption of unknown location will impact them and how they can best prepare;
- planning for an event which hasn't occurred in historical time;
- foreseeing and appropriately planning for the range of possible impacts to the built environment, local, regional and national economy and psyche.

---

[1] Equivalent to volume of 800,000 Olympic size pools.

Deligne, N.I., Lindsay, J.M. & Smid, E. (2015) An integrated approach to Determining Volcanic Risk in Auckland, New Zealand: the multi-disciplinary DEVORA project. In: S.C. Loughlin, R.S.J. Sparks, S.K. Brown, S.F. Jenkins & C. Vye-Brown (eds) *Global Volcanic Hazards and Risk,* Cambridge: Cambridge University Press.

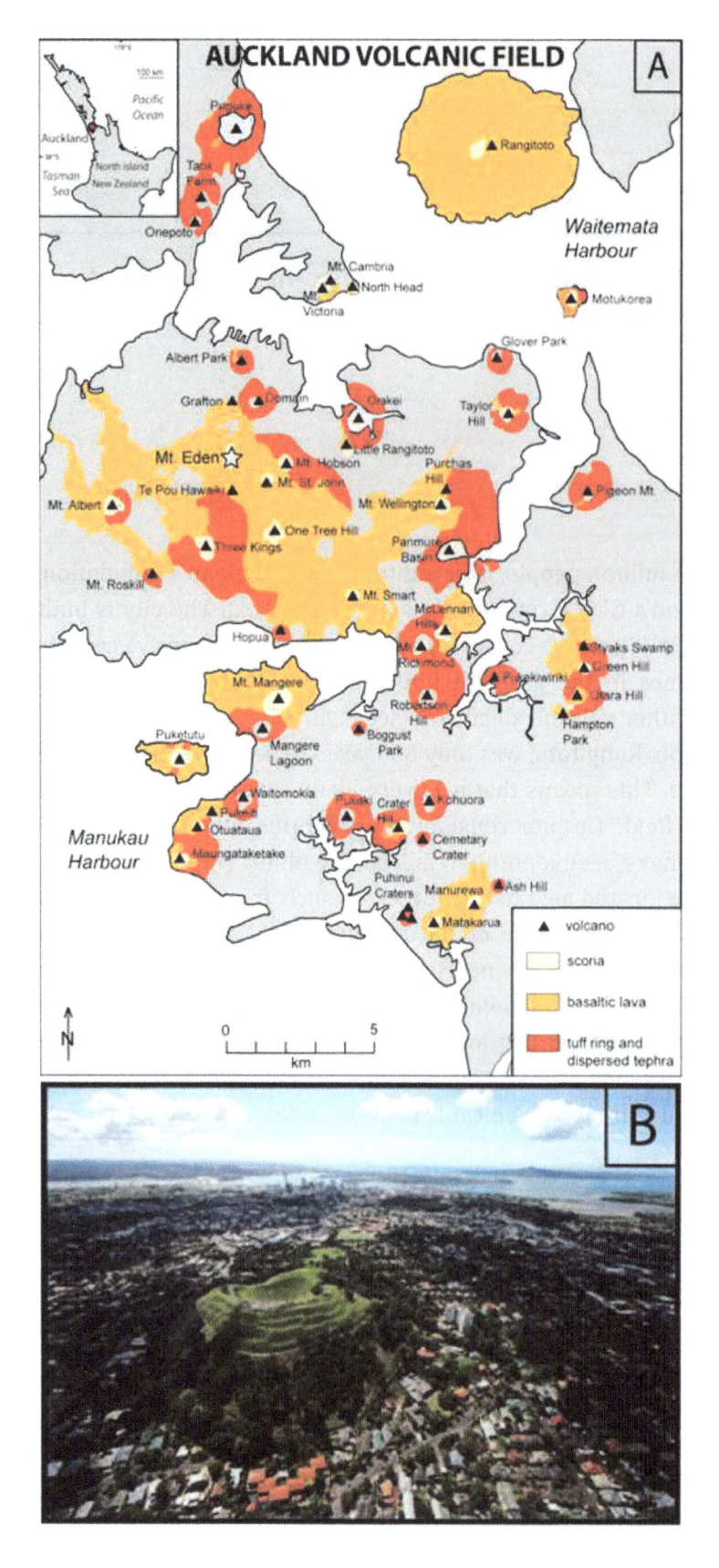

Figure 5.1 a) Map of Auckland Volcanic Field; star indicates location of Mt Eden. b) View of Mt Eden looking to the north highlighting the complete overlap of AVF and city (© Auckland Council).

5.2 DEVORA

The DEtermining VOlcanic Risk in Auckland (DEVORA) programme is a 7-year multi-agency research programme launched in November 2008. DEVORA was established following Exercise Ruaumoko, a 2007-2008 national Cabinet-lead Civil Defence exercise simulating an AVF eruption, in part to address knowledge gaps revealed by the exercise (Ministry of Civil Defence and Emergency Management, 2008). It is co-led by GNS Science (New Zealand's geologic survey) and the University of Auckland, with associated researchers at Massey University, the University of Canterbury and Victoria University of Wellington. It is funded by these organisations, the Earthquake Commission (national government), and Auckland Council (local/regional government). The DEVORA programme has a mandate to investigate the geological context of the AVF, volcanic hazards, and risk posed by the AVF, as reflected by the three themes organising the programme (Figure 5.2), listed below along with key questions:

1) Theme 1: Geological Model
    - Where is AVF magma coming from?
    - Why does it leave its source?
    - What controls the path of magma in the crust?
    - Where will the magma reach the surface?
    - What is the crust underlying the AVF made of?
    - Why is the most recent eruption the largest?
    - How fast will magma travel to the surface?
    - When will we detect the ascending magma?
2) Theme 2: Probabilistic Volcanic Hazard Model
    - What is the distribution in time of past eruptions affecting Auckland?
    - What is the likelihood and size of future eruptions affecting Auckland?
    - What are likely styles and hazards of future eruptions?
    - Where are we in the lifespan of the AVF?
    - How do we usefully calculate probabilistic volcanic hazard for Auckland?
    - What is the probabilistic volcanic hazard?
    - How intensive should the monitoring be to provide adequate warning of an AVF eruption?
3) Theme 3: Risk and Social Model for Auckland
    - Who and what are exposed to volcanic hazards in Auckland?
    - How will each hazard affect people and infrastructure?
    - How will people and organisations cope in an eruption?
    - What are the flow-on effects nation-wide from an eruption affecting Auckland?
    - How can we calculate risk to people and infrastructure?
    - What are the risks to people and infrastructure?
    - How can these risks be reduced?

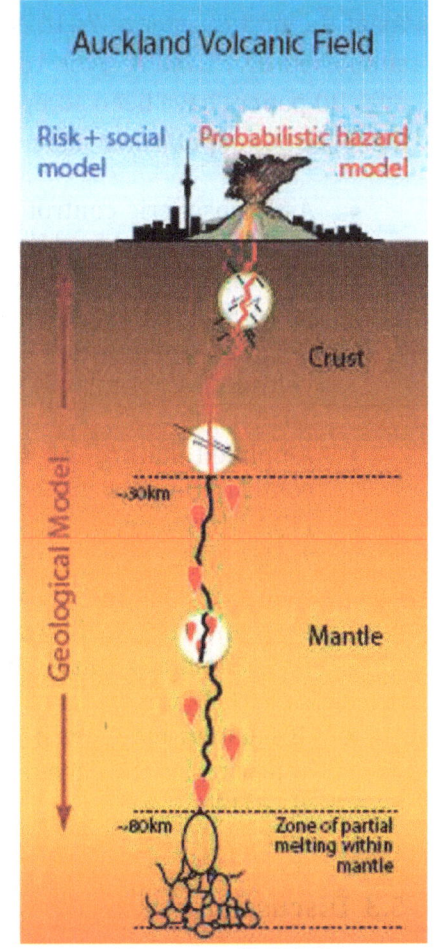

Figure 5.2 Scope of DEVORA themes.

To ensure that DEVORA outputs are useful not just scientifically but practically, government representatives sit on the DEVORA steering committee, which charts and directs DEVORA efforts. Furthermore, there is an annual research forum open to Auckland Council and Civil Defence staff and representatives from critical infrastructure and utility organisations. Here, recent findings and ongoing research are presented. This strengthens communication between scientists and decision makers, and enables policy to be informed by the most recent scientific findings. Indeed, the Auckland Volcanic Field Contingency Plan, the policy document which details response arrangements should an AVF eruption occur, has been recently reviewed and updated in close consultation with DEVORA scientists. Additionally, through DEVORA, University of Auckland students and the Auckland Civil Defence team participate in an annual informal mock eruption exercise. A longitudinal study is planned to compare public risk perception in 2008 and now, and will evaluate effectiveness of the DEVORA and associated programmes in improving public understanding of AVF hazards and risk.

As of the first quarter of 2014, seven Masters and 11 PhD projects have been at least partially supported by DEVORA, over 180 presentations have been given at scientific conferences, and over 80 papers have been accepted or published in a range of peer-reviewed scientific journals. Sample titles of published papers include:

- Asthenospheric control of melting processes in a monogenetic basaltic system: a case study of the Auckland Volcanic Field, New Zealand (McGee et al., 2013);
- Age, distance, and geochemical evolution within a monogenetic volcanic field: Analysing patterns in the Auckland Volcanic Field eruption sequence (Le Corvec et al., 2013);
- Longevity of a small shield volcano revealed by crypto-tephra studies (Rangitoto volcano, New Zealand): change in eruptive behaviour of a basaltic field (Shane et al., 2013);
- Amplified hazard of small-volume monogenetic eruptions due to environmental controls, Orakei Basin, Auckland Volcanic Field, New Zealand (Németh et al., 2012);
- LiDAR-based quantification of lava flow susceptibility in the City of Auckland (New Zealand) (Kereszturi et al., 2012);
- Some challenges of monitoring a potentially active volcanic field in a large urban area: Auckland Volcanic Field, New Zealand (Ashenden et al., 2011);
- The communication of uncertain scientific advice during natural hazard events (Doyle et al., 2011);
- Evacuation planning in the Auckland Volcanic Field, New Zealand: a spatio-temporal approach for emergency management and transportation network decisions (Tomsen et al., 2014).

5.3 Discussion

The breadth and scope of the DEVORA programme has produced not only invaluable scientific outputs that advance scientific understanding of volcanic fields, but also important and applicable information for government policy makers and risk and emergency managers. As such, DEVORA is a model for the production of scientific research for science and society, resulting in strengthened ties between scientists and practitioners. Although the location, timing, and size of the next eruption is unknown, and an AVF eruption will be unwelcomed due to its highly disruptive nature, Auckland and New Zealand will be as best prepared as possible given the high uncertainty of such an event.

References

Ashenden, C. L., Lindsay, J. M., Sherburn, S., Smith, I. E., Miller, C. A. & Malin, P. E. 2011. Some challenges of monitoring a potentially active volcanic field in a large urban area: Auckland volcanic field, New Zealand. *Natural hazards,* 59, 507-528.

Doyle, E. E., Johnston, D. M., McClure, J. & Paton, D. 2011. The communication of uncertain scientific advice during natural hazard events. *New Zealand Journal of Psychology,* 40, 39-50.

Kereszturi, G., Procter, J., Cronin, S. J., Németh, K., Bebbington, M. & Lindsay, J. 2012. LiDAR-based quantification of lava flow susceptibility in the City of Auckland (New Zealand). *Remote Sensing of Environment,* 125, 198-213.

Le Corvec, N., Bebbington, M. S., Lindsay, J. M. & McGee, L. E. 2013. Age, distance, and geochemical evolution within a monogenetic volcanic field: Analyzing patterns in the Auckland Volcanic Field eruption sequence. *Geochemistry, Geophysics, Geosystems,* 14, 3648-3665.

McGee, L. E., Smith, I. E., Millet, M.-A., Handley, H. K. & Lindsay, J. M. 2013. Asthenospheric control of melting processes in a monogenetic basaltic system: A case study of the Auckland Volcanic Field, New Zealand. *Journal of Petrology,* 54, 2125-2153.

Ministry of Civil Defence and Emergency Management. 2008. *Exercise Ruaumoko '08: Final Report* [Online]. Available: http://www.civildefence.govt.nz/memwebsite.nsf/Files/National%20Exercise%20Programme/$file/ExRuaumoko-FINAL-REPORT-Aug08.pdf.

Németh, K., Cronin, S. J., Smith, I. E. & Agustin Flores, J. 2012. Amplified hazard of small-volume monogenetic eruptions due to environmental controls, Orakei Basin, Auckland Volcanic Field, New Zealand. *Bulletin of volcanology,* 74, 2121-2137.

Shane, P., Gehrels, M., Zawalna-Geer, A., Augustinus, P., Lindsay, J. & Chaillou, I. 2013. Longevity of a small shield volcano revealed by crypto-tephra studies (Rangitoto volcano, New Zealand): Change in eruptive behavior of a basaltic field. *Journal of Volcanology and Geothermal Research,* 257, 174-183.

Statistics New Zealand. 2014. *Regional Gross Domestic Product: Year ended March 2013* [Online]. Available: http://www.stats.govt.nz/browse_for_stats/economic_indicators/NationalAccounts/RegionalGDP_MRYeMar13.aspx.

Tomsen, E., Lindsay, J. M., Gahegan, M., Wilson, T. M. & Blake, D. M. 2014. Evacuation planning in the Auckland Volcanic Field, New Zealand: a spatio-temporal approach for emergency management and transportation network decisions. *Journal of Applied Volcanology,* 3, 6.

Chapter 6

Tephra fall hazard for the Neapolitan area

W. Marzocchi, J. Selva, A. Costa, L. Sandri, R. Tonini and G. Macedonio

The Neapolitan area is one of the highest volcanic risk areas in the world, both for the presence of three potentially explosive and active volcanoes (Vesuvius, Campi Flegrei and Ischia), and for the extremely high exposure (over a million people located in a very large and important metropolitan area). Even though pyroclastic flows and lahars represent the most destructive phenomena near the volcanoes, tephra fall poses a serious threat on a wider spatial scale. Excess of tephra loading can cause building collapse, disrupt services and lifelines, and severely affect agriculture and human health. On a larger spatial scale, tephra fallout may cause a major disruption of the economy in Europe and in the Mediterranean area (Folch & Sulpizio, 2010, Sulpizio et al., 2012).

The volcanic *hazard* is the way in which scientists quantify such a kind of threat. The hazard is usually expressed in probabilistic terms in order to account for the vast irreducible (aleatory) and reducible (epistemic) uncertainties. In the past several papers focussed on the assessment of tephra fallout hazard from Neapolitan volcanoes (e.g. Barberi et al. (1990), Macedonio et al. (1990), Cioni et al. (2003), Costa et al. (2009)). These studies have combined field data of tephra deposits and numerical simulations of tephra dispersal (often considering tens of thousands of wind profiles to account for wind variability) to produce maps for the expected tephra loading in case of a specific scenario (e.g. considering one specific kind of eruption), or of a few reference scenarios at both Mount Vesuvius and Campi Flegrei.

This kind of map is still frequently used in volcanology, however, they do not represent the real volcanic hazard, because they do not consider the probability of occurrence of the specific scenarios considered, and they neglect a large part of the natural variability, such as the possibility to have eruptions of different size and from different vents. The latter is particularly important for the Campi Flegrei caldera, where the largest source of uncertainty comes from the forecast of the next eruption location. From a more technical point of view, these studies do not properly incorporate all known aleatory and epistemic uncertainties. This aspect is of primary importance in order to get a reliable volcanic hazard assessment.

The need to have a realistic volcanic hazard analysis is not only important from a scientific perspective, but it is of paramount importance for risk mitigation. Any sound (and defensible) risk assessment and mitigation plan has to be based on a reliable volcanic hazard analysis. In practical terms, the costs and benefits of any possible mitigation option have to be weighted and

Marzocchi, W., Selva, J., Costa, A., Sandri, L., Tonini, R. & Macedonio, G. (2015) Tephra fall hazard for the Neapolitan area. In: S.C. Loughlin, R.S.J. Sparks, S.K. Brown, S.F. Jenkins & C. Vye-Brown (eds) *Global Volcanic Hazards and Risk,* Cambridge: Cambridge University Press.

compared with the probability of occurrence of the wide range of possible threats, i.e. with the volcanic hazard. Any decision making based on single scenarios without considering their probability of occurrence cannot lead to any rational and defensible risk mitigation plan, in particular for high-risk areas.

The need to use the best available science for helping society to mitigate the high volcanic risk in the Neapolitan area pushed volcanologists to develop innovative tools for volcanic hazard analysis in probabilistic terms, the so-called Probabilistic Volcanic Hazard Analysis (PVHA). The attempt is to move toward hazard assessment formats that are similar to other kinds of hazards, such as, for example, the seismic hazard. Following the results of Costa et al. (2009), Selva et al. (2010) assessed tephra fallout hazard at Campi Flegrei attempting to overcome some of the limitations described above. In particular they accounted for the most important sources of uncertainty and natural variability in the eruptive processes due to many different possible scenarios (represented by a discrete number of eruptive scales, and vent positions), and statistically combining the contribution to the final PVHA from all the possible scenarios, making use of the law of total probability.

In order to provide information to the engineers to move from hazard to risk assessment we need to shape the hazard output in a way that can be easily combined with the fragility curves that represent how a building can be damaged as a function of the different intensities of the different volcanic threats (e.g. Spence et al. (2005), Zuccaro et al. (2008), Zuccaro & Leone (2011)).

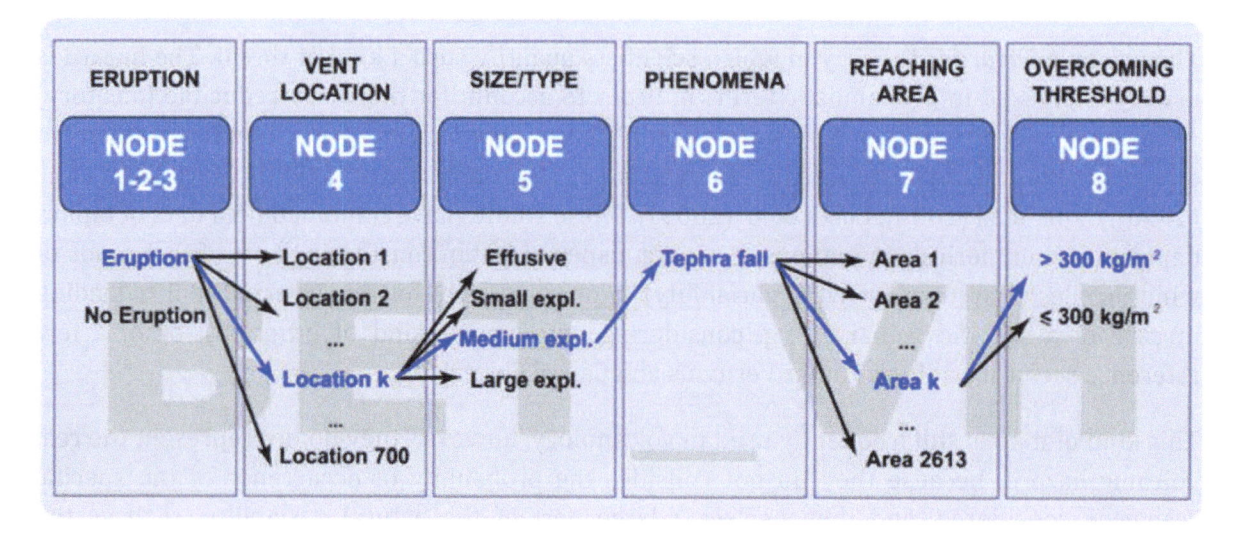

Figure 6.1 Event tree of the model BET_VH for a specific volcano to evaluate the PVHA for tephra fallout above 300 kg/m².

One of the currently adopted methodologies is based on the BET_VH tool (Marzocchi et al. (2010); https://vhub.org/resources/betvh) that performs a proper statistical mixing of the different possible scenarios, further extending the work made by Selva et al. (2010). Such an open source tool, being based on Bayesian inference modelling, properly accounts for the aleatory (intrinsic) and epistemic (linked to our limited knowledge of the eruptive process) uncertainty, propagating these two all along the different factors of PVHA. PVHA and related uncertainties are described by a probability density function instead of by a single value. This gives the interested stakeholders an idea about the confidence of the probabilities we are providing. The analysis for the volcanic hazard posed by tephra is based on an *event tree* (see Figure 6.1). An event tree is a branching graph representation of events in which individual branches are alternative steps from a general prior event, state or condition, and which evolve through time into increasingly specific subsequent events. Eventually the branches terminate in final outcomes representing specific hazards (or risks) that may occur in the future. In this way, an event tree attempts to graphically display all relevant possible volcanic outcomes in progressively higher levels of detail. Points on the graph where new branches are created are referred to as nodes. In BET_VH all uncertainties can be assessed at each level, namely on the eruption occurrence, on vent position, on the eruptive scale, on the production of tephra and on its transport, dispersal and deposition by the wind. The BET_VH tool has been used in other volcanic areas to produce a full PVHA for tephra fallout and other volcanic hazardous phenomena (Sandri et al., 2012, 2014).

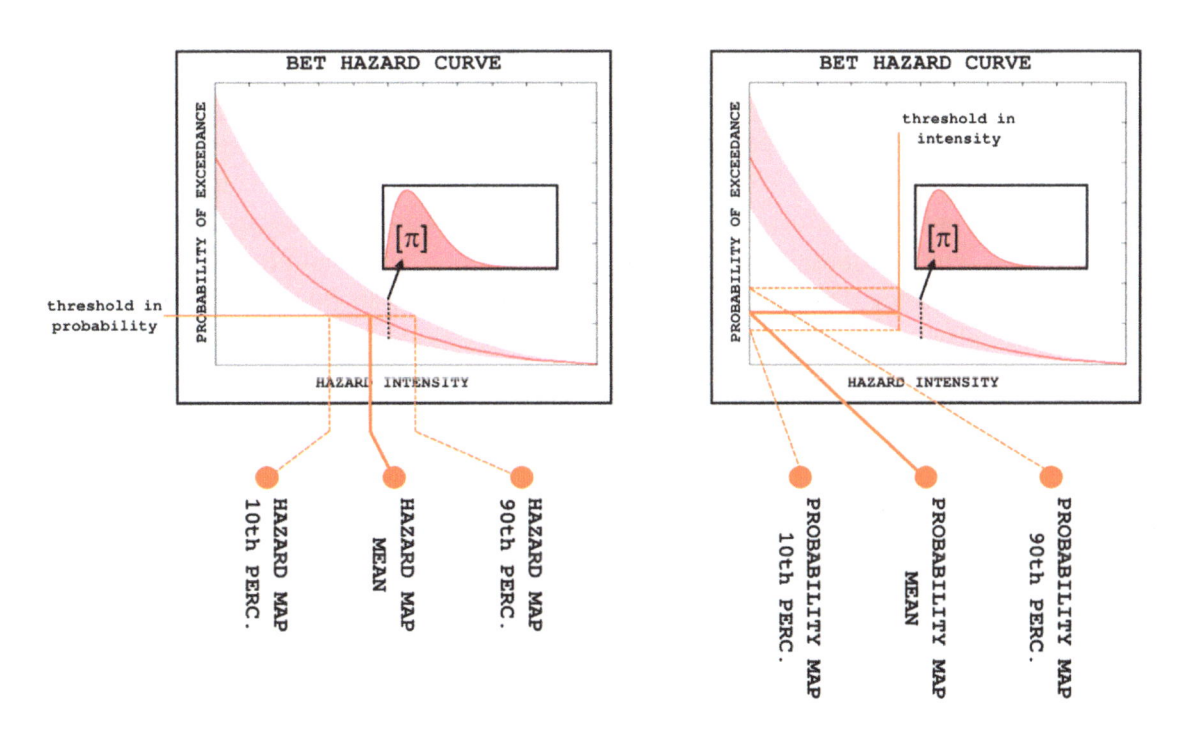

Figure 6.2 Example of hazard curve for a given target cell of the gridpoint. On the x-axis reports the different threshold of the intensity measure (tephra load in our case). On the y-axis reports, the computed exceedance probability of such intensity thresholds in a given time window and a given target position. The shaded area shows the 10 to 90th percentiles confidence interval of the hazard curve. Cutting the curves horizontally (left panels), we obtain the hazard intensity for a given exceedance probability value (basic ingredient of hazard maps). Cutting them vertically (right panel) we obtain the exceedance probability for a given intensity value (basic ingredient of probability maps). Given hazard curves at each position in a target area, maps can be produced at different levels of confidence (e.g. mean, 10th and 90th percentiles), showing the effects of epistemic uncertainties on either hazard (left panel) and probability (right panel) maps. (Modified from Selva et al. (2014).

An ongoing improvement of the method aims at performing the production of fully probabilistic hazard curves (see Figure 6.2), estimating the exceedance probability of a set of thresholds in tephra load, based on the method proposed for seismic hazard by Selva & Sandri (2013) Indeed, hazard curves represent the most complete information about the hazard, and they allow volcanologists to produce proper hazard maps at different levels of probability (SSHAC 1997), as shown in Figure 6.2. The proposed method can be used in both long (years to decades) and short (hours to weeks) perspectives (e.g. Marzocchi et al. (2008), Selva et al. (2014)). For the volcanoes threatening the Neapolitan area, several papers have already taken some steps in the direction of estimating some of the node probabilities reported in Figure 6.1 for both long- and short-term hazard. For Vesuvius, Marzocchi et al. (2004), (2008) estimated the factors probabilities of the first five nodes of the event tree of Figure 6.1, while Macedonio et al. (2008) provided an estimation of the best-guess probabilities for nodes 7 and 8. For Campi Flegrei, the probability distributions for the first five nodes have been respectively estimated by Selva et al. (2012a), Selva et al. (2012b) and Orsi et al. (2009), while Costa et al. (2009) provided an estimation of the best guess probabilities for nodes 7 and 8 in two possible vent locations (Eastern and Western parts of the caldera). Merging all these factors in a full comprehensive volcanic hazard analysis for tephra fall is one of the main goals of the ongoing research.

The results obtained so far include the PVHA for tephra fallout conditional to the occurrence of specific eruptive scenarios, i.e. the probability maps conditional to the occurrence of eruptions of specific sizes at Vesuvius (e.g. Macedonio et al. (2008)) and Campi Flegrei (e.g. Costa et al. (2009)). Figure 6.3 shows some of these maps. A significant improvement for Campi Flegrei was achieved by the proper mixing of all the possible eruptive sizes and vents, conditional to the occurrence of an eruptions, performed by Selva et al. (2010), computed by accounting for the different possible vent locations Selva et al. (2012b), eruption sizes (Orsi et al., 2009), and the probability distribution for the nodes 6, 7 and 8 (see Figure 6.4). In this respect, this kind of approach is particularly useful for large and potentially very explosive calderas, such as Campi Flegrei, for which the position of the vent is critical and it imposes a large uncertainty on the final PVHA.

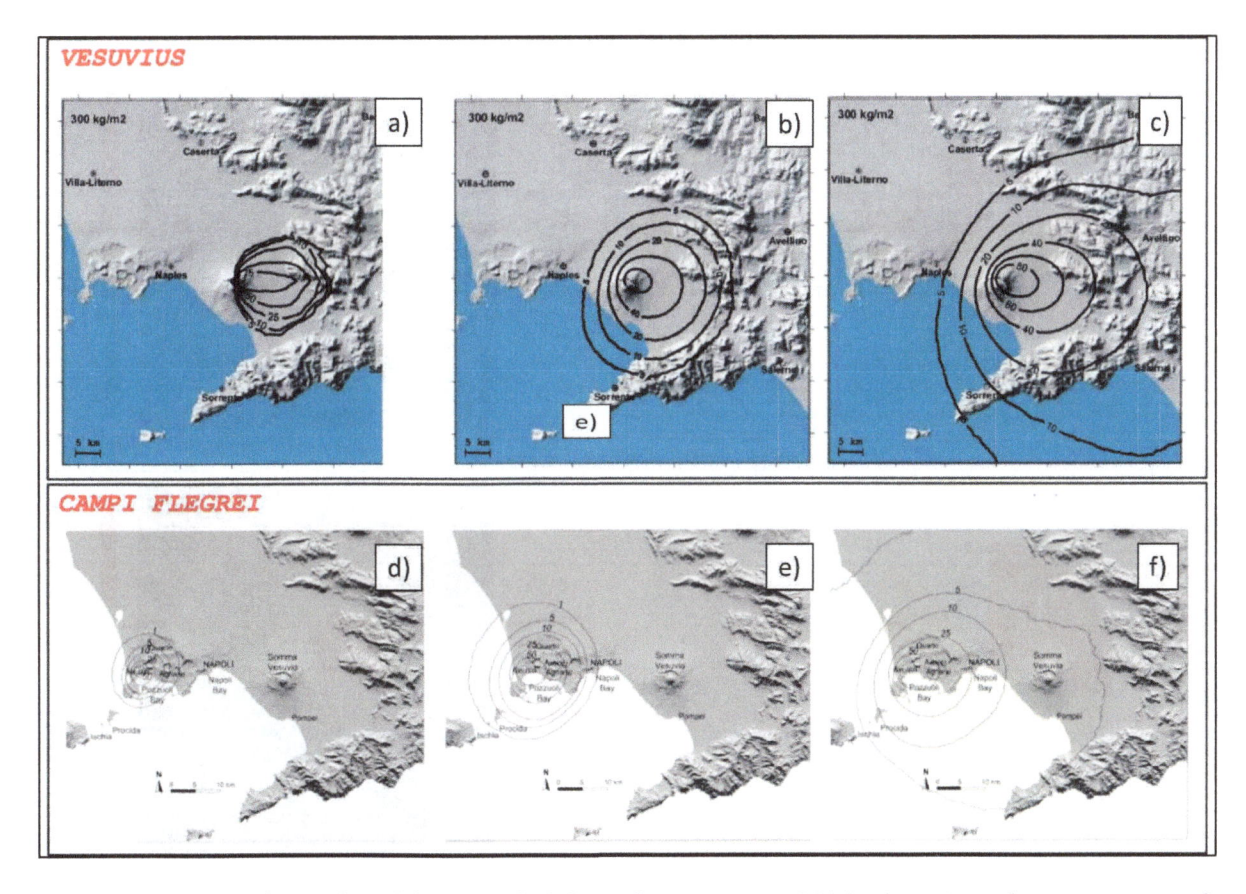

Figure 6.3 Results for tephra fallout probability of overcoming 300 kg/m² given the occurrence of an eruption of size a) Violent Strombolian, b) Subplinian and c) Plinian at Vesuvius (Macedonio et al., 2008), and of size d) low, e) medium and f) high explosive, from the eastern vent (Averno-Monte Nuovo) at Campi Flegrei (Costa et al., 2009). Each map shows the hazard footprint of the event, enabling the user to assess areas under threat.

Despite the recent significant steps ahead in achieving a full and comprehensive PVHA for tephra fall, much more work has still to be done. In two ongoing Italian projects (ByMur, 2010-2014, DPC-V1, 2012-2013), there have been attempts to provide further improvements in long-term PVHA for Vesuvius, Campi Flegrei and Ischia, by accounting for all the factors concurring to the full hazard. A preliminary merging of the full PVHA for tephra fallout posed by both Vesuvius and Campi Flegrei on the municipality of Naples is shown in Figure 6.5 (Selva et al., 2013). The variability of eruptive parameters within each size class must also be modelled, to evaluate its importance and impact on the final PVHA. The production of hazard curves, as mentioned above, is a necessary step if PVHA results are to be included into quantitative risk assessment procedures. The assessment of the epistemic uncertainty on the hazard curves represents the most complete results that we aim to achieve (Figure 6.4). The final PVHA for the municipality of Naples is planned to be ready for the end of the project ByMuR and it will consist of a hazard curves, at different level of confidence regarding epistemic uncertainties, for each cell of the grid covering the municipality of Naples.

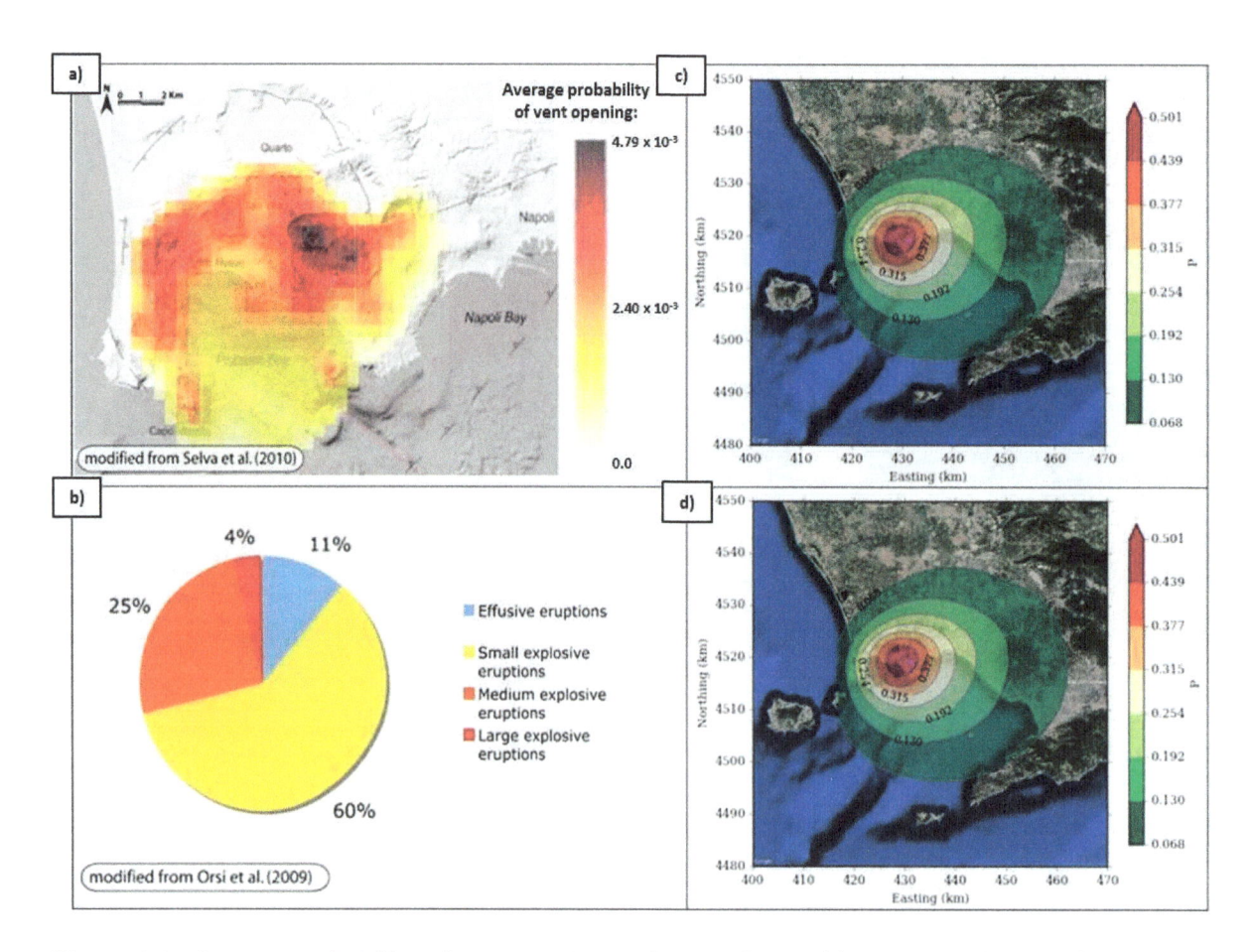

Figure 6.4 a) mean probability of vent opening at Campi Flegrei (the notation '4.7E-03' means 4.7 x 0.001=0.047); b) mean probability of possible eruptive sizes at Campi Flegrei; c) mean probability of tephra fallout and of tephra loading larger than 0 kg/m²; d) as for c) but relative to a tephra loading larger than 300 kg/m². The maps reported in panels a) and b) have been obtained by Selva et al. (2010) integrating the outcome of all possible scenarios – all possible size (panel b) and all possible vent opening (panel a) – with their own probability of occurrence.

Regarding short-term PVHA in the Neapolitan area, two research projects (the Italian DPC-V2, 2012-2014, and the EC MEDSUV 2013-2015) aim at providing quantitative improvements for Vesuvius and Campi Flegrei in order to reach its operational implementation for tephra fallout. This would represent a tool of primary importance during potential volcanic unrest episodes and for ongoing eruptions, being able to be updated frequently and accounting for the rapidly evolving situation and providing crucial information for crisis management. Theoretically, short-term PVHA should be based on sound modelling procedures stemming from frequently updated meteorological forecast and information about the crisis evolution (Selva et al., 2014). In addition, the relevance of epistemic uncertainties arising from the forecast of the future eruption dynamics and wind conditions, and from the tephra dispersal model, should be estimated.

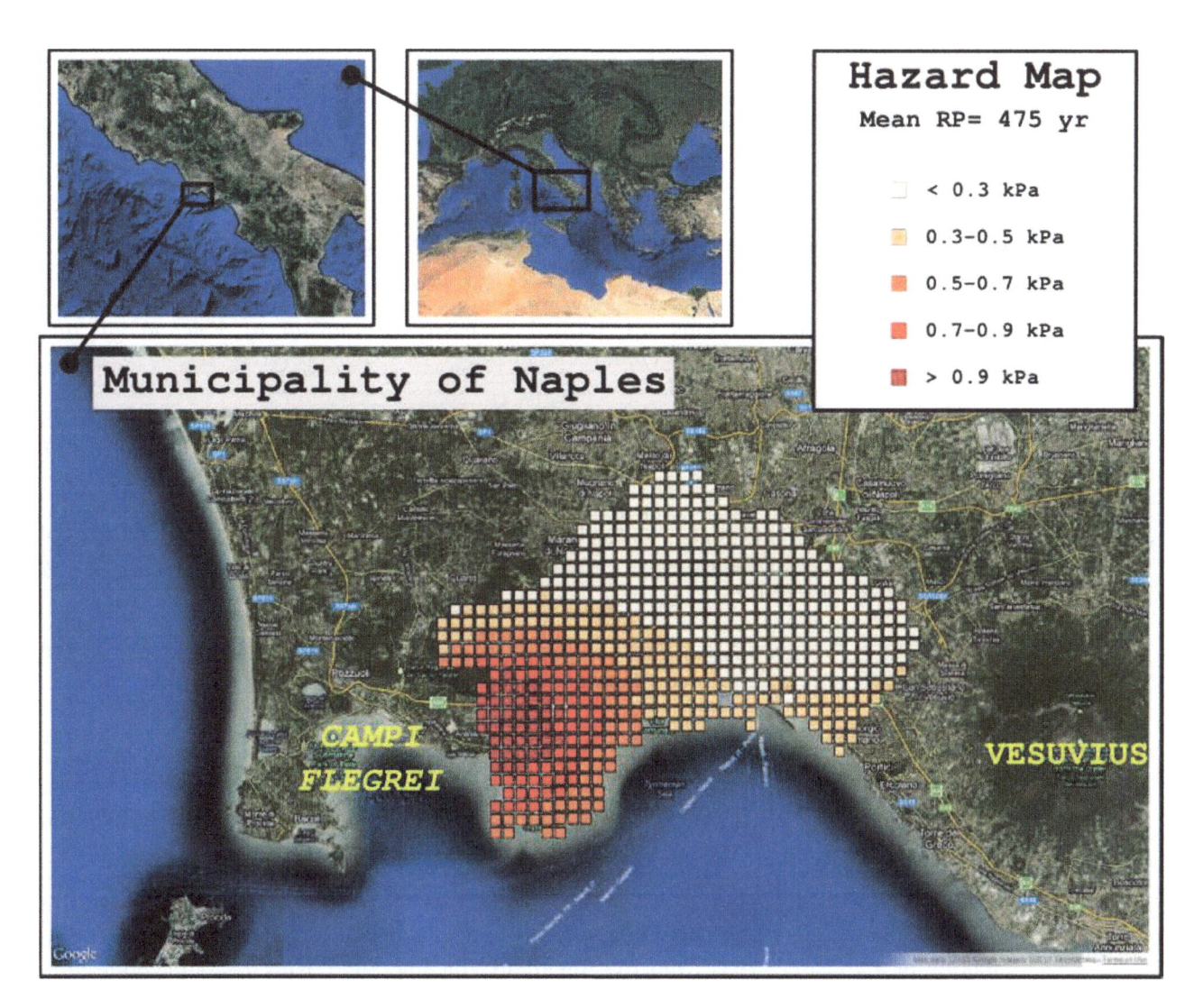

Figure 6.5 Hazard map (mean) for tephra loading with a return period of 475 years (exceedance probability threshold equal to 0.1 in 50 yr), considering both Vesuvius and Campi Flegrei on the municipality of Naples. In the legend 1 kPa stands for 1000 Pascal (or 0.1 bar).

References

Barberi, F., Macedonio, G., Pareschi, M. & Santacroce, R. 1990. Mapping the tephra fallout risk: an example from Vesuvius, Italy. *Nature,* 344, 142-144.

ByMur. 2010-2014. *Bayesian Multi-risk Assessment: a case study for the Natural Risks in the city of Naples* [Online]. Available: http://bymur.bo.ingv.it/.

Cioni, R., Longo, A., Macedonio, G., Santacroce, R., Sbrana, A., Sulpizio, R. & Andronico, D. 2003. Assessing pyroclastic fall hazard through field data and numerical simulations: example from Vesuvius. *Journal of Geophysical Research: Solid Earth (1978–2012),* 108.

Costa, A., Dell'Erba, F., Di Vito, M., Isaia, R., Macedonio, G., Orsi, G. & Pfeiffer, T. 2009. Tephra fallout hazard assessment at the Campi Flegrei caldera (Italy). *Bulletin of Volcanology,* 71, 259-273.

DPC-V1. 2012-2013. *Valutazione della pericolosità vulcanica in termini probabilistici* [Online]. Available: http://istituto.ingv.it/l-ingv/progetti/progetti-finanziati-dal-dipartimento-di-protezione-civile-1/progetti-vulcanologici-2012.

Folch, A. & Sulpizio, R. 2010. Evaluating long-range volcanic ash hazard using supercomputing facilities: application to Somma-Vesuvius (Italy), and consequences for civil aviation over the Central Mediterranean Area. *Bulletin of Volcanology,* 72, 1039-1059.

Macedonio, G., Costa, A. & Folch, A. 2008. Ash fallout scenarios at Vesuvius: numerical simulations and implications for hazard assessment. *Journal of Volcanology and Geothermal Research,* 178, 366-377.

Macedonio, G., Pareschi, M. T. & Santacroce, R. 1990. Renewal of explosive activity at Vesuvius: models for the expected tephra fallout. *Journal of Volcanology and Geothermal Research,* 40, 327-342.

Marzocchi, W., Sandri, L., Gasparini, P., Newhall, C. & Boschi, E. 2004. Quantifying probabilities of volcanic events: the example of volcanic hazard at Mount Vesuvius. *Journal of Geophysical Research,* 109.

Marzocchi, W., Sandri, L. & Selva, J. 2008. BET_EF: a probabilistic tool for long-and short-term eruption forecasting. *Bulletin of Volcanology,* 70, 623-632.

Marzocchi, W., Sandri, L. & Selva, J. 2010. BET_VH: a probabilistic tool for long-term volcanic hazard assessment. *Bulletin of volcanology,* 72, 705-716.

Orsi, G., Di Vito, M. A., Selva, J. & Marzocchi, W. 2009. Long-term forecast of eruption style and size at Campi Flegrei caldera (Italy). *Earth and Planetary Science Letters,* 287, 265-276.

Sandri, L., Jolly, G., Lindsay, J., Howe, T. & Marzocchi, W. 2012. Combining long-and short-term probabilistic volcanic hazard assessment with cost-benefit analysis to support decision making in a volcanic crisis from the Auckland Volcanic Field, New Zealand. *Bulletin of volcanology,* 74, 705-723.

Sandri, L., Thouret, J.-C., Constantinescu, R., Biass, S. & Tonini, R. 2014. Long-term multi-hazard assessment for El Misti volcano (Peru). *Bulletin of volcanology,* 76, 1-26.

Selva, J., Costa, A., Marzocchi, W. & Sandri, L. 2010. BET_VH: exploring the influence of natural uncertainties on long-term hazard from tephra fallout at Campi Flegrei (Italy). *Bulletin of volcanology,* 72, 717-733.

Selva, J., Costa, A., Sandri, L., Macedonio, G. & Marzocchi, W. 2014. Probabilistic short-term volcanic hazard in phases of unrest: a case study for tephra fallout. *Journal of Geophysical Research,* 119, 8805-8826.

Selva, J., Garcia-Aristizabal, A., di Ruocco, A., Sandri, L., Marzocchi, W. & Gasparini, P. 2013. BET_VR: a probabilistic tool for long-term volcanic risk assessment. *IAVCEI General Assembly.* Kagoshima, Japan.

Selva, J., Marzocchi, W., Papale, P. & Sandri, L. 2012a. Operational eruption forecasting at high-risk volcanoes: the case of Campi Flegrei, Naples. *Journal of Applied Volcanology,* 1, 1-14.

Selva, J., Orsi, G., Di Vito, M. A., Marzocchi, W. & Sandri, L. 2012b. Probability hazard map for future vent opening at the Campi Flegrei caldera, Italy. *Bulletin of volcanology,* 74, 497-510.

Selva, J. & Sandri, L. 2013. Probabilistic Seismic Hazard Assessment: Combining Cornell-like approaches and data at sites through Bayesian inference. *Bulletin of the Seismological Society of America,* 103, 1709-1722.

Spence, R., Kelman, I., Baxter, P., Zuccaro, G. & Petrazzuoli, S. 2005. Residential building and occupant vulnerability to tephra fall. *Natural Hazards and Earth System Science,* 5, 477-494.

Sulpizio, R., Folch, A., Costa, A., Scaini, C. & Dellino, P. 2012. Hazard assessment of far-range volcanic ash dispersal from a violent Strombolian eruption at Somma-Vesuvius volcano, Naples, Italy: implications on civil aviation. *Bulletin of Volcanology,* 74, 2205-2218.

Zuccaro, G., Cacace, F., Spence, R. & Baxter, P. 2008. Impact of explosive eruption scenarios at Vesuvius. *Journal of Volcanology and Geothermal Research,* 178, 416-453.

Zuccaro, G. & Leone, M. 2011. Volcanic crisis management and mitigation strategies: a multi-risk framework case study. Earthzine.

Chapter 7

Eruptions and lahars of Mount Pinatubo, 1991 to 2000

C.G. Newhall and R.U. Solidum

Mount Pinatubo (Philippines) – asleep for ~ 500 years – began to stir in mid-March 1991, and produced a giant eruption on 15 June 1991, second largest of the twentieth century. Only that of remote Katmai-Novarupta, Alaska in 1912 was larger. About 20,000 indigenous Aeta lived on the volcano, and ~1,000,000 lowland Filipinos lived around it. Two large American military bases, Clark Air Base and Subic Bay Naval Station, added about 40,000 Americans to those at risk. With centuries' of volcanic gas (supply) accumulated in tens of cubic kilometres of molten rock (magma), and with so many innocent people nearby, a disaster was waiting to happen.

Thick deposits from pumice-rich pyroclastic flows formed the lower slopes of the volcano and told a history of infrequent but very large eruptions - larger than any eruption in the history of modern volcano monitoring. Scientists warned that a giant eruption was possible, perhaps even likely, but none had ever been monitored, much less successfully forecast. For two months after the volcano began to stir, small earthquakes and other signs fluctuated without clear, systematic trends. The volcano was teasing the scientists, and the public was profoundly sceptical.

Against the odds, a team of scientists from the Philippine Institute of Volcanology and Seismology (PHIVOLCS), assisted by the US Geological Survey, correctly forecast a giant eruption. Evacuations that had been recommended earlier were now enforced and expanded. Over the course of a few days, small precursory eruptions escalated to a spectacular climax on 15 June that swept the whole volcano, killing virtually everything in its path. Avalanches of searingly hot ash and pumice (pyroclastic flows) filled valleys and swept over ridge crests. Tens of centimetres of ash, with weight nearly doubled by rain from simultaneous Typhoon Yunya, caused many roofs to collapse. Loss of life was relatively modest considering the population at risk and the enormous size of the events (~400 died during the eruption, and ~500 Aeta children died in evacuation camps from measles). Warnings, coupled with strong visible clues from pre-climactic eruptions, had saved nearly all of the Aeta population, plus an unknown number of lowlanders. Some damage was unavoidable, but much was also averted, especially damage to military assets and commercial jets.

Newhall, C.G. & Solidum, R.U. (2015) Eruptions and lahars of Mount Pinatubo, 1991 to 2000. In: S.C. Loughlin, R.S.J. Sparks, S.K. Brown, S.F. Jenkins & C. Vye-Brown (eds) *Global Volcanic Hazards and Risk,* Cambridge: Cambridge University Press.

What factors worked for successful mitigation of the eruption risk?

- Pinatubo, and other long-dormant volcanoes, give plenty of warning signs. The challenge is to read them correctly, and to time the warnings to be early enough for evacuation but not so early that people give up and return home. With no precedent monitoring of such a large eruption elsewhere, and no prior monitoring of Pinatubo, the scientific team just barely managed to install enough instruments, collect and interpret the data, educate those at risk, and to give the right warnings at the right time. Fortunately for all, the volcano gave scientists two months in which to work and in early June gave signs that the eruption was just days away.
- PHIVOLCS had a quick-response team that started work at Pinatubo in earliest April, and captured critical early data. The US Geological Survey also had a team of volcano scientists and technicians, experienced, fully equipped and ready to help. The latter team had been formed after the disaster at Nevado del Ruiz in Colombia just a few years earlier, and was supported by USAID's Office of Foreign Disaster Assistance. Together, the two teams accomplished what neither team by itself could have accomplished. When Nature presents an enormous challenge, rapid, joint, international responses may be necessary.
- Although public and even official scepticism was a huge challenge, trust between key scientists and officials offset that scepticism. The longer a volcano has been quiet, the less people know about it, and the more sceptical they will be. Scientists (led by the late Raymundo Punongbayan of PHIVOLCS) and officials – some who had known each other for years and some who were new-found friends – prepared for the crisis as a team and developed the trust that was needed for critical mitigation decisions. Trust in other circles – e.g. between missionaries and the indigenous people – also helped greatly.
- That the Philippines already had protocols and procedures for evacuations ahead of typhoons and floods, and even for volcanoes elsewhere in the Philippines, also helped to offset local unfamiliarity and scepticism about Pinatubo. The hierarchy of national, regional, provincial, municipal, and village civil defence worked well. A similar hierarchy of command and hazard preparedness within the military had equally beneficial effects.
- Especially because of Pinatubo's long quiescence before 1991, very few people around Pinatubo understand anything about volcanoes and their hazards. The same had been true in Armero, downriver from Nevado del Ruiz, and unfamiliarity with volcanic hazards cost residents of Armero their lives. A hard-hitting video made by the late Maurice Krafft for the International Association of Volcanology and Chemistry of the Earth's Interior (IAVCEI) was wonderfully graphic, showing quickly in images what words could not describe. This video saved many lives. Video worked where words would have failed. *(Ironically and at the same time, Maurice and his wife Katia, dissatisfied with the footage of pyroclastic flows in this video, stopped by Unzen Volcano to get better footage, and they and 41 others were sadly killed.)*
- Scientists are by training cautious about making forecasts. Invariably, they wish for more data. But during a crisis, advice must be given no matter how high the uncertainties. Ray Punongbayan and his colleagues set aside their normal caution, explained the uncertainties, and gave their best guesses. They made forecasts where others might have feared to tread. Obviously, caution and an all-out search for reliable data are important, but when Nature signals that a hazard is imminent, scientists must speak out.
- Those at risk can be diverse, and require an equal diversity of communication approaches. Some of those at risk responded best to "Trust me. Follow me". Others challenged us to convince them of the hazard. Military officers and engineers understood probability trees; others drew more from the IAVCEI video.

- Many – indeed most – of those at risk waited until the last possible moment before evacuating. Yes, many were sceptical. And yes, few people wanted to evacuate even if they knew they should. Although the plan called for evacuations at Alert Level 4, very few moved because the hazard wasn't yet in their face. Two messages by example helped. First, scientists moved themselves from the centre of Clark Air Base to the far perimeter of the base, and base commanders took notice. Second, when Americans from Clark Air Base left town, Filipinos in neighbouring towns also took notice.

Beginning during the eruption, and continuing for a more than a decade thereafter, rain-triggered volcanic mudflows (lahars) buried large areas around the foot – including many towns up to 40 km away – with an average of 5-10 m of sand and gravel. Unlike floods that come and go, lahars come ... and stay. Of roughly 6 million cubic meters of deposit on the volcano slopes, more than half was washed into the surrounding lowlands over the next 10 years. The scale of the hazard far exceeded normal sediment control measures. More than 200,000 people were "permanently" displaced, though by 2014 some have returned and built new homes on top of the lahar deposits. Within just a few years, costs of lahar damage and mitigation exceeded the ~ USD 2B damage from the eruption itself.

Again, in spite of the enormous scale of the lahar hazard, only about 400 were killed by lahars. Scientists set up high-tech warning systems with radio-telemetered rain gauges and flow meters. For lahars, the PHIVOLCS-USGS team was joined by a team from the University of the Philippines and University of Illinois-Chicago. Kelvin Rodolfo of the university team introduced the Indonesian word "lahar" which, because it is foreign, became a good educational tool. Police set up manned lahar watchpoints. More videos were shown. Time and again, warnings were sounded, towns were evacuated, and most people survived. In addition to warning systems, engineers built an elaborate set of levees (dikes) and sediment catchment structures. Early structures were too optimistic, getting filled or overrun quickly, but eventually, the increasing scale of the engineered structures matched the decreasing scale of hazard. Some of the waste was inevitable, as there was public pressure to act even before the full scale of the problem was understood. Some additional waste might be charged to politics and corruption. Debate about whether to spend for dikes or spend for relocation of towns was generally cut short, either by normal human reluctance to abandon one's home, or by lahars themselves. Much of the engineering mitigation was financed by the central government; overseas development aid financed additional studies and construction in selected watersheds.

Estimated costs for the pre-eruption scientific response (mainly, helicopter time and equipment that was destroyed and had to be replaced) were approximately USD 1.5 million; for preparation of scientists over the preceding decade ~15 million and for pre- and syn-eruption evacuations ~USD 40 million. Compare these costs to roughly 10,000 lives saved and hundreds of millions of dollars of damage averted. Clearly, maintenance of quick response teams and the warnings they gave were cost effective.

The cost-effectiveness of lahar mitigation was not as clear. Costs of scientific response were roughly USD 2M and lahar control structures plus temporary and relocation housing cost at least USD 700M of government outlays. Damage to the town of Bacolor, sandwiched by sediment control levees, should also be counted as a cost of mitigation. Savings might include other towns, e.g., Bamban, Guagua, San Marcelino, Botolan, and the large city of San Fernando, but saving them from lahars has caused substantial flooding in subsequent years. A proper cost-benefit analysis of Pinatubo lahar mitigation would be of great interest.

Preparation time for the eruption was short, the scientific team was tight and spoke with one voice, and a relatively small number of political, civil defence, and military leaders made most of the decisions. The subsequent lahar crisis was much more complicated, with more scientists, more decision makers, and more time. Before the eruption, there was no time for debate; during the lahar period, there was lots of time for debate – between scientists, among engineers and policymakers, and between citizens of one town and the next. The result was that mitigation measures during the lahar period were more controversial, and probably more expensive than they needed to be, but in the end most people and towns were protected. It wasn't perfect, and some bitterness still remains, but the overriding fact is that most people at risk survived and have been able to rebuild their lives.

Figure 7.1 Mount Pinatubo prior to the paroxysmal explosive eruption of 15 June 1991 (top) and after the eruption (bottom). Much of the edifice disappeared and became a caldera depression with a lake and many active steam vents. (V.Gempis, USAF).

References

Ewert, J. W., Miller, C. D. & Hendley, I. 1997. *Mobile response team saves lives in volcano crises. U.S. Geological Fact Sheet 064-97* [Online]. US Geological Survey. Available: http://pubs.usgs.gov/fs/1997/fs064-97/.

Newhall, C., Hendley Ii, J. W. & Stauffer, P. H. 1997. *Benefits of volcano monitoring far outweigh the costs - the case of Mount Pinatubo. U.S. Geological Survey Fact Sheet 115-97.* [Online]. Available: http://pubs.usgs.gov/fs/1997/fs115-97/.

Newhall, C. G. & Punongbayan, R. 1996. *Fire and Mud: Eruptions and Lahars of Mount Pinatubo, Philippines*, Philippine Institute of Volcanology and Seismology Quezon City.

Rodolfo, K. S. 1995. *Pinatubo and the Politics of Lahar: Eruption and Aftermath, 1991*, University of the Philippines Press and Pinatubo Studies Program, UP Center for Integrative and Development Studies.

Chapter 8

Improving crisis decision-making at times of uncertain volcanic unrest (Guadeloupe, 1976)

JC. Komorowski, T. Hincks, R.S.J. Sparks, W. Aspinall, and the CASAVA ANR project consortium[1]

8.1 Defining the problem

Scientists monitoring active volcanoes are increasingly required to provide decision support to civil authorities during periods of unrest. As monitoring techniques and their resolutions improve, the process of jointly interpreting multiple strands of indirect evidence becomes increasingly complex (Sparks & Aspinall, 2013). During a volcanic crisis, decisions typically have to be made with limited information and high uncertainty, on short time scales. The primary goal is to minimise loss and damage from any event, but social and economic losses resulting from false alarms or evacuations must also be considered (Woo, 2008). It is not the responsibility of scientists to call an evacuation or to manage a crisis; however, demands are increasing on them to assess risks and present scientific information and associated uncertainties in ways that enable public officials to make urgent evacuation decisions or other mitigation policy choices.

8.2 The 1975 - 1977 volcanic unrest at La Soufrière (Guadeloupe)

An increasing number of earthquakes were recorded and felt at La Soufrière one year prior the eruption, which began with an unexpected explosion on 8 July 1976. In the subsequent 9-month period, the volcano ejected about 2 million tonnes of old, cold volcanic ash and rocks in 26 explosions (Feuillard et al., 1983, Komorowski et al., 2005, Beauducel, 2006, Feuillard, 2011). Various volcanic gases (H_2O, minor CO_2, H_2S, SO_2) were also released during the eruption and led to moderate environmental impact with short-term public health implications (Figure 8.1), due to the presence of chlorine and fluorine in the vapour. A report that "fresh glass" was present in an ash sample, implying new magma was close to the surface, led to a major controversy among scientists that was widely echoed in the media (Fiske, 1984). With other evidence suggesting continued build-up of pressure in the volcano, this key observation – later found to be erroneous - and the uncertainty of possible transition to a devastating explosive

---

[1] https://sites.google.com/site/casavaanr/

Komorowski, J-C., Hincks, W., Sparks, R.S.J., Aspinall, W., & CASAVA ANR (2015) Improving crisis decision-making at times of uncertain volcanic unrest (Guadeloupe, 1976). In: S.C. Loughlin, R.S.J. Sparks, S.K. Brown, S.F. Jenkins & C. Vye-Brown (eds) *Global Volcanic Hazards and Risk,* Cambridge: Cambridge University Press.

eruption, led the authorities to declare an evacuation of ca. 70,000 people on 15 August, which lasted 4 to 6 months. The evacuation had severe socio-economic consequences, which persisted long after the volcanic unrest had subsided. The costs have been estimated as 60% of the total annual per capita Gross Domestic Product of Guadeloupe in 1976, excluding losses of uninsured personal assets and open-grazing livestock. There were no fatalities, but this eruption stills ranks amongst the most costly of the twentieth century (De Vanssay, 1979, Lepointe, 1999).

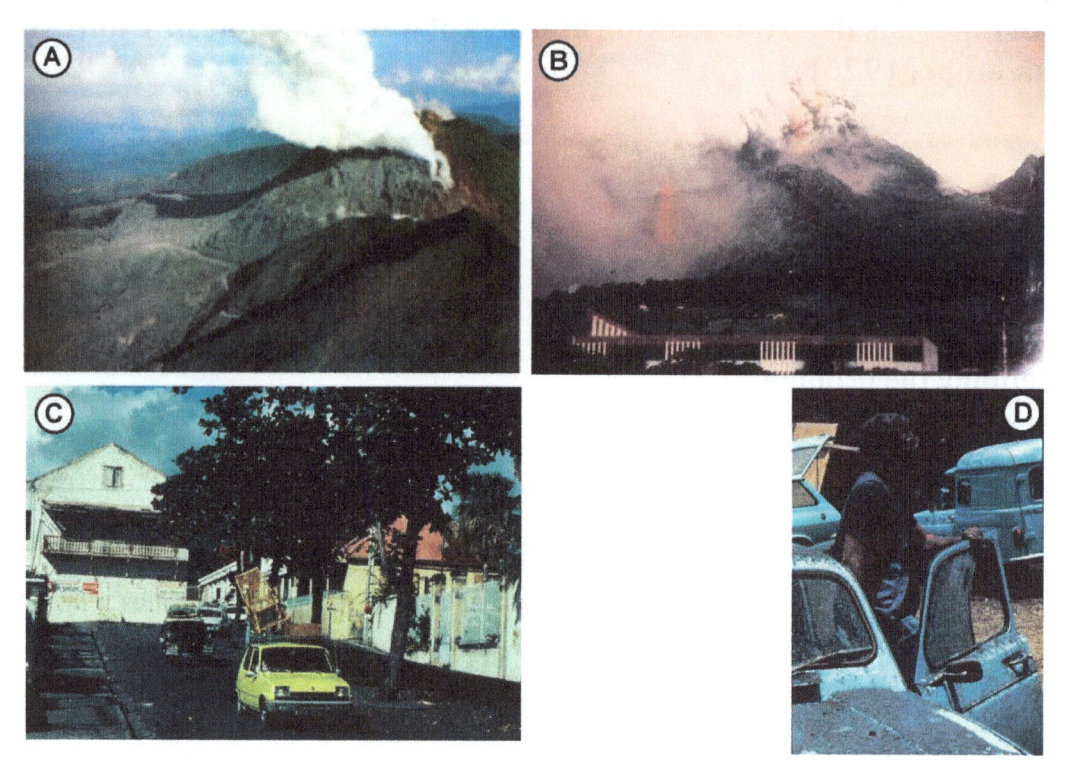

Figure 8.1 Eruptive phenomena and impact of the 1975-1977 volcanic unrest at La Soufrière (Guadeloupe). a) Gas and ash emitting fracture which opened on 8 July, photo taken before 30 August 1976 (copyright IPGP). b) Phreatic explosion and dense ash cloud, 4 October 1976 (copyright IPGP). c) People evacuating with their belongings from the towns of Saint-Claude and Basse-Terre in early September 1976 (R. Fiske). d) Ash and lapilli on car in the town of Saint-Claude from July-August 1976 explosions (R. Fiske).

8.3 Lessons learned

At La Soufrière, there was a lack of a comprehensive monitoring network prior to the 1976 crisis, limited knowledge of the eruptive history of this particular volcano, and a tendency towards caution exacerbated by the memory of past devastating Caribbean eruptions. These factors all contributed to major scientific uncertainty and a polemical publically-expressed lack of consensus and trust in available expertise (Komorowski et al., 2005, Beauducel, 2006). The combination of markedly escalating and fluctuating activity, and societal pressures, in a small island setting, made analysis, forecasting, and crisis response all highly challenging for scientists and authorities. Prior to the crisis there was no well-founded, and accepted, volcanic emergency response plan, so the authorities were compelled to resort to a "precautionary principle" approach in the face of the uncertain evidence and the absence of scientific consensus on the likely outlook.

Pre-eruption, there was a policy to move the banana export port facilities of Basse Terre to the more sheltered economic capital Pointe-à-Pitre, and the evacuation reinforced this policy. This, in turn, contributed to the ravaging of the economy of the administrative capital, Basse Terre, and to its population's bitterness and feeling of being forsaken. The evacuation is still perceived by some as having been unnecessary and an exaggerated application of the "precautionary principle". Even now, many hold to the view that much of the risk assessment was exaggerated for political reasons.

In its overseas territory context, the volcanic crisis in 1976 became a metaphor for many accumulated socio-cultural frustrations on island, and engendered a distrust of science as a possible contributor to solving such issues. The public debate at the time became polarised on issues of opposing "truths", served up and contrasted by a few strongly opinionated scientific experts, rather than focussing on how science could help constrain epistemic and aleatory uncertainty and foster improved decision-making in the circumstances. Thus this infamous crisis exemplified the need for a structured and transparent approach to evidence-based decision-making in the presence of substantial scientific uncertainty.

8.3.1 A probabilistic approach to quantifying uncertainty

Similarities of volcanic unrest interpretation with uncertainties in medical diagnosis suggest a formal evidence-based approach can be helpful, whereby monitoring data are analysed synoptically to provide probabilistic hazard forecasts. A probabilistic tool to formalise such inferences is the Bayesian Belief Network (BBN) (Bedford & Cooke, 2001). By explicitly representing conditional dependencies (relationships) between the volcanological model and observations, BBNs use probability theory to treat uncertainties in a rational and auditable manner, to the extent warranted by the strength of the scientific evidence. A retrospective analysis is given for the 1976 Guadeloupe crisis by Hincks et al. (2014), using a BBN (Figure 8.2) to provide a framework for assessing the state of the evolving magmatic system and the probability of a future eruption. Conditional dependencies are characterised quantitatively by structured expert elicitation (Aspinall, 2006, Aspinall & Cooke, 2013).

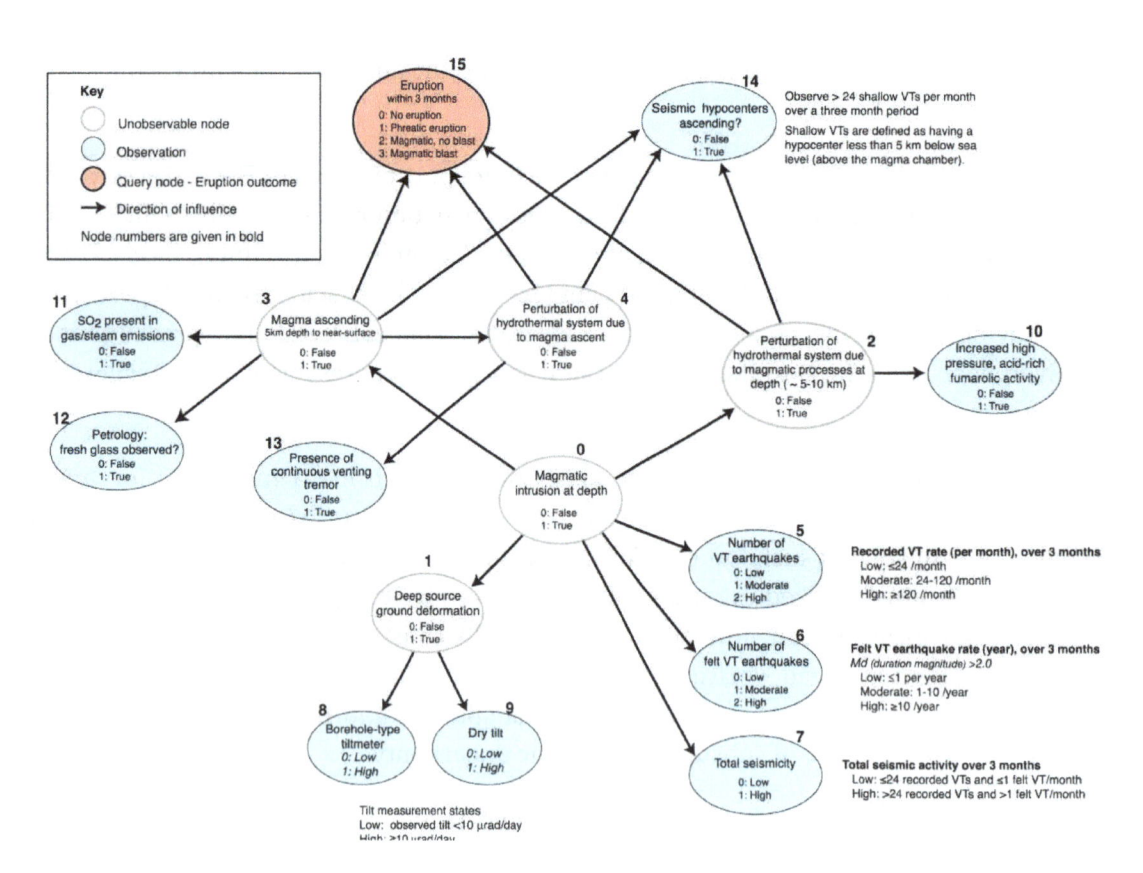

Figure 8.2 Retrospective Bayesian Belief Network (BBN) for La Soufrière (Hincks et al., 2014) showing the relationship between volcanic processes, states and observations available in 1976; used to make inferences about probabilities of future activity scenarios. Nodes represent both hidden (grey) and observable (blue) states. Arcs between nodes represent conditional dependencies (e.g. direct causal relationships or influence) and are characterised by conditional probability tables (CPTs). Arrows indicate the direction of influence. In this case, all conditional probability distributions (and associated uncertainties) were obtained by expert elicitation, the network structure being agreed by the group prior to elicitation.

Analysis of the available monitoring data suggests that at the height of the crisis the probability was high that magmatic intrusion was taking place, according with most scientific thinking at the time. Correspondingly, the probability of magmatic eruption was elevated in July and August 1976, and the signs of precursory activity were justifiably a cause for concern. However, as of 31 August 1976 collective uncertainty about the future course of the crisis was also substantial such that, of all the possible scenarios considered in the BBN, the marginally most likely outcome based on available observations was 'no eruption' (mean probability 0.5); the chance of a magmatic eruption, perhaps associated with a devastating volcanic blast, had an estimated mean probability of ~0.4 (Figure 8.3). There was, therefore, little or no evidential strength for asserting that one of these scenarios was significantly more likely than the other.

8.4 A path towards improved decision-making during crises

The analysis by Hincks et al. (2014) provides objective probabilistic expression to the volcanological narrative at the time of the 1976 crisis. Indeed a formal evidential case, such as this, would have supported the authorities concerns about public safety and their decision to evacuate. Revisiting the episode highlights many challenges for modern, contemporary decision-making under conditions of considerable uncertainty, and suggests that the BBN is a suitable framework for marshalling multiple, uncertain observations, model results and interpretations.

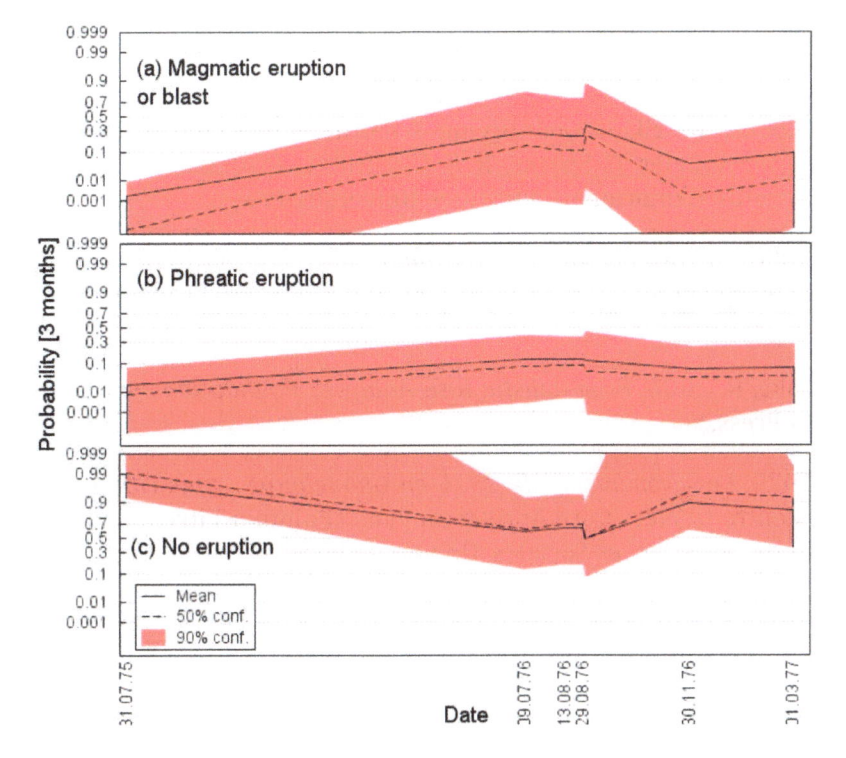

Figure 8.3 Temporal variations from July 1975 to March 1977 in BBN forecast probabilities for La Soufrière Volcano (Hincks et al., 2014), given observation states shown in the lower part of the figure: a) a magmatic eruption or magmatic blast; b) a phreatic eruption, or c) no eruption. The unbroken black line denotes the expected (mean) probability estimate and the dashed line the median, as determined by Monte Carlo re-sampling of BBN input distributions; the shaded bands show the corresponding 5-95 percentile ranges, indicating the uncertainty in the forecast probability.

More recently, mild but persistent seismic and fumarolic unrest since 1992 at La Soufrière volcano has prompted renewed interest in geologic studies, monitoring, risk modelling, and crisis response planning. Development of an advanced probabilistic formalism for decision-making could help quantify and constrain scientific uncertainty, and thereby assist public officials in making urgent evacuation decisions and policy choices should the ongoing unrest intensify in a lead-up to renewed eruptive activity.

The BBN formulation (Hincks et al., 2014) can be developed further as a tool for ongoing use in volcano observatories and can be combined with other probabilistic tools (Newhall & Hoblitt, 2002, Marzocchi et al., 2008, Marzocchi & Bebbington, 2012). This approach is complemented by a progressive quantitative hazard and risk assessment approach (CASAVA project: http://sites.google.com/site/casavaanr/home) that considers: (a) interdisciplinary determinations of infrastructural, human, systemic and cultural factors; (b) social vulnerabilities, capacity and resilience, and (c) includes also the influence of risk perception and governance issues on disaster preparedness. This new work has implications for the way monitoring should be organised for Lesser Antilles volcanoes, and for how risk-informed decision-making in crisis response and long-term strategies of volcanic risk mitigation should be formulated.

References

Aspinall, W. 2006. Structured elicitation of expert judgement for probabilistic hazard and risk assessment in volcanic eruptions. *In:* Mader, H. M. (ed.)*Statistics in Volcanology.* Geological Society of London.

Aspinall, W. & Cooke, R. 2013. Quantifying scientific uncertainty from expert judgement elicitation. *In:* Rougier, J., Sparks, R. S. J. & Hill, L. *Risk and Uncertainty Assessment for Natural Hazards.* Cambridge: Cambridge University Press.

Beauducel, F. 2006. *À propos de la polémique de Soufrière 1976* [Online]. Available: http://www.ipgp.jussieu.fr/~beaudu/soufriere/forum76.html [Accessed 21 December 2013].

Bedford, T. & Cooke, R. 2001. *Probabilistic Risk Analysis: Foundations and Methods*, Cambridge University Press.

De Vanssay, B. 1979. *Les événements de 1976 en Guadeloupe : apparition d'une subculture du désastre.* Centre Universitaire Antilles-Guyane (Pointe-à-Pitre, Guadeloupe) et Ecole des Hautes Etudes en Sciences Sociales, Université Paris 5.

Feuillard, M. 2011. *La Soufrière de la Guadeloupe: un volcan et un peuple,* Guadeloupe, Éditions Jasor.

Feuillard, M., Allegre, C., Brandeis, G., Gaulon, R., Le Mouel, J., Mercier, J., Pozzi, J. & Semet, M. 1983. The 1975–1977 crisis of La Soufrière de Guadeloupe (FWI): A still-born magmatic eruption. *Journal of Volcanology and Geothermal Research,* 16, 317-334.

Fiske, R. S. 1984. Volcanologists, journalists, and the concerned local public: a tale of two crises in the eastern Caribbean. *National Research Council, Geophysics Study Committee (eds) Explosive Volcanism. National Academy Press, Washington, DC*, 110-121.

Hincks, T. K., Komorowski, J.-C., Sparks, S. R. & Aspinall, W. P. 2014. Retrospective analysis of uncertain eruption precursors at La Soufrière volcano, Guadeloupe, 1975–77: volcanic hazard assessment using a Bayesian Belief Network approach. *Journal of Applied Volcanology,* 3:3, pp.26.

Komorowski, J.-C., Boudon, G., Semet, M. P., Beauducel, F., Anténor-Habazac, C., Bazin, S. & Hammouya, G. 2005. Guadeloupe. *In:* Lindsay, J. M., Robertson, R. E. A., Shepherd, J. B. & Ali, S. *Volcanic Hazard Atlas of the Lesser Antilles.* Seismic Research Unit of the University of The West Indies.

Lepointe, E. 1999. Le réveil du volcan de la Soufrière en 1976: la population guadeloupéenne à l'épreuve du danger. *In:* Yacou, A. (ed.)*Les catastrophes naturelles aux Antilles – D'une Soufrière à l'autre.* Paris: CERC Université Antilles et de la Guyane, Editions Karthala.

Marzocchi, W. & Bebbington, M. S. 2012. Probabilistic eruption forecasting at short and long time scales. *Bulletin of Volcanology,* 74, 1777-1805.

Marzocchi, W., Sandri, L. & Selva, J. 2008. BET_EF: a probabilistic tool for long-and short-term eruption forecasting. *Bulletin of Volcanology,* 70, 623-632.

Newhall, C. & Hoblitt, R. 2002. Constructing event trees for volcanic crises. *Bulletin of Volcanology,* 64, 3-20.

Sparks, R. S. J. & Aspinall, W. P. 2013. Volcanic Activity: Frontiers and Challenges in Forecasting, Prediction and Risk Assessment. *In:* Sparks, R. S. J. & Hawkesworth, C. J. *The State of the Planet: Frontiers and Challenges in Geophysics.* Washington: American Geophysical Union.

Woo, G. 2008. Probabilistic criteria for volcano evacuation decisions. *Natural Hazards,* 45, 87-97.

Chapter 9

Forecasting the November 2010 eruption of Merapi, Indonesia

J. Pallister and Surono

9.1 Background

Merapi volcano, Indonesia (7.542°S 110.442°E) is one of the most active and hazardous volcanoes in the world. A large population settled on and around the flanks of the volcano is at risk. Over the past century eruptions were characterised by frequent small to moderate intensity eruptions, with pyroclastic flows produced by lava dome collapse. The most recent eruption in 2010 was of unusually high intensity. In late October and early November 2010, the volcano produced its largest and most explosive eruptions since 1872, displacing about 400,000 people, and claiming nearly 400 lives.

9.2 Monitoring

A seismic network has been in place on Merapi since 1982 to identify different kinds of earthquakes that are informative about the potential for eruption. Deformation is measured using Electronic Distance Measurements (EDM) of line lengths from the flanks to reflectors near the summit, and (since 2010) also with Global Positioning Satellite (GPS) receivers. Sulfur dioxide gas (SO_2) is routinely measured at Merapi using ultraviolet absorption spectrometers. SO_2 is commonly chosen as the gas to monitor as the atmosphere normally contains only trace amounts so it is relatively easy to detect. During volcanic quiescence the SO_2 is typically emitted at less than 100 tons per day, while the emissions can double or treble during small eruptions.

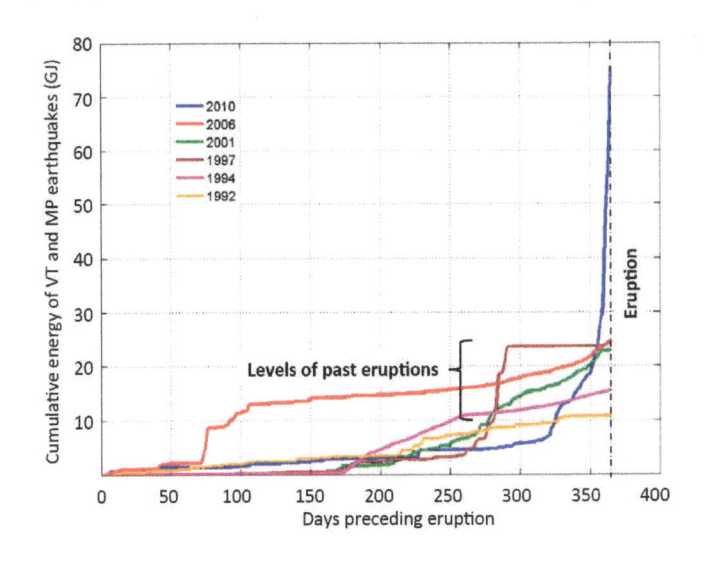

Figure 9.1 Cumulative seismic energy release of volcano-tectonic (VT) and multiphase (MP) earthquakes for eruptions of Merapi in 1997, 2001, 2006 and 26 October 2010. Modified from Budi-Santoso et al. (2013).

Pallister, J. & Surono (2015) Forecasting the November 2010 eruption of Merapi, Indonesia. In: S.C. Loughlin, R.S.J. Sparks, S.K. Brown, S.F. Jenkins & C. Vye-Brown (eds) *Global Volcanic Hazards and Risk,* Cambridge: Cambridge University Press.

9.3 Forecasting the 2010 eruption

Despite the challenges involved in forecasting the 2010 "hundred year eruption", the magnitude of precursory signals (seismicity, ground deformation, gas emissions) was proportional to the large size and intensity of the eruption. Increasing numbers of earthquakes occurred at rates of tens to hundreds of events per day in the weeks before the October 2010 eruption. While increasing seismicity is not a definitive sign of impending eruption it provides an alert of increasing potential. As is common in many volcanoes the earthquakes were located at depths between a few kilometres and the surface. In late September, high levels of CO_2 in summit fumaroles provided early warning of magmatic replenishment. In late October 2010 a series of small phreatomagmatic eruptions took place, with associated SO_2 emissions of tens of thousands of tons per day and peaks in earthquake energy. The observations of exceptionally high gas emissions and high rates of summit deformation as determined with EDM data raised concerns further. In addition and for the first time, near-real-time satellite radar imagery played a major role along with the seismic, geodetic, and gas observations in monitoring and forecasting eruptive activity during a major volcanic crisis. The satellite data documented exceptionally rapid extrusion of a voluminous summit lava dome following the initial phreatomagmatic eruptions and before the climactic eruption on 5 November. Rates of extrusion during 1-4 November were an order of magnitude greater than seen at Merapi during past eruptions, and the resulting summit lava dome quickly reached a volume of ~5 million m³, and was poised ready to collapse at the break in slope at the edge of the summit by 4 November.

The monitoring data played a key role in anticipating the major eruption of 5 November 2010. Marked escalation in summit deformation, seismic energy, SO_2 and CO_2 emissions, increased temperature of crater fumaroles, and the high extrusion rate of lava observed from satellites led to a major expansion of the evacuated zone [see Chapter 10]. The Indonesian Center of Volcanology and Geological Hazard Mitigation (CVGHM) was able to issue timely warnings of the magnitude of the eruption phases, and evacuations organised by the Indonesian National Board for Disaster Management (BNPB), provincial and local emergency managers saved an estimated 10,000 to 20,000 lives [Chapter 10].

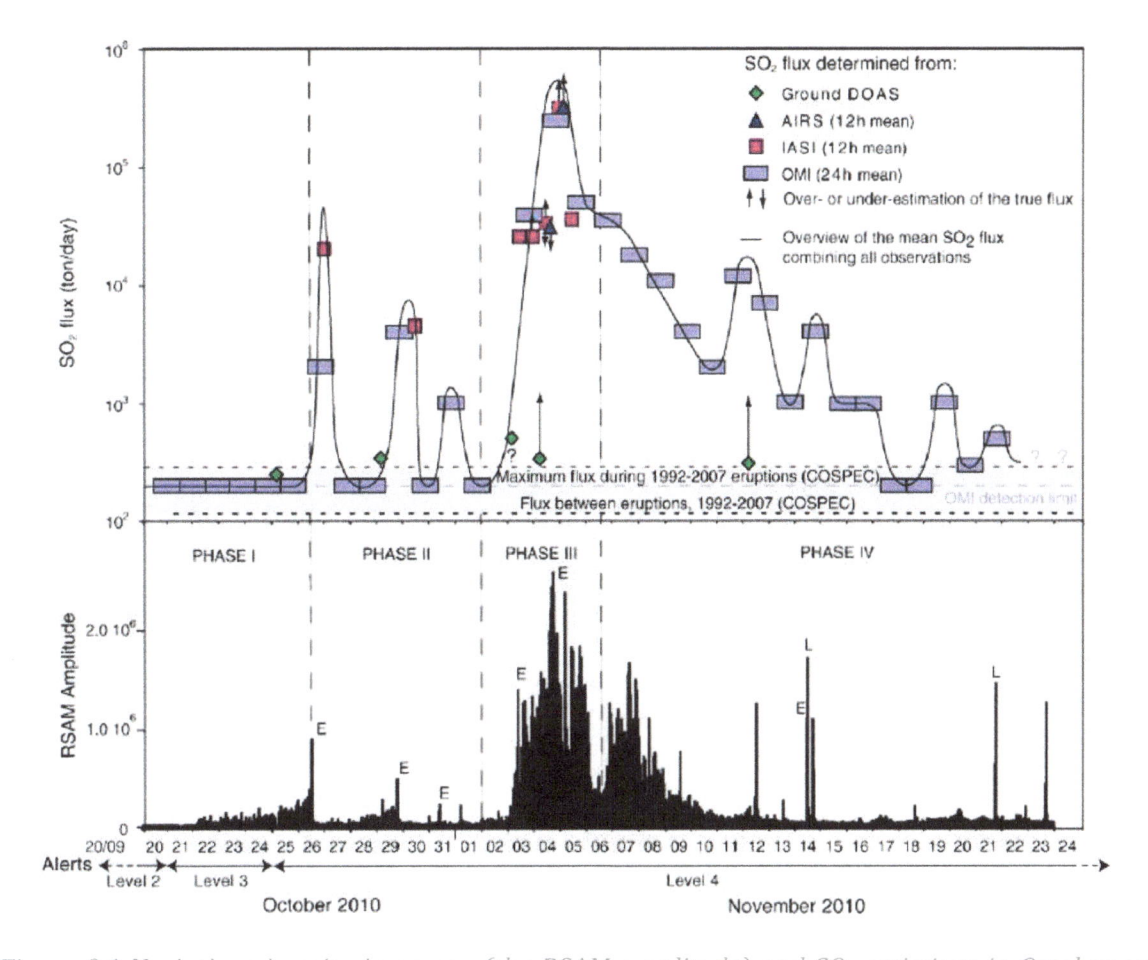

Figure 9.1 Variations in seismic energy (the RSAM amplitude) and SO$_2$ emissions in October and November 2010. Phases are phreatomagmatic explosive (I), magmatic (II), climactic (III) and waning (IV), E marks eruptions and L marks volcanic mudflows (lahars). RSAM is Real-time Seismic Amplitude Measurement, DOAS is Differential Optical Absorption Spectroscopy, satellite SO$_2$ measurements are by AIRS (Atmospheric Infrared Sounder), IASI (Infrared Atmospheric Sounding Interferometer) and OMI (Ozone Monitoring Instrument). From Surono et al. (2012)

References

Budi-Santoso, A., Lesage, P., Dwiyono, S., Sumarti, S., Jousset, P. & Metaxian, J.-P. 2013. Analysis of the seismic activity associated with the 2010 eruption of Merapi Volcano, Java. *Journal of Volcanology and Geothermal Research,* 261, 153-170.

Surono, Jousset, P., Pallister, J., Boichu, M., Buongiorno, M. F., Budisantoso, A., Costa, F., Andreastuti, S., Prata, F., Schneider, D., Clarisse, L., Humaida, H., Sumarti, S., Bignami, C., Griswold, J., Carn, S., Oppenheimer, C. & Lavigne, F. 2012. The 2010 explosive eruption of Java's Merapi volcano-A '100-year' event. *Journal of Volcanology and Geothermal Research,* 241-242, 121-135.

Chapter 10

The importance of communication in hazard zone areas: case study during and after 2010 Merapi eruption, Indonesia

S. Andreastuti, J. Subandriyo, S. Sumarti, D. Sayudi

Merapi is one of the most active volcanoes in Indonesia (2,948 m summit elevation). Eruptions during the twentieth and twenty-first centuries resulted in: 1,369 casualties (1930-1931), 66 casualties (1994) (Thouret et al., 2000), and 386 casualties (2010). The 2010 eruption had impacts that were similar to the unusually large 1872 eruption, which had widespread impacts and resulted in approximately 200 casualties (Hartmann, 1934). These casualties are considered to be a large number given the relatively sparse population in the late nineteenth century by comparison with the population density today.

The 5 November 2010 Merapi eruption affected two provinces and four regencies, including Magelang (west-southwest flank), Sleman (south flank), Klaten (southeast-east flank, and Boyolali (northern flank). The eruption led to the evacuation of 399,000 people and resulted in a total loss of US$ 3.12 billion (National Planning Agency: National Disaster Management Agency, 2011-2013).

The large number of evacuees of Merapi in 2010 was due to warnings of an unusually large eruption – a warning that was based on precursors during the months to days preceding the eruption. These precursors included large increases in seismicity and deformation of the volcano's summit, high rates of dome extrusion, increased temperature of crater fumaroles (reaching 460°C by 20 October), and an abrupt increase in CO_2 at a summit fumaroles. During the time of crisis, there was rapid escalation in rates of seismicity, deformation and rates of initial lava extrusion. All the monitoring parameters exceeded levels and rates of change observed during previous eruptions of the late twentieth century. Consequently, a Level IV warning was issued and evacuations were carried out and then extended progressively to greater distances as the activity escalated. The exclusion zone was extended from 10 to 15 and then to 20 km from Merapi's summit.

Indonesia applies four levels of warnings for volcano activity. From the lowest to highest: at Level I (Normal), the volcano shows a normal (background) state of activity; at Level II (Advisory) visual and seismic data show significant activity that is above normal levels; at Level III (Watch) the volcano shows a trend of increasing activity that is likely to lead to eruption; and at Level IV (Warning) there are obvious changes that indicate an imminent and hazardous eruption, or a small eruption has already started and may lead to a larger and more hazardous

Andreastuti, S., Subandriyo, J., Sumarti, S. & Sayudi, D. (2015) The importance of communication in hazard zone areas: case study during and after 2010 Merapi eruption, Indonesia. In: S.C. Loughlin, R.S.J. Sparks, S.K. Brown, S.F. Jenkins & C. Vye-Brown (eds) *Global Volcanic Hazards and Risk,* Cambridge: Cambridge University Press.

eruption. At Level III people must be prepared for evacuation and at Level IV evacuations are required. Figure 10.1 presents the chronology of warnings and radius of evacuations during the 2010 Merapi eruption (time increases from the bottom of the diagram upwards).

	ALERT LEVEL	DATES	RADIUS	ERUPTION
DECREASING	NORMAL	15-9-2011		
	ADVISORY	30-12-2010		
	WATCH	3-12-2010		
		4-11-2010	20 KM (11:00 UTC)	4 Nov. 17:05 UTC (16,5 km)
INCREASING		3-11-2010	15 KM (08:05 UTC)	3 Nov. 08:30 UTC (9 km)
	WARNING	25-10-2010	10 KM (11:00 UTC)	26-10-2010 (10:02 UTC)
	WATCH	21-10-2010		
	ADVISORY	20-9-2010		
	NORMAL	17-9-2010		

Figure 10.1 Chronology of warnings and radius of evacuations during the 2010 Merapi eruption (time increases from the bottom of the diagram upwards). Distances given in the eruption column show extent of pyroclastic flows.

Following the first explosive eruption on 26 October 2010 and before the climactic eruption on 5 November, a lava dome was extruded rapidly (at rates of ≥25 m³/s, Pallister et al. (2013)). Explosive eruptions took also took place and were accompanied by pyroclastic flows. The lengths of pyroclastic flows increased from 8 km (26 October 2010) to 12 km (3 November) and then to 16.5 km during the climactic eruption on 5 November.

The 2010 Merapi eruption offers an excellent lesson in dealing with eruption uncertainties, crisis management and public communication. Good decision making depends not only on good leadership, but also on the capabilities of scientists, good communication and coordination amongst stakeholders, public communication and on the capacity of the community to respond. All of these factors were in place before the 2010 eruption and contributed to the saving of many thousands of lives.

After the 2010 Merapi eruption with its large impact, revision of the hazard map was carried out to take into account the greater extent of eruption deposits and impacts compared to previous events in the twentieth century. This map is the basis for the implementation of land-use planning and it is represented by the "Map of Impacted Area by Eruption and Lahar" (Peta Terdampak Erupsi dan Lahar), shown in Figure 10.2. The map delineates three hazard zones: Hazard Zone III (directly affected area (ATL)), which includes Forest Conservation/National

Park Development areas with 'closed society settlement' (living in harmony with disaster/zero growth) and National Park and Protected Forest; Hazard Zone III indirectly affected area (ATTL), which includes National Park and Protected Forest; Hazard Zone II (not affected and intended for settlement but according to the land-use plan, highly controlled); and Hazard Zone I (area impacted by lahar). The width of restricted development in river overbank areas is decided by the Governor, and integrated into the Regency/City land-use plan.

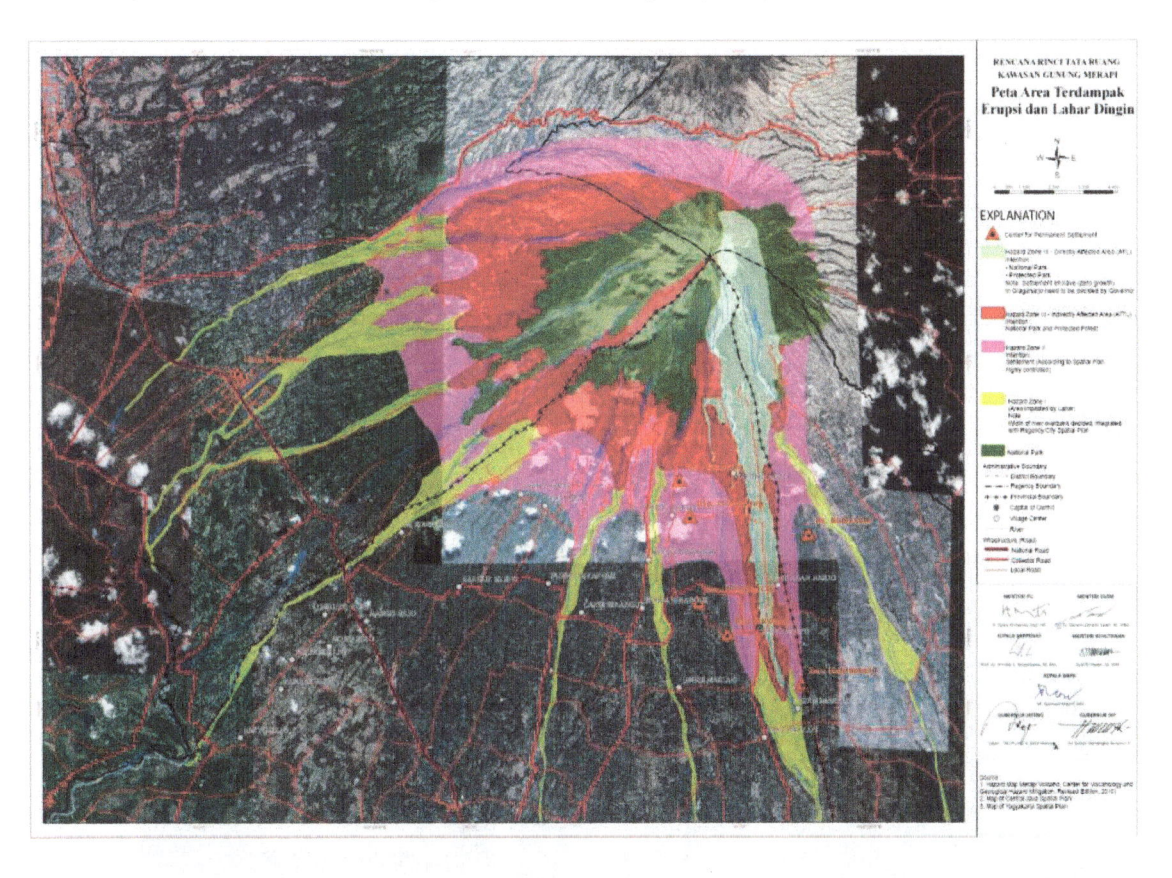

Figure 10.2 Map of Impacted Area by Eruption and Lahar (Peta Terdampak Erupsi dan Lahar) (Source: Map by Center for Volcanology and Geological Hazard Mitigation, CVGHM, Aster Landsat, courtesy Franck Lavigne).

The hazard map was approved by the Ministry of Energy and Mineral Resources, Ministry of Public Work, Ministry of Forestry, National Plan Agency, Head of National Disaster Agency, Governor of Yogyakarta and the Governor of Central Java. The map of impacted area by eruption and lahar is the basis for Merapi land-use plan and the rehabilitation and reconstruction plan. The process has been supported by Ministry Decree of the Republic Indonesia No 16, 2011 (Ministry Decree of Republic Indonesia No 16, 2011, on Team of Coordination on Rehabilitation and Reconstruction of area post disaster of Merapi Eruption, in Yogyakarta Special Province and Central Java Province).

An action plan policy for rehabilitation and reconstruction includes the land-use plan as the basis for determination of secure locations for settlementas well as the design for relocated houses, which are constructed with a risk reduction approach. The map in Figure 10.3 shows the location of temporary and permanent settlement in Sleman, Yogyakarta and the photos (Figure 10.4) show examples of permanent and temporary housing.

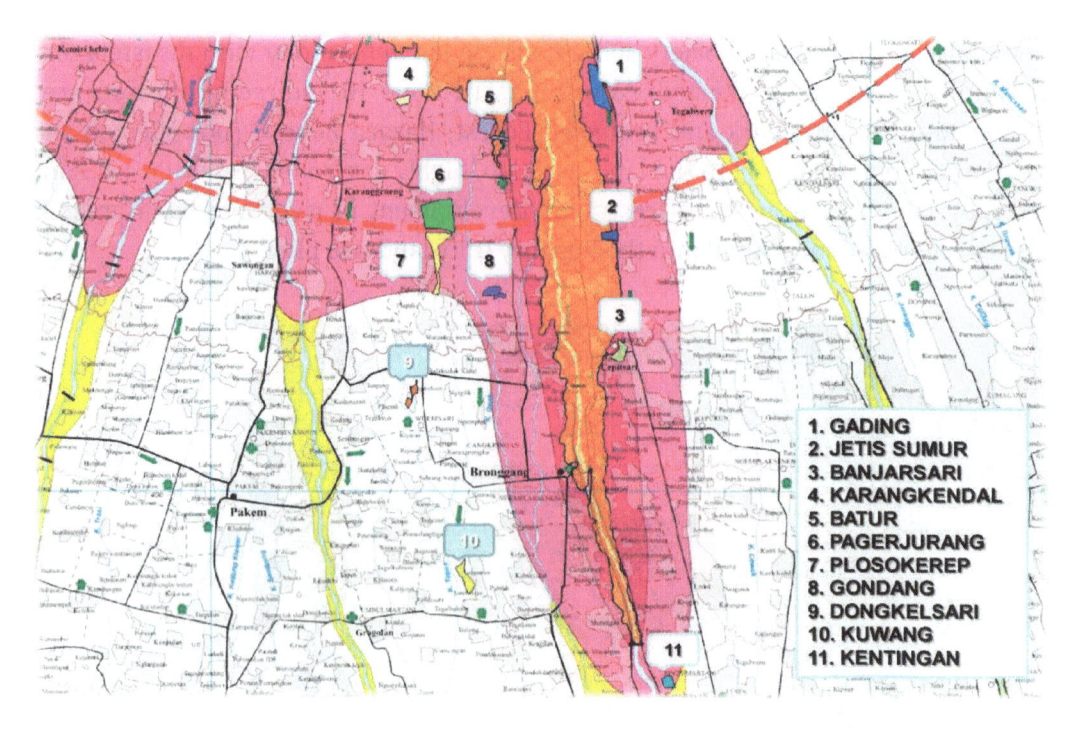

Figure 10.3 Map of Temporary and Permanent Settlement in Sleman, Yogyakarta (Source: Center for Volcanology and Geological Hazard Mitigation, CVGHM, 2011)

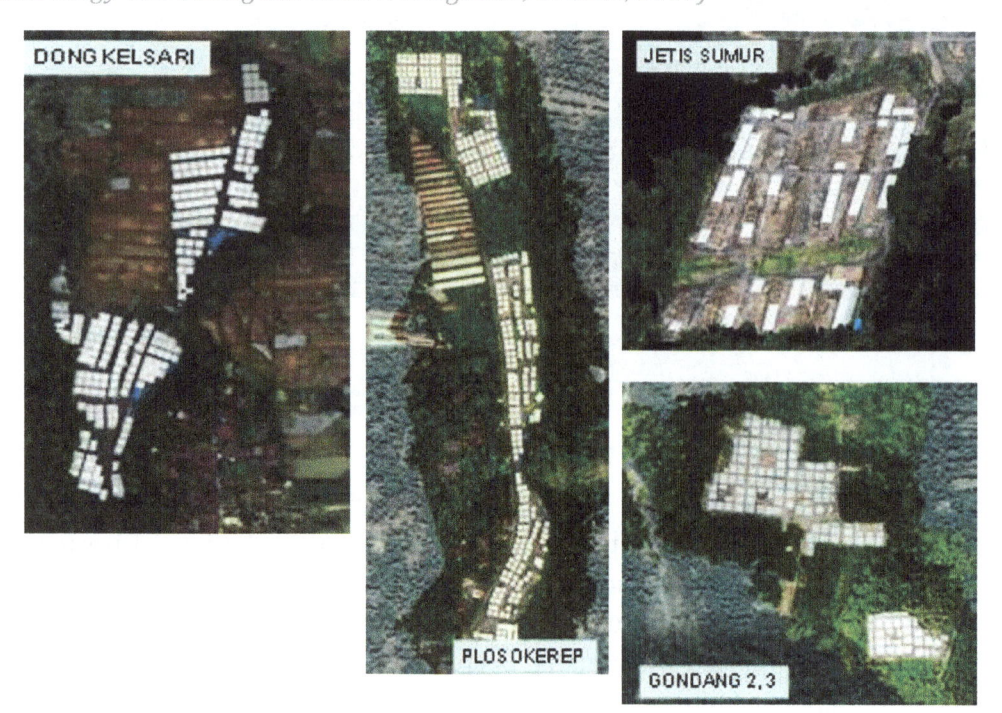

Figure 10.4 Air photo of temporary and permanent settlements of Dongkelsari, Plosokerep, Jetis-sumur, Gondang 2-3 (see map in Figure) (Source: Center for Volcanology and Geological Hazard Mitigation, CVGHM, 2012).

Impacts of Merapi eruptions on the human and cultural environment, livelihood and properties provide a lesson that in densely-populated areas around a volcano there is a need for regular review of hazard mitigation strategies, including spatial planning, mandatory disaster training, contingency planning and for regular evacuation drills. Merapi is well known for a capacity building programme named 'wajib latih' (mandatory training) required for people living near the volcano. The aim of this activity is to improve hazard knowledge, awareness and skills to protect self, family and community. In addition to the wajib latih, people also learn from direct experience with volcano hazards, which at Merapi occur frequently. However, the 2010 Merapi eruption showed that well trained and experienced people must also be supported by good management, and that training and mitigation programmes must consider not only "normal" but also unusually large eruptions (Mei et al., 2013).

References

Hartmann, M. A. 1934. *Der grosse Ausbruch des Vulkanes G. Merapi Mittel Java im Jahre 1872*, Ruygrok & Company.

Mei, E. T. W., Lavigne, F., Picquout, A., De Bélizal, E., Brunstein, D., Grancher, D., Sartohadi, J., Cholik, N. & Vidal, C. 2013. Lessons learned from the 2010 evacuations at Merapi volcano. *Journal of Volcanology and Geothermal Research,* 261, 348-365.

National Planning Agency: National Disaster Management Agency 2011-2013. Action Plan of Rehabilitation and Reconstruction , Post Disaster Area of Merapi Eruption, Yogyakarta and Central Java Province.

Pallister, J. S., Schneider, D. J., Griswold, J. P., Keeler, R. H., Burton, W. C., Noyles, C., Newhall, C. G. & Ratdomopurbo, A. 2013. Merapi 2010 eruption—Chronology and extrusion rates monitored with satellite radar and used in eruption forecasting. *Journal of Volcanology and Geothermal Research,* 261, 144-152.

Thouret, J.-C., Lavigne, F., Kelfoun, K. & Bronto, S. 2000. Toward a revised hazard assessment at Merapi volcano, Central Java. *Journal of Volcanology and Geothermal Research,* 100, 479-502.

Chapter 11

Nyiragongo (Democratic Republic of Congo), January 2002: a major eruption in the midst of a complex humanitarian emergency

J.-C. Komorowski and K. Karume

11.1 Lava flows in town: the 17 January 2002 Nyiragongo eruption

Nyiragongo is a 3470 m high volcano located in the western branch of the East African Rift in the Democratic Republic of Congo (DRC), close to the border with Rwanda. It has a 1.3 km wide summit crater that has been filled with an active lava lake since 1894. The area is affected by frequent damaging tectonic earthquakes and by permanent passive degassing of carbon dioxide (CO_2). Fatal concentrations of CO_2 can accumulate in low-lying areas, threatening the permanent population and internally displaced persons (IDPs) in refugee evacuation centres.

On 17 January 2002 fractures opened on Nyiragongo's upper southern flanks triggering a catastrophic drainage of the lava lake (Figure 11.1). An estimated 25 million cubic metres of lava erupted from many vents along the fractures, which rapidly propagated South towards and into the city of Goma located 17 km away on the shores of Lake Kivu. A small volume of lava entered the lake, which contains deep CO_2 and CH_4 (methane) gas-charged waters. This raised concerns of a potential overturn of the lake, generating lethal gas flows, but the lake was not disturbed. Nyiragongo volcano is responsible for 92% of global lava-flow related fatalities (ca. 824) since 1900. The eruption was accompanied by an unprecedented level of felt earthquakes (Allard et al., 2002, Komorowski et al., 2002/2003, Tedesco et al., 2007b).

Two main lava flows entered the city producing major devastation, and forcing the rapid exodus of most of Goma's 300,000 to 400,000 inhabitants across the border into neighbouring Rwanda. There were international concerns about the evacuation causing an additional humanitarian catastrophe exacerbating the ongoing regional ethnic and military conflict. Lava flows destroyed about 13 % of Goma, 21% of the electricity network, 80% of its economic assets, 1/3 of the international airport runway and the housing of 120,000 people. The eruption caused about 470 injuries and about 140 to 160 deaths mostly from CO_2 asphyxiation and from the explosion of a petrol station near the active hot lava flow (Komorowski et al., 2002/2003, Baxter et al., 2003).

This was the first time in history that a city of such a size had been so severely impacted by lava flows. The eruption of Nyiragongo in 1977 produced extremely fluid, fast-moving (up to 60 km/h) lava flows (Figure 11.1) that entirely covered several villages at night thus killing an estimated 600 persons, but the lava did not reach Goma.

Komorowski, J-C. & Karume, K. (2015) Nyiragongo (Democratic Republic of Congo), January 2002: a major eruption in the midst of a complex humanitarian emergency. In: S.C. Loughlin, R.S.J. Sparks, S.K. Brown, S.F. Jenkins & C. Vye-Brown (eds) *Global Volcanic Hazards and Risk,* Cambridge: Cambridge University Press.

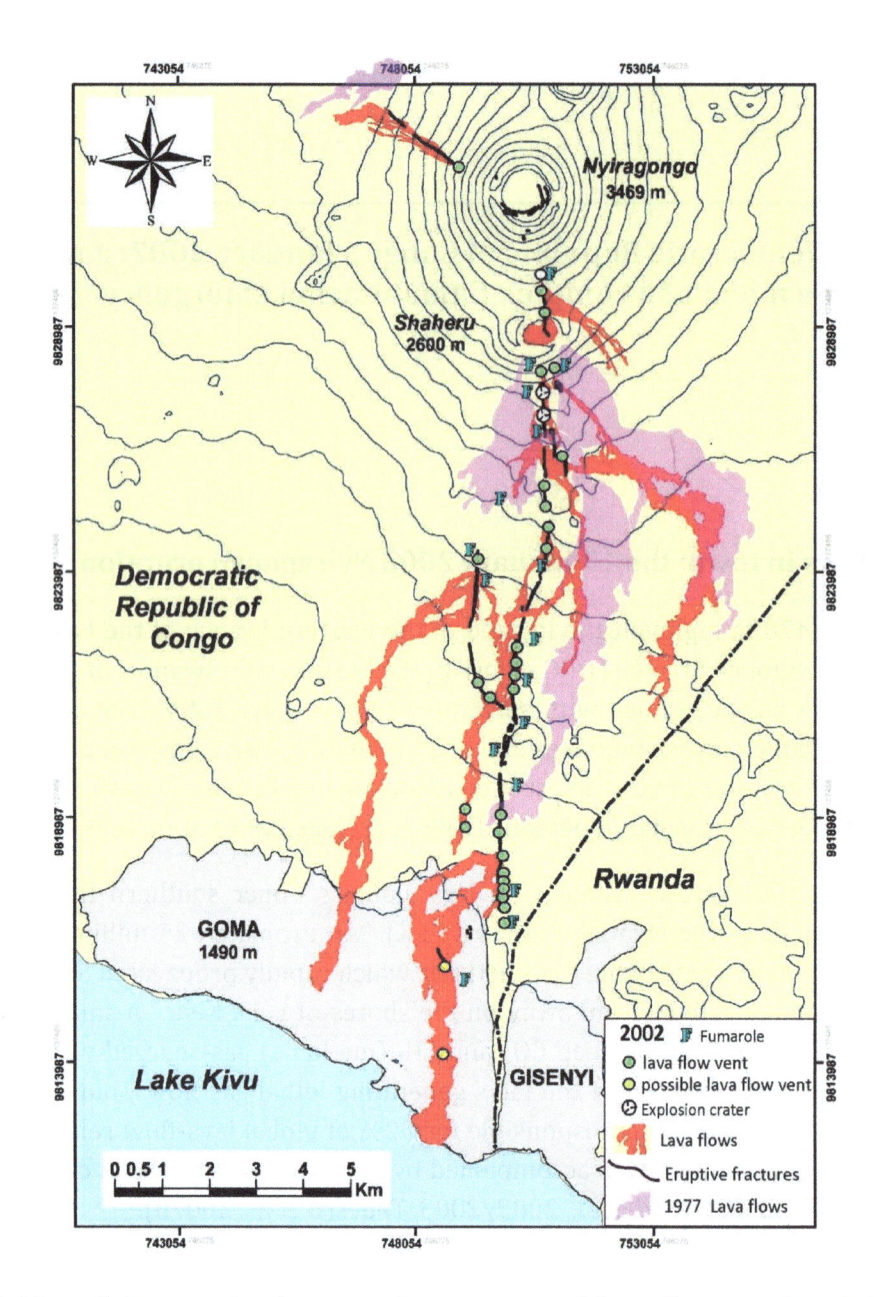

Figure 11.1 Map of the eruptive fractures, lava vents, and lava flows emplaced during the 17 January 2002 and 1977 eruptions of Nyiragongo volcano, Democratic Republic of Congo (modified from Komorowski et al. (2002/2003)).

11.2 Multiple geohazards and a complex humanitarian emergency

With its rapidly expanding demographics and a large numbers of internally displaced persons (IDPs), the city of Goma (ca. 1 million people in 2014) is one of the highest volcanic risk areas in the world. Indeed, this area is not only threatened by future lava flows from lava-lake draining eruptions of Nyiragongo, but also lava flows from Nyamuragira volcano. Long lava flows from this neighbouring volcano, which erupts on average every two years, threaten the northern shores of Lake Kivu and the town of Sake, where large numbers of IDPs shelter (Favalli et al., 2006, 2009, Chirico et al., 2009, Smets et al., 2014). Ash falls, sulfuric acid, chlorine, and fluorine-rich gases affect public health (Sawyer et al., 2008), the water supply and crops. Other hazards include major earthquakes and potentially catastrophic outbursts of CO_2 and methane gas from Lake Kivu (Schmid et al., 2005, Tassi et al., 2009) and landslides (Figure 11.3). These acute geohazards can develop in a cascading sequence and are superimposed on a decade of devastating insecurity and military conflicts in the densely populated Kivu region, that have caused a complex emergency requiring a major humanitarian effort and the largest ongoing UN peace-keeping mission (Komorowski et al., 2002/2003).

*Figure 11.2 **Top** and **middle**: Lava flows invading the city of Goma on 17 January 2002 (photo K. Mahinda, GVO, 2002); **Bottom**: Fractures propagating through the village of Monigi 14 km away from Nyiragongo near the city of Goma as magma migrated from the central conduit South towards lake Kivu (see Figure 11.4) (photo J-C Komorowski, January 23 2002).*

The eruption caused a major humanitarian emergency that further weakened the already fragile lifelines of the population in an area subjected to many years of regional instability and military conflicts. The medical and humanitarian community feared a renewal of cholera epidemics that caused a high mortality in refugee evacuations centres after the 1994 genocide. However, rapid and efficient response by relief workers from UN agencies, numerous non-governmental organisations (NGOs), and local utility agencies prevented major epidemics. Epidemiological surveillance found no major increases in infectious diseases (Baxter & Ancia, 2002, Baxter et al., 2003).

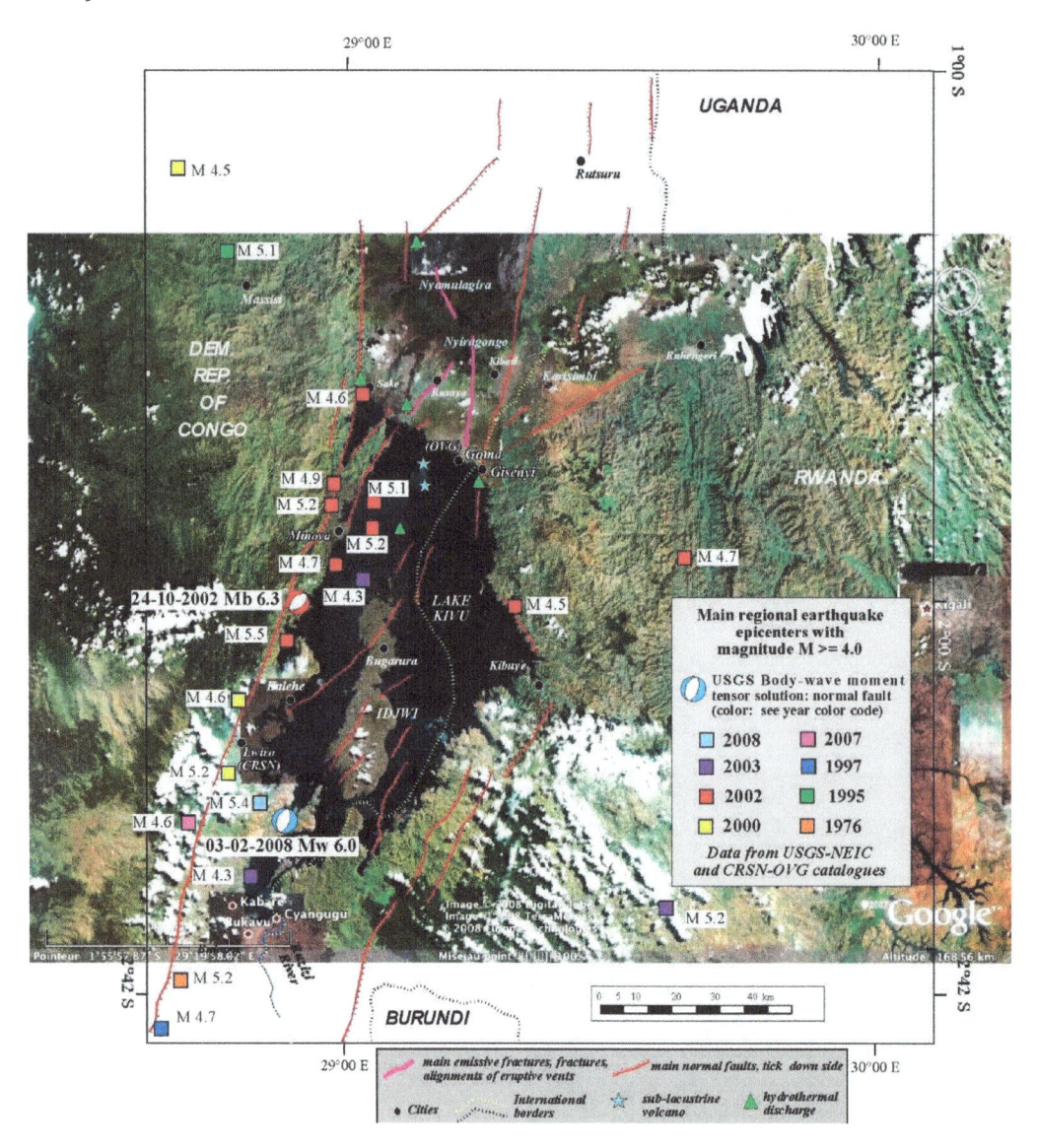

Figure 11.3 Volcanic and seismo-tectonic general setting of the lake Kivu and Virunga Volcanic Zone (Democratic Republic of Congo, Rwanda, Burundi, Uganda) in the western branch of the East African Rift System. Main normal rift faults and lake Kivu data taken from Pouclet (1977), Villeneuve, (1980), Bellon & Pouclet (1980), Ebinger (1989a, 1989b) and Kasahara et al. (1992, Degens et al.,(1973), Wong & Von Herzen (1974). Main earthquake epicenters with magnitude ≥ 4 since 1973 taken from the USGS National Earthquake Information Center (NEIC) and data from the CRSN and the OVG (Goma, RDC), satellite image background from Google Earth (modified from Komorowski et al., 2004 and references therein).

11.3 Lessons learned

Despite being in the midst of a civil war, a lack of funding, institutional support and adequate monitoring equipment, the Goma Volcano Observatory (GVO) scientists successfully made some exceptionally valuable observations about increasing fumarolic and seismic activity. Data from two distant seismic stations were interpreted, as other equipment had been vandalised during the years of military conflict. Those signs were correctly interpreted by GVO as potentially indicating that an eruption could occur, although the precise scenario of a far-reaching flank fissure eruption could not be forecast without an adequate monitoring network. Nevertheless, the GVO played a key role in the recognition of the unrest 1 year prior to the eruption and in providing expert advice to the UN authorities once the eruption began. Memory of the 1977 devastating lava flows triggered life-saving actions by villagers, including panic-less self-evacuation. This, in combination with the presence of a large humanitarian community in Goma and the advice provided by the GVO undoubtedly contributed to the low number of fatalities given the scale of the eruption (Komorowski et al., 2002/2003, Ruch & Tedesco, 2003).

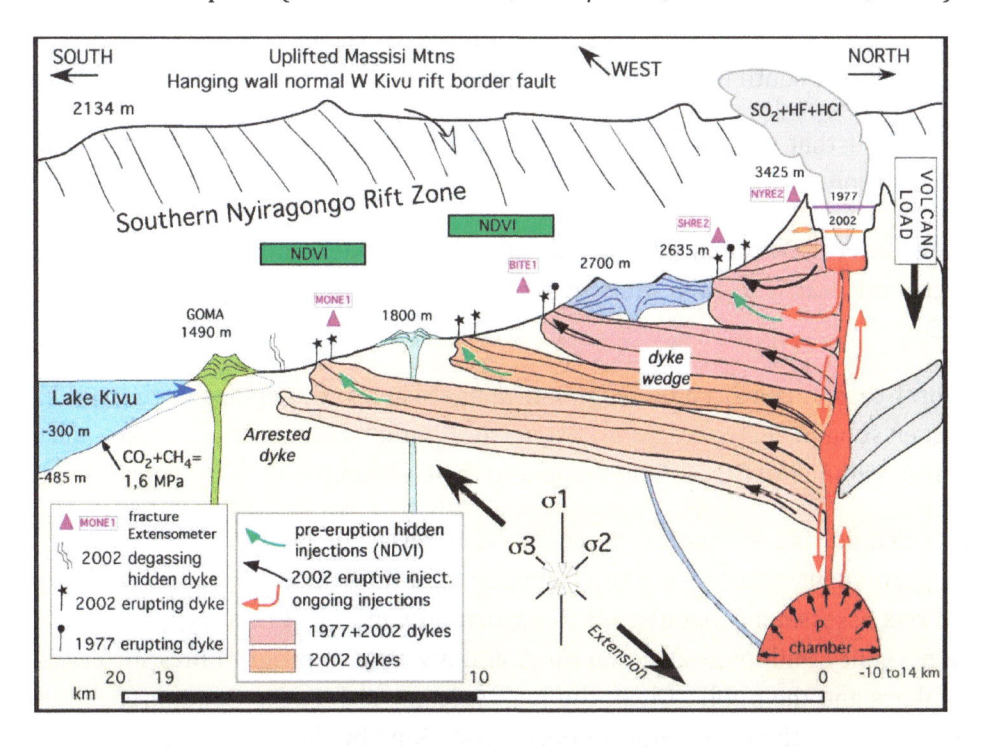

Figure 11.4 Conceptual model of the southern Nyiragongo rift zone and its volcanic activity in the context of the North Kivu tectonic rift area and gas-charged Lake Kivu. (Based on information from (Komorowski et al., 2006, Houlie et al., 2006).

For Nyiragongo, the IDNDR Decade programme had not achieved its goals as clearly stated in the 1994 Goma Declaration (Casadevall & Lockwood, 1995). The response to the 2002 Nyiragongo eruption was remarkable, with significant support provided rapidly by the international humanitarian and scientific community under the coordination and with funding from UN agencies. This support came from international and regional NGOs, government agencies, donor countries and academic research programmes (Tedesco et al., 2007a). Had the DRR goals for Nyiragongo, laid out as part of the International Decade for Natural Disaster Reduction (IDNDR) and stated in the 1994 Goma Declaration (Casadevall & Lockwood, 1995) been achieved prior to this eruption, such a complex response may have been unnecessary.

Therefore one of the first tasks of the post-crisis response to the 2002 eruption was to establish a modern operational volcano monitoring network and team. Thus, the GVO was significantly strengthened and a new multi-parameter modern monitoring system was installed gradually along with capacity-building programmes. All these efforts have significantly improved the technical and analytical capabilities of the GVO in monitoring the activity of the Virunga volcanoes (including Nyiragongo and Nyamuragira).

11.4 The way forward: the Goma Volcano Observatory

Since 2002, a new large lava lake has formed within the 1000 m deep crater and is associated with a significant sulfur, chlorine, and fluorine-rich gas plume, one of the largest in the world. The level of lava has risen slowly but continuously by at least 500 m since October 2002, resting about 400 m below the rim, and only 130 m below the level at the time of the 2002 drainage. There is considerable scientific uncertainty regarding future scenarios (Komorowski et al., 2002/2003):

1. What is the threshold level of the lava lake required to trigger another release of lava through its flanks towards the south like in 1977 and 2002?

2. How likely is it that magma will be channelled away from the summit crater through the highly fractured southern flanks of Nyiragongo?

3. In a future lateral eruption, will magma propagate faster and further towards the water-saturated ground near lake Kivu, thus increasing the likelihood of explosive eruptions within the city of Goma?

4. In the worst-case scenario, could magma propagate below the deep gas-charged basin of lake Kivu to trigger subaqueous volcanic eruptions and potential catastrophic lake-overturn events releasing large volumes of CO_2 and methane into the environment?

Successful volcanic risk mitigation depends on a series of integrated timely actions. These include: a permanent secured multi-parameter real-time monitoring network; quantitative hazard and risk assessment that quantifies uncertainty and fills knowledge gaps; early-warning systems; emergency and long-term planning; and awareness programmes for crisis managers, decision-makers and the public. Given the high volcanic risk in the Goma area, all these efforts must be further strengthened to support decision-making by the authorities.

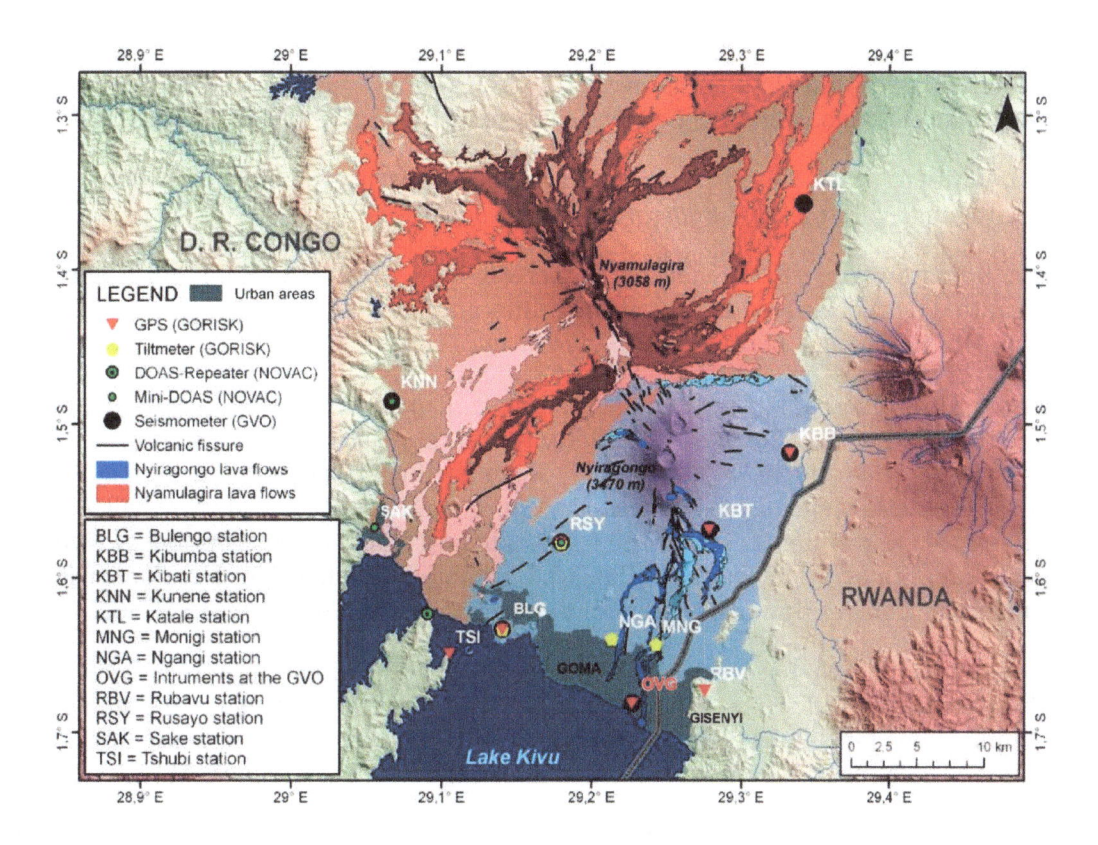

Figure 11.5 The current multi-parameter monitoring system of the Goma Volcanological Observatory (see legend for the techniques). In cooperation with foreign institutions or universities, GVO is involved in: geochemistry surveys, investigation of Mazuku (pockets of CO_2) CO_2 ground degassing, Lake Kivu chemical and physical surveys, satellite imagery and DEM mapping, geological and structural mapping, lake Kivu stability modelling, lava flow paths modelling, hazard and risk maps, ground deformation benchmarks, continuous temperature measurements and monitoring of the width of eruptive fractures. Some stations are no longer working due to equipment vandalism (GVO; WOVO). Image modified from Smets et al. (2014).

References

Allard, P., Baxter, P., Hallbwachs, M. & Komorowski, J. 2002. Nyiragongo. *Bulletin Global Volcanism Network*, 27.

Baxter, P., Allard, P., Halbwachs, M., Komorowski, J., Andrew, W. & Ancia, A. 2003. Human health and vulnerability in the Nyiragongo volcano eruption and humanitarian crisis at Goma, Democratic Republic of Congo. *Acta Vulcanologica*, 14, 109.

Baxter, P. J. & Ancia, A. 2002. Human health and vulnerability in the Nyiragongo volcano crisis Democratic Republic of Congo 2002: Final report to the World Health Organisation. *World Health Organisation*.

Casadevall, T. & Lockwood, J. 1995. Active volcanoes near Goma, Zaire - Hazards to residents and refugees. *Bulletin of Volcanology*, 57, 257-277.

Chirico, G. D., Favalli, M., Papale, P., Boschi, E., Pareschi, M. T. & Mamou-Mani, A. 2009. Lava flow hazard at Nyiragongo Volcano, DRC. *Bulletin of Volcanology*, 71, 375-387.

Favalli, M., Chirico, G., Papale, P., Pareschi, M., Coltelli, M., Lucaya, N. & Boschi, E. 2006. Computer simulations of lava flow paths in the town of Goma, Nyiragongo volcano, Democratic Republic of Congo. *Journal of Geophysical Research: Solid Earth (1978–2012)*, 111.

Favalli, M., Chirico, G. D., Papale, P., Pareschi, M. T. & Boschi, E. 2009. Lava flow hazard at Nyiragongo volcano, DRC. *Bulletin of Volcanology*, 71, 363-374.

Houlie, N., Komorowski, J., De Michele, M., Kasereka, M. & Ciraba, H. 2006. Early detection of eruptive dykes revealed by normalized difference vegetation index (NDVI) on Mt. Etna and Mt. Nyiragongo. *Earth and Planetary Science Letters*, 246, 231-240.

Komorowski, J.-C., Tedesco, D., Kasereka, M., Allard, P., Papale, P., Vaselli, O., Durieux, J., Baxter, P., Halbwachs, M., Akumbe, M., Baluku, B., Briole, P., Ciraba, H., Dupin, J.-C., Etoy, O., Garcin, D., Hamaguchi, H., Houlie, N., Kavotha, K. S., Lemarchand, A., Lockwood, J., Lukaya, N., Mavonga, G., de Michele, M., Mpore, S., Mukambilwa, M., Newhall, C., Ruch, J., Yalire, M. & Wafula, M. 2002/2003. The January 2002 flank eruption of Nyiragongo volcano (Democratic Republic of Congo): Chronology, evidence for a tectonic rift trigger, and impact of lava flows on the city of Goma. *Acta vulcanologica*, 14-15, 27-62.

Komorowski, J., Houlié, N., Kasereka, C. & Ciraba, H. Early detection of eruptive dykes revealed by Normalized Difference Vegetation Index (NDVI) on Nyiragongo and Etna volcanoes: Implications for dyke wedge emplacement, monitoring, and risk assessment. AGU Fall Meeting Abstracts, 2006 San Francisco. American Geophysical Union, 1573.

Ruch, J. & Tedesco, D. 2003. One year after the Nyiragongo Volcano alert: evolution of the communication between Goma inhabitants (populations), scientists and local authorities. *Acta Vulcanologica*, 14, 101.

Sawyer, G., Carn, S., Tsanev, V., Oppenheimer, C. & Burton, M. 2008. Investigation into magma degassing at Nyiragongo volcano, Democratic Republic of the Congo. *Geochemistry, Geophysics, Geosystems*, 9.

Schmid, M., Halbwachs, M., Wehrli, B. & Wüest, A. 2005. Weak mixing in Lake Kivu: new insights indicate increasing risk of uncontrolled gas eruption. *Geochemistry, Geophysics, Geosystems*, 6.

Smets, B., d'Oreye, N., Kervyn, F., Kervyn, M., Albino, F., Arellano, S. R., Bagalwa, M., Balagizi, C., Carn, S. A. & Darrah, T. H. 2014. Detailed multidisciplinary monitoring reveals pre-and co-eruptive signals at Nyamulagira volcano (North Kivu, Democratic Republic of Congo). *Bulletin of Volcanology*, 76, 1-35.

Tassi, F., Vaselli, O., Tedesco, D., Montegrossi, G., Darrah, T., Cuoco, E., Mapendano, M., Poreda, R. & Delgado Huertas, A. 2009. Water and gas chemistry at Lake Kivu (DRC): geochemical evidence of vertical and horizontal heterogeneities in a multibasin structure. *Geochemistry, Geophysics, Geosystems*, 10.

Tedesco, D., Badiali, L., Boschi, E., Papale, P., Tassi, F., Vaselli, O., Kasereka, C., Durieux, J., DeNatale, G. & Amato, A. 2007a. Cooperation on Congo volcanic and environmental risks. *Eos, Transactions American Geophysical Union*, 88, 177-181.

Tedesco, D., Vaselli, O., Papale, P., Carn, S., Voltaggio, M., Sawyer, G., Durieux, J., Kasereka, M. & Tassi, F. 2007b. January 2002 volcano-tectonic eruption of Nyiragongo volcano, Democratic Republic of Congo. *Journal of Geophysical Research: Solid Earth (1978–2012)*, 112.

Chapter 12

Volcanic ash fall impacts

T.M. Wilson, S.F. Jenkins, and C. Stewart

12.1 Overview

All explosive eruptions produce volcanic ash (fragments of volcanic rock < 2mm: Figure 12.1), which is then dispersed by prevailing winds and deposited as ash falls hundreds or even thousands of kilometres away. Volcanic ash suspended in the atmosphere is well known as a hazard for aviation, as was demonstrated during the 2010 eruption of Eyjafjallajökull, Iceland, which led to substantial disruption to flights in Europe and an estimated US$5 billion loss as global businesses and supply chains were affected (Ragona et al., 2011). Volcanic ash fall can also create considerable impacts on the ground. As a general rule, impacts will be more severe with increasing thickness of ash fall. Relatively thin falls (< 10 mm) may have adverse health effects for vulnerable individuals and can disrupt critical infrastructure services, aviation, agriculture and other socio-economic activities over potentially very large areas. Thick ash falls (>100 mm) may damage crops, vegetation and infrastructure, cause structural damage to buildings and create major clean-up requirements. However, they are typically confined to within tens of kilometres of the vent and, as they occur with large eruptions, are relatively rare.

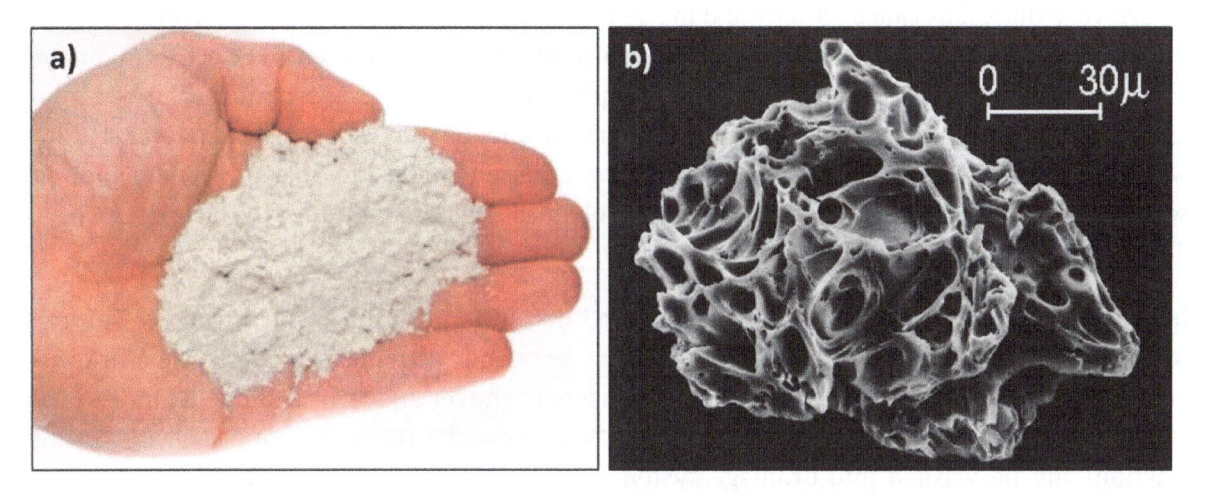

Figure 12.1 a) Rhyolitic ash produced by the eruption of Chaitén volcano, Chile, in 2008 (median grainsize: 0.02 mm). Photo: G. Wilson; b) Scanning electron microscope image of an ash particle from the eruption of Mount St Helens, USA, in 1980. Vesicles, formed as gas expanded within solidifying magma, are clearly visible. Ash is often highly abrasive because of its irregular shape and hardness. Photo: A.M. Sarna-Wojcicki, USGS.

Wilson, T.M., Jenkins, S.F. & Stewart, C. (2015) Volcanic ash fall impacts. In: S.C. Loughlin, R.S.J. Sparks, S.K. Brown, S.F. Jenkins & C. Vye-Brown (eds) *Global Volcanic Hazards and Risk,* Cambridge: Cambridge University Press.

The quantity of ash (thickness or loading) is not the only mechanism by which ash can cause damage or disruption; the surface chemistry of ash, abrasiveness, friability, ash grain size and density can all influence, or even control, how some systems or components may respond to an ash fall. The physical and chemical properties of fallen ash are largely controlled by eruptive dynamics and magma composition, although dispersal conditions, e.g. wind and rain, also play a role. Ash properties can therefore vary among different eruptions and even during the same eruption. Environmental factors such as wind and rain may lead to ash remobilisation, which may extend and/or intensify the level of impact.

12.2 Impacts

Impacts depend upon the amount of volcanic ash deposited and its characteristics (hazard), as well as the numbers and distribution of people and assets (exposure), and the ability of people and assets to cope with ash fall impacts (vulnerability). Three main zones of ash fall impact may be broadly expected, each requiring a different approach to impact management and planning: 1) Destructive and immediately life-threatening (Zone I); 2) Damaging and/or disruptive (Zone II); and 3) Disruptive and/or a nuisance (Zone III). These zones are summarised in the schematic Figure 12.2 where physical ash impacts to selected societal assets are depicted against ash deposit thickness – which generally decreases with distance from the volcano source. The more severe impacts depicted in Zones I and II may not occur in smaller magnitude eruptions because there is insufficient ash fall. A further impact zone (Zone IV) could be applied to areas that extend beyond those receiving ash fall, as aviation disruption from airborne ash can occur in areas where no ash falls on the ground. Below, we elaborate on the impacts shown in Figure 12.2. See Chapter 3 and references therein for more detailed information on each point:

Impacts to people: Exposure to volcanic ash fall rarely endangers human life directly, except where very thick falls cause structural damage (e.g. roof collapse) or indirect casualties such as those sustained during ash clean-up operations or in traffic accidents. Short-term effects commonly include irritation of the eyes and upper airways and exacerbation of pre-existing asthma; serious health problems are rare (Horwell and Baxter (2006); see Chapter 13 and www.ivhhn.org). Affected communities can also experience considerable direct and indirect social impacts, for example, impaired psychological functioning due to factors such as disruption of livelihoods and consequent anxiety.

Impacts to critical infrastructure: Damage and disruption of critical infrastructure services from ash fall impacts can substantially affect socio-economic activities. Electricity networks are vulnerable, mainly due to ash contamination causing flashover and failure of insulators (Wilson et al., 2012). Ash can also disrupt transportation networks through reduced visibility and traction; and be washed into drainage systems. Wastewater treatment systems that have an initial mechanical pre-screening step are particularly vulnerable to damage if ash-laden sewage arrives at the plant. Suspended ash may also cause damage to water treatment plants if it enters through intakes or by direct fallout (e.g. onto open sand filter beds). In addition to direct impacts, system interdependence is a problem. For example, air- or water-handling systems may become blocked by ash leading to overheating or failure of dependent systems. Specific impacts depend strongly on network or system design, typology, ash fall volume and characteristics, and the effectiveness of any applied mitigation strategies (Wilson et al., 2012).

Impacts to agriculture: Fertile volcanic soils commonly host farming operations; ash falls can be beneficial or detrimental to soil depending on the characteristics of the ash (particularly with respect to its surface composition and the soil (Cronin et al., 1998). The time of year in the agricultural production cycle strongly determines the level of impact (Cook et al., 1981). For example, ripe crops close to harvest are particularly vulnerable to contamination, pollination disruption and damage. Under very thin ash falls (< 1 mm) crops and pastures can suffer from acid damage or reduced UV light and, with increasing thicknesses, plants may be broken or buried and soil potentially smothered. Thick ash falls (>100 mm) typically require soil rehabilitation, e.g. thorough mixing or removal, to restore agricultural production (Wilson et al., 2011). For livestock, ash falls may cause starvation (damaged or smothered feed), dehydration (water sources clogged with ash), tooth wear, deaths from ingesting ash along with feed and (more rarely) acute or chronic fluorosis if ash contains moderate to high levels of bioaccessible fluoride.

Impacts to buildings: The load associated with an ash fall can cause the collapse of roofing material (e.g. sheet roofs), the supporting structure (e.g. rafters or walls) or both and, under sufficiently great loads (>> 100 mm), the entire building may collapse (Blong, 1984, Spence et al., 2005). Non-engineered, long-span and low-pitched roofs are particularly vulnerable to collapse, potentially under thicknesses of around 100 mm. Under thinner ash falls (< 100 mm), structural damage is unlikely although non-structural elements such as gutters and overhangs may suffer damage. Ash falls with increased moisture content, as a result of rain for example, will impart a greater load so that resulting damage is more likely. Building components and contents may also be damaged from ash falls due to ash infiltration into interiors, with associated abrasion and corrosion.

Impacts to the economy (including clean-up): Economic losses may arise from damage to physical assets, e.g. buildings, or reductions in production, e.g. agricultural or industrial output (Munich-Re, 2007). Most economic activities will be impacted, even indirectly, under relatively thin (< 10 mm) ash falls, for example through disruptions to critical infrastructure. Losses may even result from precautionary risk management activities, e.g. covering water supplies, business closure or evacuations. During or after an ash fall, clean-up from roads, properties, and airports is often necessary to restore functionality, but the large volumes make it time-consuming, costly and resource-intensive

12.3 Hazard and risk ranking

The wide geographic reach of volcanic ash falls, and their relatively high frequency, makes them the volcanic hazard most likely to affect the greatest number of people. However, forecasting how much volcanic ash will fall, where it will fall, when it will fall, and with what characteristics is a major challenge. Probabilistic volcanic ash hazard maps [see Chapter 6], developed using ash dispersal models and statistical analyses of likely eruption styles, frequencies, magnitudes and wind conditions, take a step towards robustly quantifying ash fall hazard. Most importantly, the probabilistic nature of such hazard maps means that some of the epistemic uncertainties associated with forecasting how much ash will be erupted, to what height and into what wind conditions are accounted for by simulating many thousands of possible ash fall footprints. By aggregating these difference scenarios, areas of relatively high and low hazard can be identified. A probabilistic ash hazard model was developed for 190 active volcanoes in the Asia-Pacific region, home to 25% of the world's volcanoes and over two billion inhabitants, and the average frequency of ash falls that exceed critical impact thicknesses estimated on a location-by-location, rather than volcano-by-volcano basis (see Jenkins et al. (2012)). By multiplying these hazard estimates with freely available exposure data (in this case LandScan population density) and a proxy for human vulnerability (the UN Human Development Index), a crude 'risk' score could then be established (Figure 12.3). This offers an insight into the relative risk across the region, building on the probabilistic ash hazard maps. By disaggregating the score, the key risk driver can be identified, which may suggest how risk can best be reduced. For example, Tokyo's risk is dominated by the high cumulative hazard (54 active volcanoes lie within 1000 km), Jakarta's risk is dominated by population exposure and Port Moresby's risk by the vulnerability. While this approach is useful at large regional to global scales, it should never replace a local risk assessment, which should use more detailed knowledge of the volcano and local analyses of societal assets and vulnerability to produce a robust assessment.

Ash fall impacts | 285

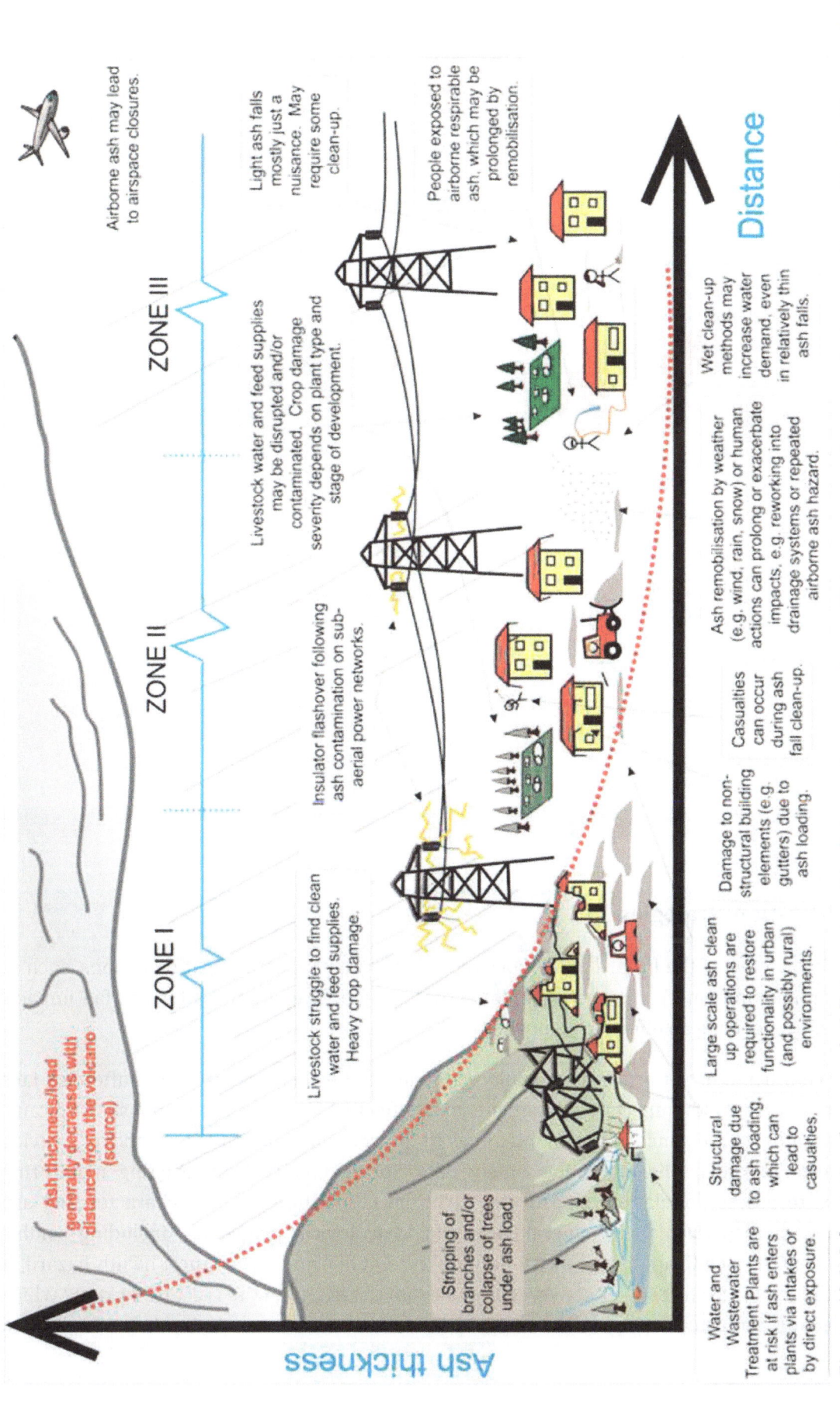

Figure 12.2 Schematic of some ash fall impacts with distance from a volcano. This assumes a large explosive eruption with significant ash fall thicknesses in the proximal zone and is intended to be illustrative rather than prescriptive. Three main zones of ash fall impact are defined: 1) Destructive and immediately life-threatening (Zone I); 2) Damaging and/or disruptive (Zone II); 3) Disruptive and/or a nuisance (Zone III).

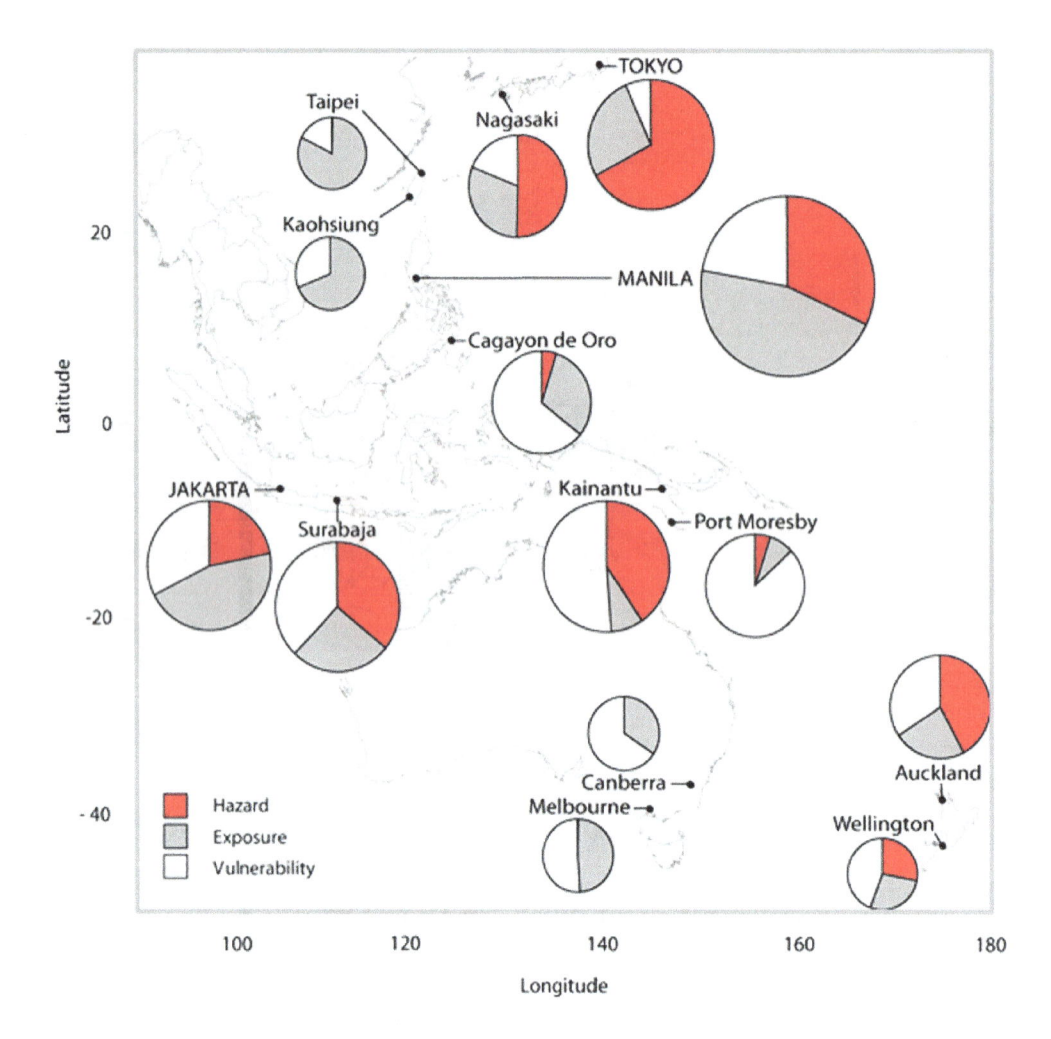

Figure 12.3 Relative risk scores (shown by circle size) and the contributions of the three factors towards the overall risk (a product of hazard, exposure and vulnerability) for cities in the Asia-Pacific region. Hazard is taken as the estimated frequency of thin (≥ 1mm) ash falls; Exposure: the population density; and Vulnerability: a composite of education, life expectancy and standard of living (the UN Human Development Index).

12.4 Mitigation strategies

Greater knowledge of the hazard and associated impact can support mitigation actions, such as crisis planning and emergency management activities. Poor preparedness for ash fall impacts can be costly, delaying an effective response.

A major concern to both the affected populace and authorities before, during and after ash falls are the potential health impacts. Key elements of an effective public health response to volcanic ash fall include: surveillance of health outcomes to inform public health advice and/or provide reassurance to the public; obtaining timely data on air quality from existing monitoring networks to assess population exposure to airborne respirable ash; and characterising ash samples with respect to their mineralogical and toxicological properties, including soluble element content [see Chapter 13]. As a relatively rare public and agricultural health hazard, it can be difficult for agencies to effectively communicate the extent of the risk and to know which ash collection and analysis methods are appropriate. The International Volcanic Health Hazard

Network (IVHHN) is invaluable in this role (www.ivhhn.org). Other preparedness activities should include:

- providing stakeholders with access to specific and relevant preparedness and post-event response/recovery information. Communication regarding the hazard and recommended mitigation steps should be transparent, repeated and from multiple trusted and authoritative sources. (e.g. www.gns.cri.nz/Home/Learning/Science-Topics/Volcanoes/Eruption-What-to-do/Ash-Impact-Posters);
- effective and timely warnings. Ashfall warnings are now standard procedures in many countries, such as in the United States, Japan and New Zealand;
- facilitating appropriate protective actions, e.g. sealing buildings, shutting down vulnerable systems, etc;
- development of clean-up plans that prioritise critical areas or lines of communication and identification of volcanic ash disposal sites and procedures;
- mutual support or continuity agreements between municipal authorities, critical infrastructure organisations and businesses, which can facilitate greater access to resources to deal with ash fall events.

While volcanic eruptions cannot be prevented, the exposure and vulnerability of the population to their impacts may, in theory, be reduced, through the considerable tasks of hazard and risk assessment, improved land use planning, risk education and communication and increasing economic development.

References

Blong, R. J. 1984. *Volcanic Hazards. A Sourcebook on the Effects of Eruptions,* Australia, Academic Press.

Cook, R. J., Barron, J., Papendick, R. I. & Williams, G. 1981. Impact on agriculture of the Mount St. Helens eruptions. *Science,* 211, 16-22.

Cronin, S., Hedley, M., Neall, V. & Smith, R. 1998. Agronomic impact of tephra fallout from the 1995 and 1996 Ruapehu Volcano eruptions, New Zealand. *Environmental Geology,* 34, 21-30.

Horwell, C. J. & Baxter, P. J. 2006. The respiratory health hazards of volcanic ash: a review for volcanic risk mitigation. *Bulletin of Volcanology,* 69, 1-24.

Jenkins, S., Magill, C., Mcaneney, J. & Blong, R. 2012. Regional ash fall hazard I: a probabilistic assessment methodology. *Bulletin of Volcanology,* 74, 1699-1712.

Munich-Re 2007. Volcanism - Recent findings on the risk of volcanic eruptions. *Schadenspiegel,* 1, 34-39.

Ragona, M., Hannstein, F. & Mazzocchi, M. 2011. The impact of volcanic ash crisis on the European Airline industry. *In:* ALEMANNO, A. (ed.) *Governing Disasters: The Challenges of Emergency Risk regulations.* Edward Elgar Publishing.

Spence, R., Kelman, I., Baxter, P., Zuccaro, G. & Petrazzuoli, S. 2005. Residential building and occupant vulnerability to tephra fall. *Natural Hazards and Earth System Science,* 5, 477-494.

Wilson, T., Cole, J., Stewart, C., Cronin, S. & Johnston, D. 2011. Ash storms: impacts of wind-remobilised volcanic ash on rural communities and agriculture following the 1991 Hudson eruption, southern Patagonia, Chile. *Bulletin of Volcanology*, 73, 223-239.

Wilson, T. M., Stewart, C., Sword-Daniels, V., Leonard, G. S., Johnston, D. M., Cole, J. W., Wardman, J., Wilson, G. & Barnard, S. T. 2012. Volcanic ash impacts on critical infrastructure. *Physics and Chemistry of the Earth, Parts A/B/C,* 45, 5-23.

Chapter 13

Health impacts of volcanic eruptions

C.J. Horwell, P.J. Baxter and R. Kamanyire

13.1 Overview

Volcanoes emit a variety of products which may be harmful to human and animal health. Some cause traumatic injury or death and others may trigger diseases, particularly in the respiratory and cardiovascular systems, or mental health problems. The impact on health is related to the style of eruption and type of volcano. Effusive eruptions tend to emit gases and aerosols, which may damage the respiratory system, and lava flows which rarely kill but may cause thermal injuries and mental stress due to the threat of loss of property. Explosive eruptions kill, injure and potentially trigger disease via a multitude of hazards ranging from proximal impacts related to production of fragmented rock and more distal impacts from ash, gas and secondary effects.

13.2 Injury agents

Injury and death are caused by a range of volcanic hazards (e.g. Auker et al. (2013)), which can be summarised by their impact on the body:

1) *Mechanical injury where the body is crushed.* Explosive eruptions may produce large volumes of fragmented rock, which range in size from boulders to fine ash. Mechanical injury/death occurs from a range of volcanic processes relating to the ejection of material and its transport through air or water (lahars, rock avalanches, ballistics). In 1985, the eruption of Nevado del Ruiz volcano, Colombia, led to glacial melt mixing with ash/rock deposits to form a lahar which buried 23,000 people downstream in the town of Armero. Roof collapse is also a common crushing injury, from the weight of ashfall, particularly on flat roofs [see Chapter 12]. Occasionally those proximal to the volcano may be buried by deposits or suffer asphyxiation from inhalation of particles.

2) *Thermal injury (burns) caused by hot volcanic emissions.* These take the form of pyroclastic density currents (PDCs) and surges (composed of searing gas, ash and rocks), lava flows and hydrothermal waters (which are used for recreational bathing). On Montserrat, West Indies, most of those killed during the Soufrière Hills eruption died on 25 June 1997 when PDCs and surges swept into the exclusion zone, where locals had returned to maintain their farms. Even survivors at the margins of the surge zone suffered serious burns from walking across the hot surge deposits to safety (Loughlin et al., 2002). In most PDC-related deaths, severe burns to the

Horwell, C. Baxter, P. & Kamanyire, R. (2015) Health impacts of volcanic eruptions. In: S.C. Loughlin, R.S.J. Sparks, S.K. Brown, S.F. Jenkins & C. Vye-Brown (eds) *Global Volcanic Hazards and Risk,* Cambridge: Cambridge University Press.

skin (cutaneous) and airways (resulting in pulmonary oedema) cause immediate mortality, or delayed mortality from respiratory complications and infection (Baxter et al., 1982).

3) *Toxicological effects where emissions react with the body.* Gases, ash and aerosols may be inhaled or ingested. A range of potentially-toxic elements may leach from particles. Cases of poisoning have been associated primarily with high levels of bioaccessible fluorine in ash, particularly for livestock which may ingest large quantities of ash during grazing (Cronin et al., 2003). The surfaces of the mineral particles themselves may be reactive in the lung, particularly if the ash is rich in crystalline silica or iron (Horwell et al., 2007, 2012). Potentially toxic elements, in particular fluorine, may present issues in some eruptions if ash contaminates water supplies. However, experience has shown that more common problems include ash blocking and restricting access to livestock drinking water, causing drinking water to become unpalatable and causing water shortages during cleanup operations (Stewart et al., 2006, Wilson et al., 2013).

4) *Electrical impact.* Lightning, generated from friction of particles in the ash plume, may strike people directly or trigger fires.

13.3 Airborne volcanic emissions

Gases. Volcanoes emit hazardous gases (e.g. CO_2, SO_2, H_2S, HF, HCl & radon). Gas exposures may occur during and following eruptions, and during periods of quiescence, and may be proximal or distal to the vent, depending on the size of eruption. Most gas-related deaths occur due to carbon dioxide or hydrogen sulfide pooling in depressions near the volcano, but large eruptions may generate mega-tonnes of SO_2 which can be transported globally and potentially trigger acute respiratory diseases, such as asthma, in exposed populations. Following the 2010 Eyjafjallajökull and 2014 Holuhraun eruptions, the potential for a large, effusive Icelandic eruption, such as the Laki eruption of 1783, is considered a major risk to Europe and is ranked as one of the highest priority risks in the UK National Risk Register with concerns that sulfur dioxide, sulfate aerosols and other gases may have substantial health and environmental impacts. Chronic dental fluorosis has been observed in rural residents of Ambrym island, Vanuatu, and linked to volcanic degassing and contamination of rain-fed drinking water supplies by the volcanic plume (Allibone et al., 2012).

Ash. Whilst ash may cause skin and eye irritation, the primary concern for humans is ash inhalation; the style of eruption and composition of the magma govern the size and composition of the particles which, in turn, control the pathogenic potential of those particles when inhaled. The most hazardous eruptions are those generating fine-grained, crystalline silica rich ash, as silica has the propensity to cause chronic lung disease.

Explosive eruptions generate inhalable ash through fragmentation of magma; the fine particles travel in a plume and, depending on the size of the eruption, ash may fall over wide areas causing disruption and anxiety to populations [see Chapter

Figure 13.1 Ash remobilisation in Yogyakarta following the 2104 Kelud eruption. Photo: Tri Wahyudi

12]. A recent World Health Organization report found that acute and chronic exposures to particles from ambient air pollution, such as PM$_{2.5}$ which can penetrate deep into the lungs, increase both mortality and morbidity (World Health Organisation, 2013). In the volcanic setting, inhalation of fine ash may trigger asthma and other acute respiratory diseases in susceptible people, but chronic effects have not been adequately studied. Active public health precautionary measures will always be needed to protect the population from heavy exposure. Some volcanoes mass-produce crystalline silica in lava domes – viscous lava piles which grow within volcanic craters - which are prone to collapse, generating clouds of fine-grained silica-rich ash. Strict controls to minimise population exposure may be needed because, in industrial settings, crystalline silica causes silicosis, an irreversible and potentially fatal lung disease, and is also classified as a lung carcinogen. This type of eruption, if long lived, may produce frequent ash falls leaving local populations periodically exposed to potentially hazardous levels of ash which, over time, may place them at higher risk of developing silicosis, although presently no cases have been recorded. At Soufrière Hills, Montserrat, West Indies the eruption, which began in 1995 and has lasted for over 15 years, generated dome-collapse ash composed of up to ~25 wt.% crystalline silica (Baxter et al., 2014, Horwell et al., 2014). Stringent and costly clean-up measures after ash falls were maintained by the UK government to protect the population. The first risk assessment of its kind, by Hincks et al., (2006), found that those potentially at greatest risk of developing silicosis, if protective measures were not undertaken, were outdoor workers e.g., gardeners, and children.

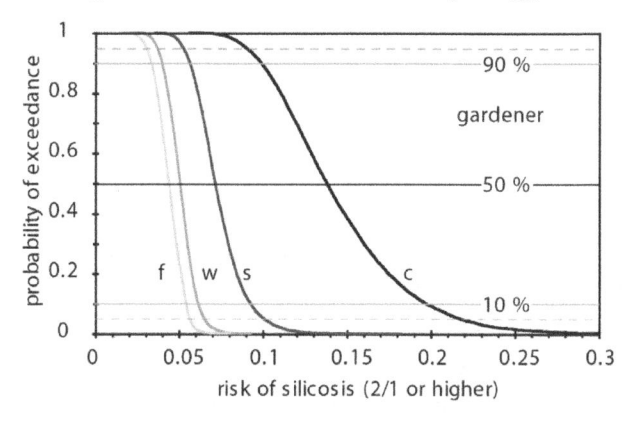

Figure 13.2 Probability of exceedance curve for risk of silicosis (classification ≥2/1) for gardeners, calculated from simulated cumulative exposures (see Hincks et al. 2006 for key to curves, which are for specific Montserrat locations).

13.4 Secondary effects

Large populations brought together in evacuation camps may contract diseases through poor sanitation. Some evacuees may suffer mental stress and other psychological disorders related to displacement and violence is also possible. Widespread ashfall or gas impact (acid rain) may lead to crop failure, loss of livestock and contamination of water supplies which, in turn, may trigger famine and related diseases. Livestock may starve due to smothering of feed and/or if feeding is impaired due to excessive tooth wear as ash is highly abrasive. Ingestion of ash by livestock can also cause fatalities due to intestinal blockages (Wilson et al., 2013). Heavy ashfall can cause roof collapse and is also slippery, making clean-up and driving hazardous. Infrastructure may be impacted, affecting primary healthcare responses [see Chapter 12].

13.5 Hazard impact and response planning

Planning for ash falls and gas release at volcanoes in states of unrest is an essential part of volcano crisis management. Concerns about the health effects of ash and gases may, in the public perception, exceed even volcanologists' warnings on the risks of death from PDCs. Public health officials must be involved in eruption planning and in the response, and work closely with scientists monitoring the volcanic activity. Planning should include setting up and maintaining airborne particulate (PM$_{10}$ and PM$_{2.5}$) monitoring networks (or gaining permission to utilise existing urban networks) so that timely data can be obtained to assess population exposure to airborne respirable ash. National regulatory or WHO guideline limits for particulate pollution (24 hour) are likely to be exceeded for as long as ash is visibly present in the air or on the ground. During and post-eruption, syndromic surveillance of acute respiratory health symptoms is also helpful for informing public health advice and providing reassurance to the public.

The International Volcanic Health Hazard Network (www.ivhhn.org), the umbrella organisation for volcanic health-related research and dissemination, has produced pamphlets and guidelines on volcanic health issues (such as preparing for ashfall) for the public, scientists, governmental bodies and agencies. IVHHN has also developed protocols for rapid characterisation of ash (such as particle size, crystalline silica content, leachate chemistry and basic toxicology) giving timely information to hazard managers during, or soon after, an eruption, to facilitate informed decision-making on health interventions. These analyses have been carried out following recent crises at Rabaul, Eyjafjallajökull, Grímsvötn, Chaitén and Merapi volcanoes (see e.g., Horwell et al. (2013)). IVHHN Expert Members are currently researching the effectiveness of various health interventions used during volcanic crises, such as types of respiratory protection.

It is essential to determine levels of crystalline silica as an urgent priority after a heavy ash-fall, by sending ash samples to a laboratory with experience of undertaking the analysis (e.g. through IVHHN). If raised levels are suspected, confirmation will be needed, which is best done by sending split samples for analysis in different laboratories and under strict scientific protocols. Specialist advice on risk assessment may be needed for reassuring the population and providing guidance on measures to reduce exposure to ash to safe limits, particularly for outdoor workers who may be most exposed, as well as children, who may be most susceptible to developing silicosis. Regular measurement of personal exposure to the ash will need to be undertaken for risk assessment purposes by an experienced team of occupational hygienists.

Even without significant concentrations of crystalline silica, there will be many who suffer acute respiratory symptoms on exposure tovolcanic emissions (gas, ash and aerosol) in the inhaled air. People with asthma and chronic respiratory conditions are most likely to be adversely affected. The public will need advice on limiting their exposure and officials will need to institute measures to remove ash deposits in public areas. Ash and gas may affect areas hundreds of kilometres away from the volcano, and cross national borders, raising public anxiety over air pollution. Fears may also arise over the presence of fluorine and other toxic elements in erupted ash, and the impacts on the environment and animal health, even with fine or sparse deposits from dispersing plumes. Most laboratories are not used to the analyses required to assess the toxic hazard and the main danger is from alarmist, but erroneous, results being disseminated to the public, politicians and media demanding rapid answers.

13.6 Long-lasting versus short-lived eruptions

The most disruptive types of eruption are the continuous, open vent eruptions (e.g., Eyjafjallajökull 2010 which lasted 6 weeks, or Kilauea, which has been ongoing since 1983) or the long-lasting dome growth and collapse type (e.g., Montserrat 1995 to present). On

Montserrat, most people abandoned the island, with the health risk of the ash playing an important part in their decisions; keeping the accumulating ash deposits clear of populated areas to minimise the risks has been a huge and highly costly undertaking (see Baxter et al. 2014, where most of the evidence-base on the health risks of volcanic ash may be found). These contrast with the major one-off eruptions such as Mount St Helens, 1980, and Pinatubo 1991, where visible, re-suspendable ash deposits may persist for many months or even a few years but are not replenished by repeated episodes of emissions or continuous venting of ash and gases with plumes being persistently blown over populated areas by prevailing winds. Another scenario arises where one-off heavy ash falls occur in semi-arid areas, for example, in Patagonia after the large eruptions of Hudson in 1991 and Puyehue Cordón-Caulle in 2011 (Wilson et al., 2013). Abandonment of farm areas and continuing problems with poor air quality in settlements on the steppe occurred due to huge ash deposits downwind of the Andean volcanoes. These deposits continue to be re-suspended by very strong winds, causing economic losses compounded by anxieties over the human health effects of such high, repeated exposures.

An important recent example of the effects of continuous, small scale eruptions in densely populated areas was at the port of Rabaul, Papua New Guinea which, for 2 years (2007- 2008), was subjected to exposure to gases (mainly SO_2), and freshly erupted fine ash from the Tavurvur cone every day for 6 months until the seasonal prevailing winds moved away from the populated area of 70,000 people. In the second year, a drought lasted for 3 months in the middle of the 6 month eruption period; the disruption nearly closed the main hospital, closed schools and many adults and children experienced asthma symptoms, causing the authorities to consider evacuating Rabaul Town. Fortunately, the eruption subsequently stopped and ordinary life resumed. IVHHN analyses, using the rapid analysis protocol, found the ash to be relatively coarse grained and low in crystalline silica (Le Blond et al., 2010). Rabaul is an active caldera and other populated calderas could experience similar small-scale but high impact eruptions due to their persistence, including disrupting infrastructure and transport, especially in modern cities (a potential example is Naples and the currently quiescent Campi Flegrei; see Chapter 6).

Further reading

Baxter, P.J. 2000. Impacts of eruptions on human health. In: H. Sigurdsson, B. Houghton, H. Rymer, J. Stix and S. McNutt (eds) Encylopedia of Volcanoes. Academic Press. pp. 1417.

Hansell, A.L. and Oppenheimer, C., 2004. Health hazards from volcanic gases: a systematic literature review. *Archives of Environmental Health* 59, 628-639.

Horwell, C.J. and Baxter, P.J., 2006. The respiratory health hazards of volcanic ash: a review for volcanic risk mitigation. *Bulletin of Volcanology* 69, 1-24.

References

Allibone, R., Cronin, S. J., Charley, D. T., Neall, V. E., Stewart, R. B. & Oppenheimer, C. 2012. Dental fluorosis linked to degassing of Ambrym volcano, Vanuatu: a novel exposure pathway. *Environmental Geochemistry and Health*, 34, 155-170.

Auker, M. R., Sparks, R. S. J., Siebert, L., Crosweller, H. S. & Ewert, J. 2013. A statistical analysis of the global historical volcanic fatalities record. *Journal of Applied Volcanology*, 2, 1-24.

Baxter, P. J., Bernstein, R. S., Falk, H. & French, J. 1982. Medical aspects of volcanic disasters: an outline of the hazards and emergency response measures. *Disasters,* 6, 268-276.

Baxter, P. J., Searl, A. S., Cowie, H., Jarvis, D. & Horwell, C. J. 2014. Evaluating the respiratory health risks of volcanic ash at the eruption of the Soufrière Hills Volcano, Montserrat, 1995-2000. *In:* Wadge, G., Robertson, R. & Voight, B. *The Eruption of Soufrière Hills Volcano, Montserrat from 2000-2010.* London: Geological Society of London.

Cronin, S. J., Neall, V., Lecointre, J., Hedley, M. & Loganathan, P. 2003. Environmental hazards of fluoride in volcanic ash: a case study from Ruapehu volcano, New Zealand. *Journal of Volcanology and Geothermal Research,* 121, 271-291.

Horwell, C., Baxter, P., Hillman, S., Calkins, J., Damby, D., Delmelle, P., Donaldson, K., Dunster, C., Fubini, B. & Kelly, F. 2013. Physicochemical and toxicological profiling of ash from the 2010 and 2011 eruptions of Eyjafjallajökull and Grímsvötn volcanoes, Iceland using a rapid respiratory hazard assessment protocol. *Environmental Research,* 127, 63-73.

Horwell, C., Hillman, S., Cole, P., Loughlin, S., Llewellin, E., Damby, D. & Christopher, T. 2014. Controls on variations in cristobalite abundance in ash generated by the Soufrière Hills Volcano, Montserrat in the period 1997 to 2010. *Geological Society, London, Memoirs,* 39, 399-406.

Horwell, C. J., Fenoglio, I. & Fubini, B. 2007. Iron-induced hydroxyl radical generation from basaltic ash. *Earth and Planetary Science Letters,* 261, 662-669.

Horwell, C. J., Williamson, B. J., Le Blond, J. S., Donaldson, K., Damby, D. E. & Bowen, L. 2012. The structure of volcanic cristobalite in relation to its toxicity; relevance for the variable crystalline silica hazard. *Particle and Fibre Toxicology,* 9, 44.

Le Blond, J. S., Horwell, C. J., Baxter, P. J., Michnowicz, S. A., Tomatis, M., Fubini, B., Delmelle, P., Dunster, C. & Patia, H. 2010. Mineralogical analyses and in vitro screening tests for the rapid evaluation of the health hazard of volcanic ash at Rabaul volcano, Papua New Guinea. *Bulletin of Volcanology,* 72, 1077-1092.

Loughlin, S., Baxter, P., Aspinall, W., Darroux, B., Harford, C. & Miller, A. 2002. Eyewitness accounts of the 25 June 1997 pyroclastic flows and surges at Soufrière Hills Volcano, Montserrat, and implications for disaster mitigation. *Geological Society, London, Memoirs,* 21, 211-230.

Stewart, C., Johnston, D., Leonard, G., Horwell, C., Thordarson, T. & Cronin, S. 2006. Contamination of water supplies by volcanic ashfall: a literature review and simple impact modelling. *Journal of Volcanology and Geothermal Research,* 158, 296-306.

Wilson, T. M., Stewart, C., Bickerton, H., Baxter, P., Outes, A., Villarosa, G. & Rovere, E. 2013. Impacts of the June 2011 Puyehue-Cordón Caulle volcanic complex eruption on urban infrastructure, agriculture and public health. Wellington: GNS Science.

World Health Organisation 2013. Review of evidence on health aspects of air pollution – REVIHAAP project: final technical report. Online http://www.euro.who.int/__data/assets/pdf_file/0004/193108/REVIHAAP-Final-technical-report-final-version.pdf.

Chapter 14

Volcanoes and the aviation industry

P.W. Webley

There are over 1500 volcanoes around the globe. Of these, 247 have been active, some with multiple eruptions, since the start of commercial airline travel in the 1950s. Guffanti et al. (2010) provide a document on all the volcanic ash – aviation encounters from 1953–2009. They classify the level of encounter by a severity index of 0-5, with no encounters at level 5 and only nine at level 4. Several of the most significant encounters at level 4 occurred in the 1980s. The first two were from the 1982 eruption of Mt. Galunggung volcano, where a B-747 aircraft lost all four engines at an altitude of 11 km above sea level and approx. 150 km from the volcano (Hanstrum & Watson, 1983) and in 1989 from the Redoubt volcanic eruption, Alaska, USA, where another B-747 encountered an ash cloud approx. 150 km from the volcano, at 7.6 km above sea level (Casadevall, 1994).

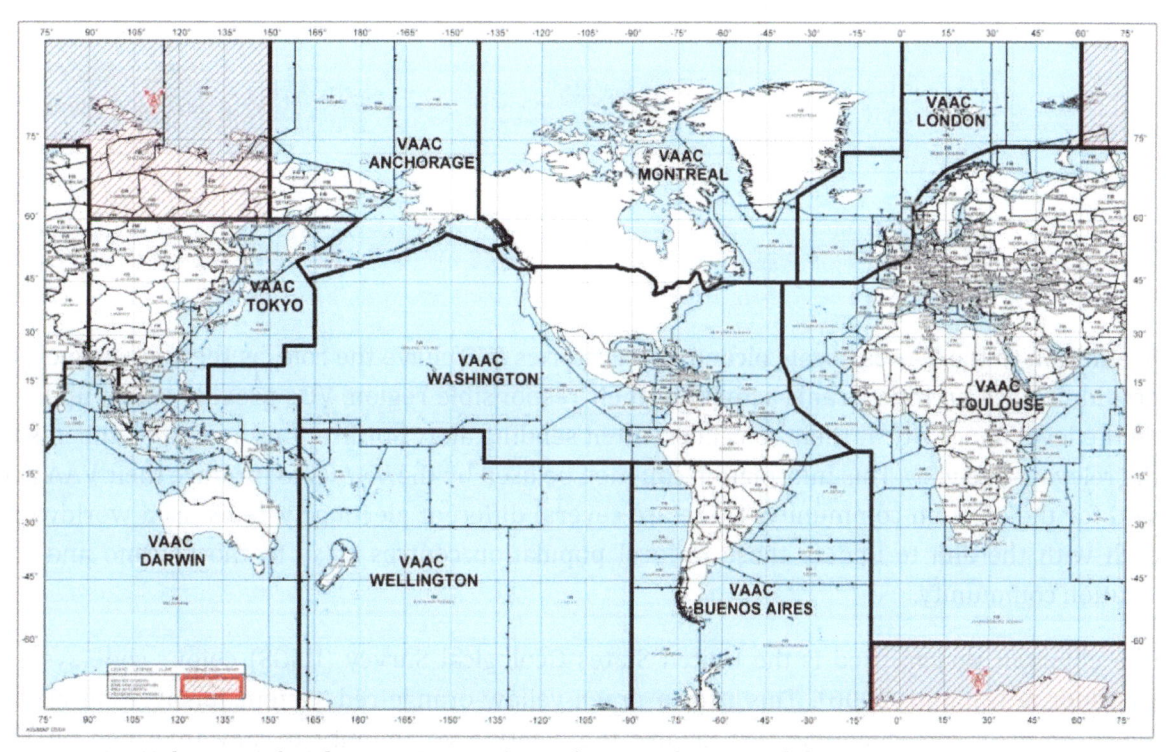

Figure 14.1 Volcanic Ash Advisory Center (VAAC) area of responsibility map, black lines represent the boundaries of each VAAC. For volcanoes in their area, each VAAC will produce the VAA and VAG for the aviation community. (International Civil Aviation Organisation).

Along with those from the 1991 eruption of Mount Pinatubo (Casadevall et al., 1996), lead the International Civil Aviation Organization (ICAO) to set up nine volcanic ash advisory centres or

Webley, P.W. (2015) Volcanoes and the aviation industry. In: S.C. Loughlin, R.S.J. Sparks, S.K. Brown, S.F. Jenkins & C. Vye-Brown (eds) *Global Volcanic Hazards and Risk,* Cambridge: Cambridge University Press.

VAACs (International Civil Aviation Organisation ICAO, 2007). These nine centres each have own areas of responsibility, see Figure 14.1 and are maintained by the local weather service. The nine VAACs are: Anchorage, Montreal, London, Toulouse, Tokyo, Washington, Darwin, Wellington and Buenos Aires. Each of these produce volcanic ash advisories (VAA) and volcanic ash graphics (VAG) for the aviation community, see example in Figure 14.2, and link to the aviation industry through local meteorological watch offices, which produce the Significant Meteorological Information (SIGMET) statements for the aviation.

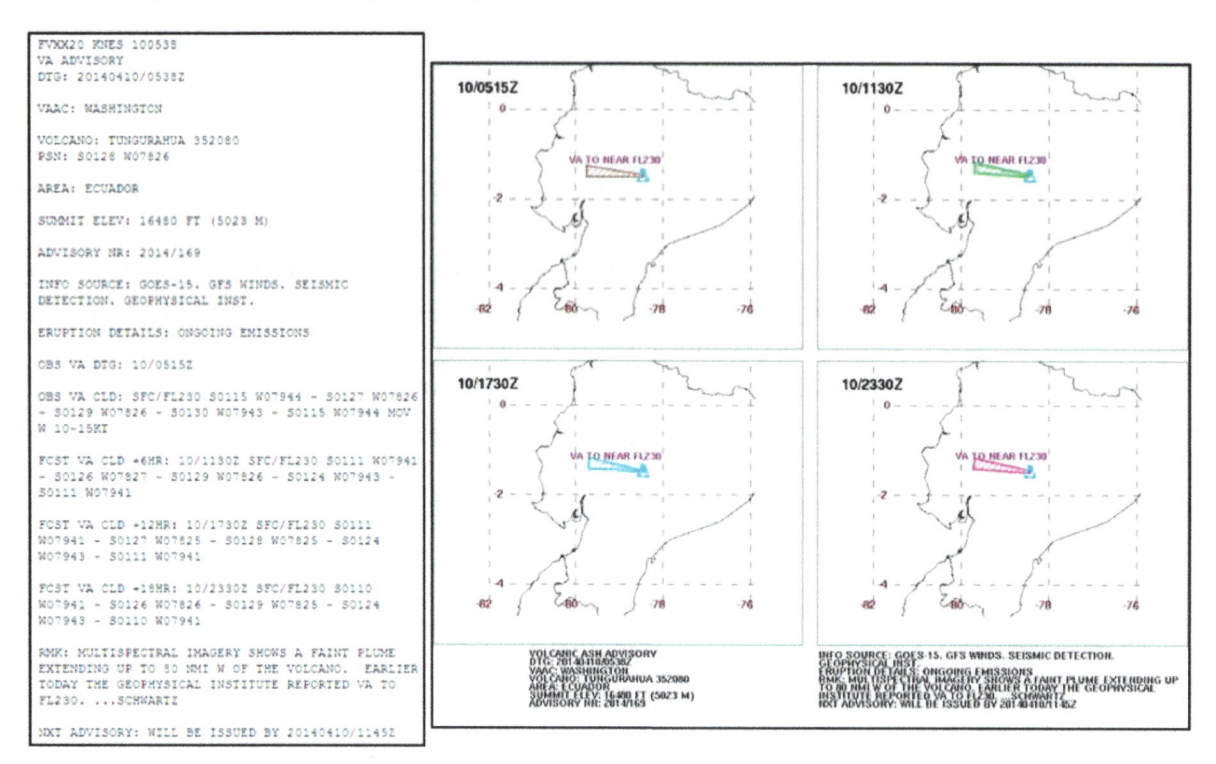

Figure 14.2 Example Volcanic Ash advisory (VAA) and volcanic ash graphic (VAG) for Tungurahua volcano on April 10, 2014 at 05:38 UTC (Z) as produced at the Washington VAAC. From NOAA Satellite and Information Service: NESDIS; http://www.ssd.noaa.gov/VAAC/ARCH14/archive.html

In addition to the VAACs, local volcano observatories (VO) have the role, as the state agency, to provide advice on the volcanic activity in their responsible region. VOs provide status updates on the level of activity at their volcanoes, often sending alert notifications and daily updates to the relevant agencies. This information can then be used by the VAACs to produce their VAA and VAG for the aviation community. There are several different alerting systems used worldwide, each with the aim to update those in local population centres close to the volcano and the aviation community.

One common system used is the United States Geological Survey (USGS) colour code system (Gardner & Guffanti, 2006). This uses a green-yellow-orange-red system for aviation alerts, which with its corresponding text, allows the aviation community to stay informed on the activity levels of the volcano, see Figure 14.3. The system in Figure 14.3 is in accordance with recommended ICAO procedures and is currently used by USGS led volcano observatories (USGS, 2014) as well as the Kamchatka Volcanic Eruption Response Team (Kamchatka Volcanic Eruption Response Team KVERT, 2014) and GeoNet in New Zealand (Geonet, 2014)

Volcano Alert Levels Used by USGS Volcano Observatories

Alert Levels are intended to inform people on the ground about a volcano's status and are issued in conjunction with the Aviation Color Code. Notifications are issued for both increasing and decreasing volcanic activity and are accompanied by text with details (as known) about the nature of the unrest or eruption and about potential or current hazards and likely outcomes.

Term	Description
NORMAL	Volcano is in typical background, noneruptive state or, *after a change from a higher level*, volcanic activity has ceased and volcano has returned to noneruptive background state.
ADVISORY	Volcano is exhibiting signs of elevated unrest above known background level or, *after a change from a higher level*, volcanic activity has decreased significantly but continues to be closely monitored for possible renewed increase.
WATCH	Volcano is exhibiting heightened or escalating unrest with increased potential of eruption, timeframe uncertain, OR eruption is underway but poses limited hazards.
WARNING	Hazardous eruption is imminent, underway, or suspected.

Aviation Color Code Used by USGS Volcano Observatories

Color codes, which are in accordance with recommended International Civil Aviation Organization (ICAO) procedures, are intended to inform the aviation sector about a volcano's status and are issued in conjunction with an Alert Level. Notifications are issued for both increasing and decreasing volcanic activity and are accompanied by text with details (as known) about the nature of the unrest or eruption, especially in regard to ash-plume information and likely outcomes.

Color	Description
GREEN	Volcano is in typical background, noneruptive state or, *after a change from a higher level*, volcanic activity has ceased and volcano has returned to noneruptive background state.
YELLOW	Volcano is exhibiting signs of elevated unrest above known background level or, *after a change from a higher level*, volcanic activity has decreased significantly but continues to be closely monitored for possible renewed increase.
ORANGE	Volcano is exhibiting heightened or escalating unrest with increased potential of eruption, timeframe uncertain, OR eruption is underway with no or minor volcanic-ash emissions [ash-plume height specified, if possible].
RED	Eruption is imminent with significant emission of volcanic ash into the atmosphere likely OR eruption is underway or suspected with significant emission of volcanic ash into the atmosphere [ash-plume height specified, if possible].

Figure 14.3 USGS volcanic activity alert-level notification system, with volcano alert levels from normal, through advisory and watch to warning and aviation colour code from green to red, adapted from Gardner and Guffanti (2006).

Additionally, to these operational groups, many other organisations have put together global meetings such as the World Meteorological Organization-International Union of Geology and Geophysics (WMO-IUGG) first and second workshops on ash dispersal forecast and civil aviation in 2010 and 2013 (World Meteorological Organisation WMO, 2013). Also, ICAO has assembled working groups and task forces such as the 2010-2012 International Volcanic Ash Task Force (IVATF) (International Civil Aviation Organistion ICAO, 2014). This task force was brought together as a focal point and coordinating body of all work related to volcanic ash being carried out by ICAO at global and regional levels.

Four meetings were held from 2010-2012, where members consisted of the VAAC's, aviation community, representatives for the volcano observatories and regulatory bodies. The IVATF provided a summary of recommendations (International Volcanic Ash Task Force IVATF, 2012) centred on science, airworthiness, air traffic management and international airways volcano watch coordination. These recommendations on aviation safety and volcanic watch operations will continue under the ICAO international airways volcano watch operations group (IAVWOPSG) (International Civil Aviation Organisation ICAO, 2014).

Globally, there can be many volcanoes active and potentially hazardous to the aviation industry. Therefore, the VAACs and local volcano observatories work closely together to provide the most effective advisory system and ensure the safety of all those on the ground and in the air. Through the release of VAA, VAG and observatory information notices then timely advisories of the ongoing activity is able to reach the relevant organisation and reduce the potential hazard and provide the best tools to mitigate the risk to all.

References

Casadevall, T. J. 1994. The 1989–1990 eruption of Redoubt Volcano, Alaska: impacts on aircraft operations. *Journal of Volcanology and Geothermal Research,* 62, 301-316.

Casadevall, T. J., Delos Reyes, P. & Schneider, D. J. 1996. The 1991 Pinatubo eruptions and their effects on aircraft operations. *In:* NEWHALL, C. & PUNONGBAYAN, R. S. (eds.) *Fire and Mud: Eruptions and Lahars of Mount Pinatubo, Philippines.*

Gardner, C. A. & Guffanti, M. C. 2006. US Geological Survey's Alert Notification System for Volcanic Activity: US Geological Survey Fact Sheet 2006-3139. Online at pubs.usgs.gov/fs/2006/3139

Geonet. 2014. *Aviation Colour Codes* [Online]. Online. Available: info.geonet.org.nz/display/volc/Aviation+Colour+Codes

Guffanti, M., Casadevall, T. J. & Budding, K. E. 2010. *Encounters of aircraft with volcanic ash clouds: A compilation of known incidents, 1953-2009,* US Department of Interior, US Geological Survey.

Hanstrum, B. & Watson, A. 1983. A case study of two eruptions of Mount Galunggung and an investigation of volcanic eruption cloud characteristics using remote sensing techniques. *Australian Meteorological Magazine,* 31, 131-77.

International Civil Aviation Organisation Icao 2007. Manual on Volcanic Ash, Radioactive Material and Toxic Chemical Clouds. 2nd ed.

International Civil Aviation Organisation Icao. 2014. *International Airways Volcano Watch Operations Group* [Online]. Available: www.icao.int/safety/meteorology/iavwopsg/Pages/default.aspx

International Civil Aviation Organistion Icao. 2014. *International Volcanic Ash Task Force* [Online]. Available: www.icao.int/safety/meteorology/ivatf/Pages/default.aspx

International Volcanic Ash Task Force Ivatf. 2012. *Summary of the accomplishments of the International Volcanic Ash Task Force* [Online]. Available: www.icao.int/safety/meteorology/ivatf/Documents/IVATF.Summary.of.Accomplishments.pdf

Kamchatka Volcanic Eruption Response Team Kvert. 2014. *KVERT: Aviation Color Codes* [Online]. Online. Available: www.kscnet.ru/ivs/kvert/color_eng.php

Usgs. 2014. *USGS Volcanic Activity Alert Notification System* [Online]. Available: volcanoes.usgs.gov/activity/alertsystem/

World Meteorological Organisation Wmo. 2013. *2nd IUGG-WMO workshop on Ash dispersal forecast and aviation* [Online]. Online. Available: www.unige.ch/sciences/terre/mineral/CERG/Workshop2.html.

Chapter 15

The role of volcano observatories in risk reduction

G. Jolly

Volcanic risk reduction is a partnership between science, responding agencies and the affected communities. A critical organisation in the volcanic risk reduction cycle is a volcano observatory (VO), which is an institute or group of institutes whose role it is to monitor active volcanoes and provide early warnings of future activity to the authorities. For each country, the exact constitution of a VO may differ, dependent on the legislative framework for disaster risk reduction and scientific advice to government. For example, in the USA, the Alaska Volcano Observatory is a joint programme of the United States Geological Survey (USGS), the Geophysical Institute of the University of Alaska Fairbanks (UAFGI), and the State of Alaska Division of Geological and Geophysical Surveys (ADGGS), whereas in New Zealand, GNS Science has sole responsibility under the country's Civil Defence Emergency Management Act to provide warnings on volcanic activity and hence provides the function of a volcano observatory.

The responsibilities of a VO also differ from country to country. In some nations, a volcano monitoring organisation may be responsible only for maintaining equipment and ensuring a steady flow of scientific data to an academic or civil protection institution, who then interpret the data or make decisions. In other jurisdictions, the VO may provide interpretations of those data and undertake cutting edge research on volcanic processes. In most cases a VO will provide volcanic hazards information such as setting Volcanic Alert Levels and issuing forecasts of future activity, and in some instances, a VO may even provide advice on when civil actions should take place such as the timing of evacuation. Some of the VOs have responsibility for multiple volcanoes, whereas others may only monitor and provide advice on a single volcano. In some countries an academic institute may fulfil both the monitoring and research function for a volcano.

This wide range of potential roles and responsibilities demonstrates the importance of a VO function, but also shows that there is no single template for the constitution of a VO. However, it is critically important that governments recognise the need for volcano monitoring, provide adequate resourcing and have clear definitions of roles for VOs, academic institutions, civil protection agencies and other key players for the pathway for issuing warnings.

A critical VO role is to provide information to Volcanic Ash Advisory Centres (VAACs). It is stated under regulations of the International Civil Aviation Organisation that states should maintain VOs that monitor pre-eruptive activity and eruptions themselves and provide

Jolly, G. (2015) The role of volcano observatories in risk reduction. In: S.C. Loughlin, R.S.J. Sparks, S.K. Brown, S.F. Jenkins & C. Vye-Brown (eds) *Global Volcanic Hazards and Risk,* Cambridge: Cambridge University Press.

information on the activity to VAACs, Meteorological Watch Offices and air traffic control authorities.

There are over 100 VOs around the world to monitor ca. 1551 volcanoes considered to be active or potentially active. Many of the VOs are members of the World Organisation of Volcano Observatories (WOVO; www.wovo.org). WOVO is a commission of the International Association of Volcanology and Chemistry of the Earth's Interior (IAVCEI) that aims to co-ordinate communication between VOs and to advocate enhancing volcano monitoring around the globe. WOVO is an organisation of and for VOs of the world and has three co-leaders for each of the following regions: Asia-Pacific, Americas, Europe/Africa. WOVO organises or co-sponsors meetings, workshops and conference sessions that focus on issues of importance for VOs. It also co-ordinates information exchange between the VOs. One of the main recent roles of WOVO has been to enhance communication between VOs and Volcanic Ash Advisory Centres.

To be able to monitor their volcanoes effectively, VOs potentially have a very wide suite of tools available to them. However, the range of the capability and capacity of VOs globally is enormous. Many active volcanoes have no monitoring whatsoever, whereas some VOs in developed countries may have hundreds of sensors on a single volcano [see Chapter 19]. This leads to major gaps in provision of warnings of volcanic activity, particularly in developing countries.

Monitoring programmes typically include: tracking the location and type of earthquake activity under a volcano; measuring the deformation of the ground surface as magma intrudes a volcano; sampling and analysing gases and water being emitted from the summit and flanks of a volcano; observing volcanic activity using webcams and thermal imagery; measurements of other geophysical properties such as electrical conductivity, magnetism or gravity. VOs may have ground-based sensors measuring these data in real-time or they may have staff undertaking campaigns to collect data on a regular basis (e.g. weekly, monthly, annually). Some VOs may also the capability to collect and analyse satellite data.

Volcano seismicity is the fundamental backbone for early warnings of eruption. Magma and fluid movement inside volcanoes create a variety of seismic signals. A typical progression of seismicity preceding and eruption starts with rock-breaking earthquakes or volcano-tectonic signals as magma starts to move upwards inside the volcano. As magma, gases or hydrothermal fluids get forced through cracks, the earthquakes change their character reflecting resonance or repetitiveness of the source of the signals. Eruptions and their products (pyroclastic density currents or lahars) also produce diagnostic signals.

There are many different ways of measuring the shape of the earth; traditional surveying methods such as levelling or electronic distance measurement can be used on a campaign basis, although most VOs now use continuous data collection using Global Navigation Satellite Systems or electronic tilt to provide real-time or near real-time sub-millimetre measurements of the location of points around a volcano. Satellite measurements [Chapter 17], particularly Synthetic Aperture Radar, also allow wide spatial coverage of a deforming volcano although this imagery normally has a return period of weeks to months.

Volcanoes emit many different gas species as the magma rises from depth; the magma also interacts with hydrothermal systems, groundwater or surface water and can change the chemistry and physical properties of existing water bodies. Measuring these changes can provide early warning of magma on the move.

The main gas species that are associated with magma are water, sulfur dioxide and carbon dioxide, although many other gases can be measured, especially halogens, and other carbon and sulfur species. Scientists have devised a wide range of techniques to measure these gases, especially sulfur dioxide and carbon dioxide. Water vapour is very difficult to measure as there is already an abundance in the atmosphere. Very few gas monitoring techniques provide real-time data at a comparable rate to measurements of seismicity or ground deformation, although this is a rapidly developing field and over the next 5-10 years, it is likely that such instrumentation will exist.

Changes in water chemistry and properties are relatively simple to measure in real-time, such as the temperature of a crater lake or the pH of a hot spring, although volcanic environments are commonly highly acidic and/or hot and thus it is difficult to maintain sensors for extended periods as they can be destroyed very easily.

Satellite observations can help with both gas emissions and physical properties of water bodies. For example, the Ozone Monitoring Instrument (OMI) onboard the NASA Aura satellite has the ability to monitor large emissions of sulfur dioxide; a variety of satellite platforms provide thermal monitoring with varying degrees of temporal or spatial resolution. However, not all VOs have either the capability or capacity to access or interpret satellite information.

Observations of volcanic activity are critical during eruptive activity to provide information to the VO on how an eruption is progressing. For example, observations of ash cloud heights are vital for VAACs so they can accurately model the dispersion of ash for aviation. Oftentimes, VOs have web cameras at various points around a volcano to provide visual or thermal imagery back to the monitoring scientists. In some cases, ground-based radar can image eruption clouds. Remote imagery is especially important if the VO is located at some distance from a volcano.

Definition of what constitutes an appropriate level of monitoring has received some attention over the last few years (Ewert et al., 2005, Miller & Jolly, 2014). One approach that has been used is to assess broadly the risk associated with each volcano [see Chapter 23], and allocate more resources to those volcanoes that pose a higher risk. This approach has been extended in this submission to GAR15. In some countries, it may be difficult to provide monitoring for all the volcanoes in its jurisdiction; however, consideration should be made of maintaining at least some minimal monitoring for the high-risk volcanoes through all periods of quiescence. Volcanoes exhibit fluctuations in their background levels of activity and it is difficult to recognise what constitutes unrest that may lead to an eruption if the VO does not understand the long-term behaviour of the volcano fully.

One aspect of VOs that is often underestimated is ensuring that a VO has sufficient human resource to develop and maintain monitoring expertise. Oftentimes, VOs have the ability to purchase capital items, or equipment is provided in an emergency situation through aid donors, but if there is insufficient staffing and/or an ongoing operational budget, the monitoring can quickly fall into disrepair in the recovery period after an eruption. It is important that VOs are resourced sustainably, so that they can maintain at least a minimum level of monitoring through periods with little or no activity, and that they have the ability to call on additional support during crises to bolster a long-term core capability. This can be achieved by partnerships with other organisations either in-country, for example, academic institutions that may be able to provide students for routine monitoring tasks such as collecting ash, or overseas, for example, GNS Science has a long-term partnership with Vanuatu Meteorology and Geohazards Department. In both cases, there has to be clear understanding of the responsibilities of the VO and any other organisations assisting the VO or undertaking research on the volcano (Newhall et al., 1999).

The process for mitigating risk differs from country to country, but in essence, communities need to understand their risk and take action to mitigate risk (by avoiding, minimising or accepting the risks). This is commonly illustrated through a series of steps in a risk management cycle, such as risk reduction (e.g. using planning legislation to prevent people from living in high risk areas), readiness (e.g. having contingency plans in place so that different parts of the community know what they should do in a crisis), response (e.g. reacting to increased activity by evacuating areas) and recovery (e.g. cleaning up after ashfall). VOs play a role in all aspects of risk management.

VOs are often involved in outreach activities in times of volcanic quiet so that the authorities and the communities can better understand the potential risk from their volcano(es); this may also involve regular exercising with civil protection agencies to test planning for eruption responses.

During the lead up to an eruption, VOs may provide regular updates on activity which inform decisions on evacuations or mitigation actions to reduce risk to people or to critical infrastructure. For example, power transmission companies may choose to shut off high voltage lines if there is a high probability of ashfall. They may also assist organisations in developing contingency plans by providing possible eruption scenarios.

During an eruption, VOs will then provide up-to-date information about the progression of activity. For an explosive eruption, information might include the duration, the height that ash reaches in the atmosphere and areas being impacted on the ground. This can inform decisions such as search and rescue attempts or provide input to ash dispersion forecasts for aviation.

After an eruption has ceased, VOs can aid recovery through advice about ongoing hazards such as remobilisation of ash deposits during heavy rainfall. They can also assist or undertake valuable research on the eruption through collection of time-perishable data. This can lead to better understanding of the volcano so that future responses can be fine-tuned.

The role of VOs is critical in reducing risk from volcanoes, both on the ground and in the air. Volcanic risk reduction can only improve if VOs are adequately resourced by national government. If adequate volcano monitoring is established ahead of a volcanic crisis, the VO can provide a wide range of information to responding agencies and to the potentially affected communities, both in quiescence and during response; ultimately, this results in better preparedness and enhanced safety of people and infrastructure.

References

Ewert, J. W., Guffanti, M. C. & Murray, T. 2005. An Assessment of the Volcanic Threat and Monitoring Capabilities in the United States: Framework for a National Volcano Early Warning System, Open File Report 2005-1164. Online.

Miller, C. A. & Jolly, A. D. 2014. A model for developing best practice volcano monitoring: a combined threat assessment, consultation and network effectiveness approach. *Natural Hazards,* 71, 493-522.

Newhall, C., Aramaki, S., Barberi, F., Blong, R., Calvache, M., Cheminee, J.-L., Punongbayan, R., Siebe, C., Simkin, T., Sparks, R. S. J. & Tjetjep, W. 1999. Professional conduct of scientists during volcanic crises. *Bulletin of Volcanology,* 60, 323-334.

Chapter 16

Developing effective communication tools for volcanic hazards in New Zealand, using social science

G. Leonard and S. Potter

16.1 Background

Social science plays an increasing and valuable role in volcanic Disaster Risk Management (DRM); social science research methods are now used globally to investigate and improve the links amongst volcanology, emergency mangement and community resilience to volcanic hazards. The biennial IAVCEI Cities on Volcanoes Conferences, each hosted by an international city at risk from volcanic hazards, held its eighth meeting in Yogyakarta (Indonesia) in September 2014. These meetings attract large attendances of social and physical scientists as well as emergency managers and DRM practitioners. By incorporating social science methodologies, information derived from volcano monitoring and data interpretation can be used in the most effective way possible to reduce the risk of volcanic hazards to society.

A range of New Zealand researchers at universities, and the government earth science research institute GNS Science, have been conducting applied social research focussed around natural hazards for nearly 20 years, spearheaded by studies of the impacts of the 1995/96 eruptions of Ruapehu volcano. In 2006 the national Joint Centre for Disaster Research was established, a joint venture between Massey University School of Psychology and GNS Science. It includes researchers from other universities and agencies and undertakes multi-disciplinary applied teaching and research aimed at gaining a better understanding of the impacts of disasters on communities, improving the way society manages risk, and enhancing community preparedness, response and recovery from the consequences of hazard events. Researchers also focus on the effective communication of likelihoods for volcanic eruption forecasts (Doyle et al., 2014). Three projects are highlighted here as examples of volcanic hazard focussed research within this collaborative national social science framework.

16.2 Development of a revised Volcanic Alert Level system

The communication of scientific information to stakeholders is a critical component of an effective Volcano Early Warning System. Volcanic Alert Level (VAL) systems are used in many countries as a tool to communicate complex volcanic information in a simple form, from which response decisions can be made. Communication tools such as these are required to meet the needs of a wide range of stakeholders, including central government, emergency managers, the aviation industry, media and the public. They are also required to be usable by the scientists who determine the levels based on volcano observations and interpretation of complex monitoring data.

Leonard, G. & Potter, S. (2015) Developing effective communication tools for volcanic hazards in New Zealand, using social science. In: S.C. Loughlin, R.S.J. Sparks, S.K. Brown, S.F. Jenkins & C. Vye-Brown (eds) *Global Volcanic Hazards and Risk,* Cambridge: Cambridge University Press.

A recent research project by Potter et al. (2014) involved the exploration of New Zealand's 20-year old VAL system. For the first time globally, a new VAL system was developed based on a robust qualitative ethnographic methodology, which is commonly used in social science research (e.g. Patton (2002)). The research involved interviews of scientists and stakeholders, document analysis, and observations of scientists over three years at GNS Science as they set the VAL during multiple unrest and eruption crises. The data resulting from the interviews underwent thematic analysis, which involves grouping comments made by participants into themes. The findings were triangulated against the document analysis and observation data to produce a draft new VAL system. The draft system then went through multiple iterations with stakeholders and scientists, until a final version acceptable to all interested parties was formed.

The new VAL system, which is presented in Figure 16.1, was integrated into the Ministry of Civil Defence and Emergency Management's Guide to the National Civil Defence and Emergency Management plan in 2014. For more information on New Zealand's VAL system, visit www.geonet.org.nz/volcano. The methodology utilised in this trans-disciplinary research is applicable worldwide, and potentially could be used to develop warning systems for other hazards.

New Zealand Volcanic Alert Level System

Volcanic Alert Level	Volcanic Activity	Most Likely Hazards
Eruption 5	Major volcanic eruption	Eruption hazards on and beyond volcano*
Eruption 4	Moderate volcanic eruption	Eruption hazards on and near volcano*
Eruption 3	Minor volcanic eruption	Eruption hazards near vent*
Unrest 2	Moderate to heightened volcanic unrest	Volcanic unrest hazards, potential for eruption hazards
Unrest 1	Minor volcanic unrest	Volcanic unrest hazards
0	No volcanic unrest	Volcanic environment hazards

An eruption may occur at any level, and levels may not move in sequence as activity can change rapidly.

Eruption hazards depend on the volcano and eruption style, and may include explosions, ballistics (flying rocks), pyroclastic density currents (fast moving hot ash clouds), lava flows, lava domes, landslides, ash, volcanic gases, lightning, lahars (mudflows), tsunami, and/or earthquakes.

Volcanic unrest hazards occur on and near the volcano, and may include steam eruptions, volcanic gases, earthquakes, landslides, uplift, subsidence, changes to hot springs, and/or lahars (mudflows).

Volcanic environment hazards may include hydrothermal activity, earthquakes, landslides, volcanic gases, and/or lahars (mudflows).

*Ash, lava flow, and lahar (mudflow) hazards may impact areas distant from the volcano.

This system applies to all of New Zealand's volcanoes. The Volcanic Alert Level is set by GNS Science, based on the level of volcanic activity. For more information, see geonet.org.nz/volcano for alert levels and current volcanic activity, gns.cri.nz/volcano for volcanic hazards, and getthru.govt.nz for what to do before, during and after volcanic activity. Version 3.0, 2014.

Figure 16.1 New Zealand's new Volcanic Alert Level system. The most up-to-date system is always accessible via www.geonet.org.nz/volcano.

16.3 Lahar hazard mitigation at Mt Ruapehu

Research into public awareness of, and response to, lahar warnings at one of New Zealand's major ski areas situated on the active Ruapehu volcano has been conducted annually for over a decade. Lahars have travelled through the ski area in multiple eruptions in the last 50 years (e.g., Figure 16.2 image). Visitors are required to evacuate from lahar-prone valleys immediatly following a siren and voice announcement automatically triggered by eruption sensors. They have as little as two minutes to move to safety and individual and group behaviour amongst visitors and ski area staff must be immediate, decisive and correct. Social research includes the following:

(a) Annual assessment of public and staff responses to simulated events, including truly 'blind' exercises where both staff and the public are un-aware that the warning is an exercise. This has full support from the tourism company operating the ski area, and the Department of Conservation with primary risk management responsibility for the world heritage status national park within which Ruapehu sits.
(b) Awareness surveys of volcanic hazards, recall of education material and messages, and correct actions to take in a warning.
(c) Organisational psychology research into staff behaviour and training needs analysis for specific roles.
(d) Analysis of potential education media and contact points to improve public response to warnings.
(e) Surveys of the demographics of the public who continue to not respond to warnings during exercises, to further direct educational resources.

All of this has indicated potential actions that could be taken to improve future responses, such as increasing ski area staff training (Christianson, 2006) and improving hazard signage (Leonard et al., 2008). It has also lead to technical improvements in hardware performance, audibility and messaging.

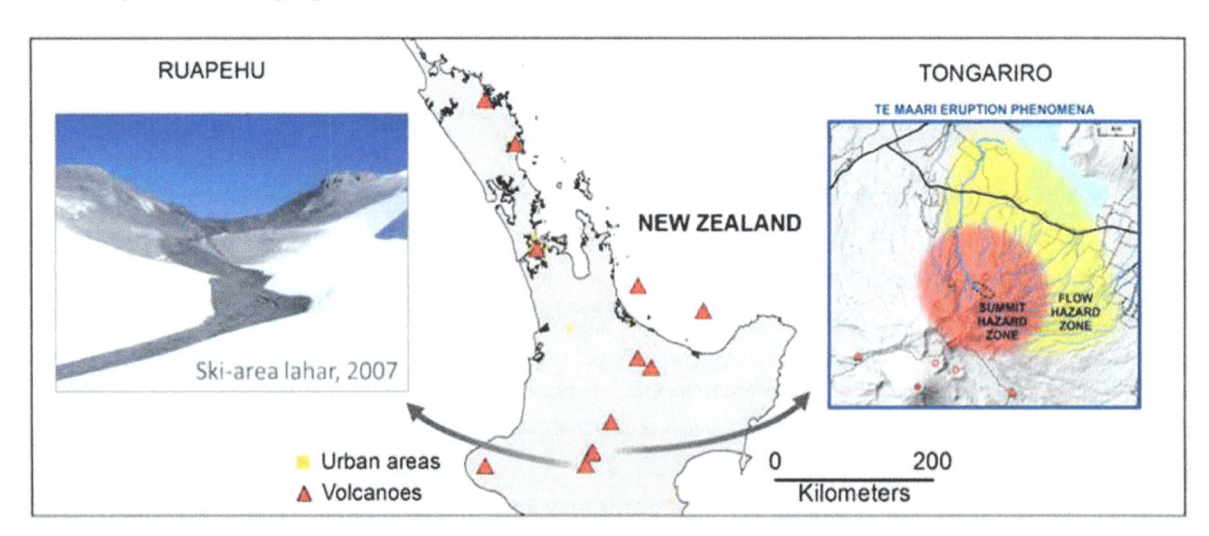

Figure 16.2 Volcanoes in New Zealand, including a photo of a ski-area lahar at Ruapehu, and the hazard map for Tongariro. The comprehensive Tongariro hazard map can be found at http://gns.cri.nz/Home/Learning/Science-Topics/Volcanoes/Eruption-What-to-do/Hazard-maps.

By repeating this research annually and tracking perceptions of visitors through time in response to real events, the communication tools continue to be improved. Surveys demonstrate that tourists appreciate that the hazard is monitored, warned for and that education materials are visible – leading to strong industry support. A design and communications research project is currently underway to direct new education initiatives to the specific demographics of people seen not responding to warnings.

16.4 Tongariro hazard maps

Social research into the perceptions of volcanic hazards and education materials supported the creation of a new volcanic crisis hazard map for eruptions at Mt. Tongariro in 2012 (Figure 16.2; Leonard et al. (2014)). The area impacted by the eruptions included a section of the popular

Tongariro Alpine Crossing walking track, which has nearly 100,000 people passing annually within less than 3 km of the 2012 vent. Requirements of tourists, concessionaires and local residents were considered alongside scientific modelling and geological information, as well as core messages from emergency management agencies, to produce an effective collaborative communication product.

The crisis map had to accommodate several complex issues:

(i) background hazard maps are used across the many potentially active vents during non-eruptive periods, but these may not match crisis hazard maps and scenarios with very elevated probability compared to the background;

(ii) the scientists' need for conservatism while constraining hazards that were initially in conflict with more probable short-term hazards in time-sensitive situations;

(iii) hazards tend to grade away spatially and should ideally be shown in a gradual probabilistically defined way, but maps need to be simple;

(iv) messaging covers several severe hazards and actions, needing to be a balance between simplicity to achieve high awareness and not clutter the map, but enough detail to be meaningful; and

(v) the visual representation of elements (i) through (iv) on a single piece of paper that can be quickly and correctly comprehended.

Ongoing social research results from Tongariro (Coomer and Leonard, 2005) and Ruapehu were applied to help guide an optimum solution in the face of these issues. International research, especially around the effective presentation of hazard maps (Haynes et al., 2007), and the development of trust amongst scientists, emergency managers and the public (e.g. Barclay et al. (2008), Johnston et al. (1999), Paton et al. (2008)) was also applied.

References

Tongariro background and crisis hazard map latest versions can be found in full at: gns.cri.nz/Home/Learning/Science-Topics/Volcanoes/Eruption-What-to-do/Hazard-maps

Barclay, J., Haynes, K., Mitchell, T., Solana, C., Teeuw, R., Darnell, A., Crosweller, H. S., Cole, P., Pyle, D. & Lowe, C. 2008. Framing volcanic risk communication within disaster risk reduction: finding ways for the social and physical sciences to work together. *Geological Society, London, Special Publications,* 305, 163-177.

Christianson, A. N. 2006. *Assessing and improving the effectiveness of staff training and warning system response at Whakapapa and Turoa ski areas, Mt. Ruapehu.* MSc, University of Canterbury.

Coomer, M. & Leonard, G. 2005. *Tongariro crossing hazard awareness survey: public perceptions of the volcanic hazard danger*, Institute of Geological & Nuclear Sciences.

Doyle, E. E., Mcclure, J., Johnston, D. M. & Paton, D. 2014. Communicating likelihoods and probabilities in forecasts of volcanic eruptions. *Journal of Volcanology and Geothermal Research,* 272, 1-15.

Haynes, K., Barclay, J. & Pidgeon, N. 2007. Volcanic hazard communication using maps: an evaluation of their effectiveness. *Bulletin of Volcanology,* 70, 123-138.

Johnston, D. M., Lai, M. S. B. C.-D., Houghton, B. F. & Paton, D. 1999. Volcanic hazard perceptions: comparative shifts in knowledge and risk. *Disaster Prevention and Management,* 8, 118-126.

Leonard, G. S., Johnston, D. M., Paton, D., Christianson, A., Becker, J. & Keys, H. 2008. Developing effective warning systems: ongoing research at Ruapehu volcano, New Zealand. *Journal of Volcanology and Geothermal Research,* 172, 199-215.

Leonard, G. S., Stewart, C., Wilson, T. M., Procter, J. N., Scott, B. J., Keys, H. J., Jolly, G. E., Wardman, J. B., Cronin, S. J. & Mcbride, S. K. 2014. Integrating multidisciplinary science, modelling and impact data into evolving, syn-event volcanic hazard mapping and communication: A case study from the 2012 Tongariro eruption crisis, New Zealand. *Journal of Volcanology and Geothermal Research.*

Paton, D., Smith, L., Daly, M. & Johnston, D. 2008. Risk perception and volcanic hazard mitigation: Individual and social perspectives. *Journal of Volcanology and Geothermal Research,* 172, 179-188.

Patton, M. Q. 2002. *Qualitative Research and Evaluation Methods,* Thousand Oaks, CA, Sage.

Potter, S. H., Jolly, G. E., Neall, V. E., Johnston, D. M. & Scott, B. J. 2014. Communicating the status of volcanic activity: revising New Zealand's volcanic alert level system. *Journal of Applied Volcanology,* 3, 1-16.

Chapter 17

Volcano monitoring from space

M. Poland

17.1 Defining the problem

Unlike many natural hazards, volcanoes usually give warnings of impending eruptions that can be detected from hours to years prior to any hazardous activity (Sparks et al., 2012). The Eyjafjallajökull eruption, for example, was preceded by several discrete episodes of subsurface magma accumulation that highlighted the potential for future eruption (Gudmundsson et al., 2010). Once it begins, an eruption can last for decades, during which time the changing conditions of associated hazards, such as ash plumes and lava flows, must be continuously assessed. Unfortunately, the resources and infrastructure needed to conduct ground-based monitoring of a volcano - especially those located in remote areas of Earth that might still have the potential to impact air traffic, like in the north Pacific (Figure 17.1) are extreme, and less than 10% of the world's volcanoes are monitored in any systematic way (Bally, 2012). Space-based methods offer a means of bridging this monitoring gap.

Figure 17.1 2009 eruption of Sarychev Peak, Kuril Islands, seen from the International Space Station (courtesy NASA).

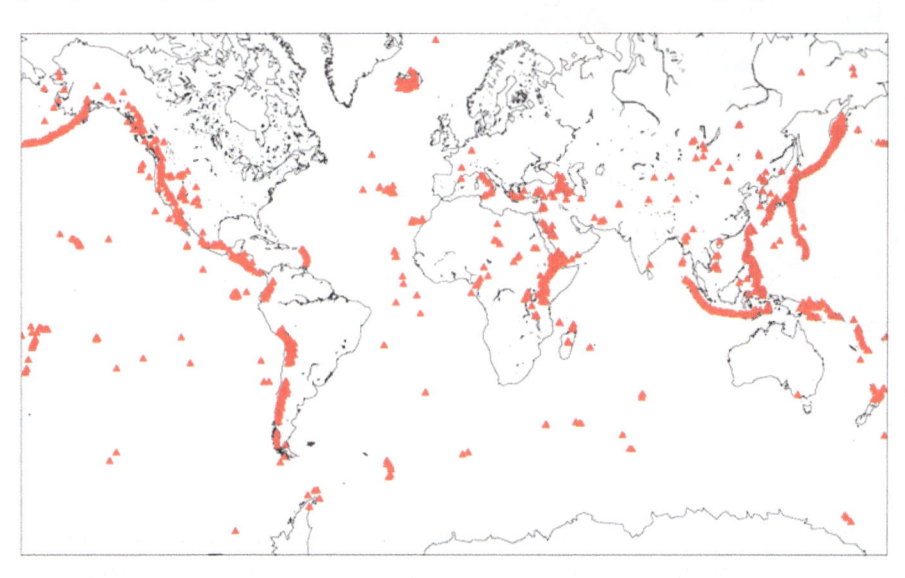

Figure 17.2 Global map of volcanoes that have erupted within the past 10,000 years (volcano locations from Siebert et al. 2010).

Poland, M. (2015) Volcano monitoring from space. In: S.C. Loughlin, R.S.J. Sparks, S.K. Brown, S.F. Jenkins & C. Vye-Brown (eds) *Global Volcanic Hazards and Risk,* Cambridge: Cambridge University Press.

17.2 Early warnings

Prior to eruption, most volcanoes provide an indication that magma is ascending towards the surface - most notably through seismicity, ground deformation, thermal anomalies and elevated gas emissions (Sparks et al., 2012). While earthquake monitoring remains rooted to the terrestrial domain, other expressions of rising magma can be detected from space. Ground deformation, for example, can be mapped by synthetic aperture radar interferometry and sometimes occurs even before earthquake swarms, providing warning of restless volcanoes that might erupt within months to years (Dzurisin, 2003). Unlike ground-based monitoring networks, satellites acquire data over broad swaths, enabling regional surveillance of volcanoes. Observations of entire volcanic arcs - for example, the Andes of South America (Pritchard & Simons, 2004) - have identified numerous volcanoes that are deforming but not erupting (Figure 17.3). Remote sensing therefore provides leverage for more time- and cost-effective deployment of ground-based resources to the volcanoes that are most likely to erupt and expose population and infrastructure to hazards.

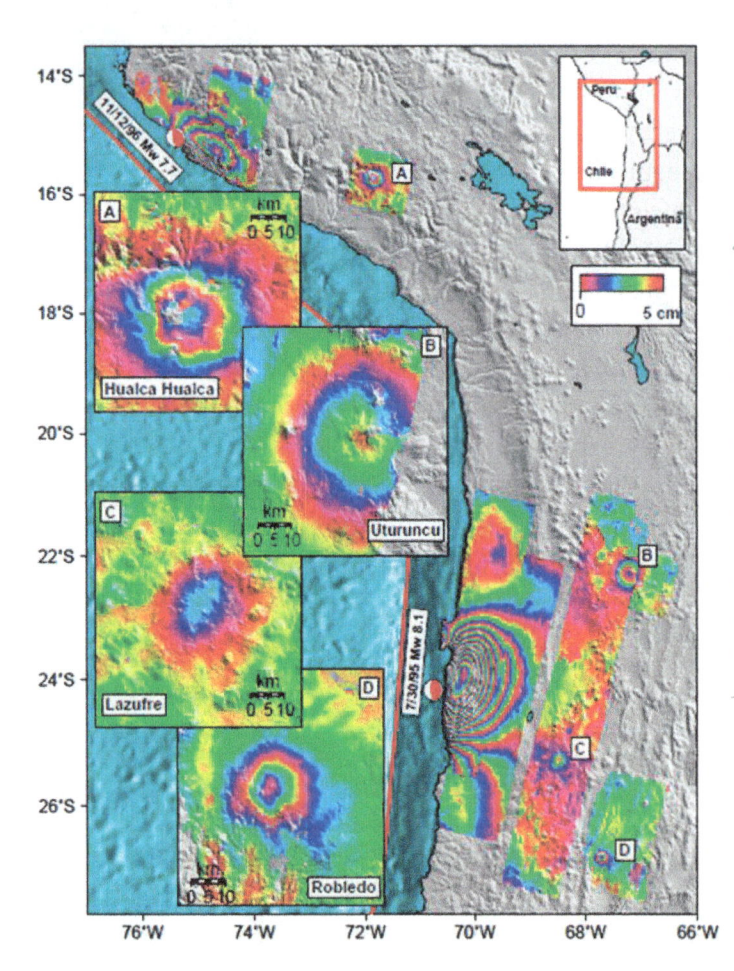

Figure 17.3 Colour contours of ground deformation draped over shaded relief from three subduction zone earthquakes along the coast and four volcanic centres in Peru, Bolivia, Argentina, and Chile. Each contour corresponds to 5 cm of deformation in the radar line-of-sight direction. Inset maps show higher-resolution interferograms at the four centres of active deformation. Reference map in upper right corner places study area in regional context. Modified from Pritchard and Simons [2004].

Remote-sensing data, primarily from meteorological satellites, also often provide the first indication of unrest or eruptive activity at a volcano. Thermal emissions are tracked by a number of satellite systems, many of which acquire images multiple times per day of the same area on the ground. Near-real-time automated analysis of thermal data is an important alerting tool for eruption onsets and is currently being applied at volcanoes around the world, especially in remote areas (Dehn et al., 2000, Wright et al., 2004). The Alaska Volcano Observatory relies on meteorological sensors for detecting thermal anomalies that can represent precursory heating or lava extrusion, ash clouds, and gas emissions across the 52 historically active volcanoes of the several-thousand-kilometre-long Aleutian volcanic arc, which, although sparsely populated, is heavily traversed by passenger and cargo aircraft (Schneider et al., 2000). Similarly, daily remote sensing of Kamchatkan and Kurile volcanoes by Russian volcanological authorities is in many cases the sole means of surveillance of dozens of volcanoes that can threaten trans-Pacific aircraft (Neal et al., 2009). At volcanoes that are not monitored by ground-based instruments, satellite data provide the best indication of eruptions that may pose hazards to air travel.

17.3 Tracking volcanic hazards

Once an eruption is in progress, timely and repeated satellite data are critical for tracking the evolution of volcanic hazards. Synthetic Aperture Radar (SAR), with meter-scale resolution and ability to "see" through clouds and ash, can detect changes on the surface that might be obscured from thermal/optical satellite data and ground observers. Near-real-time analysis of SAR imagery aided the decision to maintain evacuation zones around Merapi, Indonesia, in 2010 - a decision that likely saved several thousand lives when that volcano experienced a large eruption (Pallister et al., 2013). Tracking ash clouds is best accomplished using space-based observations (Pavolonis et al., 2013) and can be used with forecast models to warn downwind communities of impending ash fall and alert air traffic of ash location (Figure 17.4). Worldwide, Meteorological Watch Offices and Volcano Ash Advisory Centres rely on satellite data for issuing SIGMETs and Volcanic Ash Advisories, which report and forecast ash distribution. The evolution of thermal anomalies in satellite data can track lava and pyroclastic flows, and volcanic gas emissions - an under-appreciated hazard to downwind communities - can be imaged from orbit with better spatial resolution and at lower cost than from the ground (so2.gsfc.nasa.gov/), allowing for automated alerts of volcanic plumes.

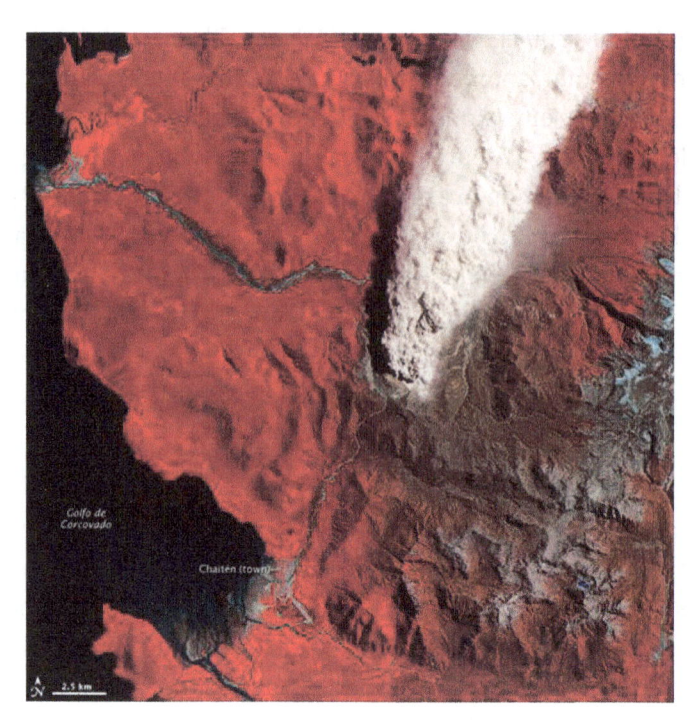

Figure 17.4 False-colour visible-infrared image acquired by the Aster satellite over Chaiten volcano, Chile, on 19 January 2009. Vegetation is red, bare or ash-covered ground is grey/brown, ash is light brown, and water is blue. Image shows a thick plume of ash and gas extending to the NNE of the volcano. Image courtesy of NASA (http://earthobservatory.nasa.gov/IOTD/view.php?id=36725).

17.4 Bridging the gap

A variety of satellite sensors have repeatedly demonstrated their ability to monitor volcanic unrest, detect eruption onsets, and track eruptive hazards; nevertheless, remote sensing is not yet a globally operational tool for volcano monitoring - some data are costly and not available in near-real time, and resources needed utilise satellite imagery are lacking in many countries. In addition, volcano-monitoring operations are rarely 24/7, so real-time alerting systems, in addition to data management and visualisation software, are required to exploit the volume of potentially available data. To bridge this monitoring, low- or no-cost data are needed with low temporal latency and adequate spatial resolution. In addition, data management systems and capacity-building must be developed to ensure broad use of those data for volcano hazard mitigation and good collaboration across international boundaries where hazards pose more than a local risk. The implementation of this vision, shared throughout the geohazards community, is underway through a number of multi-national projects dedicated to use of remote sensing data for natural hazards risk reduction, including:

- The 2012 "International Forum on Satellite EO and Geohazards," which articulated the vision for volcano monitoring from space (www.int-eo-geo-hazard-forum-esa.org/);
- The Geohazard Supersites and Natural Laboratories initiative, which aims to reduce loss of life from geological disasters through research using improved access to multi-disciplinary Earth science data (supersites.earthobservations.org/);
- The European Volcano Observatory Space Services (EVOSS), which has the goal of providing near-real-time access to gas, thermal, and deformation data from satellites at a number of volcanoes around the world (www.evoss-project.eu/);
- The Disaster Risk Management volcano pilot project of the Committee on Earth Observation Satellites (CEOS), which is designed to demonstrate how free access to a diversity of remote sensing data over volcanoes can benefit hazards mitigation efforts

Ultimately, realisation of this vision will depend on commitments from national governments and space agencies to make these data available for disaster risk reduction purposes.

References

Bally, P. 2012. Scientific and technical memorandum of the international forum on satellite EO and geohazards, 21–23 May 2012. *Santorini, Greece. doi,* 10, 5270.

Dehn, J., Dean, K. & Engle, K. 2000. Thermal monitoring of North Pacific volcanoes from space. *Geology,* 28, 755-758.

Dzurisin, D. 2003. A comprehensive approach to monitoring volcano deformation as a window on the eruption cycle. *Reviews of Geophysics,* 41.

Gudmundsson, M. T., Pedersen, R., Vogfjörd, K., Thorbjarnardóttir, B., Jakobsdóttir, S. & Roberts, M. J. 2010. Eruptions of Eyjafjallajökull Volcano, Iceland. *Eos, Transactions American Geophysical Union,* 91, 190-191.

Neal, C., Girina, O., Senyukov, S., Rybin, A., Osiensky, J., Izbekov, P. & Ferguson, G. 2009. Russian eruption warning systems for aviation. *Natural hazards,* 51, 245-262.

Pallister, J. S., Schneider, D. J., Griswold, J. P., Keeler, R. H., Burton, W. C., Noyles, C., Newhall, C. G. & Ratdomopurbo, A. 2013. Merapi 2010 eruption—Chronology and extrusion rates monitored with satellite radar and used in eruption forecasting. *Journal of Volcanology and Geothermal Research,* 261, 144-152.

Pavolonis, M. J., Heidinger, A. K. & Sieglaff, J. 2013. Automated retrievals of volcanic ash and dust cloud properties from upwelling infrared measurements. *Journal of Geophysical Research: Atmospheres,* 118, 1436-1458.

Pritchard, M. E. & Simons, M. 2004. Surveying volcanic arcs with satellite radar interferometry: The central Andes, Kamchatka, and beyond. *GSA Today,* 14, 4-11.

Schneider, D., Dean, K., Dehn, J., Miller, T. & Kirianov, V. Y. 2000. Monitoring and analyses of volcanic activity using remote sensing data at the Alaska Volcano Observatory: case study for Kamchatka, Russia, December 1997 *In:* Mouginis-Mark, P. J., Crisp, J. A. & Fink, J. H. *Remote Sensing of Active Volcanism, AGU Mongraph 116.*

Siebert, L., Simkin, T. & Kimberley, P. 2010. *Volcanoes of the World, 3rd edn,* Berkeley, University of California Press.

Sparks, R., Biggs, J. & Neuberg, J. 2012. Monitoring volcanoes. *Science,* 335, 1310-1311.

Wright, R., Flynn, L. P., Garbeil, H., Harris, A. J. & Pilger, E. 2004. MODVOLC: near-real-time thermal monitoring of global volcanism. *Journal of Volcanology and Geothermal Research,* 135, 29-49.

Chapter 18

Volcanic unrest and short-term forecasting capacity

J. Gottsmann

18.1 Background

Most volcanic eruptions are preceded by a period of volcanic unrest that perhaps is best defined as the deviation from the background or baseline behaviour of a volcano towards a behaviour which is a cause for concern in the short-term because it might prelude an eruption (Phillipson et al., 2013).

Although it is important that early on in a developing unrest crisis scientists are able to decipher the nature, timescale and likely outcome of volcano reawakening following long periods of quiescence there are still major challenges when assessing whether unrest will lead to an eruption in the short-term or wane with time.

18.2 Analysis of volcanic unrest

An analysis of 228 cases of reported volcanic unrest between 2000 and 2011 (Phillipson et al. (2013); Figure 18.1) recognises five primary observational (predominantly geophysical and geochemical) indicators of volcanic unrest:

Ground deformation: Restless volcanoes often undergo periods of ground uplift or subsidence driven for example by pressure changes in their magma reservoir or overlying geothermal reservoir. In some cases pressure increase may break the ground surface. Ground deformation is generally recorded by ground or space-borne techniques [see also Chapter 17].

Degassing: Plumes of gas may be released from craters or other vents (fumaroles) on a volcanic edifice craters. Alternatively the amount of gas released may increase or the chemical composition of gases may change over time. Ground and space-borne techniques are usually applied to monitor degassing behaviour [see also Chapter 17].

Changes at a crater lake: These changes include variations in lake temperature, lake levels, level of water chemistry, lake colour and gas release and are generally recorded using ground-based or air-borne techniques.

Thermal anomaly: Anomalous temperature changes of the ground or of fumarolic gases can be recorded by ground-based, air or space-borne sensors [see also Chapter 17].

Gottsmann, J. (2015) Volcanic unrest and short-term forecasting capacity. In: S.C. Loughlin, R.S.J. Sparks, S.K. Brown, S.F. Jenkins & C. Vye-Brown (eds) *Global Volcanic Hazards and Risk,* Cambridge: Cambridge University Press.

Seismicity: The movement of magma, fluids and gas can cause seismic signals at restless volcanoes as does the breaking of rock from stress increases at depth. Particular seismic wave forms are generated from such processes which may provide clues as to what is driving unrest at a particular volcano. Seismic observations are generally made on the ground.

The same study also recognises five idealised classes of volcanic unrest based on the temporal behaviour of these five most-commonly reported unrest indicators which can be depicted in unrest timelines and whether or not eruptive behaviour resulted from the unrest. These classes of unrest include reawakening, prolonged, pulsatory, sporadic and intra-eruptive unrest. An example of pulsatory unrest is shown in Figure for the case of Cotopaxi volcano in Ecuador. This volcano underwent a non-eruptive period of unrest during 2001 and 2003. Pulsatory unrest consists of episodes of unrest activity (lasting for days) separated by intervals of days to weeks without activity. In contrast, prolonged unrest is often expressed by long-term (years to decades) ground deformation, which may only be identifiable at volcanoes with a long-term geodetic monitoring network or satellite remote sensing.

Phillipson et al. (2013) also showed that unrest episodes at different types of volcanoes have different median unrest durations before the start of an eruption. At stratovolcanoes they last for a few weeks while at calderas their median length is two months. Shield volcanoes have the longest median unrest duration at 5 months. However, volcanoes with long periods of quiescence between eruptions do not necessarily undergo long periods of unrest before their next eruption.

To improve the knowledge-base on volcanic unrest, a globally validated protocol for the reporting of volcanic unrest (Newhall and Dzurisin, 1988) and archiving of unrest data is needed (Venezky and Newhall, 2007). Such data are important for the short-term forecasting of volcanic activity amid technological and scientific uncertainty and the inherent complexity of volcanic systems.

Figure 18.1 Location maps of 228 volcanoes with reported unrest between January 2000 and July 2011. Green circles show volcanoes with unrest not followed by eruption within reporting period, while red triangles show those with eruption.

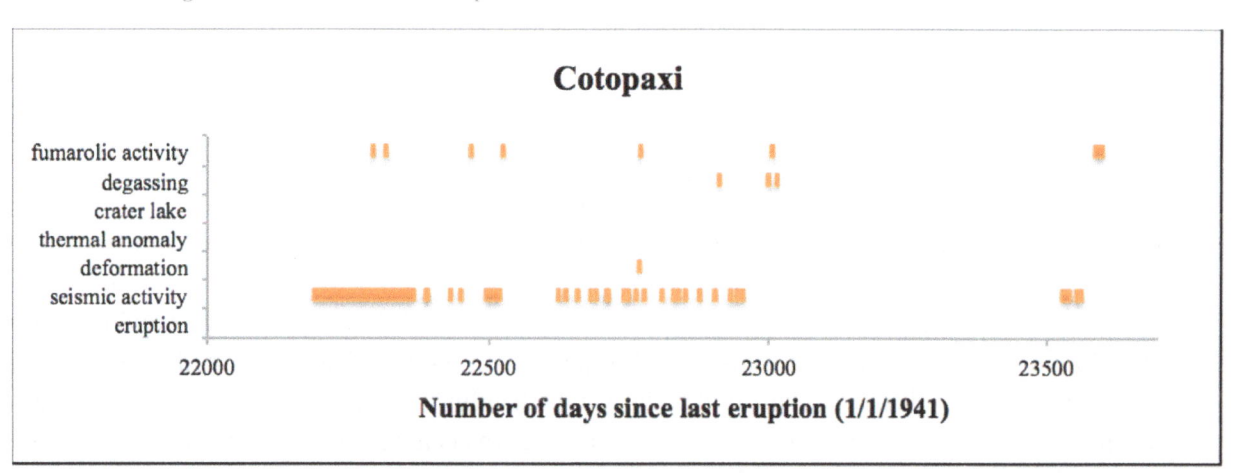

Figure 18.2 An example of a pulsatory unrest timeline using five key unrest indicators from Cotopaxi volcano in Ecuador from 27/3/2001 to 21/11/2005 (modified from Phillipson et al. (2013)). This episode of intense unrest did not culminate in an eruption.

18.3 Short-term forecasting capacity

Forecasting the outcomes of volcanic unrest requires the use of quantitative probabilistic models (Marzocchi et al., 2008, Marzocchi and Bebbington, 2012, Sobradelo et al., 2014) to address adequately intrinsic (epistemic) uncertainty as to how a unrest process may evolve as well as aleatory uncertainty regarding the limited knowledge about the process. Probabilistic forecast models applied in modern volcanology follow event tree structures which allow conditional probabilities to be attributed to different possible future eruptive or non-eruptive scenarios in an evolving unrest crisis.

An example of individual notes of an event tree structure of the HASSET (Sobradelo et al., 2014) probabilistic forecasting tool is shown in Figure . This event tree includes unrest scenarios that culminate in an eruption but also those that do not. For the probabilistic assessment of outcomes of volcanic unrest it is particularly important to assess a number of scenarios regarding the causes of unrest such as magma movement, geothermal excitation, tectonic activity or other processes. It is crucial to discriminate between unrest caused by internal triggers (magma movement) or by external triggers (regional tectonics), which ultimately condition the outcome and further development of unrest.

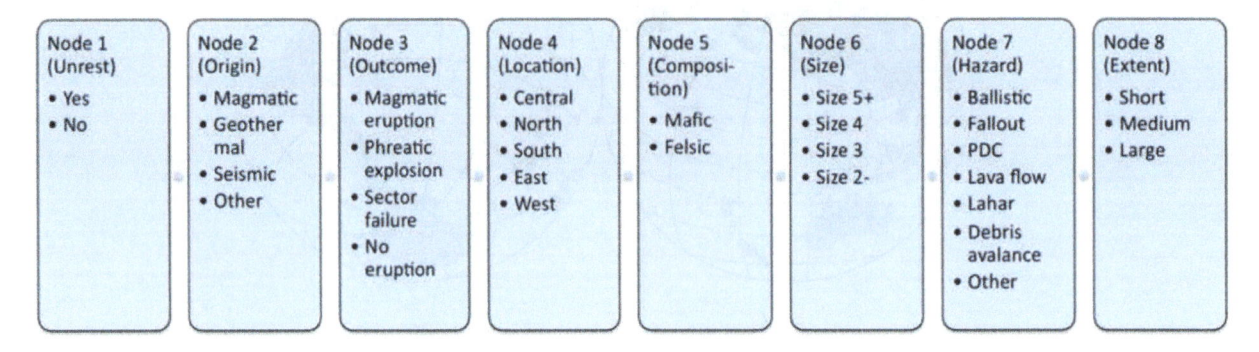

Node 1 (Unrest)	Node 2 (Origin)	Node 3 (Outcome)	Node 4 (Location)	Node 5 (Composition)	Node 6 (Size)	Node 7 (Hazard)	Node 8 (Extent)
• Yes • No	• Magmatic • Geothermal • Seismic • Other	• Magmatic eruption • Phreatic explosion • Sector failure • No eruption	• Central • North • South • East • West	• Mafic • Felsic	• Size 5+ • Size 4 • Size 3 • Size 2-	• Ballistic • Fallout • PDC • Lava flow • Lahar • Debris avalance • Other	• Short • Medium • Large

Figure 18.3 Event tree structure of the HASSET probabilistic forecasting tool (from Sobradelo et al. (2014)) formed by eight individual nodes and corresponding mutually exclusive and exhaustive branches. By the condition of independence of the nodes, the probability of a particular volcanic scenario, as a combination of branches across nodes, is the product of the individual probabilities of occurrence of each branch in that scenario.

Probabilistic forecasts may influence selection of appropriate mitigation actions based on informed societal or political decision-making. Properly addressing uncertainties is particularly critical for managing the evolution of a volcanic unrest episode in high-risk volcanoes, where mitigation actions require advance warning and incur considerable costs (Marzocchi and Woo, 2007). A major evacuation over a period of 4 months in excess of 70,000 individuals on Guadeloupe in the French West Indies in 1976 [see also Chapter 8] was initiated as a result of abnormal levels of volcanic background activity, which culminated in a series of eruptions of hot gas, mud and rock, before waning. Fortunately, no life was claimed by the activity, however, the estimated cost of the unrest was about US$ 1bn in current currency. Ninety percent of these costs were incurred by the evacuation, rehabilitation and salvage of the French economy. This in turn suggests that had the outcome of the unrest on Guadeloupe been predicted correctly the cost of the unrest would have been almost negligible. At the same time it is now acknowledged that the "proportion of evacuees who would have owed their lives to the evacuation, had there been a major eruption, was substantial" (Woo, 2008).

An improvement of the knowledge base on causes and consequences of volcanic unrest is shared the geohazards community. A number of multi-national projects are dedicated to working towards a better understanding of volcanic unrest including:

- The VUELCO project (www.vuelco.net), a European Commission-funded project on volcanic unrest in Europe and Latin America.

- The European Commission-funded MEDSUV (http://med-suv.eu) and FUTUREVOLC (http://futurevolc.hi.is) projects as part of the Geohazard Supersites and Natural Laboratories initiative (supersites.earthobservations.org/).
- The WOVOdat database, which enables the comparison of volcanic unrest data using time-series and geo-referenced data from volcano observatories worldwide in common and easily accessible formats (www.wovodat.org).

References:

Marzocchi, W. & Bebbington, M. S. 2012. Probabilistic eruption forecasting at short and long time scales. *Bulletin of Volcanology,* 74, 1777-1805.

Marzocchi, W., Sandri, L. & Selva, J. 2008. BET_EF: a probabilistic tool for long-and short-term eruption forecasting. *Bulletin of Volcanology,* 70, 623-632.

Marzocchi, W. & Woo, G. 2007. Probabilistic eruption forecasting and the call for an evacuation. *Geophysical Research Letters,* 34.

Newhall, C. G. & Dzurisin, D. 1988. Historical unrest at large calderas of the world. *US Geological Survey Bulletin*, 1-1108.

Phillipson, G., Sobradelo, R. & Gottsmann, J. 2013. Global volcanic unrest in the 21st century: an analysis of the first decade. *Journal of Volcanology and Geothermal Research,* 264, 183-196.

Sobradelo, R., Bartolini, S. & Martí, J. 2014. HASSET: a probability event tree tool to evaluate future volcanic scenarios using Bayesian inference. *Bulletin of Volcanology,* 76, 1-15.

Venezky, D. & Newhall, C. 2007. WOVOdat Design Document; The Schema, Table Descriptions, and Create Table Statements for the Database of Worldwide Volcanic Unrest.

Woo, G. 2008. Probabilistic criteria for volcano evacuation decisions. *Natural Hazards,* 45, 87-97.

Chapter 19

Global monitoring capacity: development of the Global Volcano Research and Monitoring Institutions Database and analysis of monitoring in Latin America

N. Ortiz Guerrero, S.K. Brown, H. Delgado Granados and C. Lombana Criollo

19.1 Background

Volcanic eruptions can cause loss of life and livelihoods, damage critical infrastructure and have long-term impacts, including displaced populations and long-lasting economic implications. Many factors contribute to disasters from natural hazards. One of these is the institutional capacity to enable hazard assessment for pre-emergency planning to protect populations and environments, provide early warning when volcanoes threaten to erupt, to provide forecasts and scientific advice during volcanic emergencies, and to support post-eruption recovery and remediation. Volcano observatories play a critical role in supporting communities to reduce the adverse effects of eruptions [Chapter 15]. Their capacity to monitor volcanoes is thus a central component of disaster risk reduction.

The resources are not available for extensive monitoring of all 596 historically active volcanoes. The availability of resources varies on local, national, regional and global scales, resulting in highly variable monitoring levels from volcano to volcano. Some countries have observatories dedicated to volcano monitoring, others monitor from within larger organisations, and still others have no permanent monitoring group. Individual volcanoes may have large comprehensive monitoring networks of multiple monitoring systems whilst a neighbouring volcano is unmonitored.

It is therefore vital to understand the monitoring capacity at local, national, regional and global scales to establish how well volcanoes are monitored, the distribution of monitoring equipment, the human resources, experience and education and the instrumental and laboratory capabilities. To this end a database has been developed: Global Volcano Research and Monitoring Institutions Database (GLOVOREMID).

Ortiz Guerrero, N., Brown, S.K., Delgado Granados, H. & Lombana Criollo, C. (2015) Global monitoring capacity: development of the Global Volcano Research and Monitoring Institutions Database and analysis of monitoring in Latin America. In: S.C. Loughlin, R.S.J. Sparks, S.K. Brown, S.F. Jenkins & C. Vye-Brown (eds) *Global Volcanic Hazards and Risk,* Cambridge: Cambridge University Press.

19.2 GLOVOREMID

In 2011 IAVCEI funded the development of VOMODA (Volcano Monitoring Database), whose main purpose was to obtain a realistic diagnosis of volcano monitoring and training of the human resources working on volcanological research and monitoring institutions (VRMI) in Latin America. In 2013, VOMODA was adopted and adapted for worldwide use as GLOVOREMID. The Global Volcano Model (GVM) supports this work. It is currently in both Spanish and English. This database will contribute to improving communication and cooperation between scientists and technicians responsible for volcano monitoring and may help to reduce the effects of volcanic crises. GLOVOREMID can be accessed online via http://132.248.182.158/glovoremid/.

19.2.1 Database development

The structure of GLOVOREMID was designed using a relational model. This consists of a set of tables and links that maintain information related to: volcanoes, VRMI, instrumentation and human resources responsible for volcanic surveillance. The development of the tables and relations in GLOVOREMID was completed under the normalisation method, which is a process of organizing data to minimise redundancy (Kendall & Kendall, 2010).

In order to achieve compatibility of GLOVOREMID with other existing volcanological databases, principally the Volcanoes of the World database of the Smithsonian Institution (VOTW4.0, Siebert et al. (2010)) and the Large Magnitude Explosive Volcanic Eruptions database (LaMEVE, Crosweller et al. (2012)) of the Volcano Global Risk Identification and Analysis Project (VOGRIPA), the same volcano identification codes are used and relevant data were transferred from these databases into GLOVOREMID.

For the development and implementation of GLOVOREMID KumbiaPHP Framework (Comunidad KumbiaPHP, 2012), PHP language and MYSQL engine were used. Model View Controller (MVC) was used for the architectural pattern giving a natural code organisation (De la Torre, 2010). All views were developed with HTML5 and JAVASCRIPT. The system works as follows: a query comes from browser to controller; the controller interacts with the model that is able to make data transactions directly to the engine database. Finally, the controller sends data in order to visualise it using a view (Figure 19.1).

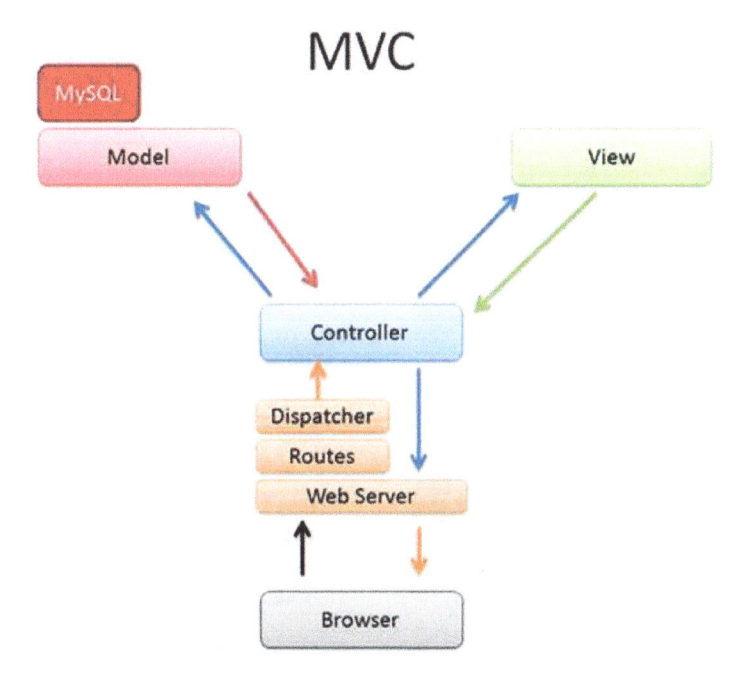

Figure 19.1 Model view controller pattern in GLOVOREMID.

GLOVOREMID is hosted on a server at the Instituto de Geofísica (UNAM). After development, the VRMI data were collected and entered into the database. Multiple users can be authorised and those working within VRMI are being given access. It is these users who are responsible for data updates. GLOVOREMID is anticipated as a global, sustainable database, accessible to and updated by those involved with volcano research and monitoring, to allow better communication and collaboration between scientists, to highlight knowledge gaps and areas where funding, training and equipment should be prioritised and perhaps even facilitate the sharing of equipment with un- or under-monitored regions as activity develops. GLOVOREMID is in the early stages of population globally, but, as it is expanding from VOMODA, is well populated for Latin America.

19.3 Monitoring in Latin America

VOMODA was developed as part of the IAVCEI project "Weaknesses and strengths in Latin America facing volcanic crises: a research for improvement of national capabilities", and hence focussed on countries of Latin America.

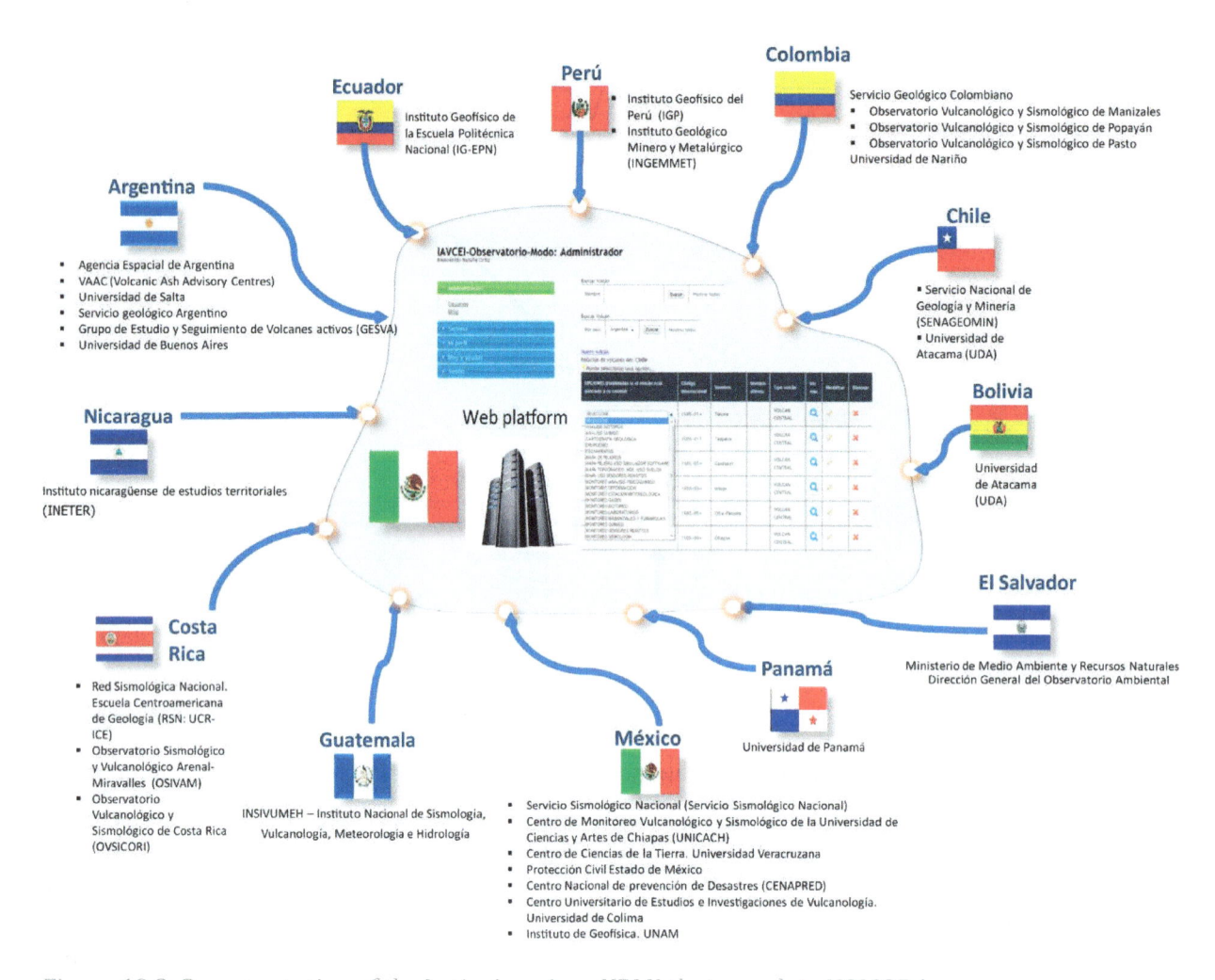

Figure 19.2 Representation of the Latin American VRMI that populate VOMODA.

Volcanoes with known or suspected Holocene activity, as recorded in VOTW4.0, are included in the database. Additional as yet unidentified volcanoes, or volcanoes with few studies or infrequent activity may also require further research or monitoring. Where volcanoes lie on the border between two countries some may be monitored by one or more VRMI. The VRMI responsible for the volcanoes are identified and were contacted to join the database and provide monitoring information.

There are many methods for monitoring volcanoes, many of which are widespread. On local scales institutions may favour particular monitoring methods or derive their own methods using the resources available to them. The database allows the recording of many types of instrumentations and methods, and can be expanded to include new methods.

To determine the monitoring level for each volcano three main lines of monitoring were chosen: seismology, deformation and gas. Monitoring levels were chosen of 0-5 based on the use of these three methods. A volcano with no seismic, deformation or gas monitoring is classed at Level 0. Level 1 is assigned when using only seismic stations. Level 2 is assigned when the volcano is monitored with seismic stations and at least one deformation station, and increasing levels represent increasing deformation and gas stations. A very well monitored volcano is that of Level 5, indicative of seismology, deformation and gas monitoring through multiple stations

(Figure 19.3). For example, Level 5 could represent a volcano with a seismic network, GPS station, EDM line, SO$_2$ and CO$_2$ monitoring.

Monitoring levels of volcanoes in Latin America

Monitoring Level	Lines of Monitoring
Level 0	Not monitored
Level 1	Seismology
Level 2	Seismology + Deformation (1)
Level 3	Seismology + Deformation (1) + Gases (1)
Level 4	Seismology + Deformation (2) + Gases (1)
Level 5	Seismology + Deformation (2) + Gases (2)

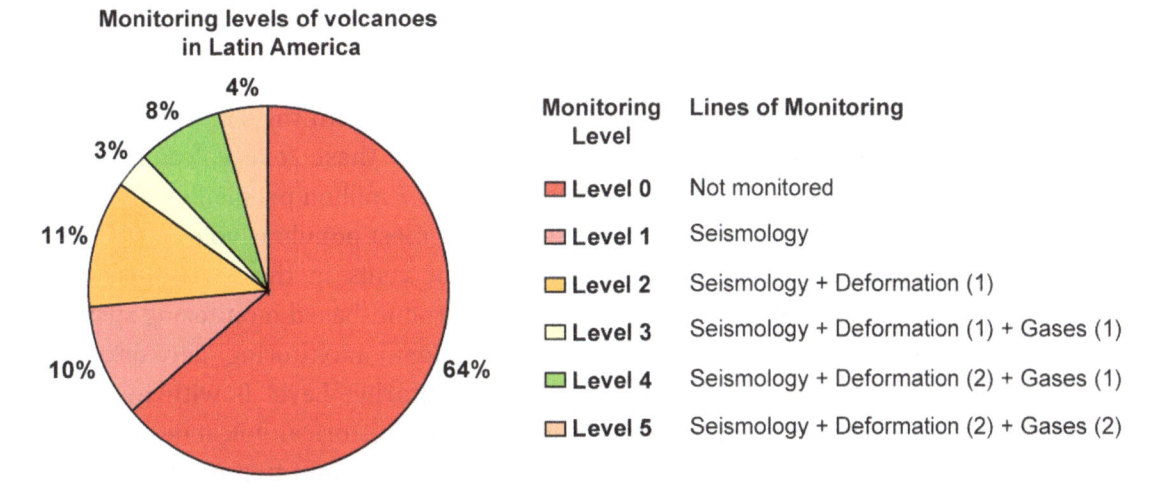

Figure 19.3 Monitoring levels for 314 Holocene volcanoes in Latin America and assignment of monitoring levels.

There are 314 Holocene volcanoes in Latin America (across the regions of Mexico and Central America and South America). Of these, 159 have confirmed eruptions recorded during the Holocene in VOTW4.0, 113 of which have confirmed historical activity. It is intuitive that a correlation between the age of eruptions and the monitoring level may exist.

Table 19.1 The number of volcanoes across Latin America that classify with each level of monitoring.

Monitoring level	Number of Latin American volcanoes	% of total Latin American volcanoes	% of monitored Latin American volcanoes
0	202	64%	-
1	30	10%	27%
2	35	11%	31%
3	10	3%	9%
4	24	8%	21%
5	13	4%	12%

There are 202 Latin American volcanoes that classify as Level 0 (i.e. unmonitored; Table 19.1). Mexico and Chile have the largest number of unmonitored volcanoes (32 and 50, respectively). These countries host the largest number of volcanoes in Latin America. 64% of Chilean volcanoes and 82% of Mexican volcanoes are unmonitored. Three countries have no *on-site* monitoring at any of their volcanoes: Argentina, Bolivia and Honduras, however, volcanoes on the Chile-Bolivia and Chile-Argentina borders are monitored. Neither Bolivia nor Honduras has recorded Holocene eruptions, with the exception of volcanoes along the Bolivia and Chile border. Four volcanoes in Argentina, excluding those on the border with Chile, have confirmed eruptions as recently as 1988.

In Latin America, 86% of unmonitored volcanoes have no recorded historic eruptions. However, 30 unmonitored volcanoes have 95 eruptions recorded between 1505 and 2008 AD (Table 19.2). These eruptions ranged in magnitude from VEI 0-5, with four volcanoes producing five large explosive VEI ≥4 eruptions in this time (Cerro Azul in Ecuador, VEI 5 eruption of 1916; Michoacán-Guanajuato in Mexico with two VEI 4 eruptions in 1759 and 1943; Carrán-Los Venados in Chile with the VEI 4 eruption of 1955; Chaitén in Chile with the VEI 4 eruption of 2008). There are populations living within 100 km distance of these four volcanoes, with Population Exposure Indices (PEI) of 2-7 [see Chapter 4]. Over 5.7 million people live within 10 km of Michoacán-Guanajuato volcanic field, ranking this as the most populous volcano (10 km) worldwide, however this is due to the wide distribution of vents in the ~50,000 square kilometre volcanic field. This volcanic field currently has no ground-based monitoring systems specifically designed for volcano monitoring, however regional monitoring networks are available. A further four Latin American volcanoes are Monitoring Level 0 with historical activity and high PEI levels of 5 to 7. Most of the unmonitored historical volcanoes have no hazard classification [see Chapter 22], with just three Hazard Level I and two Hazard Level II volcanoes; risk levels are unclassified for those 25 unmonitored historically active volcanoes with no hazard classification, whilst the majority of classified volcanoes in this group fall in the Risk Level I category (Table 19.2).

Table 19.2 Latin American volcanoes with Monitoring Levels of 0 (unmonitored) shown with their hazard level (see CS19) and PEI (see CS1). The top section shows those volcanoes with a classified hazard level; the historically active volcanoes are shown in **bold** and the warming of the background colours indicates increasing risk levels. Those volcanoes with no hazard classification are shown in the lower section, where historically active volcanoes are shown in section U-HHR; volcanoes with a Holocene record but no historical activity are shown in U-HR; the number of volcanoes with no confirmed Holocene records are shown under U-NHHR.

CLASSIFIED

Hazard III							
Hazard II						Yucamane	
Hazard I		**Cerro Azul; Lautaro; Wolf**					

UNCLASSIFIED

U – HHR	**Robinson Crusoe; Bárcena; Socorro**	**Huanquihue Group; Putana; Olca-Paruma; Pinta; Viedma; Fueguino; Sumaco; Burney, Monte; Arenales; Darwin; Marchena; Irruputuncu; Tromen; Llullaillaco; Reclus; Santiago**	**Chaitén; Carrán-Los Venados**				
U-HR	Aliso	Aguilera; Antillanca Group; Cayutué-La Viguería; Ecuador; Infiernillo; Longaví; Nevado de; Palei-Aike Volcanic Field; Yanteles	Caburgua-Huelemolle; Huambo; Soche; Sollipulli	Andahua-Orcopampa; Cumbres, Las; Quimsachata; Romeral	Chacana	Acatenango	
U-NHHR	5 volcanoes	80 volcanoes	8 volcanoes	12 volcanoes	14 volcanoes	24 volcanoes	4 volcanoes
	PEI 1	PEI 2	PEI 3	PEI 4	PEI 5	PEI 6	PEI 7

Hazard III column (right): Chichinautzin
Hazard II column: Atitlán
U-HR PEI 5: Cofre de Perote; Malinche, La; Tecuamburro
U-HR PEI 6: Jocotitlán; Naolinco Volcanic Field; Zitácuaro-Valle de Bravo
U-HR PEI 7: Almolonga; Michoacán-Guanajuato; Nejapa-Miraflores

There are 112 Latin American volcanoes that are monitored using seismic, gas or deformation stations. Thirty volcanoes (10% of Latin American volcanoes, Table 19.1) classify as Monitoring Level 1, including 11 with no recorded historical activity (Table 19.3). About half of these are in well populated regions and 11 classify at Hazard Levels II-III. Of the monitored volcanoes, 35 classify as Monitoring Level 2 making a combination of seismic and deformation monitoring the most popular choice. Further detail is available in the database regarding the type of deformation studies being used (e.g. INSAR, GPS, EDM). Forty-seven volcanoes are classified at Monitoring Levels 3 – 5, indicating that all three monitoring methods are used at 15% of the Latin American volcanoes. Just three countries have monitoring levels of 3-5 at over 50% of their *monitored* volcanoes: Mexico (86% of monitored volcanoes here – 6 out of 7), Costa Rica (75% - 6 out of 8) and Colombia (62% - 8 out of 13), indicating that several lines of monitoring are used here and that where monitoring is used in these countries, it is comprehensively undertaken.

Table 19.3 The number of volcanoes with and without historic activity in Latin America. Percentage is percentage of each age group.

Monitoring level	Number and % of Latin American volcanoes with historical activity		Number and % of Latin American volcanoes with no historical activity	
	Number	%	Number	%
0	30	27%	172	86%
1	19	17%	11	5%
2	25	22%	10	5%
3	10	9%	0	0%
4	18	16%	6	3%
5	11	10%	2	1%

Just 13 volcanoes throughout Latin America are at the highest monitoring level (Level 5) with seismic stations and two or more deformation and gas analysis techniques. All but Cuicocha in Ecuador and Cerro Machín in Colombia have recorded historical activity, but both have recent signs of unrest including elevating lake temperatures at Cuicocha (Gunkel et al., 2008) and seismic activity at Cerro Machín. Well-monitored Latin American volcanoes (Monitoring Levels (ML) 3-5) have low to high PEI levels and hazard and risk levels of I to III; however, most ML5 volcanoes have high levels of hazard and risk, and fatalities in the historic record (Table 19.4).

The largest numbers of monitored volcanoes are located in Chile, representing just 36% of volcanoes in this country. Countries with high proportions of monitored volcanoes are Colombia (87%), Costa Rica (80%) and Ecuador (53%), with monitoring levels ≥1 (Figure 19.4). Colombia and Ecuador also have the highest number and highest proportion of volcanoes at Monitoring Level 5; however, only about half of the historically active volcanoes of Ecuador are monitored (Figure 19.5). Four Latin American countries have monitoring at all historically active volcanoes: Colombia, Costa Rica, El Salvador and Nicaragua.

Table 19.4 Well-monitored Latin American volcanoes (Monitoring Levels 3-5; ML3 in green, ML4 in purple, ML5 in black) shown with their hazard level and PEI. The top section shows those volcanoes with a classified hazard level; the warming of the background colours indicates increasing risk levels. Those volcanoes with no hazard classification are shown in the lower section, where historically active volcanoes are shown in section U-HHR; volcanoes with a Holocene record but no historical activity are shown in U-HR; the number of volcanoes with no confirmed Holocene records are shown under U-NHHR.

		PEI 1	PEI 2	PEI 3	PEI 4	PEI 5	PEI 6	PEI 7
CLASSIFIED	Hazard III			Reventador	Cerro Bravo; Colima; Cotopaxi; Tungurahua	Irazú; Turrialba; Guagua Pichincha; Nevado del Ruiz	Galeras	
CLASSIFIED	Hazard II		Fernandina; Planchón-Peteroa; Antuco; Chillán, Nevados de; Copahue; Láscar;	Rincón de la Vieja; Ubinas		Santa Ana; Popocatépetl		
CLASSIFIED	Hazard I		Sierra Negra, San Pedro	Maipo	Arenal; Puracé; El Misti	Poás		
UNCLASSIFIED	U–HHR		Callaqui; Descabezado Grande; Cerro Hudson; Mentolat; Ticsani; Tinguiririca		Chichón, El; Cumbal; Miravalles; Nevado del Huila; San Martín	Ceboruco		
UNCLASSIFIED	U-HR		Corcovado; Maca; Melimoyu			Azufral; Machín	Nevado de Toluca; Cuicocha	
UNCLASSIFIED	U-NHHR		1 volcano					

Analysis of the data provided for VOMODA in 2012 shows that with just 13% and 20%, respectively, of Colombian and Costa Rican volcanoes being unmonitored and 100% of their historically active volcanoes having some monitoring, these countries are proportionally top for having at least minimal monitoring standards at their recognised Holocene volcanoes. Coupled with the monitoring of over 50% of their volcanoes at Levels 3-5, these countries show the most comprehensive monitoring regimes. With 200 unmonitored volcanoes throughout Latin America, including 30 unmonitored historically active volcanoes, resources may be required to better equip the region for anticipation and monitoring of volcanic activity.

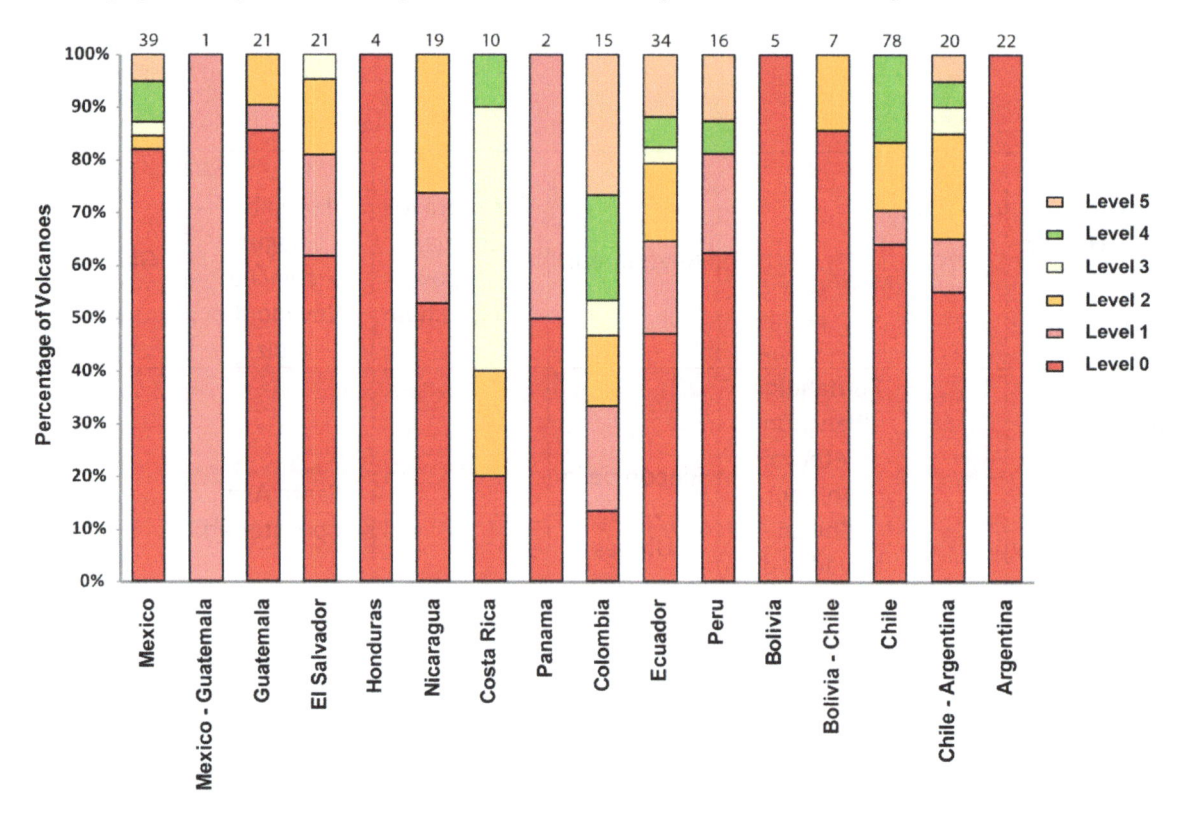

Figure 19.4 The percentage of all volcanoes in each Latin American country classified at Monitoring Levels 0-5. Data provided in 2012.

19.4 Conclusions

Efforts are underway to populate GLOVOREMID for a global dataset of VRMI and instrumentation. Further work and international cooperation with the global volcanological community is required to expand this database and the analysis of the data contained within it. Ultimately, an aim is to allow continuous data updates and to embed GLOVOREMID in other global volcanic databases in order to perform ongoing analyses of volcanic activity and monitoring.

GLOVOREMID allows a comparison between the number of active volcanoes and the investment in monitoring resources for each country. In combination with the Hazard Levels and Population Exposure Index it can be used to investigate the monitoring of high-risk volcanoes as global data are collated. The database will encourage cooperation between volcano monitoring institutions by facilitating the exchange of expertise in monitoring techniques as well as lessons learned from managing previous volcanic crises.

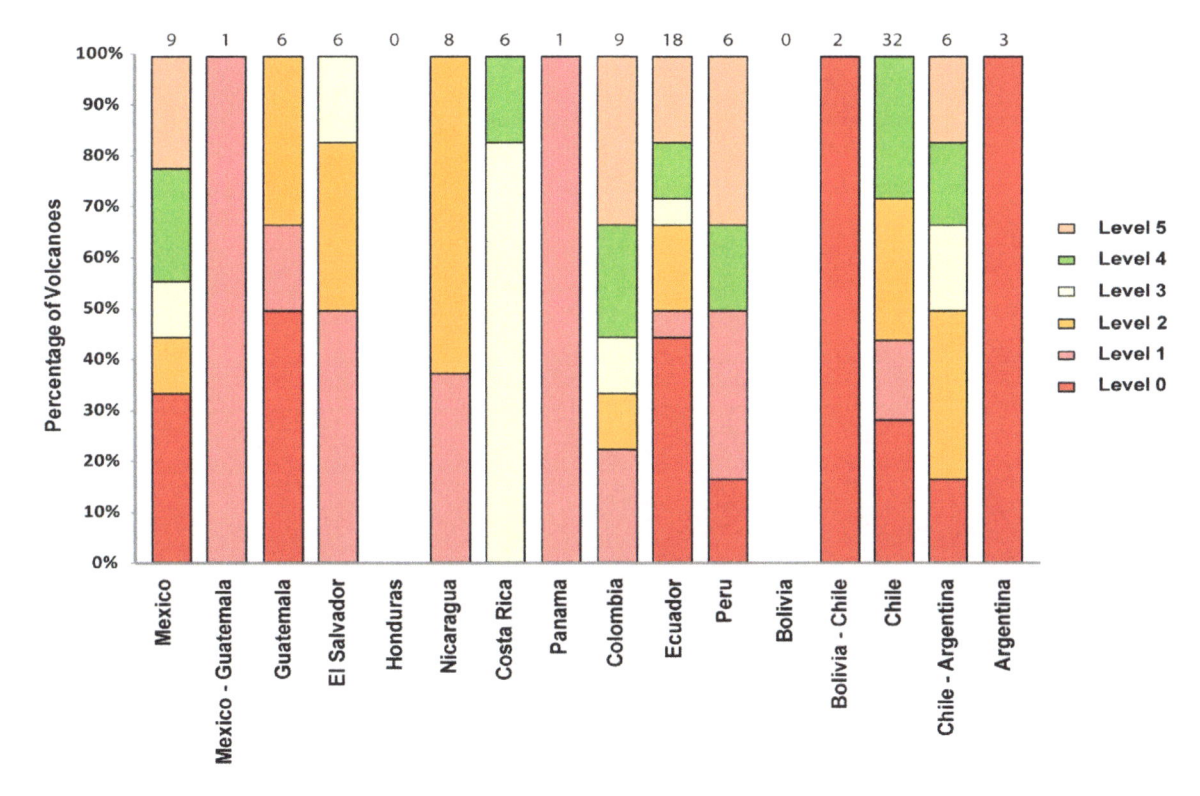

Figure 19.5 Monitoring levels of historically active volcanoes through Latin America.

References

Comunidad KUMBIAPHP. 2012. *Manual de KumbiaPHP Framework Beta 2. Capitulo 1 - Introducción–Cómoimplementar MVC* [Online]. Available: www.kumbiaphp.com

Crosweller, H. S., Arora, B., Brown, S. K., Cottrell, E., Deligne, N. I., Guerrero, N. O., Hobbs, L., Kiyosugi, K., Loughlin, S. C. & Lowndes, J. 2012. Global database on large magnitude explosive volcanic eruptions (LaMEVE). *Journal of Applied Volcanology,* 1:4, pp.13.

De la Torre, C. 2010. *Guia de arquitectura N-Capasorientada al dominio .NET 4.0. Topic: MVC Pattern,* España, Microsoft Iberica.

Gunkel, G., Beulker, C., Grupe, B. & Viteri, F. 2008. Hazards of volcanic lakes: analysis of Lakes Quilotoa and Cuicocha, Ecuador. *Advances in Geosciences,* 14, 29-33.

Kendall, K. & Kendall, J. 2010. *System Analysis and Design, 8/e,* New Jersey, Prentice Hall.

Siebert, L., Simkin, T. & Kimberley, P. 2010. *Volcanoes of the World, 3rd edn,* Berkeley, University of California Press.

Chapter 20

Volcanic hazard maps

E.S. Calder, K. Wagner and S.E. Ogburn

20.1 Introduction

Generating hazard maps for active or potentially active volcanoes is recognised as a fundamental step towards the mitigation of risk to vulnerable communities (Tilling, 2005). The responsibility for generating such maps most commonly lies with government institutions but in many cases input from the academic community is also relied on. Volcanic hazard maps communicate information about *a suite* of hazards including tephra (ash) fall, lava flows, pyroclastic density currents, lahars (volcanic mudflows) and debris avalanches (volcanic landslides). The hazard footprint of each of these depends, to a first order, on whether they are erupted into the atmosphere (and therefore dominated by transport in the atmosphere), or whether they form flows which travel along the ground surface away from the volcano. For each hazard type, the magnitude (volume) and intensity (discharge rate) of the event also determines the extent of the footprint. Tephra fall differs from the other hazards in that it can have proximal-to-regional and in extreme cases, global effects. The other hazard types characteristically affect the environs of the volcano, with the most mobile types, lahars and pyroclastic density currents, capable of reaching distal drainages over 100 km from the volcano.

It is of critical importance to understand that a wide variety of methods are currently employed to generate hazard maps, and that the respective philosophies on which they are based are equally diverse, as well as to acknowledge the notion that one model cannot fit all situations. Some hazard maps are based solely on the distribution of prior events as determined by the geology, others take into account estimated recurrence intervals of past events, or use computer simulations of volcanic processes to gauge potential future extents of impact. Increasingly, computational modelling of volcanic processes is combined with geological information and statistical models in order to develop fully probabilistic hazard maps.

20.2 Types of volcanic hazard maps currently in use

A preliminary review of hazard maps has recently been carried out by the authors. The review was based on 120 hazard maps, which were available either in print form, or electronically from legitimate sources on the internet, such as government institution websites. The hazard maps have been categorised into five main families depending on the type of information incorporated in the map and how it is conveyed (Figure 20.1).

Calder, E.S., Wagner, K. & Ogburn, S.E. (2015) Volcanic hazard maps. In: S.C. Loughlin, R.S.J. Sparks, S.K. Brown, S.F. Jenkins & C. Vye-Brown (eds) *Global Volcanic Hazards and Risk,* Cambridge: Cambridge University Press.

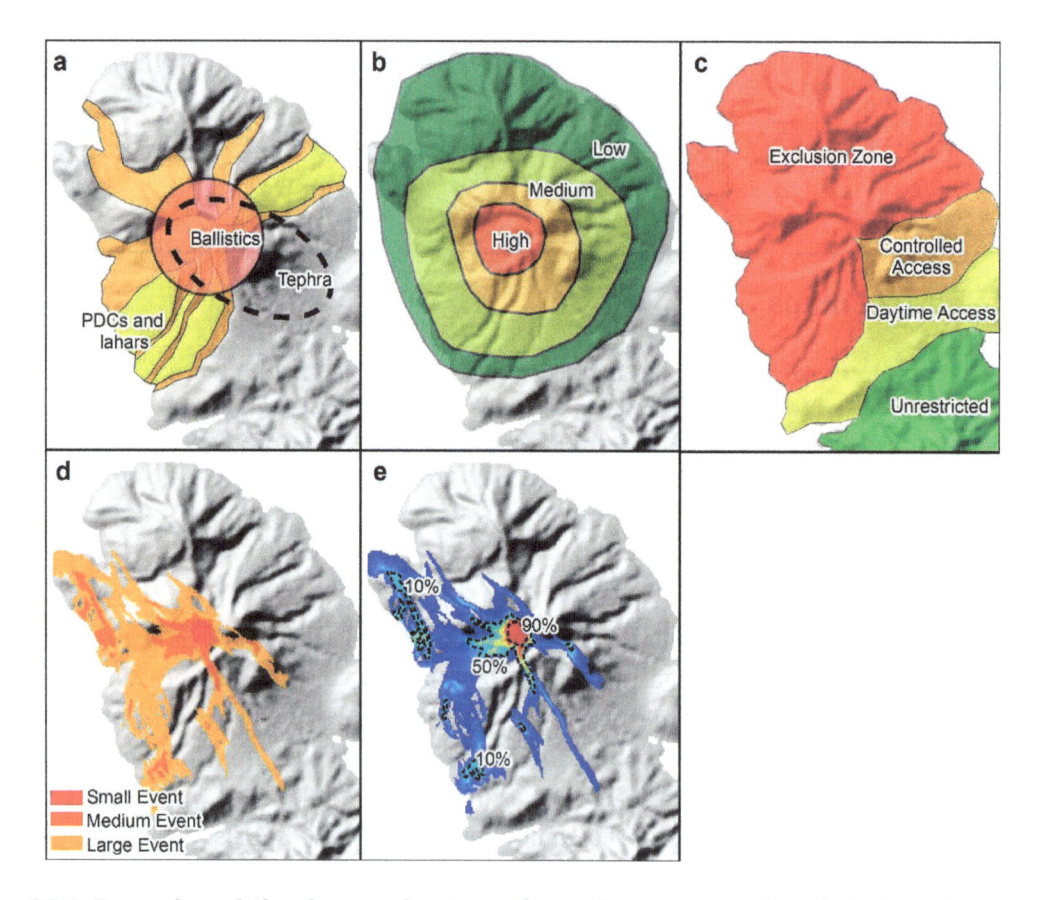

Figure 20.1 Examples of the five predominant hazard map types found during the review. a) geology-based map, b) integrated qualitative map, c) administrative map, d) modelling-based map and e) probabilistic map. These examples are not from a real volcano, they are based on a synthetic topography in order to demonstrate the variability in appearance of such maps for the same topography.

Geology-based maps: Mapped hazard footprints are based directly on the past occurrence of specific types of events. An important limitation of this type of map is that the geological record is an incomplete catalogue of events so that the distribution and extent may reflect previous events, but not all possible future events. Furthermore, the geological record can also be biased by preferentially preserving deposits from larger eruptions and because some deposits of very violent eruptions, such as those formed by volcanic blasts, are easily eroded.

Integrated qualitative maps: All available hazard information is amalgamated, resulting in simple, often concentric-type, hazard zones. The source of the information may be geology and/or modelling. These maps may be more effective for communication because they are simple. Relative hazard is communicated qualitatively [see Chapter 16 Tongariro].

Modelling-based hazard maps: Involve scenario-based application of simulation tools often for a single hazard type.

Probabilistic hazard maps: Maps based usually on the study of a single hazard using stochastic application of computer simulations. The principal limitations are that these maps deal with a single hazard, are sometimes complex to interpret or communicate and include uncertainties associated with the simulation tool or model input parameters [see Chapter 6 Vesuvius].

Administrative maps: These maps are not designed to show hazard distribution, but instead combine hazard levels with administrative needs and are constructed specifically to aid in emergency management. These maps usually inherently contain information about hazard distribution, but the geoscience content may be somewhat opaque.

Based on the review, the hazards of most widespread concern, as indicated by frequency of occurrence on hazards maps are: lahars, pyroclastic density currents (PDCs), tephra fall, ballistics, lava flows, debris avalanches, and monogenetic eruptions (Figure 20.2a). Seventy-five percent of maps include lahars and/or PDCs and 63% include tephra. Less than half include lava and/or debris avalanches, while less than 10% include hazards associated with unknown source locations, such as monogenetic eruptions. Those maps based solely on the geologic history of the area are significantly more common (63%) than all other map types (Figure 20.2b). Integrated qualitative maps make up 17% of hazard maps. Hazard maps indicate likelihood, in some form, to show the relative degree of hazard affecting the map area. The likelihood of impact can be expressed quantitatively or qualitatively, explicitly or generally. It is noteworthy that 83% of all hazard maps use simple qualitative "high-med-low" designations to indicate level of probability of impact. Such designations, however, are open to wide interpretations.

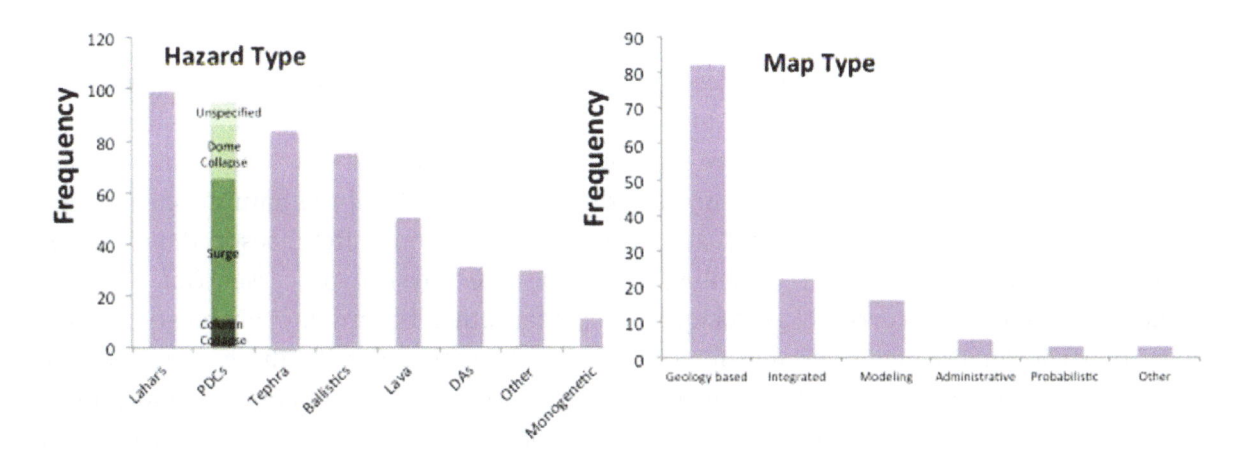

Figure 20.2 a) Types of hazards in the 120 maps reviewed, including: lahars, PDCs, tephra fall, lava flows, debris avalanches and monogenetic volcanism. PDCs were further distinguished based on specific type (column collapse, surge, dome collapse, or unspecified). b). Hazard maps can be subdivided into categories based on how and what information is conveyed. Those based solely on the geologic history of the area are significantly more common (63%) than all other map types. Map complexity increases to the right as the number of maps in that category decreases.

20.3 Modelling and uncertainty quantification in hazard maps

The computational models used for hazard mapping comprise two main types: (i) complex fluid dynamics and solid mechanics models that attempt to capture as much of the underlying physics of a process as possible; (ii) empirical, or abstracted, models that capture the essence of a complex process. Most commonly it is the latter type of models that are used for hazard mapping (e.g. Iverson et al. (1998), Bonadonna (2005). Simulations are used to indicate the outcome of an eruptive scenario, or set of scenarios (i.e. applied deterministically); or, less frequently, uncertainty is taken into account through probabilistic application of the models (e.g. Bonadonna (2005); Wadge (2009)). Assessing the types of models suitable for use in generating probabilistic hazard maps relies on our understanding of the physical processes involved, but also on our appreciation of aspects of the real phenomena that are not sufficiently captured in models. Models that can be relatively quickly run, in stochastic mode, and are coupled with digital elevation models of volcanic topography or atmospheric wind data, are being increasingly tested and employed in the generation of probabilistic hazard maps during real volcanic crises. Forward modelling applications are still largely at an experimental stage, but ongoing developments of both appropriate models and methodologies pose exciting new opportunities which will likely become more commonplace (e.g. Bayarri et al. (2009)). An increase in the application of computational models to understand potential hazards, and their use in probabilistic hazard mapping, is also intricately bound with discussions on model suitability and inherent uncertainty.

20.4 Vision for future efforts

The volcanology community currently lacks a coherent approach in dealing with hazard mapping but there is general consensus that improved quantification is desirable. Harmonisation of the terminology is needed to improve communication both within volcanology and with stakeholders. In particular, successful approaches must address and quantify uncertainty related to (i) the incompleteness and bias of the geological record and the extent to which it represents possible future outcomes; (ii) the fact that analyses based on empirical models rely on *a priori* knowledge of the events; and (iii) the ability of complex computational models to adequately represent the full complexity of the natural phenomena. The variation in currently utilised approaches results in part from differences in the extent of understanding and capability of modelling the respective physical processes (for example tephra fall hazards are currently better quantified than other hazards). Probabilistic hazard maps, in particular, are highly variable in terms of what they represent. Yet there is the need for probabilistic approaches to be fully transparent; they are used to communicate and inform stakeholders, for whom an understanding of the significance of the uncertainties involved is also crucial. A recent initiative through the newly formed IAVCEI Commission on Volcanic Hazards and Risk, will focus on hazard mapping. The effort aims to undertake a comprehensive review of current practices with a view toward:

- constructing a framework for a classification scheme for hazard maps;
- promoting harmonisation of terminology;
- defining good practices for hazard maps based on experiences of usage.

Clearly, the needs of today's stakeholders for more quantitative information about hazards and their associated uncertainties also drives the need for further research efforts in priority areas. In particular, sources of scientific advancement that would aid in the production of a new generation of more robust, quantitative, accountable and defendable hazard maps would be:

- improved methods for probabilistic analysis, especially for lahar and PDC hazards;
- methods for undertaking hazard assessments for volcanic centres from which we have sparse data;
- uncertainty quantification;
- handling 'Big Data' generated by computational modelling;
- handling uncertainty in digital elevation models and evolving volcanic topography over time;
- forecasting of extreme events and their consequences;
- communicating probabilities associated with hazard and risk;
- approaches for multi-hazard, multi-scenario probabilistic modelling.

These are research problems that require multi-disciplinary expertise to solve. There is consensus that the basic foundation on which any hazard analysis should be undertaken is the establishment of an understanding about a volcano's evolution and previous eruptive behaviour through time, based on combined field geology, dating and geochemical characterisation of the products. However, bringing together experts in modelling and statistical analysis with field scientists is then key. Our ability to achieve tangible advances in probabilistic volcanic hazard analysis hinges on the effective use of advanced modelling and statistical methods, and handling of massive and/or complex data. Dealing with such data requires fundamental advances in mathematical, statistical and computational theory and methodology but also requires training a new generation of scientists that are adapted to cross-disciplinary research environments.

20.5 Glossary of hazard map types

Administrative hazard maps: A type of map used for disaster management that takes into account local infrastructure, land use and populations in addition to information about possible hazard distribution.

Geology-based hazard maps: Indicates hazards based on the distribution of past eruptive products. Can also include information about recurrence rates.

Integrated hazard maps: All available hazard information is amalgamated, resulting in simple, often concentric-type, hazard zones. The information on which these are based can include field distributions as well as modelling. Levels of hazard are usually expressed qualitatively.

Modelling-based hazard maps: Involve scenario-based application of simulation tools often for a single hazard type.

Probabilistic hazard maps: Based on probabilistic application of hazard models (models can be empirical to fully geophysical). Levels of hazard can be expressed quantitatively.

Less common, but also in use are the following terms:

Hazard-specific maps: Considers only one hazard type in one map.

Multi-hazard maps: Considers multiple hazard types in one map.

Nested hazard maps: A type of scenario map indicating the possible distribution of eruptive products of a similar type of event (e.g. lahars), but for scenarios with varying magnitudes or intensities. The distributions are therefore nested within each other.

Rapid-response hazard maps: Generated by ascertaining the distribution of past eruptive products, rapidly (either remotely or in the field) in response to a period of unrest or impending crisis at a volcano where previously eruptive activity is not established or has not previous been well characterised.

Scenario maps: Provide information about the distribution of eruptive products, based on explicit event scenarios that may be considered likely. If levels of hazard are expressed quantitatively they can be considered conditional.

20.6 Summary

The large majority of hazard maps currently in use by government institutions around the globe are geology-based hazard maps, constructed using the distribution of prior erupted products. Such maps are based on the study of the volcano, and provide a wealth of information about its capabilities. An important limitation though, is that the distribution of previous events (even if known in their entirely), does not represent all possible future events. Increasingly, computer simulations of volcanic processes are used to augment the knowledge gained by geology, to gauge potential areas and extents of impact of future events. The hazards of most widespread concern, as indicated by frequency of occurrence on hazards maps are: lahars, pyroclastic density currents, and tephra fall. Currently, tephra hazards (which can have the most widespread effects and far-reaching economical impacts) are the best quantified. Lahars and pyroclastic density currents both have more localised impacts but do account for far greater loss of life, infrastructure and livelihoods. These hazard types present greater challenges for modelling, and as a result quantitative hazard analysis for lahar and pyroclastic density currents lags behind that for tephra fall.

References

Bayarri, M., Berger, J. O., Calder, E. S., Dalbey, K., Lunagomez, S., Patra, A. K., Pitman, E. B., Spiller, E. T. & Wolpert, R. L. 2009. Using statistical and computer models to quantify volcanic hazards. *Technometrics,* 51, 402-413.

Bonadonna, C. 2005. Probabilistic modelling of tephra dispersion. *In:* MADER, H. M., COLES, S. G., CONNOR, C. B. & CONNOR, L. J. (eds.) *Statistics in Volcanology.* London: Geological Society of London.

Iverson, R. M., Schilling, S. P. & Vallance, J. W. 1998. Objective delineation of lahar-inundation hazard zones. *Geological Society of America Bulletin,* 110, 972-984.

Tilling, R. I. 2005. Volcano hazards. *In:* MARTI, J. & ERNST, G. (eds.) *Volcanoes and the Environment.* Cambridge: Cambridge University Press.

Wadge, G. 2009. Assessing the pyroclastic flow hazards from dome collapse at Soufriere Hills Volcano, Montserrat. *In:* SELF, S., LARSEN, G., ROWLAND, S. K. & HOSKULDSSON, A. (eds.) *Studies in Volcanology: The Legacy of George Walker, Spec. Publ. IAVCEI.* London: Geological Society of London.

Chapter 21

Risk assessment case history: the Soufrière Hills Volcano, Montserrat

W.P. Aspinall and G. Wadge

21.1 Introduction

Volcanic hazard and risk at Soufrière Hills Volcano, Montserrat (SHV) has been assessed in a consistent and quantitative way for over 17 years (1997-2014), during highly variable eruptive activity involving andesitic lava dome growth (Wadge & Aspinall, 2014). This activity has placed serious stresses and constraints on the Montserrat population: about 12,000 people lived on this small Caribbean island prior to the start of the eruption in July 1995 and now (2014) this has stabilised at just over 4,000 souls. Over the years following 1995, a series of five very active dome growth episodes produced many pyroclastic flows, explosions and lahars, whose net effect was to destroy the main town, Plymouth, and most infrastructure, forcing people to leave Montserrat or live only in the northern part of the island. In June 1997, nineteen people were killed when a dome collapse pyroclastic flow caught a number of persons inside the exclusion zone.

The risks faced by the people of Montserrat from volcanic activity are the responsibility of the UK government, and hazard and risk assessment work on Montserrat has been carried out by a Scientific Advisory Committee on Montserrat Volcanic Activity (SAC) (and the predecessor Risk Assessment Panel) appointed by them, working in collaboration with the Montserrat Volcano Observatory (MVO). While the administrative basis of the SAC has changed, the quantitative risk assessment methodology for enumerating risk levels (Aspinall et al., 2002, Aspinall & Sparks, 2002), has been kept the same since 1997 to ensure comparability of findings from one assessment to the next. In a protracted eruption crisis, continuity in scientific inputs to decision-making is essential: any major change in concepts, modelling or assumptions could entail large differences in evaluated risk levels and hence engender doubts for officials and confusion in the minds of the public. This series of multiple, repeated quantitative volcanic hazard and risk assessments must be unique in volcanology.

In the case of Montserrat, by 'volcanic risk' we mean the probability that a person will be harmed by some volcanic hazard within some specified timeframe; assessing other risks and losses, such as damage to buildings or infrastructure, have had only a limited consideration in terms of framing scientific advice.

Aspinall, W.P. & Wadge, G. (2015) Risk assessment case history: the Soufrière Hills Volcano, Montserrat. In: S.C. Loughlin, R.S.J. Sparks, S.K. Brown, S.F. Jenkins & C. Vye-Brown (eds) *Global Volcanic Hazards and Risk,* Cambridge: Cambridge University Press.

21.2 Assessment methods and their effectiveness

Comprehensive risk assessments for Montserrat were first undertaken by a Risk Assessment Panel in December 1997, following the fatalities and then a series of violent Vulcanian explosions. Thereafter, whenever activity changed significantly assessments were updated. The SAC came into being in 2003, superseding the risk panel, and further developed risk assessment methods, using the following sources of information in regular meetings, usually every six months:

- MVO data on current activity at the SHV
- knowledge of other dome volcanoes
- computer models of hazardous volcanic processes
- formalised elicitations of probabilities of future hazards scenarios
- probabilistic event trees
- Bayesian belief networks
- census data on population numbers and distribution
- Monte Carlo modelling of risk levels faced by individuals and society.

Knowledge elicitation about hazard scenarios and analysis of judgements by the Classical Model method (Cooke, 1991) have been used to formulate probabilistic forecasts of future hazardous events, typically over the following 12 months. These scenario forecasts have been used in Monte Carlo simulations to quantify risk exposures of individuals and the population as a whole (Aspinall, 2006).

Hazard scenarios allow volcanologists to visualise - and describe to the authorities and the public - the potential occurrence and dangers of future events. Probabilities of occurrence of hazards, with associated confidence limits, are evaluated by judging factors that make such events likely to happen. This evaluation is done by expert elicitation, informed by experience of previous events at the volcano, its current activity state, by model simulations, and by discussion of precedents elsewhere at similar volcanoes. The elicitation process can be stimulating but also burdensome, so generally only a restricted number of scenarios or outlooks are considered (typically 10 to 30). The process may thus be limited to major hazards: i.e. explosion; large dome collapse, or lateral blast, and event probabilities are usually evaluated for each, for one year ahead. However, if a significant event occurs then the conditions under which these probabilities were judged will have changed, so a risk assessment is only valid up to the time of the next 'significant event'. Hazard and risk updating is needed at that point.

The accuracy of SAC hazard forecasts has been tested using the Brier Skill Score. Although this metric has some limitations, it is used in weather forecasting and we have adopted the analysis method for volcanic forecasts. We have tested 110 scenario probabilities against actual outcomes: seventy-five of these can be termed life-critical, and for 83% of these a positive Brier Skill Score was achieved. This shows that, in the overwhelming majority of cases, our experts outperformed an 'uninformed' baseline probabilistic forecast, demonstrating the value of the process in supporting the civil authorities and the people of Montserrat to mitigate risk. As yet, no hazard has happened that was a surprise to the scientists, and none has exceeded our scenario envelopes in terms of size or intensity.

In the main, the SAC risk assessment implementation is distinct from day-to-day observatory operations - where the duty is to provide immediate hazard advice - but it is closely linked with and relies on MVO inputs. While this may seem an unusual separation of responsibilities, we believe it has worked well on Montserrat because the SAC brings to bear a separate, independent pool of expertise and long-term experience. This approach offers a deeper analytical perspective on issues of future hazards, complementing the monitoring competencies of MVO. This said, in future it is intended that MVO staff will take on more of the tasks of quantitative risk assessment.

21.3 Expressing risk to the public

After deriving event scenario probabilities and their uncertainties by elicitation and then quantifying risks and their uncertainties by Monte Carlo modelling, the SAC assessments present the level of risk in several ways:

(i) A Preliminary Statement and the SAC Report generally state whether, overall, risk in the populated part of Montserrat has gone up, down or stayed the same.
(ii) Societal risk is expressed quantitatively as a curve of the probability of exceeding a given total number of fatalities, for different total numbers.
(iii) Individual risk is given as an annualised probability of death (from the volcano) for any person living in a specific area.
(iv) Added occupational risk, due to the volcano, is given for people working under certain conditions in specific areas.

Societal risk is presented as an *F-N* graph (probability of *N* or more potential fatalities in a given time plotted against *N*) and is useful for comparing situations from different periods and for assessing mass casualty scenarios. For instance, in Figure 21.1 the *F-N* curve for May 2003 (red), together with a second curve (green), showed how societal risk could be reduced by extending the evacuated area. For natural disaster risk comparison purposes, rudimentary *F-N* curves for hurricane and earthquake risk on Montserrat are also shown.

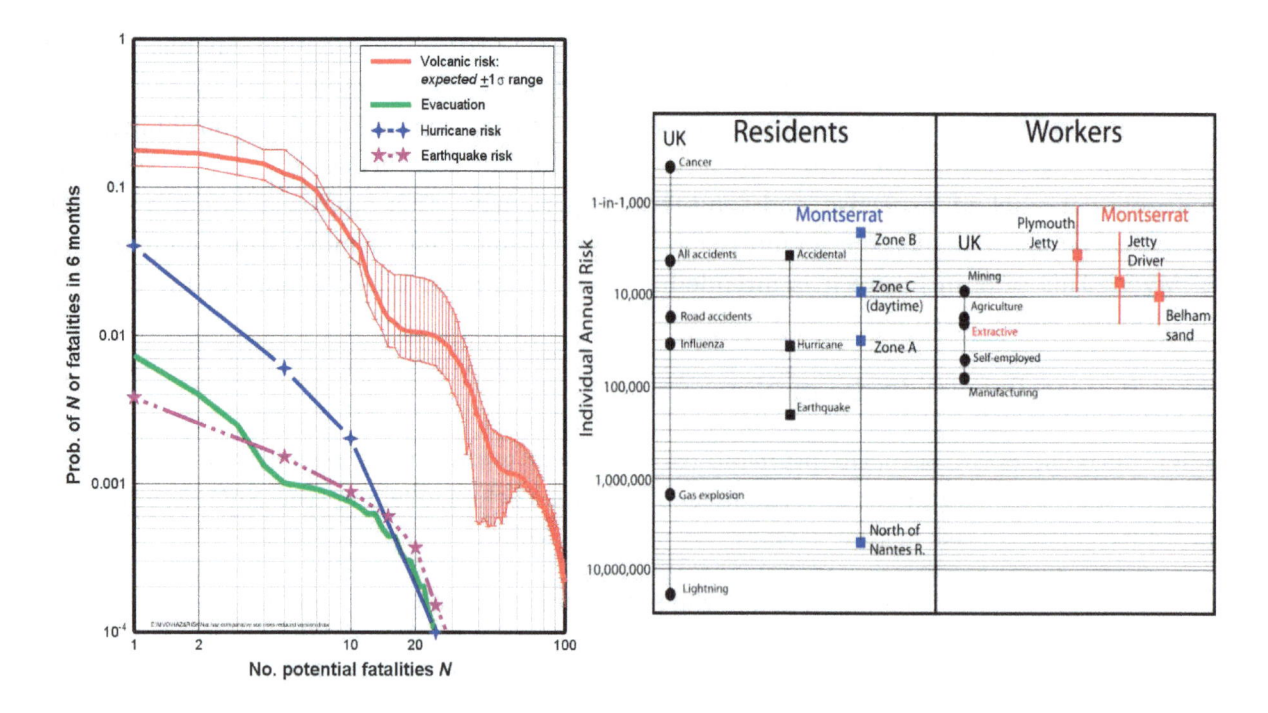

Figure 21.1 Montserrat quantitative risk assessments: (left) societal risk F-N curve for population exposure (red, with uncertainty estimates), and the risk reduction possible if the evacuation area is extended (green), plus background hurricane (blue) and earthquake risks (mauve); (right) summary plot of annualised individual risk exposures for residents and workers (latter plotted in red, with indicative uncertainties on risk estimates). Modified from Figures 24.8 and 24.9 of Wadge & Aspinall (2014).

Uncertainty analysis of the Monte Carlo societal risk modelling furnishes confidence bounds on these alternative volcanic risk *F-N* curves (Figure 21.1), indicating that meaningful risk reductions could be achieved by basic mitigation measures. In contrast with typical linear *F-N* plots for industrial risks, the volcano societal risk curves exhibit marked changes of slope and humps. These arise because an erupting volcano has a variety of ways of doing harm to a population, with differing levels of hazard intensity and spatial extent potentially causing different total magnitudes of casualty numbers, all with different probabilities of occurrence. Such complexity in volcanic *F-N* curve serves to make decision-taking about public safety very much more challenging than in the industrial risks case, especially when the risk findings are convolved with the substantial scientific uncertainties associated with volcano forecasting.

Individual risk can be a more influential measure for expressing exposure in that many people, but not all, base their own responses to threats on their own personal risk level. The SAC has used several measures for individual risk, but mainly we calculate the 'individual risk per annum' (IRPA) metric, expressed in odds form (e.g. a 1-in-1000 chance of being killed in the next year). Graphical comparison of IRPA values on a logarithmic ladder with some relevant published risk levels for everyday hazards and occupations to provide context (Figure 21.1) has proved popular with the authorities on Montserrat. The Zones used to calculate individual risk are defined geographically by the Hazard Level System managed by MVO (www.mvo.ms). The positioning of the zone boundaries has implications not just for risk levels but also evacuation actions and emergency management. There is a rudimentary feedback process, from the SAC and MVO through the civil authorities, between the level of hazard estimated and the resultant risks ensuing for a given set of boundaries (Wadge et al., 2008).

Subjectively, the general perception of whether the risk has gone up or down is reflected by changes both in calculated societal risk levels and in individual risk estimates. However, criteria for public safety and tolerable levels of volcanic risks are not established in Montserrat (or elsewhere); this vacuum is exacerbated in a crisis by inevitable tensions between individuals' acceptance of elevated risk and society's wish to avoid casualties.

The SAC risk assessments are made public within a few weeks. There are two reports: a Summary Report of about 4-5 pages and a much longer, Full Report, giving all the technical analysis. They are available at www.mvo.ms.

21.4 Government response to risk assessments

How have the authorities used the SAC risk assessments to guide actions in protecting the people of Montserrat? Because volcanic dangers at SHV are dominated by pyroclastic flows and surges, the spatial boundary of the probable extent of any future flow is the overriding concern for risk management. Generally, such actions appear to have been responsive to SAC risk level assessments, but it is difficult to evaluate this in detail (Wadge & Aspinall, 2014). This is partly because the MVO has responsibility for day-to-day guidance and recommends changes in the Hazard Level, the main cue for mitigation measures. Also, there is no formal feedback from the authorities to the SAC in terms of the reasoning about decisions to change zone boundaries. While many such decisions have been taken over the years, few if any can be directly linked to specific SAC advice, but most were consonant with levels of risk assessed by the scientists.

While the aim of the SAC has been to keep the nature and standard of advice consistent from one assessment meeting to the next, changing conditions at the volcano and changing requirements from the authorities inevitably dictated some alteration to the way advice was formulated and presented by the SAC. This said, during recent periods several pyroclastic flows have travelled further towards the populated areas and could have led easily to injuries or deaths had there not been updated official access restrictions in response to the risk assessment advice. Other risk-based decision approaches, such as cost-benefit or probabilistic criteria to inform evacuations have not, thus far, formed part of risk decision-making in Montserrat.

21.5 Summary

Since its inception in 2003, the SAC's principal role in efforts to mitigate danger from the Soufrière Hills volcano has been to consider likely future behaviour patterns of the volcano on the basis of observations, data, modelling and expert knowledge and to communicate our collective scientific understanding of processes driving the evolving eruption and limitations in that understanding. Related advances in probabilistic volcanic hazard and risk assessment methodologies have been achieved, and our appraisals of the volcano's state and the chances of potential impacts on the population by eruptive activity have represented the substantive science input to decision-making by the authorities and the provision of risk information to the public. In the face of scientific uncertainty and other challenges associated with living with an active volcano, risk-informed hazard mitigation measures have ensured there has been no fatality caused by the volcano during all the long years of eruptive activity and dangerous events since the tragic events of June 1997.

References

Aspinall, W. 2006. Structured elicitation of expert judgement for probabilistic hazard and risk assessment in volcanic eruptions. *In:* Mader, H. M. (ed.)*Statistics in Volcanology.* Geological Society of London.

Aspinall, W., Loughlin, S., Michael, F., Miller, A., Norton, G., Rowley, K., Sparks, R. & Young, S. 2002. The Montserrat Volcano Observatory: its evolution, organization, role and activities. *Geological Society of London Memoirs,* 21, 71-92.

Aspinall, W. P. & Sparks, R. S. J. 2002. Montserrat Volcano Observatory: volcanic risk estimation - evolution of models. MVO Open File Report 02/1.

Cooke, R. M. 1991. *Experts in uncertainty: opinion and subjective probability in science*, Oxford University Press.

Wadge, G. & Aspinall, W. 2014. A review of volcanic hazard and risk assessments at the Soufrière Hills Volcano, Montserrat from 1997 to 2011. *In:* Wadge, G., Robertson, R. & Voight, B. *The Eruption of Soufriere Hills Volcano, Montserrat, from 2000 to 2010: Geological Society Memoirs, Vol. 39.* Geological Society of London.

Wadge, G., Aspinall, W. P. & Barclay, J. 2008. Risk-based policy support for volcanic hazard mitigation. Report commissioned by Foreign and Commonwealth Office.

Chapter 22

Development of a new global Volcanic Hazard Index (VHI)

M.R. Auker, R.S.J. Sparks, S.F. Jenkins, W. Aspinall, S.K. Brown, N.I. Deligne, G. Jolly, S.C. Loughlin, W. Marzocchi, C.G. Newhall and J.L. Palma

22.1 Background

Globally, more than 800 million people live in areas that have the potential to be affected by volcanic hazards, and this number is growing [Chapter 4]. The need for informed judgements regarding the global extent of potential volcanic hazards and the relative threats is therefore more pressing than ever. There is also an imperative to identify areas of relatively high hazard where studies and risk reduction measures may be best focussed. Various authors have tackled this task at a range of spatial scales, using a variety of techniques. At some well-studied volcanoes, the geological record has been used in combination with numerical modelling to create probabilistic hazard maps of volcanic flows and tephra fall [Chapter 6 and 20]. Such sources of information can be hugely beneficial in land use planning during times of quiescence and in emergency planning during times of unrest. Unfortunately, creating high-resolution probabilistic hazard maps for all volcanoes is not yet feasible. There is therefore a need for a methodology for volcanic hazard assessment that can be applied universally and consistently, which is less data- and computing-intensive. The aim of such an approach is to identify, on some objective overall basis, those volcanoes that pose the greatest danger, in order that more in-depth investigations and disaster risk reduction efforts can then be focused on them.

22.2 Previous methods

An index-based approach to volcanic hazard assessment involves assigning scores to a series of indicators, which are then combined to give an overall hazard score. Indicators typically include measures of the frequency of eruptions, the relative occurrence of different kinds of eruptions and their related hazards, the footprints of these hazards, and eruption size. Indices are well suited to the problem of volcanic hazard assessment, as they allow the decomposition of the complex system into a suite of volcanic system controls and simple quantitative variables and factors that jointly characterise threat potential.

Ewert (2007) presented an index-based methodology for assessing volcanic threat (the combination of hazard and exposure) in the USA, to permit prioritisation of research, monitoring and mitigation. The study formed part of the development of the National Volcano

Auker, M.R., Sparks, R.S.J., Jenkins, S.F., Aspinall, W., Brown, S.K., Deligne, N.I., Jolly, G., Loughlin, S.C., Marzocchi, W., Newhall, C.G. & Palma, J.L. (2015) Development of a new global Volcanic Hazard Index (VHI). In: S.C. Loughlin, R.S.J. Sparks, S.K. Brown, S.F. Jenkins & C. Vye-Brown (eds) *Global Volcanic Hazards and Risk*, Cambridge: Cambridge University Press.

Early Warning System (NVEWS). The NVEWS method scores and sums twelve hazard indicators, with the result then assigned to one of five levels. Whilst representing a significant improvement on past indices, the use of a largely binary scoring system obscures much of the complexity of volcanic hazard. For example, the occurrence of Holocene lava flows and Holocene pyroclastic flows are assigned the same weighting, though the latter is far more hazardous, particularly from a loss of life perspective (Auker et al., 2013, Witham, 2005). The inclusion of indicators for historical unrest and unsatisfactory treatment of missing data make application of the NVEWS system problematic in many jurisdictions outside of the USA, where data may be scarce or absent.

Aspinall et al. (2011) developed a method for volcanic hazard assessment for application to the World Bank's Global Facility for Disaster Reduction and Recovery (GFDRR) priority countries (16 developing countries). The GFDRR method uses eight indicators to assess hazard; uncertainty indicators, used to describe the quality of information, are attached to six of the eight indicators. The summed hazard and uncertainty scores are each assigned to one of three levels.

The time period over which volcanic hazards are being assessed is not explicitly stated in the NVEWS and GDFRR methods. However, hazard assessments are more valuable if the time is specified and indeed the informativeness of the results depends on the time-frame. For example, an assessment over ten years will likely yield a very different outcome to an assessment over 10,000 years, in which very large but rare eruptions may be the dominant peril. Both the NVEWS and GFDRR methods are based on the entire Holocene eruption record. The new method developed here gives more weight to recent activity patterns, as these are more likely to indicate the character of future eruptions.

22.3 Development of a new methodology

Here, we address limitations in the NVEWS and GFDRR approaches and build on their strong points to develop an improved volcanic hazard assessment approach, which we call the Volcanic Hazard Index (VHI). We aim to assess global volcanic hazard for the next 30-year period using our indicator-based VHI method on a volcano-by-volcano basis.

The main data source is the Volcanoes of the World 4.0 (VOTW4.0) database (Siebert et al., 2010) from which we only consider 'confirmed' eruptions. There is evidence for severe under-recording of events over the entire Holocene period, which diminishes towards the present (Deligne et al., 2010). The fatalities database of Auker et al. (2013) is used in conjunction with VOTW4.0 to provide evidence that justifies indicator choices and their weightings.

22.3.1 Indicator choices

Indicators used in both the NVEWS and GDFRR methods provide a useful starting point for development of an improved methodology. These are:

- eruption frequency
- eruption magnitude
- pyroclastic flow occurrence
- mudflow (lahars and jökulhlaups) occurrence
- lava flow occurrence.

Eruption magnitude and frequency are intuitive indicators of volcanic hazard. The Volcanic Explosivity Index (VEI) is based on the size of eruptions according to volume of tephra produced; higher VEI eruptions have larger footprints and are thus more hazardous. Modal VEI and largest recorded VEI are used as magnitude indicators, aiming to capture a volcano's 'typical' and 'extreme' eruption size, based on what is known from the record. A future eruption greater than the largest recorded VEI cannot of course be excluded.

With regard to eruption frequency, given identical volcanoes producing eruptions of identical style, magnitudes and intensities, if one erupts twice as often as the other then the former is twice as hazardous as the latter. This notion suggests that eruption frequency should be used multiplicatively rather than additively, as was done in previous methods. Taking AD 1900 as base year, because recording of eruptions is almost complete after this date (Furlan, 2010), four frequency classes are defined: active (one or more eruptions since AD 1900); semi-active (historical (post-AD 1500) eruptions with or without unrest recorded since AD 1900 or Holocene (pre-AD 1500) eruptions and unrest since AD 1900); semi-dormant (Holocene (pre-AD 1500) eruptions with no post-1900 unrest or no Holocene eruptions but recorded unrest); fully-dormant (no Holocene eruptions or unrest recorded since AD 1900). Unrest is identified from Bulletin Reports which accompany VOTW4.0, and is defined subjectively as activity above background levels (for example, the presence of fumaroles does not constitute unrest, however descriptions of periods of intensified emissions does).

The fatalities database of Auker et al. (2013) can be used to infer the relative hazard posed by different eruption phenomena. Pyroclastic flows and mudflows have caused the greatest proportions of fatalities (44% and 22%, respectively), and are therefore deemed very hazardous. Lava flows have caused a relatively small proportion of fatalities (1%) but are recorded in approximately 30% of eruptions. As such, their economic impact could be large and an indicator is used to reflect this. The occurrence of pyroclastic flows, mudflows and lava flows should form part of a volcano's hazard assessment when these are common hazards (rather than rare, extreme events). We define a phenomenon as a significant contributor to overall hazard potential if it has occurred in 10% or more of a volcano's eruptions.

Indicators should capture distinct components of volcanic hazard, and problems known to affect volcanic eruption data, namely under recording, should be compensated for. Testing for the concurrence and under-recording of pyroclastic flows, mudflows, and lava flows was undertaken to identify any potential issues; no corrections were required. Further, the relationship between the occurrence of pyroclastic flows, mudflows and lava flows, and modal and maximum VEI was investigated. However, these aspects are not simply correlated, and thus VEI cannot be used as a catch-all representation of hazard. Separate indicators for pyroclastic flows, mudflows, and lava flows are required. For example, pyroclastic flows due to dome collapse and volcanic blasts can occur in eruptions with quite low VEI, and mudflow occurrence is dependent on external factors such as rainfall and the presence of crater lakes, which are unrelated to VEI.

22.3.2 Time dependence

Exploration of the global under-recording of volcanic eruptions shows two significant improvements in recording completeness, at approximately AD 1500 and AD 1900 (at which point the record becomes largely complete; Furlan (2010)). These findings can inform the length of time over which data are drawn for the hazard assessment, referred to henceforth as the 'counting period', and are useful in characterising each volcano's 'recent' history. The ideal counting period start date would be AD 1900 because of the near completeness of the volcanic eruption record after this date. However, some volcanoes have not erupted frequently enough for this time period to be representative of their recurrence statistics. Consequently we develop a simple approach where the definition of 'recent', and thus the length of the counting period, is specific to each volcano. For active volcanoes (those with at least one year in eruption recorded since AD 1900) a sliding scale is used. The counting period for an individual volcano begins in the year defined by the equation: AD Year = $1500 + \left[\left(\frac{N}{113}\right) \times 400\right]$, where N is the number of years in which the volcano is recorded as erupting between AD 1900 and AD 2013. AD 2013 is the end year of the counting period for active volcanoes. AD 1500 is used as the base year for active volcanoes' counting periods because of the significant improvement in recording at this time. The entire Holocene is used as the counting period for semi-active, semi-dormant and fully-dormant volcanoes, to maximise the amount of eruption data available for hazard assessment.

The modal VEI and pyroclastic flow, mudflow and lava flow occurrence indicators are each scored based on eruptions within the counting period only. The maximum recorded VEI is calculated using data for all Holocene eruptions (regardless of the volcano's frequency status), to maximise the likelihood of capturing the volcano's extreme events.

22.3.3 Hazard indicator scores

Modal and maximum recorded VEI are used as a reference in assigning numerical scores to the pyroclastic flow, mudflow, lava flow, and eruption frequency indicators. Comparison of the percentage of fatalities caused by the three phenomena with the percentage of fatalities caused by eruptions of each VEI suggests that pyroclastic flows and eruptions of VEI 4 generate similar impacts; pyroclastic flows have caused 44% of fatalities, whilst VEI 4 eruptions have caused 37%. This comparison leads to assigning a score of 4 to pyroclastic flow occurrence, when 10% or more of eruptions within the counting period have recorded pyroclastic flows. The total number of fatalities caused by mudflows is 50% those caused by pyroclastic flows, and by lava flows 2%. These proportions yield scores of 2 and 0.1 for mudflows and lava flows, respectively, when 10% or more of eruptions within the counting period have these flows recorded. A score of 0 is given for each of the pyroclastic flow, mudflow and lava flow indicators if the relevant phenomenon is recorded in fewer than 10% of eruptions.

Eruption frequency is used multiplicatively rather than additively in our method. As such, scoring of the frequency indicator need not be proportional to the scores of the other indicators; the only requirement is that active volcanoes are scored highest and fully-dormant volcanoes lowest, and that the scores for the four frequency status classes are proportional to each other in terms of their representation of hazard. A score of 1 is used for fully-dormant volcanoes, 1.5 is used for semi-dormant volcanoes, 2 for semi-active volcanoes, and a sliding scale from 2 to 3 is used for active volcanoes, calculated using $2+\left(\frac{N}{113}\right)$, where N is the number of years in which the volcano is recorded as erupting since AD 1900.

22.3.4 Calculating the hazard score

Scores for the indicators are combined to give a volcano-specific hazard score using the following conceptual structure:

[eruption frequency × ('frequent' characteristics of volcano's eruptions)] + extreme characteristics

This can be expressed in terms of the aforementioned indicators as:

[frequency status score × (modal VEI + PF score + mudflow score + lava flow score)] + maximum recorded VEI

with indicators for modal VEI, pyroclastic flow occurrence, mudflow occurrence and lava flow occurrence calculated using data for eruptions from the counting period only; maximum recorded VEI is calculated using all available Holocene data.

Maximum VEI is the only indicator score not multiplied by the frequency status score. Testing showed that multiplicative use of the maximum recorded VEI score gave distorted results and very high overall scores for some volcanoes that have particularly large maximum recorded VEIs. Simply adding on the maximum recorded VEI score moderates this propensity, and dilutes the weight associated with infrequent extreme eruptions.

The full method is as follows:

Indicator	Class	Criteria	Scoring
Eruption frequency	Fully dormant	- No time in eruption recorded since AD 1900 and No recorded unrest since AD 1900	1
	Semi-dormant	- No Holocene eruptions but unrest recorded since AD 1900 Or - Holocene (pre-AD 1500) eruptions but no recorded unrest since AD 1900	1.5
	Semi-active	- Holocene (pre-AD 1500) eruptions and unrest since 1900 Or - Historical (AD 1500-1900) eruptions with or without unrest since AD 1900	2
	Active	- One or more years with eruptions recorded since AD 1900	$2+\left(\dfrac{N}{113}\right)$ where N is the number of years in which the volcano is recorded as erupting since AD 1900
Pyroclastic flow occurrence	Pyroclastic flows are a significant hazard	Pyroclastic flows are recorded in 10% or more of eruptions occurring partially or fully within the volcano's counting period	4

Indicator	Class	Criteria	Scoring
Mudflow occurrence	Pyroclastic flows are not a significant hazard	Pyroclastic flows are recorded in fewer than 10% of eruptions occurring partially or fully within the volcano's counting period	0
	Mudflows are a significant hazard	Mudflows are recorded in 10% or more of eruptions occurring partially or fully within the volcano's counting period	2
	Mudflows are not a significant hazard	Mudflows are recorded in fewer than 10% of eruptions occurring partially or fully within the volcano's counting period	0
Lava flow occurrence	Lava flows are a significant hazard	Lava flows are recorded in 10% or more of eruptions occurring partially or fully within the volcano's counting period	0.1
	Lava flows are not a significant hazard	Lava flows are recorded in fewer than 10% of eruptions occurring partially or fully within the volcano's counting period	0
Modal VEI	N/A	The modal VEI of eruptions recorded with a known VEI within the volcano's counting period is X. A minimum of four such eruptions are required. Where there is no mode, the mean is used	X
Maximum recorded VEI	N/A	The greatest VEI of any eruption recorded within the volcano's Holocene eruptive history is Y	Y

22.3.5 Applicability and data constraints

For many volcanoes in VOTW4.0, data are scarce. The hazard assessment methodology requires enough data for scores to be assigned to all components of the index algorithm. The minimum amount of data required to apply the index is four or more eruptions within the volcano's counting period with a known VEI.

Volcanoes with insufficient data, i.e. those with fewer than four eruptions of known VEI within the counting period are unclassified as a classification would be accompanied by such large uncertainties as to make this irrelevant.

There are 328 volcanoes that satisfy this threshold for sufficient data to apply the scoring method fully. Such volcanoes are termed 'classified'. For the 1,223 volcanoes that do not meet these data requirements, termed 'unclassified', an alternative method that includes an assessment of the uncertainty related to lack of data must be used.

The calculation of VHI can be skewed by the inclusion of an unrepresentative modal VEI or prevented by the description of few events due to the persistent nature of activity at some volcanoes. Typically, activity separated by less than three months of apparent repose is considered an ongoing eruption. This persistent activity typically comprises small explosions or continuous effusions with infrequent bursts of larger activity. In VOTW4.22, the VEI for a persistent period of activity is taken as the maximum attained in that time. This can skew the calculation of the modal VEI upwards and result in an anomalously high VHI. These issues are accounted for through the identification of long eruptions, specifically those where the number of years in eruption in the counting period is greater than twice the number of eruptions in that period. Detailed bulletin reports accompanying VOTW4.22 have been studied to better constrain common activity and individual events. This approach will be improved and refined to ensure objectivity.

22.4 VHI application for GAR15

VHI scores have been determined for the Holocene volcanoes of VOTW4. The scores for classified volcanoes range from 0 to 30. The volcanoes' hazard scores are divided into the three hazard levels: VHI Level I (Scores 0 to <8; 134 volcanoes), VHI Level II (Scores 8 to <16; 106 volcanoes), and VHI Level III (Scores 16+; 88 volcanoes). These hazard levels are used to reflect the semi-qualitative nature of some of the data used and the approximations employed in creating and applying the method.

Ultimately, the Hazard Levels must be combined with measures of exposure in order to make statements about risk. For example, VHI Level I volcanoes may cause huge impacts if located sufficiently close to vulnerable populations or infrastructures. At this stage both the VHI and PEI are semi-quantitative, being presented as levels based upon numerical values, and risk cannot therefore be calculated as a strict product of hazard and exposure. However, plots of hazard levels against the Population Exposure Index (PEI, see Chapter 4) for each volcano in each country provide a useful visualisation of a risk matrix. This matrix is derived using a qualitative assessment of the product of the VHI and PEI, amended to consider the potential impact of hazardous phenomena within highly populated areas regardless of the hazard level. Risk is defined at three levels, I, II and III with increasing risk, shown by the warming of the colours (Figure 22.1).

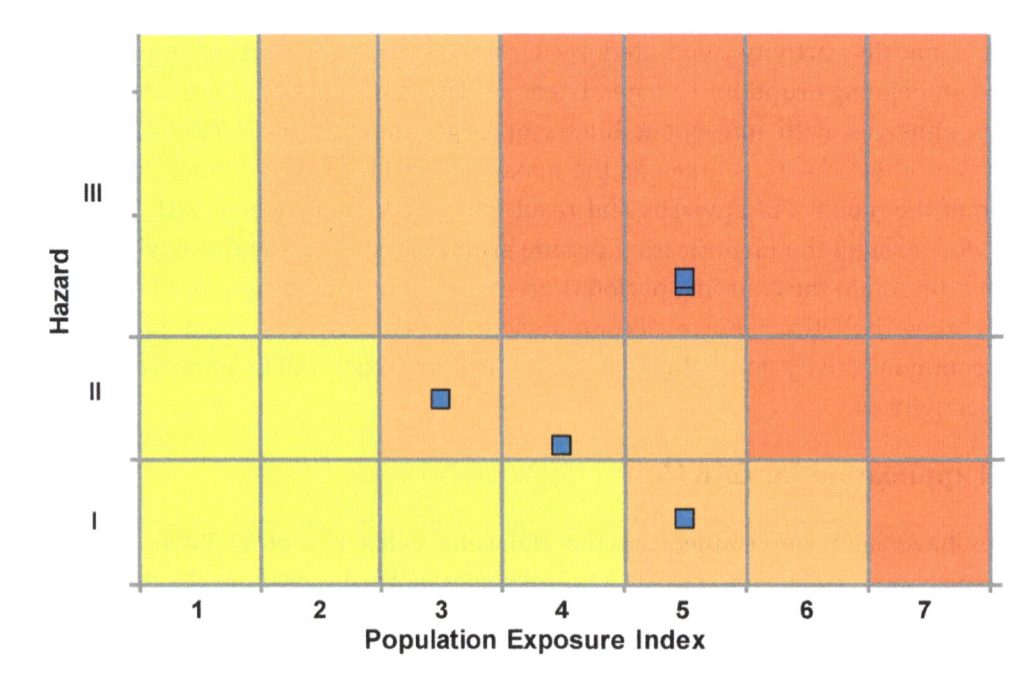

Figure 22.1 An example matrix combining VHI and PEI levels to indicate level of risk. Each volcano is represented by a point plotted using its hazard score and PEI level. The warming of the colouring of the matrix squares represents increasing risk (Risk Level I is yellow; Risk Level II is orange; Risk Level III is red).

The granularity within the matrix prevents detailed assessment and the matrix is therefore intended as a tool for the relative ranking of volcanoes. It should not be seen as a quantitative tool, as it comprises two ordinal rating scales which could be considered qualitative descriptors. This should not be used to undertake further calculations.

Unclassified volcanoes are presented with the PEI and are grouped according to their eruption record to permit an indication of known activity and for the use of PEI for an approximation for risk.

This globally applied assessment of VHI, PEI and ultimately risk does not substitute for focussed, local assessments. The PEI, for example, considers the population within concentric circles around a volcano, though in reality the exposed population will be governed by a number of factors (e.g. topography, which can shield a population on one side of the volcano and channel hazardous flows towards populations on the other). The impact on the human population is also determined by vulnerability, which is not considered here. The assessment of risk is based on these broad hazard and exposure assessments and therefore does not capture the full complexity of the situation. However, the ranking of volcanoes using this method can help identify volcanoes where monitoring and mitigation resources may need to be focussed and where localised hazard and risk assessments may be a priority.

This information is presented in the Country Profiles report (Appendix B), where the distribution of volcanoes across Hazard Levels, PEI and Risk Levels is briefly discussed for each country. Volcanoes which, at present, lack sufficient data to properly constrain their Hazard Levels should receive attention.

References

Aspinall, W., Auker, M., Hincks, T., Mahony, S., Nadim, F., Pooley, J., Sparks, R. & Syre, E. 2011. Volcano hazard and exposure in GFDRR priority countries and risk mitigation measures-GFDRR Volcano Risk Study. *Bristol: Bristol University Cabot Institute and NGI Norway for the World Bank: NGI Report,* 20100806, 3.

Auker, M. R., Sparks, R. S. J., Siebert, L., Crosweller, H. S. & Ewert, J. 2013. A statistical analysis of the global historical volcanic fatalities record. *Journal of Applied Volcanology,* 2, 1-24.

Deligne, N. I., Coles, S. G. & Sparks, R. S. J. 2010. Recurrence rates of large explosive volcanic eruptions. *Journal of Geophysical Research: Solid Earth,* 115, B06203.

Ewert, J. W. 2007. System for ranking relative threats of U.S. volcanoes. *Natural Hazards Review,* 8, 112-124.

Furlan, C. 2010. Extreme value methods for modelling historical series of large volcanic magnitudes. *Statistical Modelling,* 10, 113-132.

Siebert, L., Simkin, T. & Kimberley, P. 2010. *Volcanoes of the World, 3rd edn,* Berkeley, University of California Press.

Witham, C. 2005. Volcanic disasters and incidents: a new database. *Journal of Volcanology and Geothermal Research,* 148, 191-233.

Chapter 23

Global distribution of volcanic threat

S.K. Brown, R.S.J. Sparks and S.F. Jenkins

23.1 Calculating threat

Within the country profiles (Appendix B) individual volcanoes are ranked by risk; however, it would also be beneficial to understand the total volcanic threat borne by each country.[1] We therefore develop two measures of volcanic threat[2] to enable country ranking. The measures variously combine the number of volcanoes in the country, the size of the total population living within 30 km of volcanoes and the mean hazard score, which is calculated for each country from the relevant volcano hazard scores (VHI). We develop and use a 'Pop30' score, which calculates the number of persons, using Landscan 2011 (Bright et al., 2012) data, within a given country living within 30 km of one or more volcanoes with known or suspected Holocene activity. Note that 30 km is chosen as most fatal incidents that are caused directly by volcanic hazards fall within this distance of volcanoes [see Chapter 4]. VPI_{30}, supplied by VOTW4.0 (Siebert et al., 2010) based on the analysis of Ewert & Harpel (2004) and Siebert et al. (2008), is specific to a volcano and thus cannot be used in place of Pop30 as this would double count persons living within 30 km of neighbouring volcanoes.

We first develop a simple measure of volcanic threat to life country by country based on the number of active volcanoes, an estimate of exposed population and the mean hazard index of the volcanoes. The sum of this measure (Measure 1) for all countries is itself a simple measure of total threat and so the distribution of threat between countries can be evaluated and they can be placed in rank order using a normalised version of Measure 1. However, this measure of threat distribution can be misleading because an individual country may vary considerably in the proportion of its population that is exposed to the volcanic threat. Volcanic threat is very much higher in relation to its economy and population in a small island nation with an active volcano than in larger countries even if they have many volcanoes. Nation states vary greatly in their populations from, for example, China with 1.3 billion people (<1% exposed) to St. Kitts and Nevis in the Caribbean with only 54,000 people (100% exposed). Thus we need a measure of threat that reflects its importance to each country. Here we develop a measure (Measure 2) that rates the importance of volcanic threat in each country based on the proportion of the

---

[1] The phrase "country" is used here to denote both countries and some territories, e.g. overseas territories are classed separately to the nation state.
[2] We use threat rather than risk to describe these measures. Threat is defined here as the combination of hazard and exposure. Risk requires assessment of vulnerability, which has many different influences. Some jurisdictions can have high threat but low risk because steps have been taken to reduce the vulnerability (for example through having a well managed and equipped volcano observatory).
Brown, S.K., Sparks, R.S.J. & Jenkins, S.F. (2015) Global distribution of volcanic threat. In: S.C. Loughlin, R.S.J. Sparks, S.K. Brown, S.F. Jenkins & C. Vye-Brown (eds) *Global Volcanic Hazards and Risk,* Cambridge: Cambridge University Press.

population that is exposed: numbers of volcanoes and the total exposed population are not included in the calculation.

There are some caveats and limitations to our measures. Clearly the measures are not a full evaluation of risk and in particular do not take account of vulnerability. In general populations in high income countries are less vulnerable to loss of life than in low-income countries for a wide variety of reasons. Thus a risk measure might usefully include measures of vulnerability to natural hazards, such as GDP, the Human Development Index (HDI) and the World Risk Index (WRI). For volcanoes these general indicators of vulnerability might not be adequate; for example a measure specific to volcanic hazard should include the existence and resourcing of a volcano observatory. There was not time in this study to explore possible ways that our measures might be combined with vulnerability indicators. If the measures of country volcanic threat were to be combined with vulnerability measures there would be an issue of how to weight the vulnerability indices relative to the hazard and exposure data.

This global assessment of volcanic threat must be understood as a tool for relative ranking based on coarse global data. This approach cannot substitute for focussed local assessments of hazard and risk, as vital information such as topography, which exerts strong controls on hazard emplacement and population exposure, cannot be incorporated into our assessments at present.

23.2 Data completeness

The assessment of threat per country is partially dependent on the hazard classification for the constituent volcanoes. About 20% of the world's volcanoes have been assigned a hazard score, VHI, on the basis of their eruption records [see Chapter 22 and individual country profiles for results]. The use of these classified volcanoes to inform global threat distribution limits the number of countries that can be analysed, with approximately half of the countries having no classified volcanoes.

Hazardous phenomena and eruption size are somewhat associated with volcano morphology, as it is the nature of eruptions which largely determines volcano structure. The volcano type can therefore be used to provide a very approximate indicator of the hazard level at unclassified volcanoes. All volcanoes are grouped into similar types, as indicated by their morphology (the classification of types is adapted from Jenkins et al. (2012), Table 23.1), and the mean hazard scores of the classified volcanoes of each volcano type can be used as proxies for the unclassified volcanoes.

Table 23.1 Volcano type classification modified after Jenkins et al. (2012).

Volcano type group	Includes VOTW4.0 volcano types
Caldera(s)	Caldera, Caldera(s), Pyroclastic shield
Large cone(s)	Complex, Compound, Somma, Stratovolcano, Stratovolcano(es), Volcanic Complex
Shield(s)	Shield, Shield(s)
Lava dome(s)	Lava dome, Lava dome(s)
Small cone(s)	Cinder cone, Cinder cones, Cones, Cone, Crater rows, Explosion craters, Fissure vent(s), Lava cone, Maar, Maar(s), Pyroclastic cone(s), Scoria cones, Tuff cones, Tuff rings, Volcanic field
Hydrothermal field	Hydrothermal field, Hydrothermal field(fumarolic)
Submarine	Submarine
Subglacial	Subglacial

Substitution of proxy VHI scores at unclassified volcanoes in practice introduces rather limited uncertainty with most of these volcanoes being scored over a narrow range, with the key drivers of threat ranking being the number of volcanoes and the size of the population within 30 km.

The following measures therefore use a combination of data from classified and unclassified volcanoes. The percentage of volcanoes per country which are classified and on which the ranking is partially controlled by is presented to provide a sense of data quality.

23.3 Volcanic threat to life by country (Measure 1)

A measure of overall threat in a country is obtained using the following equation:

$$Overall\ threat = mean\ VHI\ x\ number\ of\ volcanoes\ x\ Pop30$$

The sum of the resultant scores for all countries with active volcanoes is an indicator of total global volcanic threat. The countries are normalised by this total and ranked as a percentage of the total global threat:

$$\frac{Overall\ threat\ score}{Total\ global\ threat\ score} x\ 100$$

Indonesia scores the highest level of threat and accounts for about two thirds of the total score (Table 23.2) as a consequence of the number of volcanoes (142), extent of population exposure (nearly 69 people million live within 30 km of a Holocene volcano) and the number of Hazard Level II and III volcanoes. The Philippines, which has the second highest rank has just 16% of the score of Indonesia. The Philippines has a similar mean VHI to Indonesia, but has about a third of the number of volcanoes (47) and less than half the exposed population (still over 30 million people). Japan ranks third for overall threat to life, with a comparatively small exposed population of about 9 million, reflecting concentration of the population in Japan in coastal cities and communities. All countries ranked in the top ten for overall volcanic threat have exposed populations of over 4 million.

The global distribution of this volcanic threat is illustrated in Figure 23.1, where the warming of the colours indicates increasing risk rank.

Table 23.2 The top 20 countries with highest overall volcanic threat to life. The percentage classified is the percentage of volcanoes in the country which have a classified VHI. The normalised percentage represents the country's threat as a percentage of the total global threat.

% classified	Rank	Country	Normalised %	% classified	Rank	Country	Normalised %
40	1	Indonesia	66.0	18	11	Papua New Guinea	0.4
17	2	Philippines	10.6	37	12	Nicaragua	0.4
38	3	Japan	6.9	33	13	Colombia	0.4
10	4	Mexico	3.9	0	14	Turkey	0.4
2	5	Ethiopia	3.9	50	15	Costa Rica	0.3
17	6	Guatemala	1.5	0	16	Taiwan	0.2
31	7	Ecuador	1.1	8	17	Yemen	0.2
43	8	Italy	0.9	14	18	Chile	0.2
14	9	El Salvador	0.8	29	19	New Zealand	0.2
5	10	Kenya	0.4	0	20	China	0.2

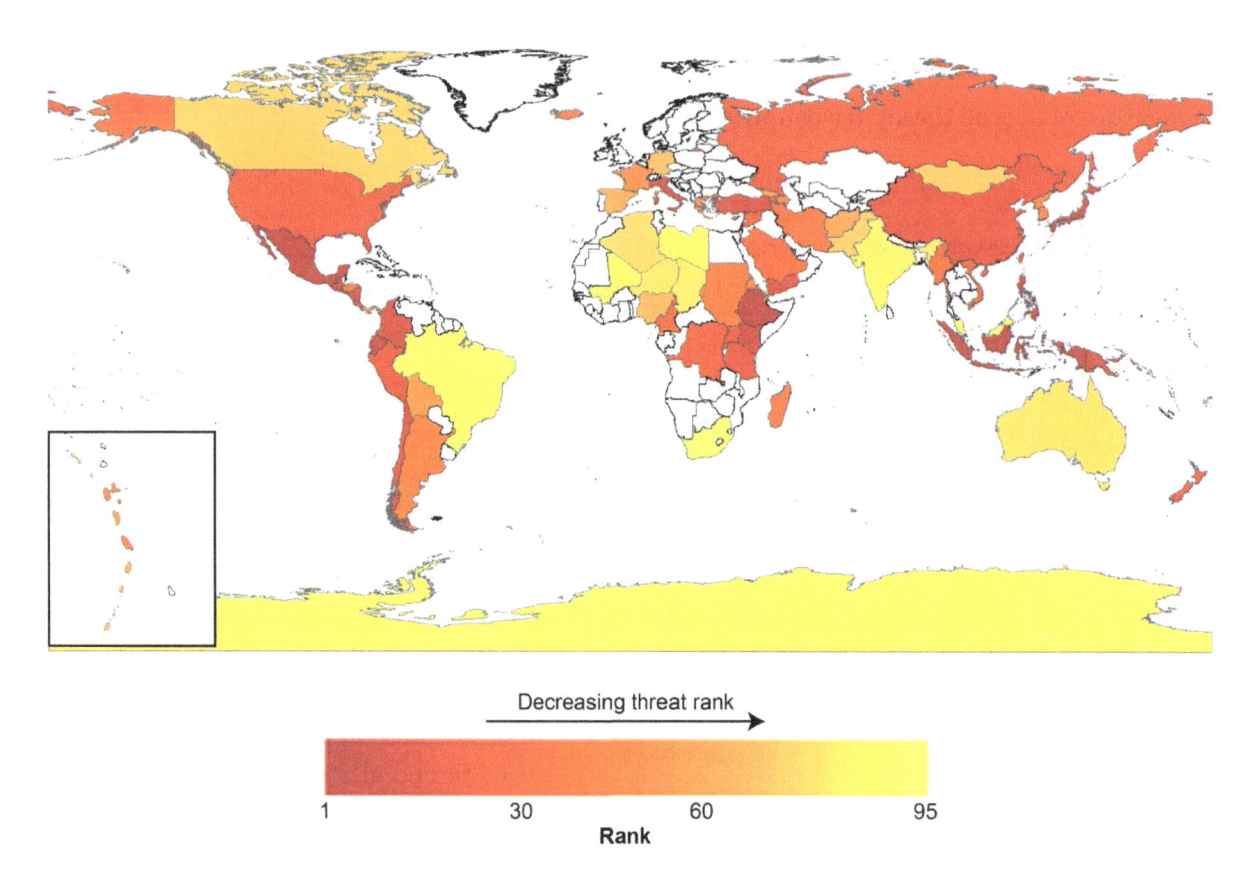

Figure 23.1 Global distribution of volcanic threat to life. Inset map shows the West Indies.

23.3.1 Distribution of volcanic threat and fatalities

Auker et al. (2013) undertook an analysis of fatality distributions, and found that Indonesia, Melanesia and the Philippines have had the highest number of fatalities (with the largest ten disasters removed). The regions considered in Table 23.3 are amended from the standard regions of VOTW4.0, to correspond with those used in Auker et al. (2013) incorporating the volcanic threat data for only those countries in which fatalities are recorded. Indonesia's history of fatal incidents corresponds well with the overall volcanic threat, and indeed ten regions only change in rank by a maximum of two positions, indicating a reasonable correlation between the overall threat and occurrence of fatalities.

Table 23.3 Regional ranking of volcanic fatalities (from Auker et al. (2013) and the threat measure.*The regions used here comprise only the countries or territories named, allowing for comparison of ranks with the fatality data. The percentage of fatalities per region with the largest ten disasters removed is shown (Auker et al., 2013).

Overall threat rank	Region* (Country)	Fatalities rank	% of fatalities
1	Indonesia (Indonesia)	1 (=)	38
2	Philippines and China (Philippines, SE China)	3 (-1)	10
3	Japan (Japan)	6 (-3)	8
4	Mexico and Central America (Costa Rica, El Salvador, Guatemala, Mexico, Nicaragua)	4 (0)	10
5	Africa and Red Sea (Cameroon, DRC, Ethiopia, Tanzania)	9 (-4)	3
6	South America (Chile, Colombia, Ecuador, Peru)	7 (-1)	5
7	Mediterranean (Italy, Greece, Turkey)	5 (+2)	9
8	Melanesia (Papua New Guinea, Solomon Islands, Vanuatu)	2 (+6)	11
9	New Zealand to Fiji (New Zealand, Tonga)	11 (-2)	0.53
10	North America (Alaska, Canada, USA-contiguous states)	12 (-2)	0.11
11	Atlantic Ocean (Azores, Canary Islands, Cape Verde)	10 (+1)	0.90
12	Kuril Islands and Kamchatka (Russia)	14 (-2)	0.07
13	Indian Ocean (Comoros, French territories)	15 (-2)	0.05
14	Iceland (Iceland)	16 (-2)	0.02
15	West Indies (Martinique and Guadeloupe, Montserrat, St. Vincent and the Grenadines)	8 (+7)	4
16	Hawaii (Hawaii)	13 (+3)	

The correlation between overall threat and regional distribution of fatalities where the largest ten disasters are included is less clear, with just five regions being of similar rank (within 2 positions). These high fatality events can significantly alter the regional ranking, and are shown by Auker et al. (2013) to dominate the fatalities record in several regions, obscuring the record of smaller events.

23.4 Proportional threat – Measure 2

The calculation of volcanic threat in Measure 1 considers the total number of people exposed and the number of volcanoes within a country. We have developed a second measure that is independent of country size but indicates how important volcanic risk is to each country. The following measure (Measure 2) is used:

$$Proportional\ threat = \frac{Pop30}{TPop} \times Mean\ VHI$$

The countries in which volcanic threat is highly significant in terms of the proportion of population exposed are small-area nations. The top 20 countries or territories ranked most highly using this measure are dominantly countries of Central and South America and small island nations or territories. All islands of the West Indies, with the exception of the Dutch Antilles, are ranked in the top 20, as most have comparatively high mean hazard scores and significant proportions of their populations living within 30 km of a volcano. The Dutch Antilles ranks at position 24, with several non-volcanic islands located in the southern Caribbean Sea off the coast of Venezuela.

Table 23.4 The top 20 countries or territories ranked by an index of proportional threat: the product of the proportion of the population exposed per country and the mean VHI.

% Classified	Rank	Country	% Classified	Rank	Country
100	1	UK- Montserrat	17	11	Guatemala
100	2	St. Vincent & Grenadines	0	12	Sao Tome & Principe
100	3	France – West Indies	33	13	Canary Islands
0	4	St. Kitts & Nevis	50	14	Grenada
0	5	Dominica	43	15	Vanuatu
29	6	Azores	37	16	Nicaragua
0	7	St. Lucia	0	17	Samoa
0	8	UK – Atlantic	0	18	American Samoa
14	9	El Salvador	0	19	Armenia
50	10	Costa Rica	17	20	Philippines

There are some strong caveats about the rankings in Table 23.4 and the information should not be over-interpreted. As emphasised earlier the assessment is quite crude and takes no account of important local factors, including the detailed distribution of populations and, the specifics of the particular volcano in a small island state. Here it is even more important not to conflate the threat measure with risk. Many of the jurisdictions in Table 23.4 are small territories with only one volcano and so a complete assessment of risk and ranking against other jurisdictions would need to take account of many local factors that affect vulnerability. In some jurisdictions the threat can be ranked high but the risk is in fact low and vice versa; the relationship between threat and risk is now explained.

Montserrat appears at the top of the list but such a ranking would be highly misleading if the measure were used to imply high risk. The Soufriere Hills Volcano, Montserrat, has a well-established volcano observatory and the population has been relocated to the north of the island, which is now at very low risk because of the intervening topography. Thus, even though the population all live within 30 km, vulnerability and hence risk is actually very low. The volcanic threat though on Montserrat remains high, and continues to prevent re-population of areas where most people lived before the eruption, requiring the continued vigilance of a well-founded Observatory.

Indonesia and the Philippines ranked most highly for threat by Measure 1, but these countries drop in rank to 23 and 20 respectively when using Measure 2. Measure 2 cannot be used to infer either how risk is distributed globally or to rank in terms of risk, but highlights small nations with high exposure to volcanic hazards in relation to their size (Figure 23.2).

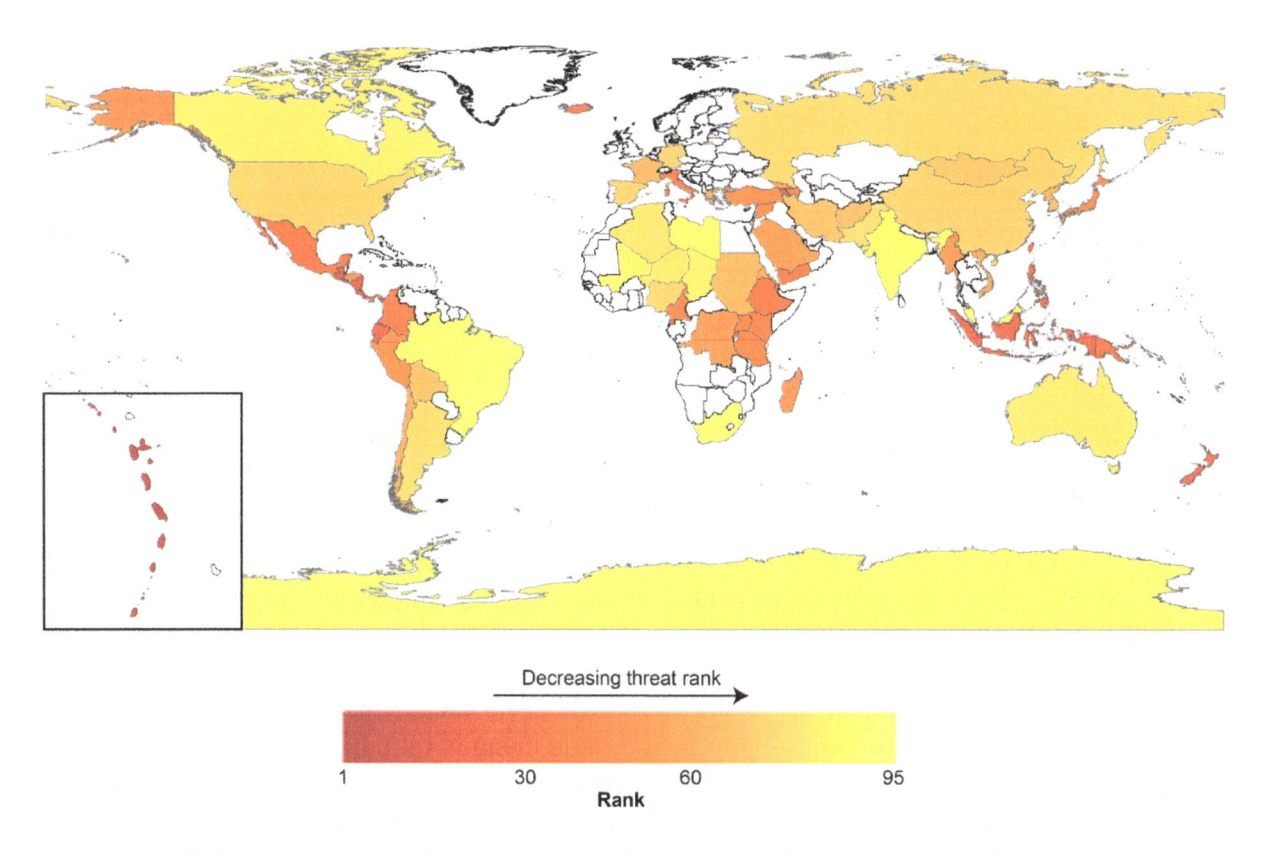

Figure 23.2 Global distribution of proportional risk. Inset map shows the West Indies.

23.4.1 Regional distribution of proportional threat

Many of the highest ranking regions for proportional threat comprise multiple small island groups: notably the small island nations and territories in the West Indies, the island groups of the Canaries, the Azores and Cape Verde in the Atlantic, and those of Fiji, Samoa and Tonga in

New Zealand to Fiji (Table 23.5). Not all of the highest ranked regions comprise small island groups. Mexico and Central America ranks highly, comprising multiple nations in which high proportions of the population are exposed. Africa and the Red Sea region also ranks highly, comprising countries that range in size from small (e.g. Sao Tome and Principe, 964 km² area (United Nations Statistics Division, 2014)) to large (e.g. Algeria, 2,381,741 km² (United Nations Statistics Division, 2014)) resulting in a range of exposed populations from less than 1% of the country's total to 97%. It is those small nations which control this region's ranking.

Table 23.5 Proportional threat as ranked by region. Note the Kuril Islands region is not included due to the absence of population data. The percentage shows the percentage risk of the top ranked region: e.g. Indonesia has about 3% of the proportional risk of the West Indies.

Proportional threat rank	Region	% of top region	Proportional threat rank	Region	% of top region
1	West Indies	100	10	Philippines & SE Asia	4
2	Mexico & Central America	35	11	Indonesia	3
3	Atlantic Ocean	32	12	Japan, Taiwan, Marianas	3
4	Africa & Red Sea	17	13	Iceland & Arctic	2
5	New Zealand to Fiji	14	14	Alaska	<1
6	Melanesia & Australia	9	15	Hawaii & Pacific	<1
7	Mediterranean & West Asia	9	16	Kamchatka & Mainland Asia	<1
8	Middle East & Indian Ocean	8	17	Canada & Western USA	<1
9	South America	5	18	Antarctica	-

23.5 Discussion

There are numerous methods available for the classification and determination of global volcanic threat. Here we only consider threat to life. The two ranking systems adopted here are shown in Table 23.6 in full.

Measure 1 allows the identification of those countries with the highest overall level of threat to life due to a combination of large numbers of people living within 30 km of an active volcano, large numbers of volcanoes and high hazard scores. Indonesia by far has the highest level of volcanic threat worldwide, with about 30% of the population living close to volcanoes. To better understand the importance of volcanic risk to individual countries, the calculation of the proportional threat is independent of the country size and number of volcanoes (Measure 2). This highlights those countries where large portions of their population live within close proximity of volcanoes – chiefly small island nations and territories where the population and volcanoes share small areas.

The differences in threat rank illustrate how whilst many countries could be expected to suffer large losses in absolute terms as shown by a high rank using Measure 1, it is the small island nations where the relative social and economic losses could be much larger (Measure 2).

Table 23.6 All countries or territories ranked in order of overall risk to life (Measure 1). Ranking through Measure 2, proportional threat, is also shown. The percentage of volcanoes per country which are classified is shown.

Country	% of volcanoes in country with classified VHI	Measure 1: Overall threat to life rank	Measure 2: Proportional threat rank
Indonesia	40	1	23
Philippines	17	2	20
Japan	38	3	43
Mexico	10	4	34
Ethiopia	2	5	36
Guatemala	17	6	11
Ecuador	31	7	22
Italy	43	8	33
El Salvador	14	9	9
Kenya	5	10	42
Papua New Guinea	18	11	31
Nicaragua	37	12	16
Colombia	33	13	39
Turkey	0	14	47
Costa Rica	50	15	10
Taiwan	0	16	29
Yemen	8	17	37
Chile	14	18	62
New Zealand	29	19	26
China	0	20	74
Tanzania	10	21	45
Peru	24	22	50
Uganda	0	23	44
USA Contiguous States	19	24	75
Russia	12	25	77
DR Congo	33	26	55
Syria	0	27	46
Cameroon	20	28	41
Spain: Canary Islands	33	29	13
Portugal: Azores	29	30	6
Vietnam	0	31	61
Armenia	0	32	19
Rwanda	0	33	30
Saudi Arabia	0	34	57
Burma (Myanmar)	0	35	53
Iran	0	36	69
Madagascar	0	37	54
France: Indian Ocean	11	38	27
Iceland	50	39	28
USA: Alaska	24	40	51

Vanuatu	43	41	15
Honduras	0	42	49
Sudan	20	43	67
South Korea	0	44	66
Argentina	10	45	78
France: West Indies	100	46	3
Greece	40	47	60
North Korea	0	48	65
France: Mainland	0	49	64
Eritrea	0	50	63
Azerbaijan	0	51	38
Panama	0	52	40
Solomon Islands	25	53	35
Comoros	50	54	21
USA: Hawaii	27	55	52
Cape Verde	33	56	25
Dominica	0	57	5
Bolivia	0	58	70
Afghanistan	0	59	68
Georgia	0	60	58
Equatorial Guinea	0	61	32
Spain: Mainland	0	62	72
Samoa	0	63	17
Saint Vincent & the Grenadines	100	64	2
Saint Lucia	0	65	7
Nigeria	0	66	76
Djibouti	0	67	48
Germany	0	68	73
USA: American Samoa	0	69	18
Grenada	50	70	14
Saint Kitts and Nevis	0	71	4
Sao Tome and Principe	0	72	12
Pakistan	0	73	79
Fiji	33	74	56
Mongolia	0	75	71
Canada	0	76	84
Tonga	33	77	59
Algeria	0	78	81
UK: West Indies	100	79	1
Netherlands	0	80	24
Australia	67	81	82
UK: Atlantic	0	82	8
Niger	0	83	83
Chad	0	84	85
France: Pacific Ocean	13	85	80
India	33	86	89
Mali	0	87	86
Libya	0	88	88
USA: Marianas Islands	14	89	87
Norway	33	90	90
South Africa	0	91	91

Malaysia	0	92	92
Brazil	0	93	93
Antarctica	13	-	-
Kuril Islands	27	-	-

References

Auker, M. R., Sparks, R. S. J., Siebert, L., Crosweller, H. S. & Ewert, J. 2013. A statistical analysis of the global historical volcanic fatalities record. *Journal of Applied Volcanology,* 2, 1-24.

Bright, E. A., Coleman, P. R., Rose, A. N. & Urban, M. L. 2012. *LandScan 2011* [Online]. Oak Ridge, TN, USA. Available: http://www.ornl.gov/landscan/

Ewert, J. W. & Harpel, C. J. 2004. In harm's way: population and volcanic risk. *Geotimes,* 49, 14-17.

Jenkins, S., Magill, C., McAneney, J. & Blong, R. 2012. Regional ash fall hazard I: a probabilistic assessment methodology. *Bulletin of Volcanology,* 74, 1699-1712.

Siebert, L., Ewert, J. W., Kimberley, P. & Shilling, S. P. 2008. Population in proximity to volcanoes: a global perspective. IAVCEI 2008 General Assembly, Reykjavik, Iceland, August 17-22, 2008.

Siebert, L., Simkin, T. & Kimberley, P. 2010. *Volcanoes of the World, 3rd edn,* Berkeley, University of California Press.

United Nations Statistics Division. 2014. *UN Data A World of Information: Country Profiles* [Online]. Available: http://data.un.org/CountryProfile.aspx

Chapter 24

Scientific communication of uncertainty during volcanic emergencies

J. Marti

24.1 Summary

Forecasting potential outcomes of volcanic unrest and activity is usually associated with high levels of scientific uncertainty. Knowing whether particular volcanic unrest will end with an eruption or not implies reliance on scientific knowledge on how the volcano has behaved in the past and on how monitoring signals can be interpreted in terms of magma movement. This may be relatively straightforward in volcanoes that erupt frequently, but may be much more challenging in volcanoes with long eruptive recurrence intervals or even more in those without historical records. The dramatic consequences that wrong interpretation of volcanic unrest signals may have should persuade volcanologists to understand that communication among them during an emergency is crucial. Consensus to quantify scientific uncertainty must be reached, in order to provide the decision maker with a simple and clear forecast of the possible outcome of the volcano reactivation. Unfortunately scientific communication during volcanic emergencies is not an easy task and there is not a general agreement on how such communication should be conducted, not only among scientists, but also between scientists and other stakeholders (e.g. decision makers, media, local population). The critical questions here, as occurs with other natural hazards, are how to quantify the uncertainty that accompanies any scientific forecast and how to communicate this understanding to policy-makers, the media and the public. In addition to scientific advance in eruption forecasting, future actions in volcanology should also address improving management of uncertainty and communication of this uncertainty.

24.2 Rationale

One of the most challenging aspects in the management of volcanic emergencies is scientific communication. Volcanology is by its nature an inexact science, such that appropriate scientific communication should convey information not only on the volcanic activity itself, but also on the uncertainties that always accompany any estimate or forecast. Deciphering the nature of unrest signals (volcanic reactivation) and determining whether or not an unrest episode may be precursory to a new eruption requires knowledge of the volcano's past and current behaviour to help establish future behaviour. In order to achieve such a complex objective it is necessary

Marti, J. (2015) Scientific communication of uncertainty during volcanic emergencies. In: S.C. Loughlin, R.S.J. Sparks, S.K. Brown, S.F. Jenkins & C. Vye-Brown (eds) *Global Volcanic Hazards and Risk,* Cambridge: Cambridge University Press.

to have different groups of individuals involved in information exchange, including those from disciplines such as geology, volcano monitoring, experimentation, modelling and probabilistic forecasting. Communication is required on a level that caters for needs and expectations of all disciplines; i.e. there is a need to share a common technical language. This is particularly relevant when volcano monitoring is carried out on a systematic survey basis without continuous scientific scrutiny of monitoring protocols or interpretation of data. In an emerging unrest situation, difficulties may arise with communication among scientists and between scientists and Civil Protection officers, decision makers, media or the public, due to the different skills and degree of knowledge of volcanic phenomena.

Of particular importance is the communication link between scientists with Civil Protection agents and decision makers during evolving volcanic crises. In this case, it is necessary to translate the scientific understanding of volcanic activity into a series of clearly explained scenarios that are accessible to the decision-making authorities. Also, direct interaction between scientists and the general public is inevitable both during times of quiescence and activity. Information coming directly from the scientific community has a special influence on risk perception and on public confidence in this information. Therefore, effective volcanic emergency management requires identification of feasible actions to improve communication strategies at different levels including: scientists-to-scientists, scientists-to-technicians, scientists-to-Civil Protection, scientists-to-decision makers, and scientists-to-general public.

The main goal of eruption forecasting seeks to respond to how, where, and when an eruption will occur. To answer these questions there is an emerging recognition that probabilities should be used for characterisation of associated uncertainties. However, communicating probabilities and, in particular, uncertainty, is not an easy task, and may require a very different approach depending on who is the receptor of such information. Making forecasts on future volcanic activity follows basically the same approach as in other natural hazards (e.g. storms, landslides, earthquakes, tsunamis). However, this approach does not necessarily require the same level of understanding by the population and decision-makers. Compared to meteorologists who have much more data and observations, volcanologists have to deal with a higher degree of uncertainty, mainly derived from this lack of observational data. It is also important to consider that all volcanoes behave in a different way, so a universal model to understand behaviour of volcanoes does not exist. Each volcano has its own particularities depending on magma composition and physics, rock rheology, stress field, geodynamic environment, local geology, etc., which make them unique, so that what is indicative in one volcano may be not relevant in another. All this makes volcano forecasting very challenging and even more difficult to communicate this high degree of uncertainty to the population and decision makers.

24.3 How to communicate volcano forecast?

Significant work has been done during last years to improve communication in natural hazards (Atman et al., 1994, Morgan et al., 2002, Karelitz and Budescu, 2004, Visschers et al., 2009, Stein and Geller, 2012). In a similar way, several studies on communication during volcanic emergencies have been carried out (Newhall et al., 1999, McGuire et al., 2009, Aspinall, 2010, Donovan et al., 2012a, Donovan et al., 2012b, Marzocchi and Bebbington, 2012, Doyle et al., 2014). The common factor in all these studies is the need to communicate the uncertainty that accompanies any forecast on the future behaviour of a natural system. Most of these studies agree that probabilities are the best way to communicate scientific forecasts and, consequently, its associated uncertainties. In this sense the use of probability theory is common in natural hazards (Cooke, 1991, Colyvan, 2008, Stein and Stein, 2013) and also in volcano forecasting (Aspinall and Cooke, 1998, Marzocchi et al., 2004, Aspinall, 2006, Sobradelo and Martí, 2010, Marzocchi and Bebbington, 2012, Donovan et al., 2012c).

Probability can be defined as one measure of uncertainty. The way in which probabilities are understood depends on the degree of numeracy we have. However, during our life we are everyday confronted with situations that require making decisions, and in many cases, even if we are not aware of doing this, we use probabilities (commonly known as 'common sense') to evaluate the degree of uncertainty in a decision. A common situation were we to use probabilities, even if they are not expressed mathematically, is when making decisions based on weather forecasts (e.g. do we take an umbrella? Do we go to an outdoor festival? etc.) that we see everyday in newspapers or on TV. We do not have problems with understanding and accepting a forecast that says the probability of having rain tomorrow is high. It is not necessary for the meteorologists to indicate this in a more precise way (e.g. the probability of rain for tomorrow is of 80%); although in some countries this is communicated. The accuracy of contemporary weather forecasts is typically quite high for periods of a few days in advance. But still there may be incorrect forecasts that may cause serious trouble, for example when bad weather is predicted people may change their plans and important economic losses may result.

Making predictions on the future behaviour of a volcano basically follows the same reasoning that is behind weather forecasts (analysis of past data, monitoring of the current situation and identification of possible future scenarios). However, typically there is less familiarity about the behaviour of volcanoes than about the weather. Thus volcano forecasts are not so easily understood by the population and decision makers. Although there are some scientists who are highly experienced communicators in well established volcano observatories and teams, there is still need for taking this practise to a higher extent as meteorologists have done. This is in part due to lack of observational data and, consequently, to the limitations that volcanologists have to obtain precise probability estimates. This makes it difficult to communicate volcano forecasts in a simple language, difficulty that usually increases because population and decision-makers are not as used to listening to volcano forecasts as they are with weather forecasts. Therefore, scientific communication of volcano forecasts needs to reduce the considerable distance that today still separates the proper scientific language from its understanding by a non-scientific audience.

The uncertainty that accompanies the identification and interpretation of eruption precursors, derives from the unpredictably of the volcano as a natural system (aleatory uncertainties) and from our lack of knowledge on the behaviour of the system (epistemic uncertainties). These uncertainties can be redefined as shallow or deep (Cox Jnr, 2012, Stein and Stein, 2013) depending on the eruption frequency of the volcano. Highly active volcanoes with high eruption frequencies can be more easily predicted (i.e. they are reasonably well known) than those characterised by low eruption frequencies, respectively.

There are different ways in which probabilities (and uncertainties) can be described. These include words, numbers, or graphics. The use of words to explain probabilities seeks to offer a language that appeals to people's intuition and emotions (Lipkus, 2007). However, it usually lacks precision as it tends to introduce significant ambiguity by the use of words such as 'probable', 'likely', 'doubtful', etc, which lack precision or clear definitions. Probabilities are defined mathematically, but such descriptions may fail when the audience has a low numeracy. In the last years it has been increasingly common to use graphics to represent probabilities in natural hazards (Kunz et al., 2011, Spiegelhalter et al., 2011, Stein and Geller, 2012). The advantage of communicating uncertainties (or probabilities) visually is that we are everyday better prepared and trained to use and understand infographics. A graphic can be adapted to the aims of the communicator, stressing the importance of the context of the communication exercise and the needs and capabilities of the audience (Spiegelhalter et al., 2011). Volcanologists can adapt these modern methodologies to their needs, in order to make volcano forecasts and their intrinsic uncertainty clear enough to any potential receptor of this information.

In addition to these 'formal' or 'academic' ways to describe and communicate probabilities (and uncertainties) there are other important aspects (namely odds, regulations and culture) related to each particular society, that volcanologists should take into account for communication purposes. 'Odds' is an expression of relative probabilities that is well understood by many communities (e.g. gambling, games of chance) and can also be effective to communicate volcano forecasting if it is correctly adapted to such a goal. Regulations are not a direct communication tool but are frequently used to manage environmental and natural hazards. Some regulations are widely understood or at least accepted by the public even if they don't understand the science behind them. Regulations are not widely used in volcanic crisis management but can be useful in communication. An example is in Case Study 18 where occupational risk regulations were used to explain risk to workers. Finally, culture is of key importance in communication. The cultural diversity of societies facing volcanic threat determines that some communication approaches may work in one country or culture but not in another. Therefore, it is important to analyse and understand the particular cultural aspects of each society in order to define the best communication procedures and languages in each case.

24.4 What should be communicated?

Forecasting the future behaviour of a volcano requires good knowledge of its past behaviour, based on the analysis of the geological record and/or historical eruptive records, and a precise understanding of its current activity through monitoring systems. Will the eruption occur? What style of eruption will it be? When the eruption will occur? Where the eruption will occur? What is the size of the problem?, are the basic questions that the decision makers will surely pose to the scientist once an alert has been declared and the process of managing a volcanic emergency has begun. Usually, scientists can answer these questions with approximations (with probabilities in some cases) based on knowledge of previous cases from the same volcano, or from other volcanoes of similar characteristics, knowledge of the past eruptive history of the volcano, the degree of accuracy in the detection of warning signals (geophysical and geochemical monitoring), and knowledge about the significance of these warning signs. Giving probabilities as outcomes of volcano forecasts may be relatively easy for the scientist depending on the degree of information available, but it may be not fully understood by the decision-maker or any other receptor of such information. It is necessary to find clear and precise ways to transfer this information from scientist to decision-maker, to avoid misunderstandings and misinterpretations that could lead to incorrect management of the volcanic crisis and, consequently, to a disaster.

Volcano forecasts should be focussed on the science, just communicating precise and clear scientific advice on the potential evolution of volcanic phenomena in the most appropriate terms, in order to make it understandable to all potential receptors. Scientists may also recommend safe behaviour directly to the public, providing advice that saves people's lives (e.g. go up a hill if a lahar threatens). However, this should not imply or be confused with making decisions on how to manage a volcanic emergency (e.g. evacuation), as this belongs strictly to the decision maker.

24.5 When volcano forecasts should be communicated?

Scientific communication in active volcanic areas should be always present. This means that there should be a permanent flow of information from scientists to the population and policy-makers on the eruptive characteristics of the volcano, its current state of activity, or its associated hazards, even when volcanoes do not show signs for alarm. This is crucial preparation for when an emergency starts and things need to move much faster. However, in many cases scientific communication in hazard assessment and volcano forecasting is just restricted to volcanic emergencies. This may reduce considerably the understanding of the scientific information and its reliability due to the previous lack of knowledge on these subjects by the population and decision-makers.

When volcanic unrest starts and escalates, the origin of this unrest needs to be investigated to assess the likelihood of evolving into an eruption. As previously mentioned, the calculation of probabilities will be subjected to considerable uncertainties, as most of the data will be obtained from monitoring systems, so they will constitute indirect evidence of what could be happening inside the volcanic system. In volcanoes with a high eruption frequency comparison with previous unrest episodes will assist understanding unrest.

Past experience shows that good detection and interpretation of precursors allows prediction of what will happen with a considerable degree of confidence. This implies that scientific communication during volcanic crises needs to be constant and permanently updated with the arrival of new data. The longer is taken in making a decision the higher could be the costs incurred, as reaction time decreases and vulnerability increases. This constitutes the main worry in managing volcanic crises, WHEN to make a decision, which in most cases could be to order an evacuation. In essence, the relationship between the decrease of uncertainty in the interpretation of the warning signs of pre-eruptive processes to acceptable (reliable) levels, and the time required to make a correct decision, is a function of the degree of scientific knowledge of the volcanic process and the effectiveness of scientific communication. Therefore, scientific communication during volcanic emergencies needs to be effective from the beginning of the process, but would be significantly improved if this communication channel has already been established when the level of activity of the volcano did not represent a cause of concern.

24.6 What needs to be done?

In order to improve scientific communication during volcanic crises comparisons between communication protocols and procedures adopted by different volcano observatories and scientific advisory committees is recommended, in order to identify difficulties and best practice at all levels of communication: scientist to scientist, scientist to technician, scientist to Civil Protection, scientist to general public. Experience from the management and communication of other natural hazards should be brought in and common communication protocols should be defined based on clear and effective ways of showing probabilities and associated uncertainties. Although each cultural and socio-economic situation will have different communication requirements, comparison between different experiences will help to improve each particular communication approach, thus reducing uncertainty in communicating eruption forecasts.

References

Aspinall, W. 2006. Structured elicitation of expert judgement for probabilistic hazard and risk assessment in volcanic eruptions. *In:* MADER, H. M. (ed.) *Statistics in Volcanology.* Geological Society of London.

Aspinall, W. 2010. A route to more tractable expert advice. *Nature,* 463, 294-295.

Aspinall, W. & Cooke, R. M. Expert judgement and the Montserrat Volcano eruption. Proceedings of the 4th international conference on Probabilistic Safety Assessment and Management PSAM4, 1998. 13-18.

Atman, C. J., Bostrom, A., Fischhoff, B. & Morgan, M. G. 1994. Designing risk communications: completing and correcting mental models of hazardous processes, Part I. *Risk Analysis,* 14, 779-788.

Colyvan, M. 2008. Is probability the only coherent approach to uncertainty? *Risk Analysis,* 28, 645-652.

Cooke, R. M. 1991. *Experts in uncertainty: opinion and subjective probability in science*, Oxford University Press.

Cox Jnr, L. A. 2012. Confronting deep uncertainties in risk analysis. *Risk Analysis,* 32, 1607-1629.

Donovan, A., Oppenheimer, C. & Bravo, M. 2012a. Science at the policy interface: volcano-monitoring technologies and volcanic hazard management. *Bulletin of Volcanology,* 74, 1005-1022.

Donovan, A., Oppenheimer, C. & Bravo, M. 2012b. Social studies of volcanology: knowledge generation and expert advice on active volcanoes. *Bulletin of Volcanology,* 74, 677-689.

Donovan, A., Oppenheimer, C. & Bravo, M. 2012c. The use of belief-based probabilistic methods in volcanology: Scientists' views and implications for risk assessments. *Journal of Volcanology and Geothermal Research,* 247, 168-180.

Doyle, E. E., Mcclure, J., Johnston, D. M. & Paton, D. 2014. Communicating likelihoods and probabilities in forecasts of volcanic eruptions. *Journal of Volcanology and Geothermal Research,* 272, 1-15.

Karelitz, T. M. & Budescu, D. V. 2004. You say "probable" and I say "likely": improving interpersonal communication with verbal probability phrases. *Journal of Experimental Psychology: Applied,* 10, 25-41.

Kunz, M., Grêt-Regamey, A. & Hurni, L. 2011. Visualization of uncertainty in natural hazards assessments using an interactive cartographic information system. *Natural hazards,* 59, 1735-1751.

Lipkus, I. M. 2007. Numeric, verbal, and visual formats of conveying health risks: suggested best practices and future recommendations. *Medical Decision Making,* 27, 696-713.

Marzocchi, W. & Bebbington, M. S. 2012. Probabilistic eruption forecasting at short and long time scales. *Bulletin of Volcanology,* 74, 1777-1805.

Marzocchi, W., Sandri, L., Gasparini, P., Newhall, C. & Boschi, E. 2004. Quantifying probabilities of volcanic events: the example of volcanic hazard at Mount Vesuvius. *Journal of Geophysical Research,* 109.

Mcguire, W., Solana, M., Kilburn, C. & Sanderson, D. 2009. Improving communication during volcanic crises on small, vulnerable islands. *Journal of Volcanology and Geothermal Research,* 183, 63-75.

Morgan, M. G., Fischhoff, B., Bostrom, A. & Atman, C. J. 2002. *Risk Communication: A Mental Models Approach,* Cambridge, Cambridge University Press.

Newhall, C., Aramaki, S., Barberi, F., Blong, R., Calvache, M., Cheminee, J.-L., Punongbayan, R., Siebe, C., Simkin, T., Sparks, R. S. J. & Tjetjep, W. 1999. Professional conduct of scientists during volcanic crises. *Bulletin of Volcanology,* 60, 323-334.

Sobradelo, R. & Martí, J. 2010. Bayesian event tree for long-term volcanic hazard assessment: Application to Teide-Pico Viejo stratovolcanoes, Tenerife, Canary Islands. *Journal of Geophysical Research: Solid Earth (1978–2012),* 115.

Spiegelhalter, D., Pearson, M. & Short, I. 2011. Visualizing uncertainty about the future. *Science,* 333, 1393-1400.

Stein, S. & Geller, R. J. 2012. Communicating uncertainties in natural hazard forecasts. *Eos, Transactions American Geophysical Union,* 93, 361-362.

Stein, S. & Stein, J. 2013. How good do natural hazard assessments need to be? *GSA Today,* 23, 60-61.

Visschers, V. H., Meertens, R. M., Passchier, W. W. & De Vries, N. N. 2009. Probability information in risk communication: a review of the research literature. *Risk Analysis,* 29, 267-287.

Chapter 25

Volcano Disaster Assistance Program: Preventing volcanic crises from becoming disasters and advancing science diplomacy

J. Pallister

25.1 VDAP

The Volcano Disaster Assistance Program (VDAP) is a cooperative partnership of the USAID Office of US Foreign Disaster Assistance (OFDA) and the US Geological Survey (USGS). Founded in 1986 in the wake of the Nevado del Ruiz catastrophe wherein more than 23,000 people perished needlessly in a volcanic eruption, VDAP works by invitation to reduce volcanic risk, primarily in developing nations with substantial volcano hazards. The majority of emergency responses and capacity building projects occur in, but are not limited to, Pacific Rim nations. The single most successful VDAP operation was its response with the Philippine Institute of Volcanology and Seismology to the reawakening and subsequent eruption of Mount Pinatubo in 1991. This response alone saved 20,000 lives, including US military personnel at Clark Air Base, and a conservative estimate indicates that at least 250 million dollars in tangible assets were removed from harm's way ahead of the eruption (Newhall et al., 1997). More recently, in late 2010 VDAP assisted Indonesia's Center for Volcanology and Geologic Hazard Mitigation respond to the eruption of Merapi volcano, which saved 10,000 to 20,000 lives.

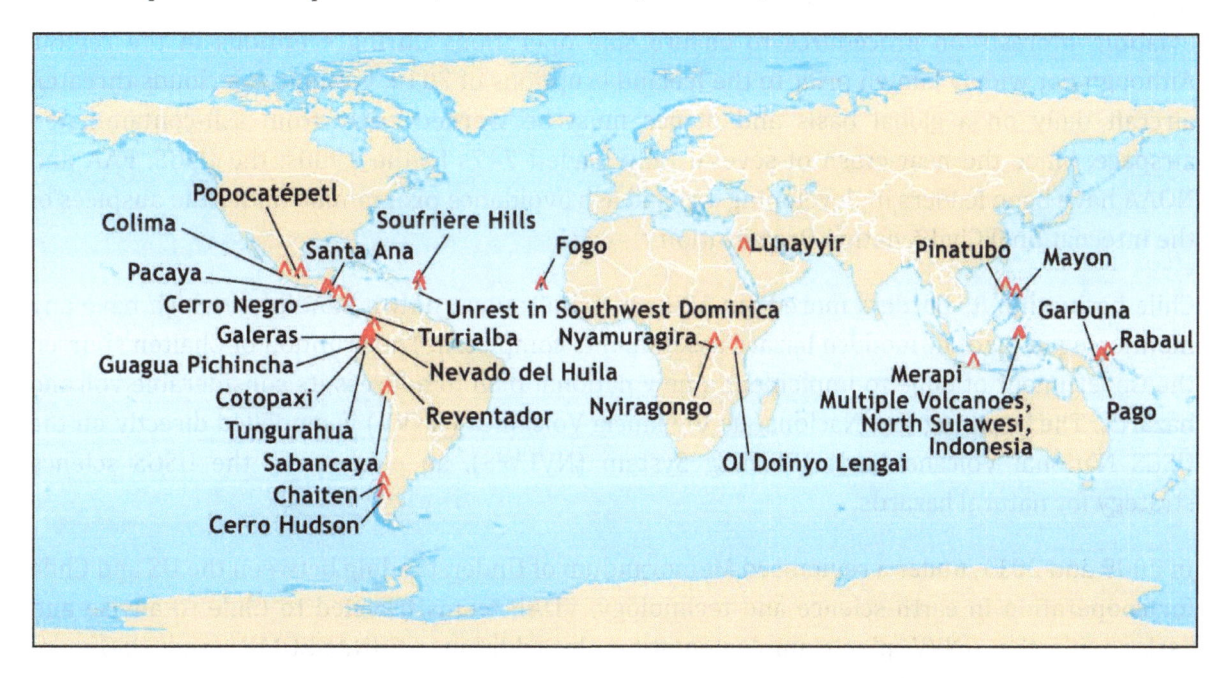

Figure 25.1 Map of VDAP deployments 1986-2012.

Pallister, J. (2015) Volcano Disaster Assistance Program: Preventing volcanic crises from becoming disasters and advancing science diplomacy. In: S.C. Loughlin, R.S.J. Sparks, S.K. Brown, S.F. Jenkins & C. Vye-Brown (eds) *Global Volcanic Hazards and Risk,* Cambridge: Cambridge University Press.

25.2 Current activities

The VDAP team is on call to respond to volcano emergencies globally with crisis response teams. Since 1986, VDAP has responded to 25 major volcano crises. In addition to this on-call activity, VDAP has conducted capacity-building projects and helped build or strengthen volcano hazards institutions in a dozen Pacific Rim countries. The VDAP approach is to work in the background to support and strengthen our partners' crisis response and hazard mitigation programmes.

25.3 Chile

On 2 May, 2008, Chaitén volcano in southern Chile suddenly re-activated after hundreds of years of dormancy and produced the largest eruption so far in the twenty-first century. A town of more than 4,000 people lying just 10 km from the volcano had to be evacuated within 48 hours of the eruption onset. A VDAP rapid-response team assisted the Chilean Servicio Nacional de Geología y Minería (SERNAGEOMIN) to deploy radio-telemetered monitoring instruments and forecast eruption hazards. Added to the disruption of lives and livelihoods on the ground, the Chaitén eruption severely disrupted air traffic in the region. In addition to the VDAP response, two experts on volcanic ash hazards to aviation from the USGS Volcano Hazards Program traveled to Argentina and Chile to advise civilian and military

Figure 25.2 Gas, ash, and steam erupt from the growing lava dome in the crater of Chaitén, southern Chile, May 26, 2008. USGS photo.

aviation interests on procedures to ensure safe operations during eruptions in the region. Although not widely known prior to the Iceland eruptions of 2010, volcanic ash clouds threaten aircraft daily on a global basis and planes must be warned away from ash-contaminated airspace. Since the near-crash of several fully loaded 747s in the 1980s, the USGS, FAA and NOAA have been leaders in developing a global ash avoidance programme under the auspices of the International Civil Aviation Organization.

Chile has within its borders more than 122 active volcanoes, only a handful of which have any monitoring in place or modern hazard assessments completed. The eruption of Chaitén spurred the Government of Chile to implement a new national plan to address its considerable volcano hazards. The resulting Red Nacional de Vigilancia Volcánica (RNVV) is modelled directly on the USGS National Volcano Early Warning System (NVEWS), an element of the USGS science strategy for natural hazards.

In 2009 and 2011, under a revitalised Memorandum of Understanding between the US and Chile for cooperation in earth science and technology, VDAP teams traveled to Chile to advise and assist with the RNVV plan's implementation. In addition, SERNAGEOMIN and USGS are continuing to work together to assess the volcanological and ecological impacts.

Figure 25.3 Group photo of VDAP team with President Michelle Bachelet, US Ambassador to Chile Paul Simons (left) and USGS Director's representative Tom Casadevall (right). Photo taken following briefing by the team on the situation at Chaitén volcano and SERNAGEOMIN's new national volcano early warning plan (Red Nacional de Vigilancia Volcánica).

25.4 Colombia

VDAP has had an ongoing collegial crisis response and capacity-building relationship with Colombia since the disaster at Nevado del Ruiz in 1985. Over the years, VDAP has worked closely with the Instituto Colombiano de Geología y Minería (INGEOMINAS) on various projects, including direct involvement in establishing the three volcano observatories now functioning in Colombia.

In 2007-2011 VDAP worked closely with the INGEOMINAS Observatory in Popayan to monitor and forecast eruptions at Huila volcano. Like Nevado del Ruiz, Huila is a large snow-and-ice-clad volcano with a history of producing exceedingly dangerous debris flows (lahars). On 20 November 2008, following a period of escalating unrest of several weeks, Huila erupted and generated a huge debris flow. Owing to accurate forecasting and good communications with downstream communities, fewer than 10 casualties occurred in an area where in 1994 an event of similar sized killed more than 1000. Most of the credit for the success of this risk mitigation effort belongs to INGEOMINAS and other involved Colombian institutions, but the effort was, and still is, substantially supported by VDAP.

Figure 25.4 USGS and INGEOMINAS personnel meet to define alert level system for Huila volcano, Colombia. February 2007. USGS photo.

Figure 25.5 Eruption-generated debris flow overran portions the town of Belalcazar, Colombia approximately 15 miles downstream of Huila volcano on 20 November 2008. Accurate forecasts and warnings triggered evacuation of the 4000 residents, saving lives and property. INGEOMINAS photo.

In 2008 a Memorandum of Understanding between the USGS and INGEOMINAS was signed, and subsequently the Director of INGEOMINAS has sought a Project Annex with VDAP to work with them to develop their volcano monitoring and analysis capabilities on all 15 Colombian volcanoes. Within the limitations of its resources, VDAP will continue to work with INGEOMINAS on volcano hazard mitigation in this important South American country.

25.5 Indonesia

Indonesia is the world's most volcanically active nation, with numerous eruptions each year and several million people living directly on the flanks of the volcanoes. Currently, VDAP's largest capacity-building project is conducted in partnership with the Indonesian Center for Volcanology and Geologic Hazard Mitigation (CVGHM). USGS collaboration with CVGHM's predecessor, the Volcanology Survey of Indonesia, dates back to the 1980s. At that time USGS helped evaluate hazards and establish monitoring at several of the highest-risk volcanoes, such as Merapi, located in the suburbs of Yogyakarta (metropolitan area population, 1.6 million). Following an absence of 20 years, at the invitation of CVGHM and with support of OFDA, VDAP returned to Indonesia in 2004 to help build a new regional volcano observatory in the North Sulawesi and Sangihe islands region. Over the succeeding five years,

CVGHM and VDAP built what is now one of the best monitoring networks in the country, with real-time seismic monitoring of the 10 active volcanoes in place and signals relayed in real-time by satellite and internet to CVGHM offices in Bandung. In 2006,

Figure 25.6 VDAP and CVGHM scientists install a seismic station to monitor volcanic activity in North Sulawesi. USGS photo.

VDAP sent a crisis response team to assist CVGHM during eruptions of Merapi volcano, which directly threatens the lives of several hundred thousand people. VDAP also provided seismic monitoring equipment, eruption forecasts, remote sensing, and a technical advisor to assist the OFDA Disaster Assistance Response Team (DART) during their response to the M 6.3 Yogyakarta earthquake, which took place during the eruption and killed 5,700.

In 2010 at the request of the President of Indonesia, a VDAP team was again deployed to assist CVGHM in their response to the largest eruption at Merapi in more than 100 years. VDAP and international partners utilised satellite radar data to "see through" clouds that obscured the volcano and delivered near-real-time analyses of changes directly to the CVGHM response team during the crisis. This information allowed CVGHM to assess the magnitude of the eruption and areas affected, thereby informing their decisions regarding the extent of evacuations needed. In addition, the on-site VDAP team provided technical assistance and monitoring equipment to replace systems destroyed in the early phase of the eruption and to expand the monitoring programme. Although ~380 fatalities were recorded, it is estimated that CVGHM warnings and prompt actions by the Government of Indonesia saved 10,000-20,000 lives.

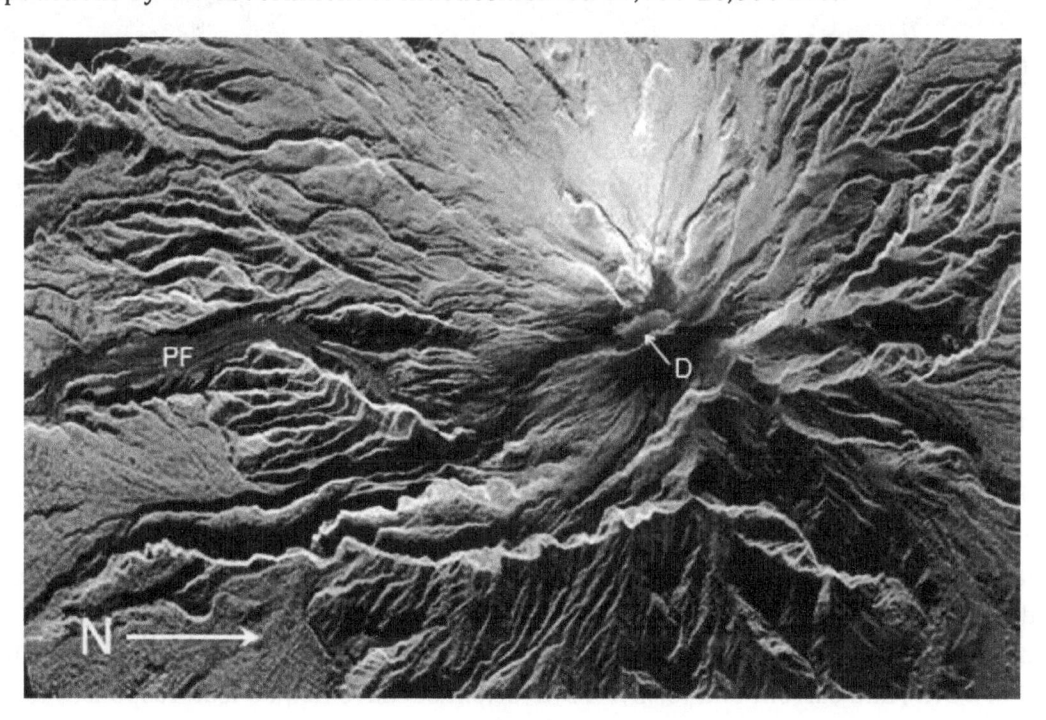

Figure 25.7 TerraSAR-X Synthetic Aperture Radar image of 4 November 2010, provided courtesy of the German Aerospace Center (DLR), showing pyroclastic flow deposits (PF) from the 26 October eruption and a new lava dome growing at the summit. Very rapid growth of the lava dome was followed by a large eruption during the night of 4-5 November.

US Ambassador Cameron Hume and the Director of the Indonesian Geology Agency, Bambang Dwiyanto, signed Annex IV to the 2006 Memorandum of Understanding between the governments of Indonesia and the US for Cooperation in Science and Technology for Natural Hazards. This Annex calls for continued VDAP assistance to CVGHM. Subsequently, VDAP was singled out among multiple international donors by CVGHM with a request to assist them in modernising volcano monitoring networks and hazard assessments in Java, where more than 100 million people live in the shadows of active volcanoes. This expansion of VDAP work was approved by OFDA and USGS in 2009. In 2010, VDAP and CVGHM completed monitoring installations in North Sulawesi, conducted joint training workshops and began the Java expansion with new installations and hazard assessments at Tangkuban Perahu volcano, the highly populated "city volcano" near Bandung, Java. In subsequent years, VDAP and CVGHM have worked together at Ijen, Raung and Dieng volcanoes in Java and at Agung volcano in Bali.

Figure 25.8 USAID Mission Director, Walter North (far left), VDAP Chief John Pallister (left centre) and US Ambassador to Indonesia Scot Marciel (centre) brief Indonesian Vice President Boediono on the 2010 eruption of Merapi and the effectiveness of the response by Indonesia's Center for Volcanology and Geologic Hazard Mitigation. USGS photo.

25.6 VDAP benefits to the USGS Domestic Program

Over the past 25 years, VDAP has served as a development and proving ground for much of the volcano monitoring technology and eruption forecasting science that is applied at US volcanoes. International experience in crisis response and risk mitigation has informed, strengthened, and helped guide development of domestic capabilities. The Scientists-in-Charge at the USGS Alaska and Cascades Volcano Observatories and the Director of the USGS Volcano Science Center are alumni of VDAP, and current and former VDAP scientists have helped lead responses to recent eruptions in Washington State, Alaska, and the US Commonwealth of the Northern Mariana Islands. The USGS plan for domestic volcano hazard mitigation (National Volcano Early Warning System (NVEWS)), outlined in the USGS Bureau Science Strategy, draws many key elements from decades of VDAP experience. VDAP serves as an enduring and productive strategic partnership between the US Departments of State and Interior. VDAP seeks to enhance US relationships with other nations through science diplomacy and to build international friendships through work toward a shared goal of saving lives and property.

Reference

Newhall, C., Hendley II, J. W. & Stauffer, P. H. 1997. *Benefits of volcano monitoring far outweigh the costs - the case of Mount Pinatubo. U.S. Geological Survey Fact Sheet 115-97.* [Online]. Available: http://pubs.usgs.gov/fs/1997/fs115-97/.

Chapter 26

Communities coping with uncertainty and reducing their risk: the collaborative monitoring and management of volcanic activity with the *vigías* of Tungurahua

J. Stone, J. Barclay, P. Ramon, P. Mothes and STREVA

Long-lived episodic volcanic eruptions share the risk characteristics of other forms of extensive hazard (such as flood, drought or landslides). They also have the capacity for escalations to high intensity, high impact events. Volcán Tungurahua in the Ecuadorian Andes has been in eruption since 1999. The management of risk in areas surrounding the volcano has been facilitated by a network of community-based monitoring volunteers that has grown to fulfil multiple risk reduction roles in collaboration with the scientists and authorities.

26.1 Inception and evolution

Renewed activity from Tungurahua (1999) prompted the evacuation, via Presidential Order, of the large tourist town of Baños and surrounding communities. Social unrest associated with the displacement and attendant loss of livelihood culminated in a forcible civil re-occupation of the land, crossing and over-running military checkpoints (Le Pennec et al., 2012). This re-occupation prompted a radical re-think of management strategy around the volcanic hazard, shifting emphasis from enforcement to communication (Mothes et al., 2015). This enabled the community to continue their way of life alongside the volcano when it is relatively quiet and to prepare for and rapidly mobilise themselves during acute activity.

To do this, a network of volunteers, formed from people already living in the communities at risk, was created with two main goals in mind: (i) to facilitate timely evacuations as part of the Civil Defence communication network, including the management of sirens, and (ii) to communicate observations about the volcano to the scientists (Stone et al., 2014). These volunteers are collectively referred to as '*vigías*' and their input provides a pragmatic solution to the need for better monitoring observations and improved early warning systems when communities are living in relative proximity to the hazard. As a part of the solution, the communities feel strong ownership and involvement with the network (Stone et al., 2014). The communication pathways, formal and informal are shown in Figure 26.1.

Stone, J., Barclay, J., Ramon, P., Mothes, P. & STREVA (2015) Communities coping with uncertainty and reducing their risk: the collaborative monitoring and management of volcanic activity with the vigías of Tungurahua. In: S.C. Loughlin, R.S.J. Sparks, S.K. Brown, S.F. Jenkins & C. Vye-Brown (eds) *Global Volcanic Hazards and Risk,* Cambridge: Cambridge University Press.

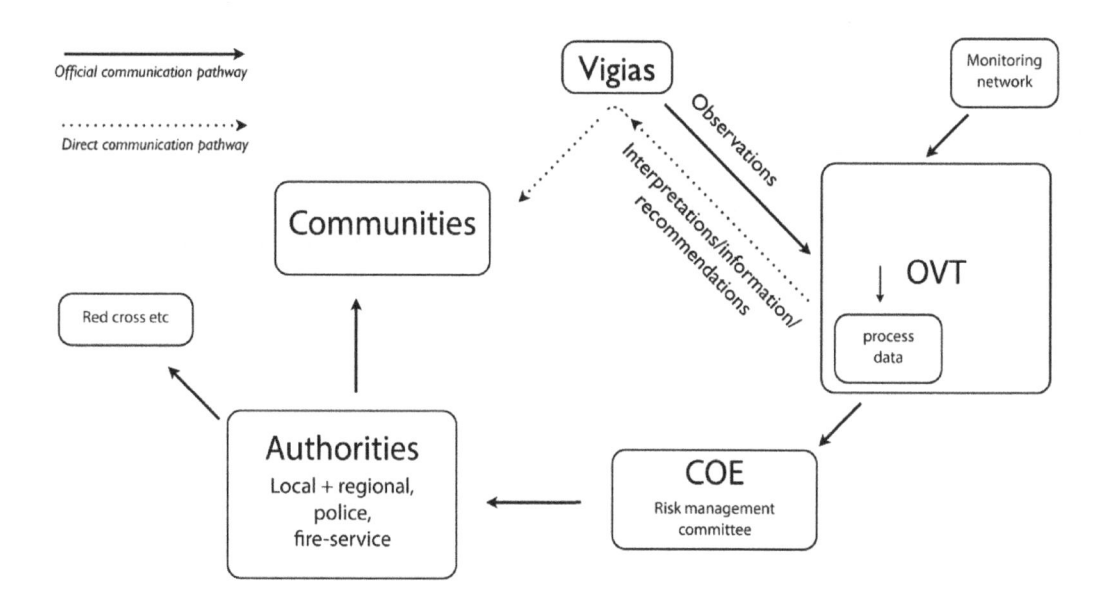

Figure 26.1 The volcanic risk communication network, with its official pathway and the more direct 'vigía mediated' pathway. Adapted from Stone et al. (2014).

26.2 Success and value of the network

The current network consists around 25 *vigías* who use radios with which they maintain daily contact with the observatory (see Figure 26.2). In theory there are up to 43 *vigías*, but not all have radios or actively take part currently. The network has been sustained and has even grown since its inception in 2000. There was a rapid expansion in numbers of *vigías* after the August 2006 eruption. This was a pivotal event, whereby lives saved in the Juive Grande area were attributed to the presence of *vigías* working with the local volcano observatory and lives lost in Palitahua were thought to be in part due to a lack of *vigías* there (Stone et al., 2014). No loss of life has been recorded in recent events in July and October, 2013 and on 1 February 2014 and this can be attributed to the prompt actions to evacuate and reduce risk via the network. Further, community trust in scientific advice and information has reformed since the events of 1999, with vigías acting as intermediaries. Some of the *vigías* now maintain the scientific monitoring equipment near their houses and make daily observations that add considerably to the sum of knowledge of the range and impact of the volcanic behaviour (Bernard, 2013, Mothes et al., 2015), often assisting with visual confirmation of inferred activity seen on the geophysical monitoring network. Apart from reducing volcanic risk, the network has been able to coordinate the response to fires, road traffic accidents, medical emergencies, thefts, assaults and to plan for future earthquakes and landslides. The economic value of allowing affected communities to remain and adapt their existing livelihoods has not, as yet, been determined, but is considered by those communities to be immeasurable.

So far, the communities have responded dynamically to the risk from the volcano, allowing them to live in close proximity and evacuating rapidly when necessary. Tungurahua is capable of producing far larger eruptions than those seen in the last 14 years (Hall et al., 1999), but the trust developed by the network should engender the capacity for action should such an eruption be forecasted, and crucially allows the people to manage their risk in the mean-time, when long-term relocation is simply not an option.

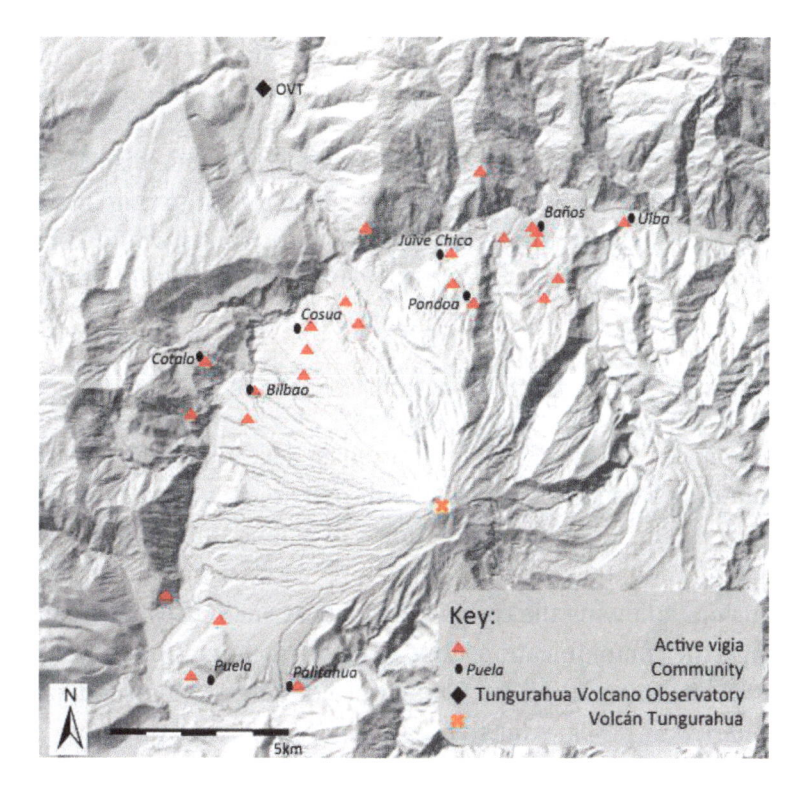

Figure 26.2 Map showing the locations of vigías relative to the volcano and communities significantly affected by volcanic hazards (adapted from Stone et al. (2014)).

26.3 Requirements of the network

Even now, the network still consists of volunteers; and the main requirement from all stakeholders is just the time needed to maintain shared goals and values. The voluntary aspect of the network is vitally important and the motivations of those involved are to help reduce risk to their communities. Nonetheless, its success is due to the willingness with which time is given by *vigías*, observatory scientists (and those in civil protection during its early years) to listen and to share. While some initial *vigías* were drawn from those already involved with Civil Defence (26%), many were also recruited by scientists due to their location relative to the volcano (21%), for their position in the community (26%) and ultimately through other *vigías* (5%). The *vigías* were given basic training from the scientists about what to observe, how to describe phenomena and how to communicate with the local observatory. The largest infrastructural investment was in a VHF radio network, upgraded by another volunteer, and the distribution of handheld radios. Radio communication is a key ingredient in developing relationships and is strictly and professionally observed: every night at 8pm, someone from civil protection calls on the joint (OVT, Civil Defence) radio system and asks the *vigías* to report in. If activity changes then communication frequency increases. Initially, if a *vigía* missed several radio checks they were told to participate properly or not be part of the team. Similarly a sense of shared pride in the role comes from the uniforms provided, initially, by civil defence.

26.4 Sustainability of the network

The network is entering into its fifteenth year; and like conventional geophysical monitoring instruments, relationships continue to function only with regular maintenance; in this instance through contact and discussion. Although the actual financial requirements are small; those that are required (maintenance of the radio network; uniforms) become important symbols to all for the value of the network; long-term neglect of this funding represents a significant threat.

The clear value that the transmission of timely messages to evacuate also reinforces the value of the *vigías and the scientists* to the wider community, providing a strong incentive to volunteers to continue. There is less evidence for whether these motivations would persist in the absence of a volcanic threat but this type of network is exceptionally well suited to extensive hazards and risks.

26.5 Risk reduction for more than 14 years

The sustained involvement of *vigías* (community-based monitoring volunteers) has allowed communities surrounding Tungurahua to live with dynamically changing risk. The network of *vigías* have greatly assisted the monitoring efforts of scientists providing visual observations and by maintaining equipment. Frequent interactions with the scientists have fostered strong trust-based relationships, allowing the *vigías* to act as intermediaries between scientists and the communities during risk communication. These activities have undoubtedly saved lives and helped to preserve livelihoods in the area. The nature of long-lived episodic volcanic eruptions, and thus their similarity to other extensive hazards, means that this type of approach could reduce risk in the case of flooding, landslides and droughts.

References

Bernard, B. 2013. Homemade ashmeter: a low-cost, high-efficiency solution to improve tephra field-data collection for contemporary explosive eruptions. *Journal of Applied Volcanology*, 2, 1-9.

Hall, M. L., Robin, C., Beate, B., Mothes, P. & Monzier, M. 1999. Tungurahua Volcano, Ecuador: structure, eruptive history and hazards. *Journal of Volcanology and Geothermal Research*, 91, 1-21.

Le Pennec, J.-L., Ruiz, G. A., Ramón, P., Palacios, E., Mothes, P. & Yepes, H. 2012. Impact of tephra falls on Andean communities: The influences of eruption size and weather conditions during the 1999–2001 activity of Tungurahua volcano, Ecuador. *Journal of Volcanology and Geothermal Research*, 217-218, 91-103.

Mothes, P., Yepes, H., Hall, M., Ramon, P., Steele, A. & Ruiz, M. 2015. The scientific-community interface over the fifteen-year eruptive episode of Tungurahua Volcano, Ecuador. *Journal of Applied Volcanology*.

Stone, J., Barclay, J., Simmons, P., Cole, P. D., Loughlin, S. C., Ramón, P. & Mothes, P. 2014. Risk reduction through community-based monitoring: the vigías of Tungurahua, Ecuador. *Journal of Applied Volcanology*, 3:11, 1-14.

Index

administrative maps, 66, 337
aerosols, 12, 55, 77, 101, 102, 108, 142, 289-92
air traffic, 4, 82, 126, 132, 295-298, 312, 313
   see aviation
airborne volcanic emissions, 290
airspace, 120, 177, 380
alert levels, 16, 26, 114, 149, 297, 299
andesite, 86, 89, 91
Aniakchak caldera, 90
ash
   see tephra
   airborne, 9, 105, 162, 177, 180, 194, 282, 290
   characteristics, 193
   deposit remobilisation, 9, 125, 193, 199, 204, 282, 290
   fall, 76, 78, 140, 160, 173-214, 302
   hazard, 55, 104, 133, 289
   inhalation, 105, 289-291
ASHFALL (Computer Program), 182-83
asthma, 55, 105, 176, 197, 282, 290-93
   see respiratory disease
Auckland Volcanic Field, 44, 88, 98, 154, 229, 233-37
   see AVF
AVF, *see* Auckland Volcanic Field
aviation, 9, 14, 26, 53, 57-8, 59, 76, 82, 104-5, 125-27, 145, 152, 194, 281, 295-98
   see air traffic

ballistics, 9, 55, 65, 104, 289, 337
basalt, 86, 89, 91,101
Bayesian
   Belief Network, 16, 48, 67, 132, 138, 174,   188, 240, 257, 344
      see BBN
BBN, 257-259
   see Bayesian Belief Network
BET_VH tool, 188-90, 240-41
   see Bayesian Event Tree Volcanic Hazard tool
ByMuR, 191, 243

caldera, 5, 89, 190, 239, 242, 293, 318
Campi Flegrei, 45, 98-9, 102, 188-93, 226, 229, 239-45
CAPRA risk modelling platform, 210, 211
carbon dioxide, 12, 52, 86, 108, 273, 290, 301
Casita volcano, 12, 107

Centre of Volcanology and Geological Hazard Mitigation, 49, 101, 124, 264, 269, 382
   see CVGHM
Chaitén volcano, 4, 92, 100, 176, 187, 281, 292, 328, 380
Chichinautzin, 225-26
civil defence, 3, 44, 75, 154, 155, 235-6, 250, 252, 299, 306, 385, 387
Colima, 89, 90
contaminated water, 13
coping capacity, 31, 32, 83, 160, 162
Cotopaxi, 318, 319
critical infrastructure, 18, 19, 53, 59, 104, 124, 176, 184, 193, 195, 201, 206, 213, 281-87, 302
crop failure, 13, 55, 102, 109, 291
CVGHM, see Centre of Volcanology and Geological Hazard Mitigation

debris avalanche, 12, 65, 107, 335, 337
decision-making during crises, 26, 28, 30-32, 48, 72-73, 132, 135, 138-39, 159-61, 255-59, 347, 372
Determining Volcanic Risk in Auckland, 44, 233
   see DEVORA
DEVORA, *see* Determining Volcanic Risk in Auckland
Disaster Risk Management, 27, 154, 305,
   see DRM
disaster risk reduction, 15, 26-30, 64, 124, 147-58, 159, 230, 299, 323, 349
disease, 13, 55, 105, 109, 276, 289-91
DRM, *see* Disaster Risk Management

earthquake, 3, 12-13, 14-16, 49, 59, 82, 87, 98-100, 108, 113, 114-18, 121, 128, 134, 249, 255, 263-64, 273, 300, 312, 346
   long period, 16, 63, 118
East African Rift valley, 5, 52, 85, 273
emergency management, 76, 83, 138, 159, 174, 179, 192, 213, 236, 299, 306, 337, 372
environmental impact, 48, 255, 290
epidemiology, 21, 140
eruption
   explosive, 5-8, 9-13, 49, 53, 59, 77, 82, 86, 89-95, 96-97, 101-2, 125, 187, 194-96, 213, 252, 263, 268, 281, 289
   style, 89-102
Etna, 8, 119, 195

Index

European Volcano Observatory Space Services, 62, 314
   see EVOSS
evacuation, 10, 14, 17, 19-21, 27-28, 33, 47, 48, 50, 52, 55, 75, 82, 99-100, 109, 111, 113 122-23, 135-37, 139, 147-150, 154-55, 249-51, 255-59, 264, 267-68, 273, 291, 302, 320, 385
event trees, 16, 31, 67, 134, 138, 160, 188, 190, 240, 241, 320, 344
EVOSS, see European Volcano Observatory Space Services
expert elicitation, 17, 134, 257, 258, 344
Eyjafjallajökull, 4, 14, 26, 76, 77,82, 116-17, 120, 123, 152, 157, 173, 184, 281, 290,292, 311

famine, 13, 55, 102, 105, 109, 111, 291
fatality, 2, 43, 109-112, 144, 227, 348, 363
flood, 11, 101, 106-7, 206
fluorine, 11, 105, 108, 255, 275, 278, 290, 292
fluorosis, 198, 283, 290
F-N curve, 21, 138, 345-46
forecasting, 3, 9, 14-17, 26-27, 30-32, 47, 48-49, 53, 57, 59, 63, 67, 72-3, 104, 113-29, 132, 139, 147, 154, 159-62, 180, 244, 249-50, 257-9, 263-65, 283, 317-21, 344, 371-75, 380-84
Fuego, 92-93
Fuji, 89, 226

Galunggung, 9, 105, 226, 295
geothermal field, 122
Global Navigation Satellite System, 118, 300
   see GNSS
Global PEI, 228
Global Volcanism Program, 14, 212
   see GVP
Global Volcano Model, 1, 83, 181, 212, 324
   see GVM
GLOVOREMID Global Volcano Research and Monitoring Institutions Database, 15, 64, 127, 323-333
GNSS, 118
   see Global Navigation Satellite System
Goma Volcano Observatory, 52, 101, 277, 278
   see GVO
grain size, 104, 199, 213, 282

ground deformation, 14, 32, 45, 49, 62, 108, 114, 119, 128, 161, 301, 312, 317
GVM, 83, 125, 156, 157, 163, 210, 324
   see Global Volcano Model
GVP, 212
   see Global Volcanism Program
GVO, 52, 101, 275, 277, 278
   see Goma Volcano Observatory

Hawaiian eruption, 91, 92
Hawaiian volcanic chain, 5, 85
hazard
   and risk ranking, 283
   assessment, 18, 69-70, 103, 104, 141-43, 180, 181, 183, 187-92, 213, 239-45, 349-57
   footprint, 18, 103, 104, 107, 226, 243, 336
   indicator scores, 352
   intensity, 175, 185, 189, 193, 204, 207-208, 346
   map, 61, 131, 192, 268-69, 308, 335-40
HDI, 21, 43, 141, 223, 229-30, 360
   see Human Development Index
Hekla, 16
HFA, 156, 157
Hudson volcano, 176, 194, 195, 293
Human Development Index, 21, 133, 141, 223, 229-30, 284, 286, 360
   see HDI
Human Vulnerability Index, 141
Hyogo Framework for Action, 15, 28, 32, 156, 159, 162
   see HFA

IAVCEI (International Association of Volcanology and Chemistry of the Earth's Interior), 1, 59, 66, 83, 125, 148, 156, 157, 210, 250, 300, 305, 325, 338
impacts to
   agriculture, 194, 195, 198, 199, 201, 203, 209, 283
   buildings, 53, 195, 197, 199, 201, 203, 283, 287
   the economy, 174, 188, 198, 200, 202, 203, 283, 359
injury agents, 55, 289
InSAR, 15, 117, 119
   see synthetic aperture radar interferometry
International Aviation Colour Code, 26, 57, 126, 149, 296-97

Index

International Charter for Space and Major Disasters, 15
International Volcanic Health Hazard Network, 13, 55, 213, 292
  see IVHHN
Ischia, 45, 98, 188, 29, 243
IVHHN, 55, 56, 213, 282, 287, 292, 293
  see International Volcanic Health Hazard Network

jökulhlaup, 118

Katmai-Novarupta, 249
Kilauea, 89, 92-93, 118, 119, 292
Krakatau (Krakatoa), 6, 12, 106, 107

La Soufrière, 27, 48, 100, 137, 148, 255-259
lahar, 2, 11, 12, 13, 18, 27, 47, 55, 61, 65, 82, 100, 107, 118, 130, 153, 154, 249-52, 265, 268, 269, 289, 300, 307-08, 337, 339, 340, 343, 375, 381
  see volcanic mudflow
Lake Nyos, 12, 108
Laki, 77, 89, 101-02, 109, 290
LaMEVE database, 8, 96, 97, 226, 324
  see Large Magnitude Explosive Volcanic Eruptions database
Large Magnitude Explosive Volcanic Eruptions database, 8, 97, 324
  see LaMEVE database
lateral volcanic blast, 12, 107
lava
  dome, 13, 15, 49, 89, 90, 92-93, 119, 138, 263, 264, 268, 343, 380, 383
  flow, 12- 13, 27, 52, 92-93, 101, 108, 154, 236, 273-79, 352
  lake, 52, 273, 274, 278
Lesser Antilles, 5, 84, 259
lightning, 13, 55, 109, 290
livestock, 13, 55, 105, 108, 109, 137, 147, 155, 196, 198, 199, 200, 256, 283, 290, 291

magma chamber, 86-88, 89, 114
Masaya, 229
Mauna Loa, 89, 90
Mayon, 10, 92
Merapi, 15, 26, 49, 50-51, 74, 82, 100, 101, 103, 106, 123, 137, 149, 152, 153, 200, 263-5, 267-71, 292, 313, 379, 382, 384

model
  based hazard maps, 337, 339
monitoring
  data, 14, 31, 87, 115, 123, 125, 147, 160, 163, 180, 257, 258, 264, 305
  ground-based, 15, 33, 59, 76, 119, 121-22, 161-3, 300, 301, 311, 312, 313, 317, 328
  network, 3. 82, 148, 256, 277, 278, 318, 386
  in Latin America, 15, 64, 127, 323-333
monogenetic, 44, 65, 88, 96, 98, 233, 236, 337
Mount Pelée, 2, 6, 106, 110
Mount Longonot, 118
Mount Ontake, 104, 124
Mount St Helens, 12, 92-3, 102, 106, 107, 176, 179, 195, 281, 293
Mount Eden, 44, 98, 234
Mount Ruapehu, 61, 118, 153, 195, 204, 205, 305, 307, 308
Mount Tongariro, 61, 131, 308

National Volcano Early Warning System, 380, 384
  see NVEWS
Nevado del Huila, 27, 381, 382
Nevado del Ruiz, 2, 6, 27, 74, 82, 101, 106, 110, 123, 250, 289, 379, 381
Ngozi, 229
Nishino-shima volcano, 92
NVEWS, 350, 380, 384
  see National Volcano Early Warning System
Nyamuragira, 274, 278
Nyiragongo, 12, 52, 101, 103, 108, 109, 273-79

Paricutin, 97, 98
PEI, 22-23, 43, 69, 142-43, 223-31, 328-31, 355-56
  see Population Exposure Index
Pelée
  see Mount Pelée
Perbakti-Gagak, 225, 226
Philippine Institute of Volcanology and Seismology, 74, 249, 250, 251, 379 (PHIVOLCS)
phreatomagmatic eruptions, 264-65

Index

Pinatubo, 8, 9, 11, 12, 47, 57, 74, 95, 99-100, 101, 103, 105, 106, 111, 119, 121, 125, 147, 152, 194, 202, 249-252, 293, 295, 379
Plinian eruption, 91, 94, 95, 189, 243
Popocatepetl, 16, 225, 226
Population Exposure Index, 22-23, 30, 33, 43, 69, 142-43, 159, 163, 223-31, 332, 355-56
    see PEI
probabilistic
    frameworks, 180
    hazard and risk assessment, 132
    hazard maps, 66, 335, 337, 338, 339
    method, 180, 187
Probabilistic Volcanic Ash Hazard Analysis, 211
    see PVAHA
Probabilistic Volcanic Hazard Analysis, 188, 240, 339
    see PVHA
pulmonary oedema, 290
Pupuke, 233
Puyehue Cordón-Caulle, 194, 195, 293
PVAHA, 211
    see Probabilistic Volcanic Ash Hazard Analysis
PVHA, 190, 240-43
    see Probabilistic Volcanic Hazard Analysis
pyroclastic
    density current, 9-10, 13, 18, 19, 43, 65, 92, 103, 104-06, 111, 130, 133, 141, 150, 155, 176, 227, 289, 300, 336
    flow, 10, 12, 13, 49, 55, 67, 92-93, 100, 105-7, 118, 138, 231, 239, 249-50, 263, 268, 313, 343, 347, 351-3
    surge, 10, 13, 55, 65, 105, 259, 337, 347

qualitative maps, 65, 66, 336, 337,
quantifying uncertainty, 257
quantitative probabilistic model, 16, 63, 132, 319

Rabaul, 195, 292, 293
Rangitoto, 44, 98, 234, 236
Redoubt, 9, 105, 295
remote sensing, 14, 15, 32, 33, 62, 96, 161-62, 311-14, 318, 383
resilience, 3, 14, 29, 22, 28, 29, 30-33, 46, 77, 113, 135-36, 138, 148, 153, 154, 156, 159-63, 174, 259, 305
respiratory disease, 55, 105, 290, 291

rhyolite, 86, 89, 91, 92-3
rift zone, 5, 277
ring of fire, 5

risk
    mitigation, 3, 15, 18, 19, 27, 31-32, 45-46, 48, 59, 74, 99, 148, 154, 160-61, 240, 259, 278, 381, 384
    reduction, 3, 21, 26-29, 30-33, 59, 64, 75, 83, 114, 124, 125, 147-58, 159-63, 299-302, 314, 323, 346, 349, 385-88

Sakurajima, 8
Salak, 225
Santorini, 89
seismic station/seismometer, 14, 15, 115, 118, 123, 124, 127-28, 161, 277, 326, 382
shield volcano, 5, 89, 90, 122, 318
silicosis, 55, 105, 141, 201, 291, 292
Sinabung, 96
Soufrière Hills Volcano, 3, 8, 16, 20, 67, 89, 99, 113, 134, 137, 138, 139-40, 343-48, 365
South Sandwich Islands, 5
stochastic model, 131, 181, 337, 338
stratovolcano, 89, 90, 113, 122, 318
strombolian eruption, 92-93, 94, 189, 243
sulfur dioxide, 12, 15, 76, 766, 100, 101, 102, 108, 119, 120, 263, 290, 301
super-eruption, 8, 97, 102
synthetic aperture radar interferometry, 312
    see InSAR

Tambora, 2, 6, 101, 102, 109, 110
Tangkubanparahu, 225
Taupo, 102
tephra, 9, 21, 45, 55, 65, 104-05, 130, 131, 175, 176, 239-45, 335, 337, 338
    see ash
Tongariro, 61, 131, 308, 309
Tourist/tourism, 4, 20, 28, 81, 104, 109, 131, 136, 151, 160, 198, 200, 204, 308, 309, 385
tsunami, 12, 14, 46, 107
Tungurahua, 2, 3, 8, 27, 75, 110, 137, 139, 155, 296, 285-88

Index

uncertainty
   aleatory, 16, 63, 132, 188, 239, 241, 257, 319, 373
   epistemic, 16, 63, 132, 188, 191, 192, 239, 241, 243, 257, 283, 319, 373
UNISDR, 111
United Nations Office for Disaster Reduction,
   see UNISDR
Unzen, 2, 12, 90, 106, 107, 110, 250

VAAC, 14, 26, 57-58, 105, 125, 126-27, 152, 153, 295-97, 299-301
   see Volcanic Ash Advisory Centre
Value of a Statistical Life, 113
   see VSL
VDAP, 74, 134, 151, 152, 156, 379-84
VEI, 5-8, 94, 110-11, 181-82, 195, 223, 226, 229, 328, 351
   see Volcanic Explosivity Index
Vesuvius, 45, 94, 98-99, 106, 179, 188-192, 208, 229, 239-45
VHI, 22-23, 69, 71, 141-45, 349-56, 359-61
   see Volcano Hazard Index
Virunga volcano, 276, 278
volcanic
   blast, 10, 12, 105-6, 107, 258, 336, 351
   gases, 10, 12, 48, 55, 76, 77, 86, 87, 108, 119, 255, 275, 289, 290, 291, 293, 300, 301, 317
   mudflow, 2, 11, 82, 100, 106, 194, 251, 265, 335, 351, 352
   see lahar
Volcanic
   Ash Advisory Centre, 9, 14, 57, 59, 105, 125, 295-300
   Explosivity Index, 5,6, 94, 177, 351
      see VEI
volcano,
   observatory, 14, 16, 18, 21, 27, 28, 32, 59, 90, 113-15, 124, 127, 128, 148, 149, 153, 161, 299-302
   polygenetic, 88
   shield, 5, 89, 90, 122, 318
   submarine, 5, 8, 84, 92-3, 96
   fissure, 76, 77, 89, 90, 107, 157, 277
Volcano Disaster Assistance Program, 74, 100, 134, 151, 379-84
   see VDAP
Volcano Hazard Index, 22-23, 69, 71, 141-45, 349-56, 359-61
   see VHI

Volcano Population Index, 142, 223
   see VPI
VPI, 223, 230, 359
   see Volcano Population Index
VSL, 113
   see Value of a Statistical Life
Vulcanian eruption, 94, 344
vulnerability, 18, 19-21, 31, 54, 111, 133, 135-38, 140, 174, 175, 177, 205-210, 213, 282-86, 360, 365, 375
   estimate, 174, 205, 206, 210, 213
   physical, 20, 138, 140, 210
   social, 19, 135l 136

White Island, 28, 139
World Organisation of Volcano Observatories, 59, 123, 124, 125, 127, 300
   see WOVO
WOVO, *see* World Organisation of Volcano Observatories
WOVOdat, 123, 163, 321

Yellowstone, 8, 89, 97, 102

Lightning Source UK Ltd.
Milton Keynes UK
UKOW07n1952290715

256055UK00002B/4/P

9 781107 111752

Jenni Barclay

Jenni Barclay